W9-CCA-577

PREFACE

This edition supersedes AC 61–21, Flight Training Handbook, dated 1965, and is intended to supplement AC 61–23B, Pilot's Handbook of Aeronautical Knowledge and appropriate FAA Flight Test Guides. It was developed by the Office of Flight Operations of the Federal Aviation Administration in cooperation with various Regional and District Offices and the FAA Academy.

The materials contained in this handbook include the skills and knowledges considered necessary to satisfy the pilot's basic needs to effectively operate present-day general aviation airplanes, and conform to the pilot's training and certification concepts established by Federal Aviation Regulations, Part 61.

This Flight Training Handbook was developed to assist (1) student pilots learning to fly airplanes, (2) certificated pilots who wish to improve their flying proficiency and aeronautical knowledge, or are preparing for additional certificates or ratings, and (3) flight instructors engaged in the instruction of both students and certificated pilots. It introduces the prospective pilot to the realm of flight and provides information and guidance to all pilots in the performance of procedures and maneuvers which have specific functions in various pilot operations. Performance standards for demonstrating the competence required by regulations for pilot certification are prescribed in the appropriate FAA Flight Test Guides.

It is realized there are different ways of teaching as well as performing certain procedures and maneuvers, and that many variations exist in the explanations of aerodynamic theories and principles. While many are well-founded, it would be impossible to explain all the different methods and concepts applied throughout the aviation industry. Therefore, this handbook takes a selective approach and adopts a uniform method and concept for purposes of simplification. The discussions and explanations reflect the most commonly used practices and principles in today's pilot operations, which are acceptable by the Federal Aviation Administration.

It is essential for persons using this handbook to also become familiar with and apply the rules, operating procedures, and other related matters contained in pertinent Parts of Federal Aviation Regulations, the Airman's Information Manual, and appropriate FAA Advisory Circulars.

Comments regarding this handbook may be directed to the U.S. Department of Transportation, Federal Aviation Administration, Flight Standards National Field Office, P.O. Box 25082, Oklahoma City, Oklahoma 73125.

Reprinted by asa PUBLICATIONS

For sale by the Superintendent of Documents, U.S. Government Printing Office
Washington, D.C. 20402

CONTENTS

CONTENTS—*Continued*

CONTENTS—*Continued*

Chapter 7—AIRPORT TRAFFIC PATTERNS AND OPERATIONS

Chapter 8—TAKEOFFS AND DEPARTURE CLIMBS

Chapter 9—LANDING APPROACHES AND LANDINGS

CONTENTS—*Continued*

Chapter 10—FAULTY APPROACHES AND LANDINGS

Chapter 11—PROFICIENCY FLIGHT MANEUVERS

CONTENTS—*Continued*

Chapter 12—CROSS-COUNTRY FLYING

Chapter 13—EMERGENCY FLIGHT BY REFERENCE TO INSTRUMENTS

Chapter 14—NIGHT FLYING

Chapter 15—SEAPLANE OPERATIONS

CONTENTS—*Continued*

Chapter 16—TRANSITION TO OTHER AIRPLANES

Chapter 17—PRINCIPLES OF FLIGHT AND PERFORMANCE CHARACTERISTICS

CONTENTS—*Continued*

FLIGHT TRAINING HANDBOOK
ILLUSTRATIONS

Chapter 1—INTRODUCTION TO FLIGHT TRAINING

Chapter 2—INTRODUCTION TO AIRPLANES AND ENGINES

Chapter 3—INTRODUCTION TO BASICS OF FLIGHT

ILLUSTRATIONS—*Continued*

Chapter 4—THE EFFECTS AND USE OF CONTROLS

Chapter 5—GROUND OPERATIONS

Chapter 6—BASIC FLIGHT MANEUVERS

Chapter 7—AIRPORT TRAFFIC PATTERNS AND OPERATIONS

Chapter 8—TAKEOFFS AND DEPARTURE CLIMBS

ILLUSTRATIONS—*Continued*

Page

Chapter 9—LANDING APPROACHES AND LANDINGS

Chapter 10—FAULTY APPROACHES AND LANDINGS

Chapter 11—PROFICIENCY FLIGHT MANEUVERS

Chapter 12—CROSS-COUNTRY FLYING

Chapter 13—EMERGENCY FLIGHT BY REFERENCE TO INSTRUMENTS

ILLUSTRATIONS—*Continued*

Page

Chapter 14—NIGHT FLYING

Chapter 15—SEAPLANE OPERATIONS

Chapter 16—TRANSITION TO OTHER AIRPLANES

Chapter 17—PRINCIPLES OF FLIGHT AND PERFORMANCE CHARACTERISTICS

ILLUSTRATIONS—*Continued*

ILLUSTRATIONS—*Continued*

CHAPTER 1

Introduction to Flight Training

The primary objective of this chapter is to briefly discuss the knowledges and skills required of a safe and proficient pilot and the physiological factors associated with flight.

Flight training, if it is to be truly effective, involves more than learning the mechanical manipulation of an airplane's flight controls. Physical or mechanical skill alone is not enough. Operational knowledge and understanding of the associated elements are particularly essential in flying, where safety is the most important factor.

The more the pilot understands the principles of flying, how to apply those principles in performing maneuvers, and how the maneuvers relate to pertinent pilot operations, the more competent that person will be as a pilot. Ground instruction (whether in a formal or informal classroom) and flight training go hand in hand. Each complements the other and results in a training program which is more meaningful and comprehensive. This handbook is based on the premise that knowledge and understanding, as well as skill, are *essential to safety in flight.*

The Instructor and Student Relationship

The primary purpose of all flight training is to develop safe, proficient pilots; the instructor expects from the student total cooperation and maximum effort toward this objective.

Learning to do something well normally involves a high degree of interest and the inner urge or drive to succeed. It makes little difference whether this interest and drive is spontaneous or a result of persuasion, but without some degree of ambition it is probable that both teaching and learning, even in an endeavor as interesting as flying, will at times become burdensome.

The indifferent or irresponsive student pilot who expects to slide effortlessly through the training program simply because it's the instructor's job to "teach," exhibits a poor attitude toward the task. Those who resent or refuse to accept constructive criticism, or who continually make excuses or blame others for all their problems, are lacking in the attitudes necessary for effective learning. In all flight training, strong motivation and positive attitudes are essential ingredients to assure reaching the objectives.

During the training period, the instructor will explain each lesson before the flight. This explanation will include *what* will be done, *why* it should be done, and *how* it should be done. Any point that is not clear should be questioned. By asking questions on the ground, considerable time can be saved for in-flight instruction. An unanswered question in the student's mind can seriously interfere with progress. Questions should be asked even at the expense of appearing ignorant. Pilots with years of experience, and thousands of flying hours, are still asking questions and still learning.

After each flight, the instructor will review the day's lesson. This is the student's chance to clear up any mistaken ideas about each element of a procedure or maneuver. It is imperative that a complete understanding be attained so that corrective action can be taken if necessary. The time to clarify any misconceptions is immediately after the flight, when any problems are still fresh in the person's mind.

The FAA certificated flight instructor had to meet broad flying experience requirements, pass rigid written and flight tests, and demonstrate the ability to apply recommended teaching techniques before being certificated (Fig. 1–1). In addition, the flight instructor's certificate must be renewed every 24 months by showing continued success in training pilots, or by satisfactorily completing a flight instructor's refresher course or a flight check designed to upgrade aeronautical knowledge, pilot proficiency, and teaching techniques.

The Role of the FAA

The FAA prescribes through Federal Aviation Regulations the requirements which must be met before a person is certificated as a pilot. These requirements include the applicant's physical condition, aeronautical experience, knowledge, and skill.

The FAA designates selected physicians (Aviation Medical Examiners) to conduct initial and periodic physical examinations and issue medical certificates as evidence that the pilot meets the medical standards appropriate to the class of pilot certificate.

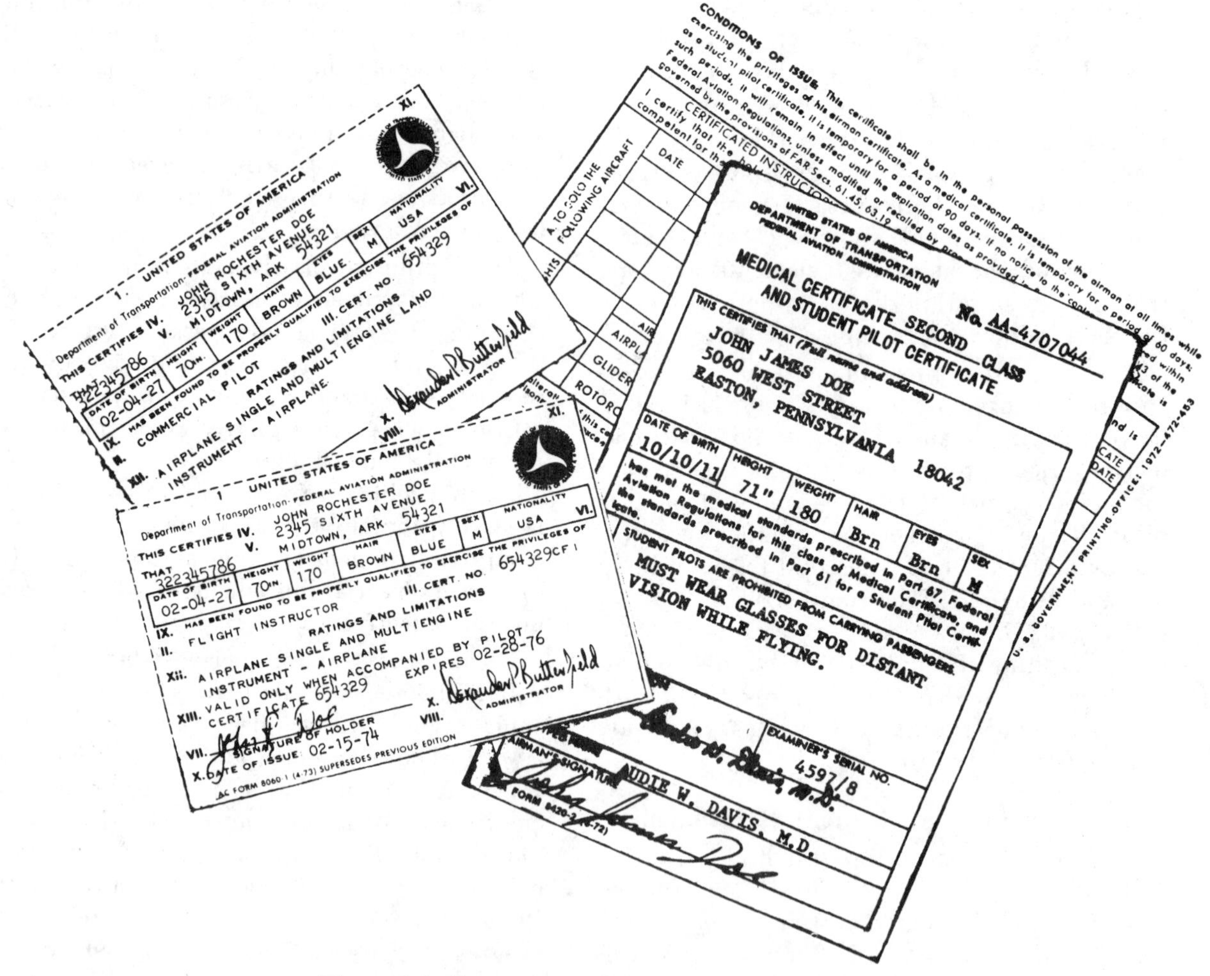

Figure 1–1 Instructor and Student Certificates

Federal Aviation Regulations, Part 61, prescribe the minimum aeronautical experience required for each grade of pilot certificate or rating. Flight time consisting of both dual instruction and solo flying are indicative of the pilot's training and experience. All flight time required for certificates or ratings must be recorded and attested to in the pilot's logbook or other reliable record as evidence that the flight training and experience have been acquired.

Regulations also prescribe the areas of aeronatical knowledge required for pilot certification. Written tests to assure that applicants meet the required aeronautical knowledge standards for certification must be passed. The *practical application* of that knowledge to typical flight situations is emphasized in the written tests.

Regulations also prescribe the pilot operations in which proficiency is required for various pilot certificates or ratings. (A pilot operation is a group of procedures and maneuvers involving skills and knowledges necessary to safely and efficiently function as a pilot.) During a certification flight check, an applicant must demonstrate a satisfactory level of skill in the required pilot operations as outlined in the FAA Flight Test Guides. The maneuvers and procedures used to demonstrate skill in the pilot operations are described in this handbook.

The ability of an applicant for an FAA private or commercial pilot certificate to perform the required pilot operations is based on the following:

1. Executing procedures and maneuvers within the aircraft's performance capabilities and limitations, including use of the aircraft's systems.
2. Executing emergency procedures and maneuvers appropriate to the aircraft.
3. Piloting the aircraft with smoothness and accuracy.
4. Exercising judgment.
5. Applying aeronautical knowledge.
6. Showing that the applicant is the master of the aircraft, with the successful outcome of a procedure or maneuver never seriously in doubt.

Pilot flight checks are administered by FAA inspectors or by FAA designated pilot examiners in the aviation industry.

The FAA considers the qualifications and performance skills of the pilots to be vital to the safety of all flight operations. All modern advances in aircraft and aircraft equipment intended to improve efficiency and safe operation are futile if the pilot is unqualified or unable to use the equipment effectively. Consequently, the certification flight tests are designed to evaluate the pilot's abilities with regard to safe and efficient pilot performance. Because evaluation through flight tests is a mere sampling of pilot ability and is compressed into a short period of time, much reliance must be placed on the flight instructor to demand proper performance throughout the entire period of training.

Practical questions and problems in the FAA tests confine their coverage primarily to considerations which have a direct application to safety. The flight tests are designed to evaluate an applicant's ability to prepare for a flight, maneuver an aircraft safely, plan and execute cross-country flights, and use radio navigation and communications procedures as necessary. All this is important if the pilot is to fit into the modern national airspace system.

To safeguard the lives and property of the flying public (as well as the nonflying public), regulatory control must be maintained over the operation of all aircraft. Without regulations for flying, confusion would prevail in the national airspace.

Federal Aviation Regulations, Part 91, contain the basic safety rules which govern flying in the United States. Every rule, every requirement, every obligation has only one main purpose—SAFETY. Factors attributable to many of the aircraft accidents can be traced to failure to adhere to certain rules of flight and good operating practices. In a democratic society, individuals must accept certain limitations for the general welfare of society. Limitations in the form of regulations and operating procedures are accepted in aviation to make flying safer for everybody and to permit expansion and development of aviation. Therefore, it is imperative that all pilots have a good working knowledge of FAR Part 1,

Definitions and Abbreviations; Part 61, Certification: Pilots and Flight Instructors; and Part 91, General Operating and Flight Rules, as well as their *amendments* as soon as they are published.

Throughout the nation the Federal Aviation Administration has approximately 85 General Aviation District Offices, often referred to as "GADOs." It is through these offices that "grass roots" contact is provided between the FAA and the aviation community. The General Aviation Inspectors in those offices are professionally trained, highly competent airmen, and are there to serve the aviation public. They are prepared to advise and assist in any way possible on aviation matters, particularly flight safety. The locations of these offices are convenient to most people and are usually found on or adjacent to an airport. The GADO inspectors make periodic visits to outlying cities within their districts for the purpose of administering written tests and flight tests, as well as conducting flight safety seminars for the aviation public.

Study Material

If good study habits are to be fully effective, the pilot in training must consider the nature and sources of information. "Hangar flying," while an impressive and pleasant pastime from which some information may be gained, is a poor means for resolving problems and gaining knowledge. All too often, exaggerated tales of wild experiences or "war stories" are related merely to impress the less experienced pilot. Frequently, such sources of information reflect subjective opinions and personal prejudices of those least qualified to give sound advice. The instructor is usually the student's prime reference source, but all pilots should obtain and use appropriate reference texts, manuals, handbooks, periodicals, technical releases, etc.

The FAA develops and makes available to the public various sources of aeronautical information, some free, others at a nominal fee. Of particular interest and value to those persons getting their start in flying are the Student Pilot Guide, VFR/IFR Exam-O-Grams, Pilot's Handbook of Aeronautical Knowledge, Airman's Information Manual, and the Private Pilot Flight Test Guide, in addition to this Flight Training Handbook (Fig. 1-2). Also, there are several excellent commercially produced reference books, and audiovisual aids. All of these, however, should be used only under the personal guidance of an instructor.

Figure 1-2 Selected Study Material

High on all the lists of reference material should be the FAA approved Airplane Flight Manual, or the Pilot's Operating Handbook issued for the airplane being flown. It will prove beneficial for all pilots to acquire a basic aeronautical reference library and to study or review specific topics from time to time.

The FAA issues Advisory Circulars to provide a systematic means for the issuance of nonregulatory material for the guidance and information of the aviation public. These circulars contain an explanation of various subjects of interest and importance to all pilots. They are issued in a numbered-subject system corresponding to the subject areas in the Federal Aviation Regulations. This makes it easy to locate both regulations and advisory material which the FAA has issued on any subject. Most of these circulars are available free of charge and can be obtained individually or by having one's name placed on FAA's mailing list for future circulars. Others may be purchased for a small fee. Complete listings and instructions for ordering are contained in the latest issue of AC 00-2, Advisory Circular Checklist, which can be obtained from the U.S. Department of Transportation, Publications Section, M 443.1, Washington, D.C. 20590.

Study Habits

Learning to fly, like all other learning, should be approached one step at a time. It is a waste of time to study advanced maneuvers before becoming familiar with the basic maneuvers. A training syllabus should be followed and each lesson studied in progression. Each step must be understood in proper sequence in order to avoid waste of time and money.

Good study habits include the practice of visualizing the instructor's explanation and those of the textbook. In addition to reading the text, a clear "mind's eye" picture should be developed in terms of airplane attitude, reaction, and performance. "Study," especially as it pertains to learning to fly, is the development of precise, meaningful images in the mind.

Good study habits in pilot training also include time spent in the cockpit reviewing checklists, identifying controls, and learning the cockpit arrangement (Fig. 1-3). The more familiar a pilot is with the airplane and its equipment, the more time can be devoted to precise airplane control. Lack of familiarity with the airplane is the cause of a high percentage of pilot-error accidents. Incomplete knowledge of aircraft equipment, systems, and capabilities often results in improper use of emergency systems, hard or gear-up landings, undershooting or overshooting the runway, fuel exhaustion, stalls, spins, etc. In a very real sense, such accidents are the direct result of poor study habits.

Figure 1-3 Good Study Habits/Cockpit Familiarization

Ground Safety

A high percentage of accidents in aviation occur during ground operations of aircraft and equipment. Failure to observe rules of safety is certain to cause problems. The old saying that "safety is no accident" is more than a cliche—it's an axiom. Observance of common sense rules is generally sufficient, but there is little chance that pilots will comply with rules, however reasonable, if they are not sincerely concerned with safety.

Pilots should be familiar with safety measures relative to fire and contamination hazards during fuel servicing, precautions for walking on parking ramps in close proximity to aircraft, taxiing and parking aircraft, entering and leaving aircraft, pre- and post-flight procedures such as the positioning of switches and controls, the use of tiedowns, and the proper installation of fuel and oil tank filler caps, etc. (Fig. 1-4). It is recommended that a checklist which includes preflight, in-flight, and post-flight operation, including emergency procedures, be immediately available and constantly used. Checking important items by rote is not an acceptable substitute for checklists. Many of the items which should be checked will be covered in more detail later in this handbook.

A flight training program that ignores, or only pays lip service to ground safety, is lacking in one of the elements requisite to good training.

Figure 1-4 Securing the Airplane After Flight

Flight Safety

The most conscientious preflight planning and careful compliance with ground safety rules will in no way compensate for a lackadaisical attitude toward safety in flight. Since safety in flight relates to a great many things, no attempt will be made to cover them all in this chapter. There are, however, certain practices relative to flight safety that must be stressed at this time. Good habits established early and continually reinforced throughout a training program are habits that the pilot is likely to follow faithfully throughout his or her flying career.

When operating in the seemingly limitless space of the sky, it is easy to forget that many other aircraft are flying in today's airspace, especially in airport traffic patterns and along busy airways (Fig. 1–5). The pilot MUST maintain a continuous vigilance for other aircraft around, above, and below. Though a pilot may have a high degree of skill in aircraft control, excellent coordination, and faultless execution of pilot operations, when one fails to maintain a constant and careful watch for other aircraft, that person is a *dangerous* pilot—dangerous not only to one's self but to passengers and all others in the immediate vicinity.

Figure 1–5 Safe Flying Requires Alertness

Safe Operating Limitations

Every pilot should be thoroughly familiar with the operating limitations or capabilities applicable to the airplane and its equipment. There can be little excuse for accidents that result from failure to observe or be aware of the airplane's flight limitations. A complete understanding of the pertinent FAA approved Airplane Flight Manual or the Pilot's Operating Handbook is important.

Just as important as the airplane's limitations, are the pilot's limitations—one cannot operate safely if the pilot exceeds his or her own capabilities. These are learned only through training and experience and an honest evaluation of one's knowledge and proficiency, as well as physical condition.

Aeromedical Factors

This section is not intended to be an indepth discussion of all the aeromedical factors with which pilots must be concerned. However, the sooner the pilot learns that at least a basic knowledge of these matters is vital to safe flight operations, the sooner adequate attention will be devoted to learning more about aeromedical problems.

Once the pilot enters the airplane, the ability to function properly is absolutely essential to safe flight. Ignorance of and indifference to the physical demands of flight can be as senseless as the lack of concern for an airplane's structural integrity. It is the responsibility of the pilot to consider the status of his or her personal health and to be informed on aeromedical facts. Advisory Circular 67–2, *Medical Handbook for Pilots*, provides much of this information.

A few of the more important medical factors dealing with a pilot's ability to safely function in the flight environment and to cope with its problems are briefly discussed here as an introduction to beginning pilots.

General Health

The person who, for whatever reason, does not feel well should not attempt to participate as a pilot in flying activities. General discomfort, whether due to colds, indigestion, nausea, overwork, lack of sleep, worry or any other bodily weakness is not conducive to safe flying. Perhaps the most insidious and common of all conditions that can result in dangerous inattentiveness, slow reactions, and confused mental processes is *excessive fatigue.* Marked fatigue is as valid a reason for cancelling or postponing a flight as an engine which is found unacceptable during an operations check.

Self-medication can be a very hazardous undertaking for pilots. Probably the best general recommendation for flyers is *abstinence* from all drugs when flying is anticipated. In some instances, the need for a particular drug or medication is an indication that the pilot's health is such that flying is automatically precluded. In other cases, it is unlikely that a pilot who is ill enough to require a drug would be well enough to fly by the time the chance of toxicity from the drug has disappeared.

Antihistamines, tranquilizers, reducing pills, barbiturates, nerve tonics, and many other over-the-counter drugs, can be lethal to the pilot in flight. It is best for those in doubt about such medication to consult a doctor, preferably an FAA designated aviation medical examiner (Fig. 1–6).

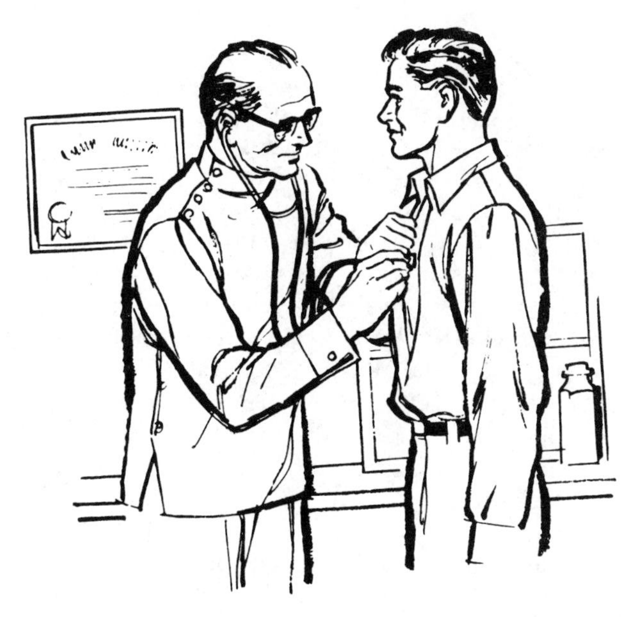

Figure 1–6 Physical Condition Affects Flight Safety

FAA's *Guide to Drug Hazards in Aviation Medicine*, AC 91.11–1, is available through the Superintendent of Documents, U.S. Government Printing Office, and is recommended for those who are interested in more complete information on this subject. The FAA *Airman's Information Manual* (Basic Flight Manual and ATC Procedures) includes additional discussions on pilot aeromedical factors. All pilots should be aware that regulations *forbid* any person from acting as a flight crewmember while using any drug that affects the person's faculties in any way contrary to safety.

Hypoxia

In simple terms, hypoxia is the result of insufficient oxygen in the blood stream. Many are prone to associate hypoxia only with flights at high altitude. While it is true that there is a progressive decrease of oxygen with an increase in altitude, there are many other conditions or situations which can and do interfere with the blood's ability to carry oxygen to the brain. Many drugs, alcohol, and heavy smoking will either diminish the blood's ability to absorb oxygen or the ability of the brain to tolerate hypoxia.

Because of wide individual variations in susceptibility to hypoxia, it is impossible to predict exactly when, where, or how hypoxia reactions will occur in each pilot. As a general rule, however, flights below 10,000 feet MSL without the use of supplemental oxygen can be considered safe, though night vision is particularly critical, and impairment of sight can occur at lower altitudes—especially for heavy smokers. The onset of hypoxia is insidious and progresses slowly. Impaired reactions, confused thinking, poor judgment, unusual fatigue, and dull headaches are typical reactions. Sometimes there is a sense of "well-being" characterized by high spirits or by the feeling that "things could not be better" (euphoria), but the individual thus involved may have little if any insight into his or her actual condition. Because of this and because many pilots never experience this false, happy, carefree feeling, it is not always a symptom one can rely upon to warn of an active or incipient hypoxic condition.

Hyperventilation

Hyperventilation is simply a matter of breathing too rapidly. This condition probably occurs with greater frequency among student pilots than is generally recognized. It is seldom completely incapacitating but it does produce one or more of the symptoms, noted later, that are disturbing if not alarming to the uniformed pilot. Therefore, it only aggravates the problem by further increasing anxiety, and thus the breathing rate.

Under conditions of stress and anxiety, a person's body reacts automatically to such stimuli whether the danger be imaginary or real. One of these automatic reactions is a marked increase in breathing rate. This results in a significant decrease in the carbon dioxide content of the blood. Carbon dioxide, of course, is needed to automatically regulate the breathing process. The common symptoms of this condition are dizziness, nausea, hot and cold sensations, tingling of the hands, legs, and feet, sleepiness and finally unconsciousness. Many of these symptoms are also common to hypoxia and some to ordinary airsickness.

It may be that many students who feel dizzy, lightheaded, or grow nauseated on their early flights are suffering from hyperventilation as well as from motion sickness. Both students and instructors should be aware of this possibility. This condition can be relieved by consciously slowing the breathing rate. Talking loudly or breathing into a bag to restore carbon dioxide will effectively slow the breathing rate.

Alcohol

There is only one safe rule to follow with respect to combining flying and drinking—*don't*. Alcohol consumed by a person is metabolized at a fixed rate by the body. This rate is not altered by the use of coffee or other popular "quack" remedies. Hangovers, whether masked by aspirin or other medication are included in the preceding admonition about flying.

Recent medical investigations of general aviation accidents indicate that alcohol has been a factor in a significant number of aircraft accidents. The inherent danger in drinking and flying apparently has not impressed some pilots. Possibly they labor under the deadly delusion that flying after a few drinks is no more dangerous than driving while in the same condition. (Even this would be a false assumption since drinking is involved in about half the fatal auto accidents investigated.)

We must first accept two simple truths. First, flying an airplane *is* more complex than the two-dimensional demands of driving a car. Second, increased altitude multiplies the intoxicating effect of alcohol on the body.

For all practical purposes, only the brain gets "drunk." When a person drinks an alcoholic beverage, the alcohol begins immediately to pass from the stomach to the bloodstream. Two ounces of bourbon will be absorbed by the bloodstream in ten minutes, four ounces in thirty minutes, and eight ounces in one and one-half hours. The alcohol is carried by the bloodstream to all parts of the body with varying effects, but the brain is really affected the most. Alcohol numbs the brain in the area where our thinking takes place, then proceeds to the area that controls ordinary body movements. Coordination is affected, eyes fail to focus, and hands lose their dexterity.

Any pilot who flies within 8 hours after the consumption of alcoholic beverages or while under the influence of alcohol, is not only dangerous but is in violation of Federal Aviation Regulations.

Carbon Monoxide

Carbon monoxide, always present in fumes from the internal combustion engine, is a colorless, tasteless and odorless gas. Even minute quantities breathed over a long period of time can have serious consequences. Its effects can be cumulative and are not easily corrected. A breath of fresh air will *not* bring early relief—several days may be required to completely rid the body of carbon monoxide. This gas has the ability to saturate the blood's hemoglobin and prevent the absorption of oxygen. The brain and body tissue must have oxygen to function and survive. Aircraft heaters designed to utilize the heat of engine exhaust gases are the usual source for this insidious danger. Be wary if there is a smell of exhaust fumes, especially if mental confusion, dizziness, uneasiness or headaches follow. If such symptoms develop, shut off the cabin heater, ventilate the cabin to the maximum extent possible, descend to lower altitudes where need for heat is less critical, and land as soon as possible for a thorough check of the source of the trouble. It is wise to then consult a doctor. Remember it may take several days to rid the body of carbon monoxide.

Though there are several types of relatively inexpensive detectors available today which may warn of unsafe conditions with respect to carbon monoxide in the cabin, they may not always be completely reliable, and their use should not lull one into a sense of false security.

Middle Ear Discomfort or Pain

Certain persons (whether pilots or passengers) have difficulty balancing the air loads on the eardrum while descending. This is particularly troublesome if a head cold or throat inflammation keeps the eustachian tube from opening properly (Fig. 1–7). If this trouble occurs during descent, the person should try swallowing, yawning, or holding the nose and mouth shut and forcibly exhaling. If no relief occurs, it may be best to climb back up a few thousand feet to relieve the pressure on the outer drum. Then a descent should be made, using these same measures. A more gradual descent may be tried, and it may be necessary to go through several climbs and descents to "stair-step" down. If a nasal inhaler is available, it may afford some relief. If trouble persists several hours after landing, an Aviation Medical Examiner should be consulted.

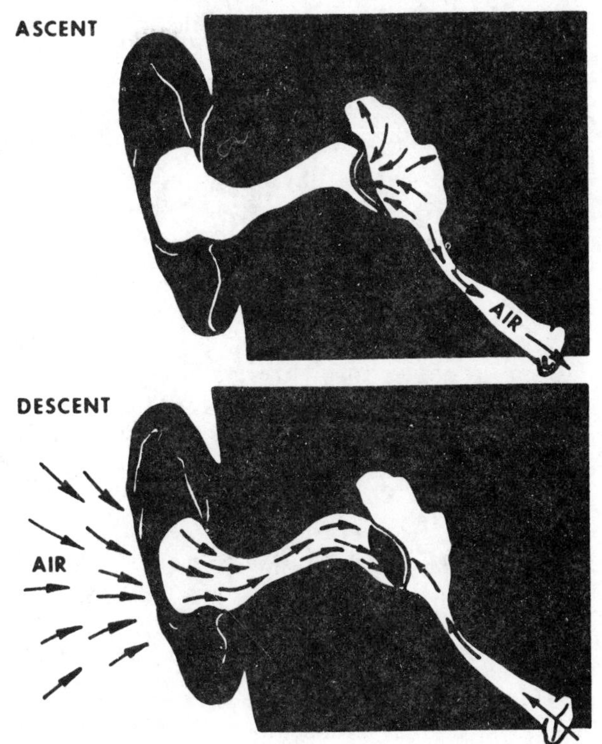

Figure 1–7 The Effect of Pressure on the Inner Ear

Disorientation (Vertigo)

The flight attitude of an airplane is generally determined by reference to the natural horizon. When the natural horizon is obscured, attitude can sometimes be maintained by reference to the surface below. If neither horizon nor surface references exist, the airplane's attitude must be determined by artificial means—an attitude indicator or other flight instruments. Sight, supported by other senses such as the inner ear and muscle sense, is used to maintain spatial orientation. However, during periods of low visibility, the supporting senses sometimes conflict with what is seen. When this happens, a pilot is particularly vulnerable to spatial disorientation. The degree of disorientation may vary considerably with individual pilots, as do the conditions which induce the problem. Spatial disorientation to a pilot means simply the inability to tell "which way is up."

Surface references or the natural horizon may at times become obscured by smoke, fog, smog, haze, dust, ice particles, or other phenomena, although visibility may be above Visual Flight Rule (VFR) minimums. This is especially true at airports located adjacent to large bodies of water or sparsely populated areas, where few, if any, surface references are available. Lack of horizon or surface reference is common on over-water flights, at night, or in low visibility conditions. Other contributors to disorientation are reflections from outside lights, sunlight shining through clouds, and light beams from the airplane's anticollision rotating beacon.

The following are certain basic steps which should assist materially in preventing spatial disorientation:

1. Before flying with less than 3 miles visibility, obtain training and maintain proficiency in airplane control by reference to instruments.
2. When flying at night or in reduced visibility, use the flight instruments.
3. Maintain night currency if intending to fly at night. Include cross-country and local operations at different airports.
4. Study and become familiar with unique geographical conditions in areas in which the flight is intended.
5. Check weather forecasts before departure, enroute, and at destination. Be alert for weather deterioration.
6. Do not attempt visual flight when there is a possibility of getting trapped in deteriorating weather.

7. Rely on instrument indications unless the natural horizon or surface reference is clearly visible.

Motion Sickness

Although motion sickness is uncommon among experienced pilots, it does occur occasionally. A person who has been its victim knows how uncomfortable it is. Most important, it jeopardizes the pilot's flying efficiency—particularly in turbulent weather and in instrument conditions when peak skill is required. Student pilots are frequently surprised by an uneasiness usually described as motion sickness. This is probably a result of combining anxiety, unfamiliarity, and the vibration or jogging received from the airplane, and usually is overcome with experience.

Motion sickness is caused by continued stimulation of the tiny portion of the inner ear which controls the pilot's sense of balance. The symptoms are progressive. First, the desire for food is lost. Then saliva collects in the mouth and the person begins to perspire freely. Eventually, he or she becomes nauseated and disoriented. The head aches and there may be a tendency to vomit. If the air sickness becomes severe enough, the pilot may become completely incapacitated.

Pilots who are susceptible to airsickness should not take the preventive drugs which are available over the counter or by prescription. These medications may make a person drowsy or depress his or her brain function in other ways. Careful research has shown that most motion sickness drugs cause a temporary deterioration of navigational skills or other tasks demanding keen judgment.

If suffering from airsickness while piloting an aircraft, open up the air vents, loosen the clothing, use supplemental oxygen, and keep the eyes on a point outside the airplane. Avoid unnecessary head movements. Then cancel the flight and land as soon as possible.

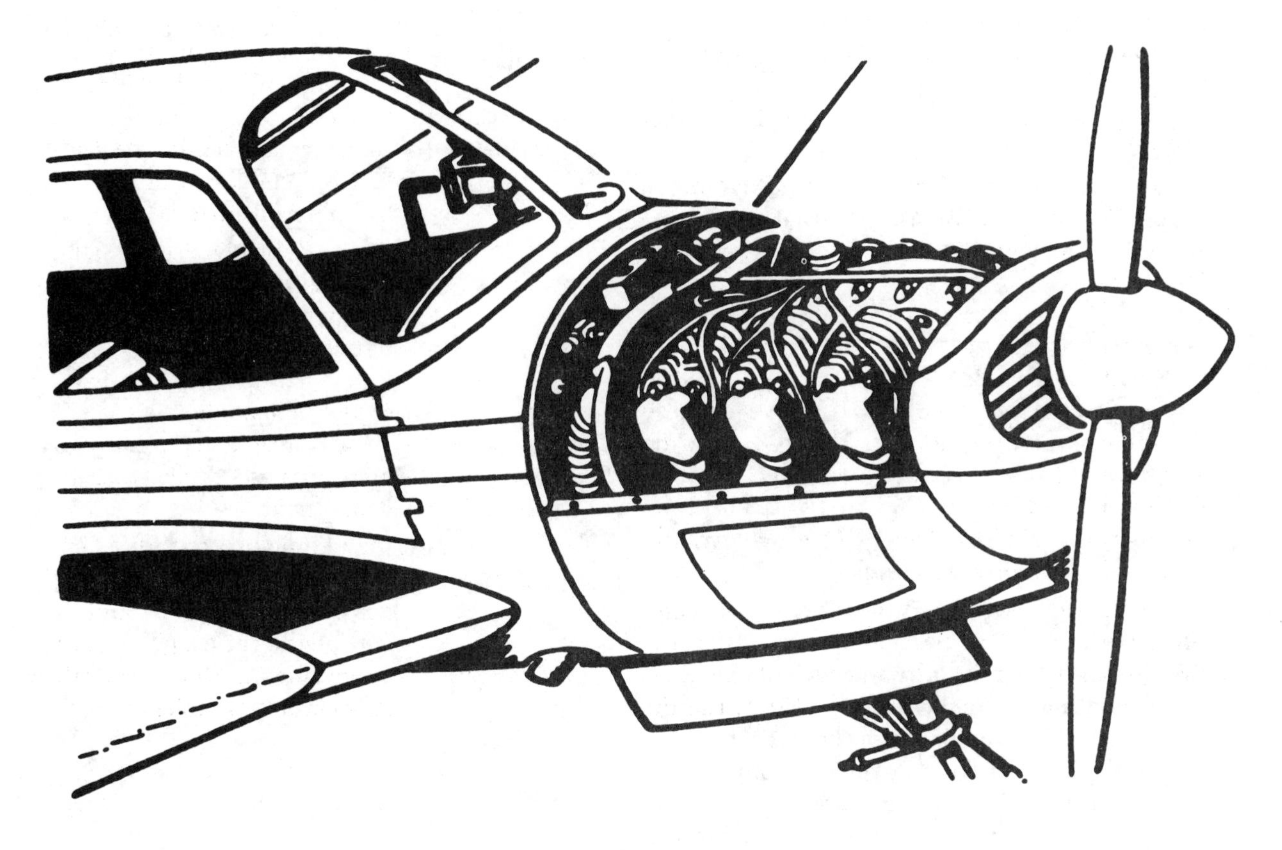

CHAPTER 2

Introduction to Airplanes and Engines

This chapter provides an introduction to the basic airplane, engine, and associated equipment. Although there is no need for the pilot to know how to disassemble or assemble an aircraft, a knowledge of the various parts is essential in understanding their purpose and use. To the experienced pilot, the discussions that follow may seem elementary; but for the beginning pilot they lay the foundation on which to build knowledge of airplane and engine operation.

As the pilot gains experience, more will be learned about how the airplane flies and how its many components operate. In this chapter, then, the objective is to identify the major parts of the airplane and engine, and briefly explain their functions and principles of operation.

The structural units of any conventional airplane are: (1) fuselage, (2) wings, (3) empennage, (4) flight controls and control surfaces—primary and auxiliary, and (5) landing or flotation gear.

When assembled, these units constitute the airplane structure or airframe (Fig. 2–1).

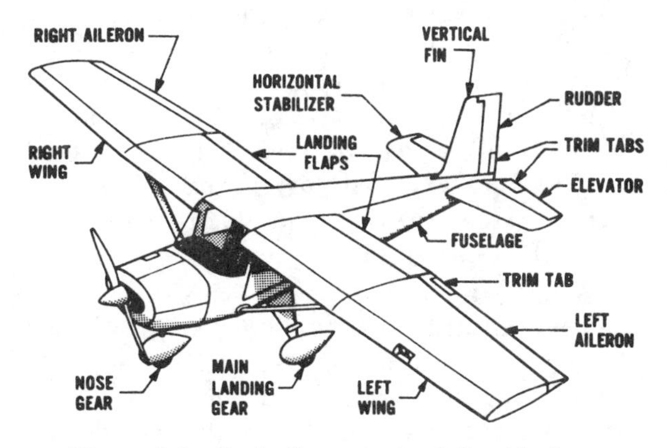

Figure 2–1 Basic Components of the Airplane

Fuselage

The fuselage is one of the principal structural units of the airplane. It houses the crew, passengers, cargo, instruments, and other essential equipment. Most present day airplanes have a fuselage made of a combination of truss and monocoque design. In the truss type construction, strength and rigidity are obtained by joining tubing (steel or aluminum) to produce a series of triangular shapes, called trusses. In monocoque construction, rigs, formers, and bulkheads of varying sizes give shape and strength to the stressed-skin fuselage (Fig. 2-2).

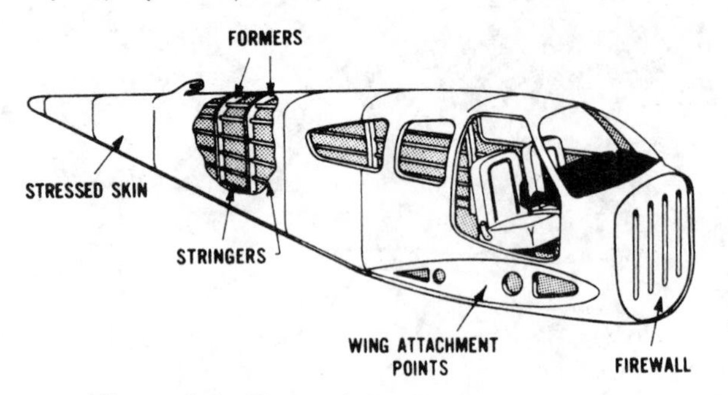

Figure 2-2 Type of Fuselage Construction

On single engine airplanes the engine is usually attached to the front of the fuselage. There is a fireproof partition between the rear of the engine and the cockpit or cabin to protect the pilot and passengers from accidental engine fires. This partition is called a firewall and is usually made of a high heat-resistant, stainless steel.

Wings

The wings are airfoils attached to each side of the fuselage and are the main lifting surfaces which support the airplane in flight. There are numerous wing designs, sizes, and shapes used by the various manufacturers. Each fulfill a certain need with respect to performance expected for the particular airplane. How the wing produces lift is explained in subsequent chapters.

Wings are of two main types—cantilever and semicantilever (Figure 2-3). The cantilever wing requires no external bracing; the stress is carried by internal wing spars, ribs, and stringers. Generally, in this type wing the "skin" or metal wing covering is constructed to carry much of the wing stresses. Airplanes with wings so stressed are called stressed skin types. Treated aluminum alloy is most commonly used as the wing covering (Fig. 2-4). The semicantilever wing is braced both externally by means of wing struts attached to the fuselage, and internally by spars and ribs.

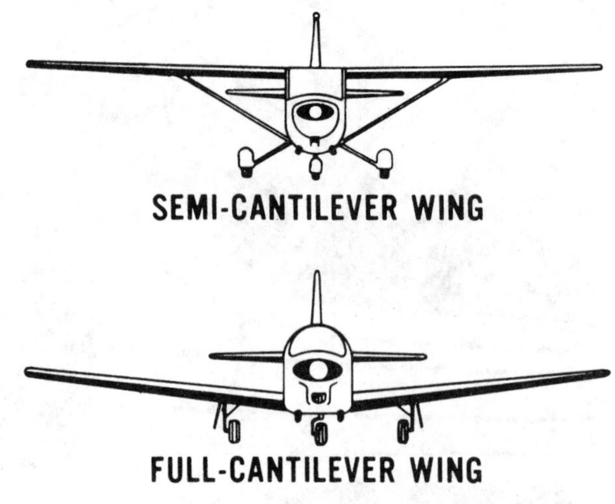

Figure 2-3 External VS Internal Wing Bracing

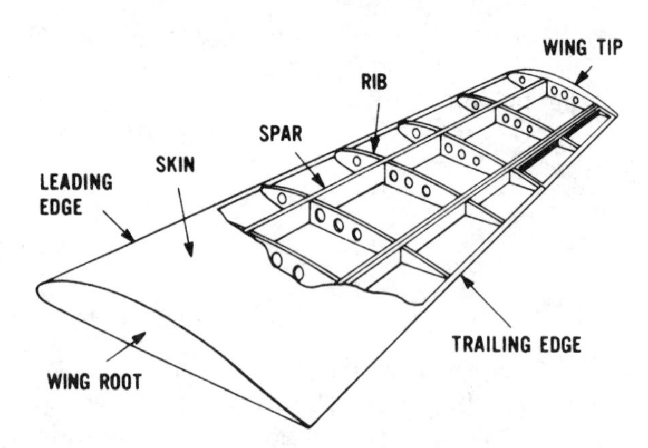

Figure 2-4 Type of Wing Construction

The principal structural parts of the wing are spars, ribs, and stringers. These are reinforced by trusses, I-beams, tubing, or other appropriate devices. The wing ribs actually determine the shape and thickness of the wing (airfoil). In most modern airplanes, the fuel tanks are either an integral part of the wing's structure, or consist of flexible containers mounted inside of the wing structure.

Empennage

Commonly known as the "tail section," the empennage includes the entire tail group consisting of fixed surfaces such as the vertical fin or stabilizer and the horizontal stabilizer; the movable surfaces including the rudder and rudder trim tabs, as well as the elevator and elevator trim tabs. These movable surfaces are used by the pilot to control the horizontal rotation (yaw) and the vertical rotation (pitch) of the airplane.

In some airplanes the entire horizontal surface of the empennage can be adjusted from the cockpit as a complete unit for the purpose of controlling the pitch attitude or trim of the airplane. Such designs are usually referred to as stabilators, flying tails, or slab tails.

The empennage, then, provides the airplane with directional and longitudinal balance (stability) as well as a means for the pilot to control and maneuver the airplane. This is further explained in subsequent chapters.

Flight Controls and Surfaces

The airplane is controllable around its lateral, longitudinal, and vertical axes by deflection of flight control surfaces. These control devices are hinged or movable surfaces with which the pilot adjusts the airplane's attitude during takeoff, flight maneuvering, and landing. They are operated by the pilot through connecting linkage by means of rudder pedals and a control stick or wheel (Fig. 2-5).

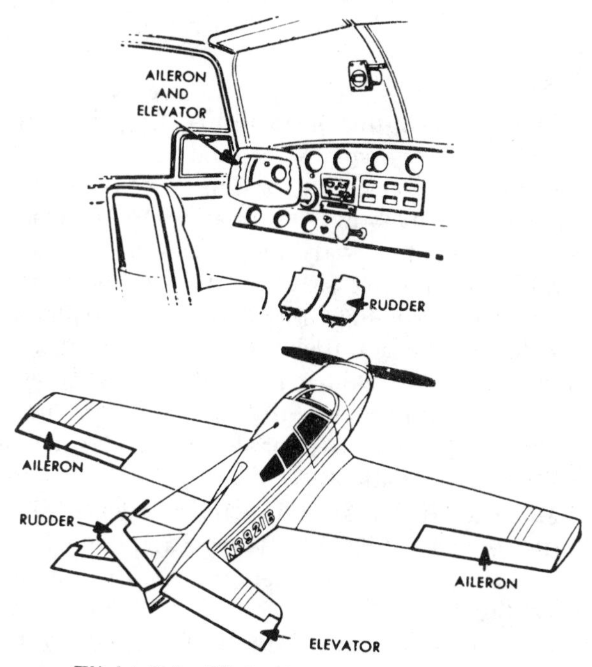

Flight 2-5 Flight Controls and Surfaces

The rudder is attached to the fixed vertical portion of the empennage—the vertical fin or vertical stabilizer. It is used by the pilot to control the direction (left or right) of yaw about the airplane's vertical axis. It is *not* used to make the airplane turn, as is often erroneously believed. This fact will be explained in the chapter on The Effect and Use of Controls.

The elevators are attached to the horizontal portion of the empennage—the horizontal stabilizer. The exception to this is found in those installations where the entire horizontal surface is a one-piece structure which can be deflected up or down to provide longitudinal control and trimming. The elevators provide the pilot with control of the pitch attitude about the airplane's lateral axis.

The movable portions of each wing are the ailerons. The term "aileron" is the French word for "little wing." They are located on the trailing (rear) edge of each wing near the outer tips. When deflected up or down they in effect change the wing's camber (curvature) and its angle of attack, and therefore change the wing's lift/drag characteristics. Their primary use is to bank (roll) the airplane around its longitudinal axis. The banking of the wings results in the airplane turning in the direction of the bank.

The ailerons are interconnected in the control system to operate simultaneously in opposite directions of each other. As the aileron on one wing is deflected downward, the aileron on the opposite wing is deflected upward. This action causes more lift to be produced by the wing on one side than the wing on the other, resulting in a controlled roll or bank.

The effect and use of the flight controls is explained in more detail in Chapter 4.

Secondary Flight Controls

In addition to the primary flight controls, there is, on most modern airplanes, a group termed "secondary controls." These include trim devices of various types, spoilers, and wing flaps.

Trim tabs are commonly used to relieve the pilot of maintaining continuous pressure on the primary controls when correcting for an unbalanced flight condition resulting from changes in aerodynamic forces or weight (Fig. 2-6). Some types also help to actuate the main control surfaces by exerting force on the main surface, thus reducing the amount of force the pilot must exert on the controls to maneuver the airplane. The trim tab is mounted on or attached to the primary control surfaces to provide easier movement or better aerodynamic balance of the surfaces.

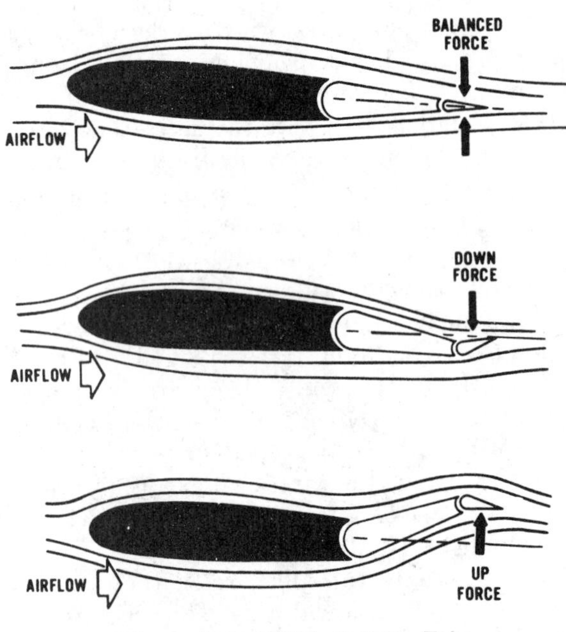

Figure 2–6 Effect of Trim Tabs

Most airplanes, except a few of the very oldest and lightest types, are equipped with trim tabs that can be controlled from the cockpit. On those other types, the tabs are manually adjustable only when the airplane is on the ground.

Spoilers, though found only on certain airplane designs and most gliders, are mounted on the upper surface of each wing. Their purpose is to spoil or disrupt the smooth flow of air over the wing to reduce the lifting force of the wing. This provides the pilot with a means of increasing the rate of descent without increasing the airplane's speed.

Wing Flaps

Wing flaps, installed on the wings of most modern airplanes, have two important functions. First, they permit a slower landing speed and, therefore, decrease the required landing distance. Second, because they permit a comparatively steep angle of descent without an increase in speed, it is possible, while making maximum utilization of the available landing area, to safely clear obstacles when making a landing approach to a small field. They may also be used to shorten the takeoff distance and provide a steeper climb path.

Most wing flaps are hinged near the trailing edges of the wings, inboard of the ailerons (Fig. 2–7). They are controllable by the pilot either manually, electrically, or hydraulically. When they are in the up (retracted) position, they fit flush with the wings and serve as part of the wing's trailing edge. When in the down (extended) position, the flaps pivot downward from the hinge points to various angles ranging up to 40°–50° from the wing. This in effect increases the wing camber (curvature) and angle of attack, thereby providing greater lift and more drag so that the airplane can descend or climb at a steeper angle or a slower airspeed.

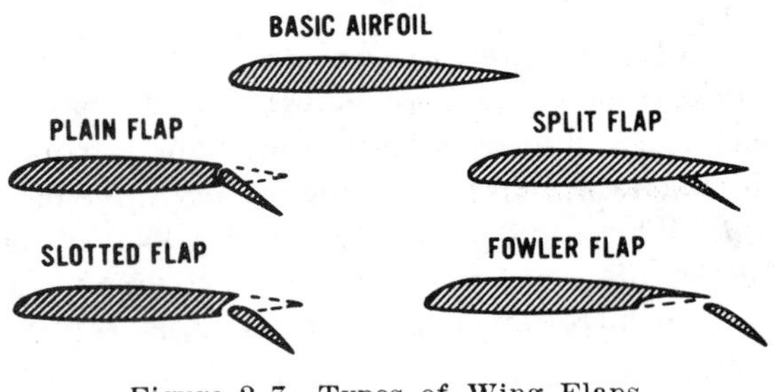

Figure 2–7 Types of Wing Flaps

Landing Gear

The main landing gear forms the principal support of the airplane on land or water. It may include wheels, floats, skis, shock-absorbing equipment, brakes, retracting mechanism with controls and warning devices, and structural members necessary to attach the gear to the primary airplane structure. The airplane also has either a tailwheel or a nosewheel, or may have a tail skid. The tailwheel or nosewheel also supports the airplane on the ground but steering and directional control are their primary functions.

Modern airplanes having the nosewheel installation are called tricycle gear airplanes. A nosewheel gear arrangement has at least three advantages:

1. It allows more forceful application of the brakes during landings at high speeds without resulting in the airplane nosing over.
2. It permits better forward visibility for the pilot during takeoff, landing, and taxiing.

3. It tends to prevent ground looping (swerving) by providing more directional stability during ground operation, since the airplane's center of gravity (CG) is forward of the main wheels. The forward CG, therefore, tends to keep the airplane moving forward in a straight line rather than ground looping.

The main landing gear assembly consists of two main wheels and struts. Each main strut is attached to the primary structure of the fuselage or the wing. With the tailwheel type of airplane, the two main struts are attached to the airplane slightly ahead of the airplane's center of gravity. In the nosewheel type, the two main struts are attached to the airplane slightly to the rear of the airplane's center of gravity.

The landing gear shock-struts may be either self-contained hydraulic units or flexible spring-like structures that support the airplane on the ground and protect the airplane's structure by absorbing and dissipating the shock loads of landing and taxiing over rough surfaces, much as they do on automobiles. Many airplanes are equipped with what is known as oleo or oleo-pneumatic struts, the basic parts of which are a piston and a cylinder. The lower part of the cylinder is filled with hydraulic fluid and the piston operates in this fluid; the upper part is filled with air. Several holes in the piston permit fluid to pass from one side of the piston to the other as the strut compresses and expands and forces the piston back and forth.

The landing gear of many light airplanes is fixed in the extended position (Fig. 2-8), while in the so-called complex airplanes it is retractable in flight.

Figure 2-8 Typical Fixed Gear Airplane

Retractable Landing Gear

Many airplanes are equipped with a retractable landing gear to reduce the drag created while it is extended during flight. Some landing gears retract rearward into the wing and some sideways into the wing (Fig. 2-9). Others retract into the fuselage, and some into the engine nacelles. The retraction and extension device may be operated either manually, hydraulically, or electrically. Warning indicators are provided to show the pilot when the wheels are down and when they are up, or if they are stuck part way. In nearly all installations, systems for emergency operation are provided.

Figure 2-9 Typical Retractable Gear Airplane

Devices used in a typical hydraulically operated landing gear retraction system include actuating cylinders, position selector valves, uplocks, downlocks, sequence valves, tubing, and other conventional hydraulic components. These units are so interconnected that they permit properly sequenced retraction and extension of the landing gear and the landing gear doors or fairing. When the landing gear control in the cockpit is moved to the UP or DOWN position, the landing gear retracts or extends as selected, by the force of hydraulic pressure which is applied to the up or down side of the gear actuator. The gear actuator applies the force required to raise and lower the gear as selected. A locking mechanism locks the gear in the desired up or down position, so that the gear will remain in that position until the pilot operates the landing gear control.

Basically, the electrically operated system is an electrically driven screw-jack for raising or lowering the landing gear. When the landing gear switch in the cockpit is moved to the UP position, the electric motor operates. Through a system of shafts, gears, adapters, an actuator screw, and a torque tube, a force is transmitted to the landing gear strut. Thus, the landing gear retracts and locks. If the switch is moved to the DOWN position, the motor reverses and the gear is moved down and locks. The sequential operation of fairing doors and landing gear is similar to that of the hydraulically operated landing gear system.

An emergency extension system installed in the airplane permits the pilot to lower the landing gear if the main electrical or hydraulic power system fails. The design configurations of some airplanes make emergency extension of the landing gear by gravity and airloads alone impossible or impractical. In such airplanes, provisions are included for forceful gear extension in an emergency. Some installations are designed so that either hydraulic fluid or compressed air provides the necessary pressure, while others use a manual system for extending the landing gear under emergency conditions.

Wheel Brakes

The brakes are used for slowing, stopping, holding, or steering the airplane. They must (1) develop sufficient force to stop the airplane in a reasonable distance; (2) hold the airplane stationary during engine run-ups at high power settings; and (3) permit steering of the airplane on the ground. Brakes installed in each main landing wheel of an airplane are actuated independently of each other by the pilot. The right-hand brake is controlled by applying toe pressure to the top portion of the right rudder pedal and the left-hand brake is controlled by pressure applied to the top portion of the left rudder pedal. This provides for simultaneous use of rudder and brakes to control the airplane's direction of movement on the ground. Some airplanes are equipped with a hand lever which, when held in the ON position, operates the individual wheel brake as rudder pressure is applied.

The independent brake system is used on most small airplanes. This type of brake system is termed "independent" because it has its own hydraulic fluid reservoir and is entirely independent of the airplane's main hydraulic system.

Independent brake systems are powered by master cylinders similar to those used in conventional automobile brake systems. The system is composed of a reservoir, two master cylinders, mechanical linkage which connects each master cylinder with its corresponding brake pedal, connecting fluid lines, and a brake assembly in each main landing gear wheel.

Each master cylinder is actuated by toe pressure on its respective pedal. The master cylinder builds up pressure by the movement of a piston inside a sealed, fluid-filled cylinder. The resulting hydraulic pressure is transmitted through the fluid line which is connected to the brake assembly in the wheel. This results in the braking action (friction) necessary to stop or slow the wheel.

The brakes may be locked for parking by a ratchet-type lock built into the mechanical linkage between the master cylinder and the brake pedal. The brakes are unlocked by application of sufficient pressure on the brake pedals to unload the ratchet.

Nosewheel Steering System

Light airplanes generally are provided with nosewheel steering capabilities through a simple system of mechanical linkage connected to the rudder pedals. Most common applications utilize push-pull rods to connect the pedals to fittings located on the pivotal portion of the nosewheel strut.

Large aircraft, with a need for more positive control, utilize a separate power source for nosewheel steering. Even though nosewheel steering system units of large aircraft differ in their construction features, basically all of them work in approximately the same manner and require: (1) a cockpit control, such as a wheel, handle, lever, or switch to allow for starting and stopping the swiveling movement of the nosewheel and to control the action of the system; (2) mechanical, electrical, or hydraulic connections for transmitting cockpit control movements to a steering control unit on the nosewheel; (3) a control unit, which is usually a metering or control valve; (4) a source of power, which is, in most instances, the airplane's hydraulic system.

Under certain conditions nosewheels vibrate and shimmy during taxiing, takeoff, or landing. If shimmy becomes excessive, it can damage the nose gear or attaching structure. Many airplanes, though, have nosewheel steering systems with built-in features to prevent the shimmy.

Aircraft Engines

Since the engine develops the power to give the airplane its forward motion, thus enabling it to fly, the pilot should have a basic knowledge of how an engine works, and how to control its power.

The engine is commonly referred to as the "powerplant." Not only does it provide power to propel the airplane, but it powers the units which furnish electrical, hydraulic, and pneumatic energy for operation of electric motors, pumps, controls, lights, radio, instruments, retractable landing gear, and flaps. In many cases the engine also provides heat for crewmembers' and passengers' comfort and for deicing equipment. In view of these varied functions it is properly referred to as an "engine" or "powerplant" rather than as a "motor."

The study of the powerplant begins with the definition of the term "internal combustion engine." Internal combustion is the process by which a mixture of fuel and oxygen is burned in a chamber from which the power can be taken directly. This type of combustion can be contrasted to external combustion, such as occurs in a steam engine, in which water is heated in one chamber and transferred as steam to another chamber for transmission of its power. The word "engine" therefore, is interpreted as meaning a machine in which heat energy (released from burning gases) is transformed into mechanical energy.

Two types of aircraft engines are in common use today. One type, used so widely in the typical training airplanes, is known as the reciprocating engine. In this type, pressures from burning and expanding gases cause a piston to move up and down in an enclosed cylinder. This reciprocating motion of the piston is transferred through a connecting rod into rotary motion by a crankshaft, splined or geared to a propeller. In the second type, the turbine jet engine generally used in military, airline, and many corporate-type airplanes, the continuous burning, expansion, and exhausting of gases in one direction pushes the engine, and therefore the airplane, in the opposite direction. Because the latter type engine is found only in airplanes with very high performance characteristics, it will not be dealt with in this handbook.

Reciprocating engines can be further classified as to the manner in which the fuel is introduced into the cylinder. In training type airplanes the usual method is by carburetion, a process of atomizing, vaporizing, and mixing gasoline with air in a unit called a carburetor, before the mixture enters the engine's cylinders. The mixture of gasoline and air is then drawn into each of the cylinders by the up-and-down moving pistons, or is forced under pressure into the cylinders by a blower or supercharger. The other method of supplying the combustible fuel is by fuel injection, whereby the gasoline is injected under pressure by a pump directly into the cylinders where it vaporizes and mixes with air. The fuel/air mixture is then fired (ignited) by timed electric ignition.

Only the simplest reciprocating type of engine and its operating principles are considered appropriate to the purpose of this handbook. For ease of description, the text speaks of "up and down" movements of the piston, but the same principles apply to engines which have the cylinders arranged horizontally or radially around the crankshaft.

The propeller, which uses the engine power to produce thrust, is discussed later in this chapter as a separate unit or accessory.

The basic parts of a reciprocating engine are the crankcase, cylinders, pistons, connecting rods, valves, spark plugs, and crankshaft (Fig. 2–10). In the head or top of each cylinder are two valves and two spark plugs. One of these valves opens and closes a passage leading from the carburetor (or induction manifold) and is called the *intake* valve. The other opens and closes a passage leading to the outside atmosphere (or exhaust manifold) and is called the *exhaust* valve. Inside each cylinder is a movable *piston* which is attached to a *crankshaft* by means of a *connecting rod.* When the rapidly expanding gases (resulting from the heat of combustion of the fuel ignited by spark plugs) push the piston down within the cylinder, it causes the crankshaft to rotate. At the same time, pistons in the other cylinders and attached to the same crankshaft are moved within their individual cylinders by the rotation of the crankshaft and go through the exact same sequence or cycle.

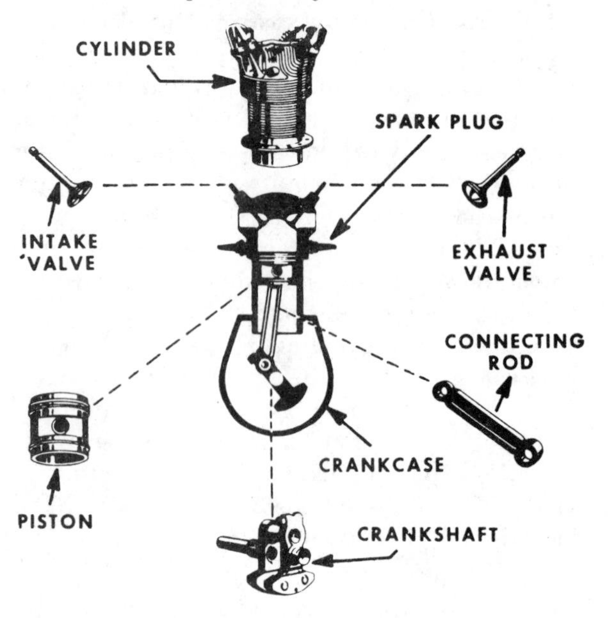

Figure 2–10 Basic Components of the Engine

Engine Cycle

The series of operations or events through which each cylinder of a reciprocating engine must pass in order to operate continuously and deliver power is called an engine *cycle.* The events that occur are intake, compression, power, and exhaust (Fig. 2–11). This cycle requires four strokes of the piston, two up and two down. Ignition of the fuel/air mixture at the end of the compression stroke adds a fifth event; consequently, the cycle of events is known as the *four-stroke, five-event* cycle principle.

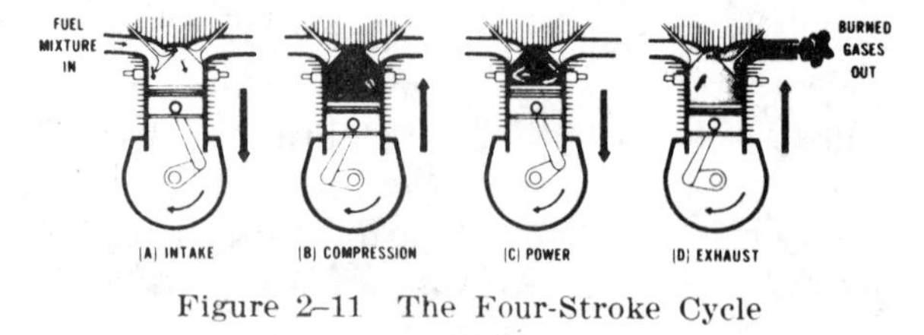

Figure 2–11 The Four-Stroke Cycle

As the piston moves downward on the *intake stroke*, the intake valve is open and the exhaust valve is closed. The piston draws air from the air intake through the carburetor, past the intake valve into the cylinder or combustion chamber. As the air passes through the carburetor, gasoline is introduced into the air flow, forming a combustible mixture. The quantity or weight of the fuel/air mixture is governed by the throttle setting selected by the pilot to regulate the amount of power the engine develops.

When the piston approaches the lower limit of its downward stroke, the intake valve closes and traps the gaseous mixture of air and fuel within the combustion chamber. Next, because both valves are closed as the piston moves upward, the fuel/air mixture is highly compressed between the piston and cylinder head when the upmost position (top dead center) is reached. This is the *compression stroke.*

At the appropriate instant, an electric spark passes across the electrodes or terminals of each spark plug in the cylinder and ignites the mixture. This third event, *ignition* takes place just slightly before the piston reaches top dead center of the compression stroke. As the mixture burns, temperature and pressure within the cylinder rise rapidly. The gaseous mixture, expanding as it burns, forces the piston downward and causes it to deliver mechanical energy to the crankshaft. This is the *power stroke.* Both valves are closed at the start of this stroke.

The energy delivered to the crankshaft during the power stroke causes the crankshaft to rotate on its bearings. Continued rotation causes the piston to move upward again. The exhaust valve, which opened during the latter part of the downward power stroke, remains open during the subsequent upward stroke and allows the burned gases to be ejected from the combustion chamber or cylinders. This stroke is the *exhaust stroke.* On the next stroke of the piston, air is again drawn into the cylinder; hence the four-stroke, five-event cycle begins again and is repeated as long as the engine is operated.

Aircraft engines are multicylinder, having four or more cylinders. Each individual cylinder has its own four-stroke cycle, but all cylinders do not pass through the sequence of events simultaneously. While one cylinder is operating on the power stroke, others are passing through the compression, exhaust, or intake strokes, timed accurately to occur in the correct sequence and at the right instant. This arrangement provides a virtual steady flow of power. Regardless of the number of cylinders on the engine, two complete revolutions of the crankshaft are required for all pistons to accomplish their four-stroke cycle.

Each event in this five-event sequence—*intake*, *compression*, *ignition*, *power*, and *exhaust*—is essential. However, it is only when ignition is added to the other four events and takes place in the proper sequence that the engine will operate. If the ignition switch is turned off, the mixture will not be ignited, there will be no power stroke, and the engine will stop. Also, if the gasoline supply is shut off, there is no gasoline in the cylinder to ignite and, therefore, no power event occurs and the engine stops. In order to start the engine initially, the crankshaft must be rotated by an outside source of power until the ignition and power event take place. This rotation is generally accomplished by an electric motor—a starter—geared to the crankshaft.

Carburetion Systems

Carburetion may be defined as the process of mixing fuel and air in the correct proportions so as to form a combustible mixture. Liquid fuel cannot be burned efficiently in an internal combustion engine; it must first be vaporized into small particles and thoroughly mixed with air. The *carburetor* performs this function.

Although detailed operation varies considerably, every carburetor operates on the same basic principle; i.e., the measurement of airflow and the metering of fuel. Every carburetor contains an air venturi to cause a decrease in air pressure (suction) which draws fuel into the airflow, and a throttle for regulating the amount of airflow. By means of the throttle, then, the pilot can manually control the fuel/air charge entering the cylinders. This, in turn, regulates the engine speed and power.

Some airplanes use a system called *fuel injection*. In this system, instead of mixing the fuel with air in a carburetor, the metered fuel is fed into injection pumps which force it under high pressure directly into the cylinders, or the intake valve passageway, where it mixes with air.

All airplane engines incorporate a device called a *mixture control*, by which the fuel/air ratio can be controlled by the pilot during flight. The purpose of a mixture control is to prevent the mixture from becoming too rich (excess fuel) at high altitudes, due to the decreasing density (weight) of the air. It is also used to "lean" the mixture during cross-country flights to conserve fuel and provide optimum power.

Fuel/Air Mixture Control

Gasoline will not burn unless it is first mixed with air (oxygen) in the proper proportion.

For the mixture to burn efficiently, the ratio of fuel to air must be maintained within a specific range. These mixtures are described either by ratio expressions or by decimal figures. Thus, 12 pounds of air and 1 pound of fuel may be described as an air/fuel ratio of 12:1.

Mixtures as rich as 8:1 and as lean as 16:1 will burn in the cylinder of an engine that develops *maximum* power with about a 12:1 ratio. Why not preset this mixture, then, and do away with manual controls? Because as altitude increases, the air becomes less dense. At 18,000 feet the air is only half as dense as at sea level; that is, a cubic foot of space contains only half as many molecules of air (oxygen). Similarly, an engine cylinder full of air at 18,000 feet contains only half as much oxygen as the same cylinder full at sea level. Meanwhile, without adjustment, the fuel is being metered at the same rate (by volume) as at lower altitudes. Consequently, as the airplane gains altitude at a fixed throttle setting, the cylinders are getting progressively less oxygen with the same amount of fuel; i.e., a richer mixture. As the airplane climbs higher and higher and the fuel/air mixture becomes richer, the excessive fuel causes the engine to lose power and to run rougher and rougher (Fig. 2–12). The mixture control, then, provides a means for the pilot to decrease fuel to compensate for this imbalance in mixture as altitude increases.

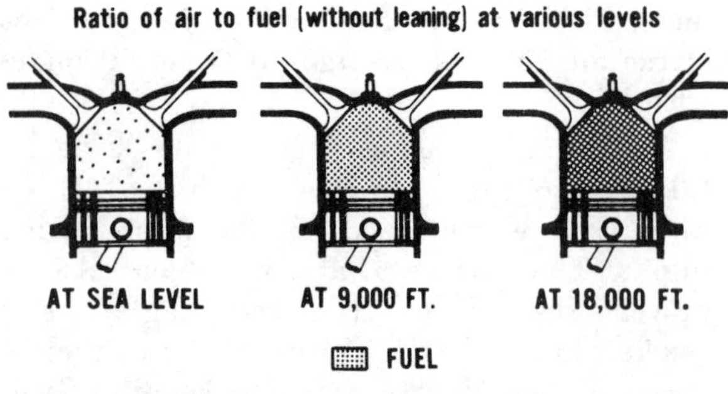

Figure 2–12 Changes in Fuel-Air Mixture with Increased Altitude

If the pilot adjusts the engine's fuel mixture precisely (according to the engine manufacturer's recommendations), there will be a noticeable saving in fuel for a given flight—with no sacrifice in airspeed. In fact, under some conditions, there might even be an airspeed gain.

Engines are more efficient when they are supplied the proper mixture of fuel and air. Flying with an over-rich mixture may induce spark plug fouling, which may shorten the life of the plugs. Proper leaning also results in a smoother running engine, and a cleaner combustion chamber with less likelihood of preignition that sometimes results from accumulated carbon deposits.

If the engine is equipped with a *float-type carburetor* but no instrument to show the fuel/air mixture ratio, this simple but satisfactory method of adjusting the mixture may be used. Lean the mixture until the engine begins to run rough, then enrich the mixture slightly until the engine runs smooth again. When this method is used with a fixed-pitch propeller at a given throttle setting in level flight, a small increase in engine RPM (revolutions per minute) and airspeed may be noted at or before the point of roughness. The mixture then should be enriched to the point of smoothness, though the RPM may stay the same or it may drop slightly. With a controllable prop and a fixed throttle setting in level flight, a slight increase in airspeed only may be noted without the RPM increase noted for fixed-pitch propellers.

Fuel injected engines usually have fuel flow indicators on the instrument panel. On these engines, adjust the mixture by setting the fuel flow according to the instructions in the airplane operating handbook or the markings on the fuel flow indicator itself.

Operating an engine with a too-lean mixture can result in serious and expensive engine damage, although severe damage rarely occurs at normal cruise power. It is much more apt to occur when operating at high power settings (above 75% of rated power for most engines) combined with too-lean mixtures which damage the engine due to overheating. Nevertheless, it has proven harmful to *over-lean* at any power setting (besides, maximum engine efficiency comes only with the proper fuel mixture). If flying with higher than normal power settings, it is permissible to lean, but this needs to be done scientifically. It needs sophisticated instrumentation and precise adjustments, to avoid engine damage. The following basic rules are recommended at all times:

1. Always enrich the mixture *before* increasing power.
2. Reset the mixture after any change in power or altitude.
3. Use full rich mixture for takeoff and climb, unless the manufacturer recommends leaning at high elevations to eliminate roughness and loss of power.

Some pilots believe that the mixture should never be leaned below 5,000 feet MSL, but this is not necessarily true. Most fuel metering devices are set on the rich side, which means that prudent leaning is permitted—in fact advisable—at almost any altitude when the engine is operated at *less* than 75% of its rated power. When a 5,000-foot restriction on leaning is recommended by the manufacturer, it usually is applicable during climbs when climb power is being used.

Ignition System

The function of the ignition system is to provide an electrical spark to ignite the fuel/air mixture in the cylinders. The ignition system of the engine is completely separate from the airplane's electrical system. The magneto type ignition system is used on most reciprocating aircraft engines. Magnetos are engine-driven self-contained units supplying electrical current without using an external source of current. However, before they can produce current the magnetos must be actuated as the engine crankshaft is rotated by some other means. To accomplish this, the aircraft battery furnishes electrical power to operate a starter, which through a series of gears, rotates the engine crankshaft. This in turn actuates the armature of the magneto to produce the sparks for ignition of the fuel in each cylinder. After the engine starts, the starter system is disengaged, and the battery no longer contributes to the actual operation of the engine. If the battery (or master) switch were turned OFF, LEFT, RIGHT, and BOTH (Fig. 2–13). With the switch in the "L" or "R" position, only one magneto is supplying current and only one set of spark plugs in each cylinder is firing. With the switch in the BOTH position, both magnetos are supplying current and both spark plugs are firing. The main advantages of the dual ignition system are:

Modern airplane engines are required by Federal Aviation Regulations to have a dual ignition system—that is, two separate magnetos to supply the electric current to the two spark plugs contained in each cylinder. One magneto system supplies the current to one set of plugs; the second magneto system supplies the current to the other set of plugs. For that reason the ignition switch has four positions: OFF, L, R, and BOTH (Fig. 2–13). With the switch in the "L" or "R" position, only one magneto is supplying current and only one set of spark plugs in each cylinder is firing. With the switch in the BOTH position, both magnetos are supplying current and both spark plugs are firing. The main advantages of the dual ignition system are:

1. Increased safety. In case one system fails, the engine may be operated on the other until a landing is safely made. Consequently, it is extremely important for each magneto to be checked for proper operation before takeoff. This should be done in accordance with the manufacturer's recommendations. Unless the manufacturer specifies otherwise, it is advisable to turn the ignition switch from the "BOTH" position to the fartherest "ON" position first, then back to "BOTH;" then to the nearest "ON" position and back to "BOTH." This sequence best assures that the magneto switch will be returned to the "BOTH" position for normal operation. Also, the performance of each ignition system will have been compared with a "BOTH" performance.
2. More complete and even combustion of the mixture, and consequently improved engine performance; i.e., the fuel mixture will be ignited on each side of the combustion chamber and burn toward the center.

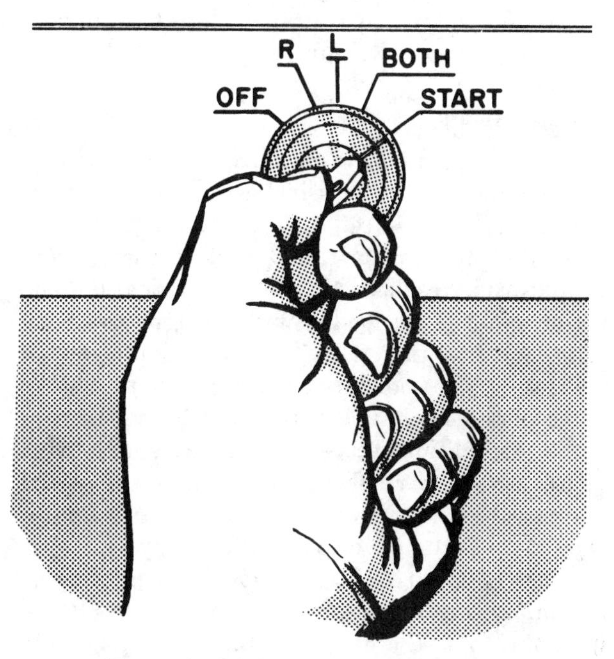

Figure 2–13 Typical Ignition Switch

Normal Combustion

Normal combustion occurs when the fuel/air mixture ignites in the cylinder and burns progressively with a normal pressure increase, producing maximum pressure immediately after the piston passes top dead center of the compression stroke (Fig. 2–14). A flame-front starts at the spark plugs and travels across the combustion chamber at a speed of approximately 70 to 100 feet per second. The velocity of the flame-front is influenced by the type of fuel, the ratio of fuel-to-air mixture, the pressure on the fuel/air mixture, and the temperature of the fuel/air mixture. The pilot has control of these items by ensuring that (1) the airplane has been serviced with the recommended grade of fuel; (2) the pressure of the fuel/air mixture is properly regulated by propeller and throttle controls; (3) the engine temperature is adequately regulated by use of the cowl flaps and by the engine speed when an internal supercharger is used; and (4) the fuel/air ratio is accurately adjusted by the mixture control. When any of these factors is not properly controlled, abnormal combustion will result.

When the fuel/air mixture is ignited by means other than the normal spark ignition, the result is abnormal combustion. This abnormal combustion is divided into two distinct types—*detonation* and *preignition.*

Detonation

When the fuel/air mixture is subjected to a combination of excessively high temperature and high pressure within the cylinder, the spontaneous combustion point of the gaseous mixture is reached. When this critical detonation point is reached, normal progressive combustion is replaced by a sudden explosion, or instantaneous combustion (Fig. 2–14). Due to the piston's position in the cylinder at the time the detonation wave starts, extremely high pressures are reached, often in excess of the structural limits of the cylinder and engine parts. Tests have proven that pressures in excess of 4,000 PSI are reached during detonation. Since these pressures are virtually instantaneous, the effect on the piston is equivalent to a sharp blow with a sledge hammer. This shattering force is what is sometimes heard in an automobile as it is accelerated rapidly. In an automobile engine this is not so serious, because it is heard and can be readily remedied by reducing the engine power. In aircraft engines it is much more serious because it is difficult to detect above other aircraft noises and corrective action may then be too late to be effective. This form of combustion causes a definite loss of power, engine overheating, preignition and, if allowed to continue, physical damage to the engine.

Although detection of detonation may be extremely difficult, the indications are an otherwise unexplained rise in cylinder head temperature, an unexplained loss of power, especially at the higher power settings, and a whitish-orange exhaust flame accompanied by puffs of black smoke.

Corrective action for detonation may be accomplished by adjusting any of the engine controls which will reduce both temperature and pressure of the fuel/air charge. Good engine operating procedures by the pilot will prevent detonation and from the pilot's standpoint, prevention is the best cure.

Preignition

Preignition is defined as ignition of the fuel prior to normal ignition, or ignition before the electrical arcing occurs at the spark plugs.

Preignition may be caused by excessively hot exhaust valves, carbon particles or spark plug electrodes heated to an incandescent or glowing state. In most cases these local "hot spots" are caused by the high temperatures encountered during detonation.

This form of abnormal combustion has the same effect on the engine as an early or advanced timing of the ignition system, and is so harmful in its effects that an engine will continue to operate normally only for a short period of time. This holds especially true if detonation and preignition are in progress simultaneously. During preignition conditions, cylinder pressures are in excess of the normal limits of the cylinder and the engine structure.

One significant difference between preignition and detonation lies in the fact that if the conditions for detonation exist in one cylinder, they may exist in all cylinders; but, preignition may exist in only one or two cylinders. This can make preignition rather difficult to detect, because of the possibility of preignition occurring in a cylinder which is not the location of the thermocouple which measures cylinder head temperature. Probably the most reliable indication is a loss of power, but this also may be difficult to determine unless the engine has a torquemeter. Another indication of preignition may be the observation of glowing carbon particles being discharged from the exhaust system.

Corrective actions for preignition include any type of engine operation which would promote cooling, such as enriching the fuel/air mixtures, reducing cylinder/manifold pressures, and properly controlling engine cowl flaps when available.

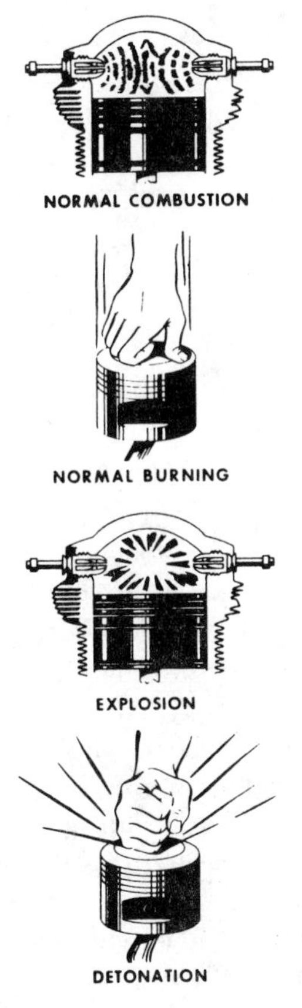

Figure 2–14 Normal Combustion VS Detonation

Carburetor Ice

The formation of ice in the carburetor has always been a problem in those airplanes equipped with a float-type carburetor. It may restrict the power output of the engine or even cause the engine to quit operating. It is not a problem in fuel injection systems because such systems use no carburetor—fuel is vaporized by other means.

In the float-type carburetor, the fuel is evaporated and vaporized immediately downstream from the throttle valve at the narrowest portion of the carburetor venturi. The effect of the evaporating fuel and decreasing air pressure causes a sharp drop in temperature within the carburetor venturi. In moist air or high relative humidities, the moisture in the air entering the carburetor condenses and, because of the lowered temperature, may result in the formation of ice. When ice forms in the carburetor, it tends to choke off the flow of air and reduce the power output, or even prevent the engine from operating.

Carburetor ice can be prevented or eliminated by raising the temperature of the air entering the carburetor venturi through the use of a heating device controlled by the pilot. The heat is usually obtained through a control valve that can be adjusted to allow heated air from the engine compartment to enter the carburetor. The pilot should apply carburetor heat whenever conditions are conducive to icing and in the manner recommended by the airplane manufacturer.

Propellers

A propeller is the unit which provides thrust to propel the airplane through the air. It consists of two or more blades and a central hub to which the blades are attached. Each blade of a propeller is an airfoil and, therefore, is essentially a rotating wing.

The power needed to rotate the propeller blades is furnished by the engine. The propeller is mounted on a shaft, which may be an extension of the crankshaft on low horsepower engines or, on high-horsepower engines it may be mounted on a propeller shaft which is geared to the engine crankshaft. In either case, the engine rotates the airfoils of the blades through the air at relatively high velocity, and in this way the propeller transforms the rotary power of the engine into thrust.

Fixed-Pitch Propeller

As the name implies, a fixed-pitch propeller has the blade pitch (blade angle) built into the propeller. For that reason the pitch angle cannot be changed by the pilot, as it can on controllable pitch propellers. Generally, this type of propeller is constructed of wood or aluminum alloy (Fig. 2–15).

Fixed-pitch propellers are designed for best efficiency at one rotational and forward speed.

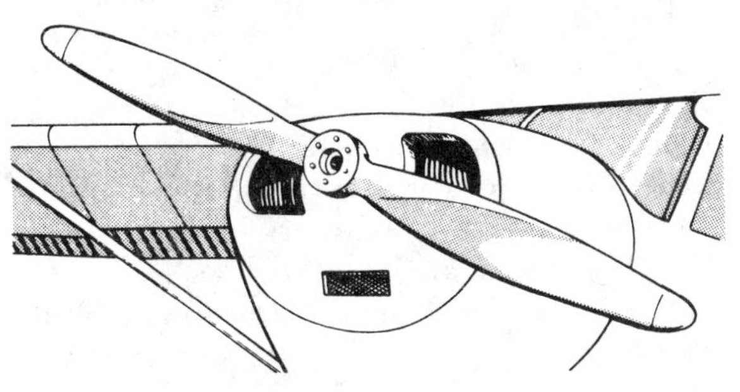

Figure 2–15 Fixed Pitch Propeller

They are designed to fit a specific set of conditions involving both the engine rotational speed and the airplane's forward speed. Any change in these conditions reduces the efficiency of both the propeller and the engine.

Constant Speed Propellers

In automatically controllable pitch propeller systems, a control device adjusts the blade angle to maintain a specific preset engine RPM without constant attention by the pilot (Fig. 2–16). For example, if engine RPM increases as the result of a decreased load on the engine, the system automatically increases the propeller's blade angle (increasing the air load) until the RPM has returned to the preset speed. A good automatic control system will respond to such small variations in RPM that, for all practical purposes, a constant RPM will be maintained. These automatic propellers are termed "constant speed" propellers.

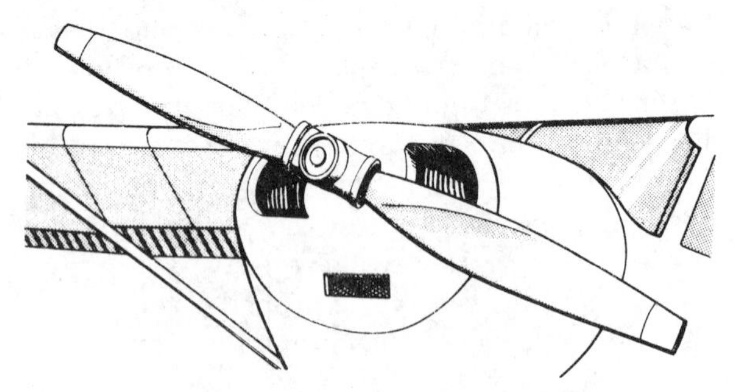

Figure 2–16 Adjustable Pitch Propeller

An automatic system consists of a governor unit which controls the pitch angle of the blades so that the engine speed remains constant. The propeller governor can be regulated by the pilot with controls in the cockpit, so that any desired blade angle setting (within its limits) and engine operating RPM can be obtained, thereby increasing the airplane's operational efficiency in various flight conditions. A low-pitch, high-RPM setting, for example, can be utilized to obtain maximum power for takeoff; then after the airplane is airborne, a higher pitch and lower RPM setting can be used to provide adequate thrust for maintaining the proper airspeed. This may be compared to the use of low gear in an automobile to accelerate until high speed is attained, then shifting into high gear for the cruising speed (Fig. 2–17).

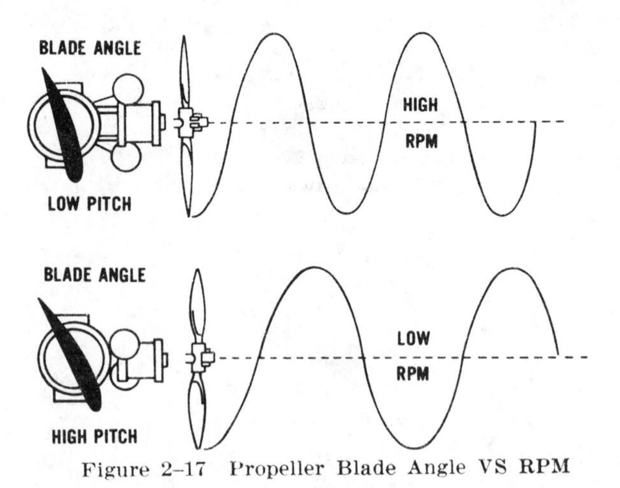

Figure 2–17 Propeller Blade Angle VS RPM

Manifold Pressure and Engine RPM

An airplane equipped with a fixed-pitch propeller has only one main power control—the throttle. In that case, the setting of the throttle will control both the amount of power and the propeller or engine RPM.

On the other hand, an airplane equipped with a constant speed propeller has two main power controls—a throttle and a propeller control. The throttle controls the engine's power output which is indirectly indicated on the manifold pressure gauge. The propeller control changes the pitch of the propeller blades and governs the RPM which is indicated on the tachometer (Fig. 2–18). As the throttle setting (manifold pressure) is increased, the pitch angle of the propeller blades is automatically increased through the action of the propeller governor system. This increase in propeller pitch proportionately increases the air load on the propeller so that the RPM remains constant. Conversely, when the throttle setting (manifold pressure) is decreased, the pitch angle of the propeller blades is automatically decreased. This decrease in propeller pitch decreases the air load on the propeller so that the RPM remains constant.

Figure 2–18 Power Instruments—RPM and Manifold Pressure

On most airplanes, for any given RPM, there is a manifold pressure that should not be exceeded. If an excessive amount of manifold pressure is carried for a given RPM, the maximum allowable pressure within the engine cylinders could be exceeded, placing undue stress on them. If repeated too frequently, this undue stress could weaken the cylinder components and eventually cause engine structural failure.

What can the pilot do to avoid conditions that would possibly overstress the cylinders? First, there must be a constant awareness of the tachometer indication (engine RPM), especially when increasing the throttle setting (manifold pressure). Then, the pilot should know and conform to the manufacturer's recommendations for power settings of a particular engine to maintain the proper relationship between manifold pressure and RPM. The combination to avoid is a *high* throttle setting (manifold pressure indication) and a *low* RPM (tachometer indication).

When both manifold pressure and RPM need to be changed significantly, the pilot can further help avoid overstress by making power adjustments in the proper order. On most airplanes when power settings are being decreased, reduce manifold pressure before RPM. When power settings are being increased, reverse the order—increase RPM first, then manifold pressure. If RPM is reduced *before* manifold pressure, manifold pressure will *automatically increase* and possibly *exceed* manufacturer's tolerances.

Turbochargers

An increasing number of engines used in general aviation airplanes are equipped with externally driven supercharger systems. These superchargers are powered by the energy of exhaust gases and are called turbochargers. With a normally aspirated (unsupercharged) engine, the maximum manifold pressure (power) that can be developed by the engine is slightly less than 30 inches of mercury (standard sea level pressure). By increasing the manifold pressure above atmospheric pressure, more fuel/air mixture can be packed into the cylinders. Consequently, supercharging the engine increases the maximum power output of the engine at sea level.

Supercharging is more important, however, at high altitudes. This is because the density of the air decreases as altitude increases, resulting in a decreased power output of an unsupercharged engine. By compressing the thin air by means of a supercharger, the turbocharged engine will maintain the preset power as altitude is increased, until the engine's critical altitude is reached. At that altitude the turbine is rotating at its highest speed and can no longer compensate for the decreasing power after that altitude is exceeded.

The turbocharger consists of a compressor to provide pressurized air to the engine, and a turbine driven by exhaust gases of the engine to drive the compressor. It is controlled automatically by a pressure controller or waste gate to maintain the manifold pressure at approximately a constant value from sea level to the engine's critical altitude. Once the pilot has set the desired manifold pressure, virtually no throttle adjustment is required with an increase of altitude until the critical altitude is reached. That altitude is reached when the exhaust gas waste gate is fully closed and the turbine is operating at maximum speed.

Instruments

Safe, economical, and reliable operation of modern airplanes is dependent upon the use of instruments. Instrumentation is basically the science of measurement. Speed, distance, altitude, attitude, direction, temperature, pressure, and engine revolutions are measured and these measurements are displayed on instrument dials in the airplane's cockpit. There are those which are referred to as "flight instruments" and others are termed "engine instruments." Only the basic instruments found in typical general aviation airplanes are discussed here (Fig. 2–19).

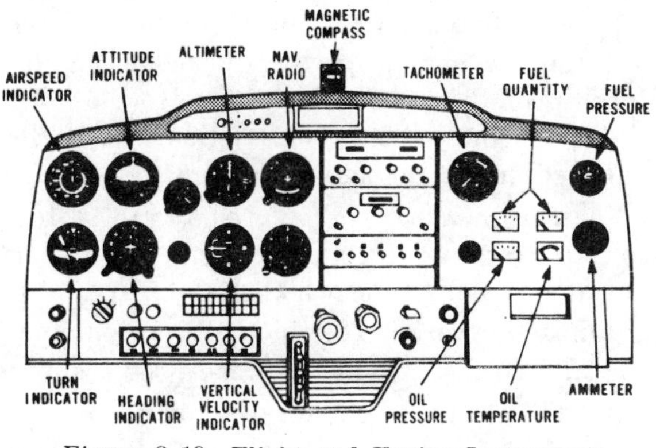

Figure 2–19 Flight and Engine Instruments

Flight Instruments

The *altimeter* measures the height of the airplane above a given atmospheric pressure level by measuring atmospheric pressure at the level of flight. Since it is the only instrument that gives altitude information, the altimeter is one of the most important instruments in the airplane. Its principle of measuring pressure is similar to that of an aneroid barometer. The altitude is presented on the instrument by multipointers and the dial is usually calibrated in units of 20, 100, and 1,000 feet.

The *airspeed indicator* is a sensitive, differential pressure gauge which indicates the speed at which the airplane is moving *through the air* (not the speed along the ground). It measures and shows promptly the difference between (1) pitot, or impact pressure of the air as the airplane moves forward, and (2) static pressure, the undisturbed atmospheric pressure at the level of flight. These two pressures will be equal when the airplane is stationary on the ground in calm air. When the airplane moves through the air, however, the pressure in the pitot line becomes greater than the pressure in the static lines. This difference in pressure is registered by the airspeed pointer on the face of the instrument, which is calibrated to show the pilot the speed of the airplane in statute miles per hour (MPH), or nautical miles per hour (Knots), or both.

The *attitude indicator* (artificial horizon), with its display of a miniature or representative airplane and a bar representing the natural horizon, is the one instrument that will give a clear picture of the flight attitude of the real airplane. The bank attitude is graduated in 10°, 20°, 30°, 60°, and 90°. The pitch attitude may or may not show the actual degree of pitch. When properly adjusted, the relationship of the miniature airplane to the horizon bar is the same as the relationship of the real airplane to the actual horizon.

The *turn and slip indicator* shows the direction and rate at which the airplane is turning and at the same time whether the airplane is slipping sideward. It is actually a combination of two instruments, a broad needle and a ball. The turn needle depends upon gyroscopic properties of precession for its indications and the ball is actuated by gravity and centrifugal force.

The turn indicator, being a gyroscopic instrument, indicates the direction and rate at which the airplane is turning about its vertical axis. It responds only to the rate of yaw. (The more modern "turn coordinator" also indicates the *rate of roll* about the airplane's longitudinal axis since it responds to both roll and yaw.) Unlike the attitude indicator, neither of these instruments gives a direct indication of the degree of bank of the airplane. However, for any given airspeed, there is a definite angle of bank necessary to maintain a coordinated turn at a given rate. The faster the airspeed, the greater the angle of bank required to obtain a given rate of turn. Thus, the turn indicator and the turn coordinator give only an *indirect* indication of the airplane's banking attitude or angle of bank during a coordinated turn.

The ball or slip indicator is a simple inclinometer consisting of a sealed, curved glass tube containing a steel ball which is free to move inside the tube. The tube is curved so that during coordinated flight centrifugal force or gravity causes the ball to rest in the lowest part of the tube centered between reference marks. When the effects of centrifugal force and gravity become unbalanced, as in a slip or skid, the ball moves away from the center of the tube.

The *magnetic compass* is a direction-seeking instrument whose basic component consists of two magnetized steel bars mounted on a float, around which is mounted the compass card. The compass card includes the letters N, S, E, and W to show the cardinal azimuth headings, and each 30° interval is represented by a number, the first and last zero of which is omitted. For example, 030° would appear as a 3, and 300° would appear as 30. Between the numbers, the compass card is graduated in 5° segments.

Since the magnetic compass is the only direction-seeking instrument in most airplanes, the pilot should be able to turn the airplane to the desired compass heading and then maintain it. There are certain inherent errors in the magnetic compass, making exact straight flight and precise turns to specific compass headings difficult to accomplish, particularly in turbulent air. The pilot should be familiar with the inherent errors and how to compensate for them. These are briefly explained in Chapter 13.

The *heading indicator* (directional gyro) is fundamentally a mechanical instrument designed to help in maintaining a magnetic heading since its indications are much more stable than the magnetic compass. However, because it has no inherent direction seeking characteristics, it must first be set to the heading shown on the compass. The calibration of its dial, though presented in an upright position for easier reading, is similar to that of the magnetic compass. The heading indicator, being gyroscopically operated, is not affected by the forces that cause the errors in the magnetic compass.

The *vertical velocity indicator* is a sensitive differential pressure gauge that senses the rate of change in static air pressure to indicate the rate at which the airplane is climbing or descending. The dial is graduated in hundreds of feet per minute.

Engine Instruments

The *tachometer* is an instrument for indicating the speed at which the engine crankshaft is rotating. The dial is calibrated in revolutions per minute (RPM). The pilot controls the RPM by use of the throttle and/or the propeller control.

The *oil pressure gauge* indicates the pressure under which oil is being supplied to the internal moving engine parts by the lubricating system. Thus, it warns the pilot of impending engine failure which may result from an exhausted oil supply, failure of the oil pump, broken oil lines, etc.

The *oil temperature gauge* indicates the temperature of the oil entering the engine so that the pilot can determine if the oil has reached the proper temperature for applying takeoff power and whether oil temperature is becoming excessive in flight. With a very low temperature the oil may be so thick that it is not able to lubricate the internal moving engine parts. With an excessively high temperature the lubricating oil tends to thin out and decreases the lubricating qualities of the oil.

Fuel quantity gauges enable the pilot to determine the amount of fuel in each of the tanks to ensure that sufficient fuel is available for the flight.

The *fuel pressure gauge* measures the difference in pressure between the fuel and the air being supplied to the engine. Thus, the instrument indicates whether fuel is being supplied and the actual pressure at which it is being forced into the engine.

The *manifold pressure gauge* indirectly indicates the power output of the engine by measuring the pressure of the air in the fuel/air induction manifold. The higher the manifold pressure, the greater the power being developed by the engine. By means of the throttle, the pilot can control the pressure in the manifold, and thus the power. By reference to this gauge, which is calibrated in inches of mercury, the power can be adjusted to specific values within the capabilities of the engine. The gauge is required on airplanes equipped with a supercharger or a constant speed propeller.

The *cylinder head temperature gauge* is an important instrument for engines capable of high compression and/or high power. It indicates the temperature of the cylinder head of the hottest cylinder (usually one of the rear ones in a horizontally-opposed or flat engine) and gives the pilot a means of determining whether the engine is operating at normal or excessive engine temperatures. Operating the engine at a temperature higher than it was designed for will cause loss of power, excessive oil consumption, and damage to the cylinder walls, pistons, and valves.

CHAPTER 3

Introduction to the Basics of Flight

The basic principles discussed in this chapter explain and establish the fact that an airplane is a stable, controllable flying machine, and why it *can* and *will* fly. To many people, flying machines seem complex and mysterious. Indeed, the technical details are complex. But behind the technology are basic principles that can be easily understood by anyone.

A complete and detailed discussion of the technology and aerodynamic principles is beyond the scope of this chapter. For those persons seeking the more advanced pilot ratings, particularly flight instructors and those who have a need or desire to learn more about aerodynamics, the last chapter in this handbook is devoted to the more technical aspects of airplane design and characteristics, and the physical forces and conditions that affect flight.

The long span of years between man's initial dream of flying and the final accomplishment was broken intermittently by the feats of the aeronauts, or balloonists. In the latter 18th century the Montgolfier Brothers of France first ascended into the air, and the observation balloon was quickly adapted to battle in the Franco-Austrian wars. The United States, also, first used the observation balloon to great advantage during the war between the states.

But it was not the same as free flight—man remained discontent as a captive dangling beneath the silk of a balloon. The pioneer pilots, Montgomery, Lilienthal, and Chanute made short flights in gliders, but they were brief and frustrating, for these men were unable to sustain flight and were forced back to earth against their will.

It was up to the Americans, Orville and Wilbur Wright to approach it scientifically, and to succeed. They built a wind tunnel, experimented with models, and learned much about wing curves, lifting forces, air resistance, and efficient wing design. Because of this careful research and intelligent study, the gulls circling the sand dunes at Kittyhawk in the winter of 1903 saw a man and a machine rise and conquer the sky for an historic 12 seconds.

The Wright brothers studied their model planes in the wind tunnel, learning the laws of physics which aided or prevented flight, and then learned how they could use the helpful forces and overcome the inhibiting forces. In similar manner, the discussions in this chapter provide the learning pilot of today with the basic facts of flight before actually beginning to fly.

Throughout the history of aviation, science and technology have produced many different types of aircraft. However, only the facts of flight that relate to the typical light plane used in general aviation will be discussed here.

Theories of Flight

A balloon rises in the air because it is filled with a gas which is lighter than the air around it. But an airplane, being heavier than the air surrounding it, cannot float as does a balloon; consequently, it must get its lift in a different manner. In effect, it must do something to the air surrounding it to make the air support its greater weight. To act on the air, the airplane must be placed in motion.

The majority of general aviation type airplanes employ a reciprocating engine to turn a propeller that "bites" into the air forcing the air backward while pulling the airplane forward. This leads our discussion to several basic laws of physics that help in explaining how an airplane flies.

The English philosopher and mathematician Sir Isaac Newton is credited with having observed in 1687 that "for every action there is an equal and opposite reaction" (Fig. 3–1). This principle applies whenever two things act upon each other, such as the air and the propeller, or the air and the wing of an airplane. In short, the statement about "action and reaction" tells us how lift and propulsion of airplanes are produced.

The only way air can exert a force on a solid body, such as an airplane's wing, is through pressure. In the 1700's, Daniel Bernoulli (a Swiss mathematician) discovered that if the velocity of a fluid (air) is increased at a particular point, the pressure of the fluid (air) at that point is decreased. The airplane's wing is designed to increase the velocity of the air flowing over the top of the wing as it moves through the air. To do this, the top of the wing is curved, while the bottom is relatively flat. The air flowing over the top travels a little farther (since it is curving) than the air flowing along the flat bottom. This means the air on top must go faster. Hence, the pressure decreases, resulting in a lower pressure (as Bernoulli stated) on top of the wing and a higher pressure below. The higher pressure then pushes (lifts) the wing up toward the lower pressure area (Fig. 3–2).

At the same time, the air flowing along the underside of the wing is deflected downward. As stated in Newton's theorem, "for every action there is an equal and opposite reaction." Thus, the downward deflection of air reacts by pushing (lifting) the wing upward (Fig. 3–2). This is like the planing effect of a speedboat or water skier skimming over the water. To increase the lift, the wing is tilted upward in relation to the oncoming air (relative wind) to increase the deflection of air. Relative wind during flight is not the natural wind, but is the direction of the airflow in relation to the wing as it moves through the air. The angle at which the wing meets the relative wind is called the *angle of attack.*

These two natural forces, pressure and deflection, produce *lift.* The faster the wing moves through the air and the greater the two forces become, the more lift is developed. Another of Newton's theorems can be used to explain just how much air deflection is needed to lift an airplane. This law of motion says

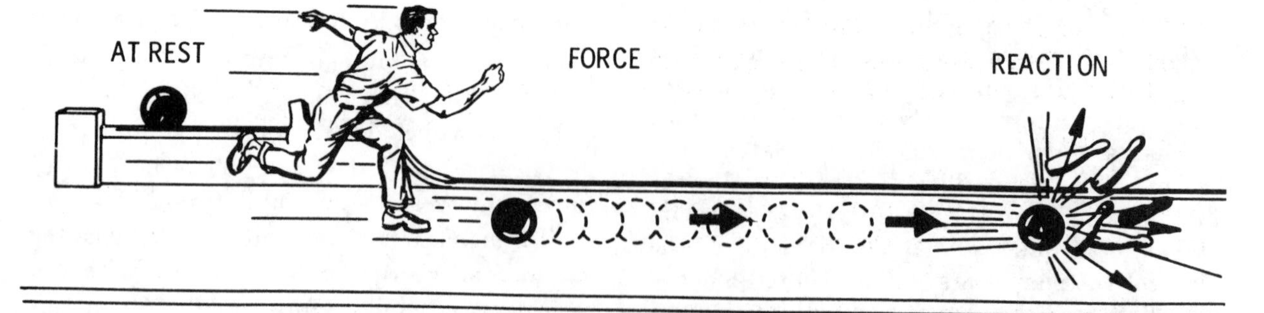

Figure 3–1 Equal and Opposite Reaction

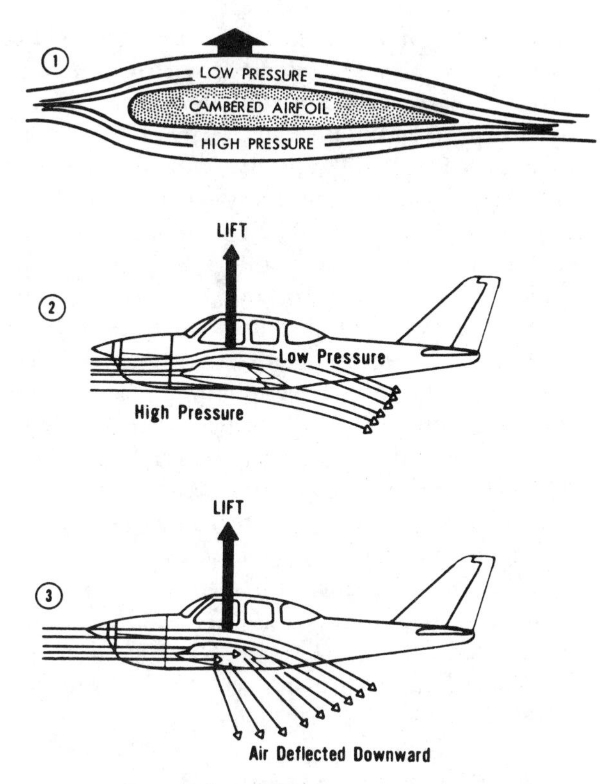

Figure 3–2 Forces Creating Lift

"the force produced will be equal to the mass of air deflected multiplied by the acceleration given to it." From this we can see the way lift, speed, and angle of attack are related.

The amount of lift needed can be produced by moving a large mass of air through the process of making a small change in velocity, or by moving a small mass of air through the process of making a large change in velocity. For example, at a high speed where the wing affects a large amount of air, only a small angle of attack is needed to deflect the air a small amount. Conversely, at low speeds during which the wing affects a lesser amount of air, a larger angle of attack is needed to deflect the air a large amount. Thus, the angle of attack at various speeds must be such that the deflection of air is adequate for the amount of lift needed.

If the airplane's speed is too slow, the angle of attack required will be so large that the air can no longer follow the upper curvature of the wing. This results in a swirling, turbulent flow of air over the wing and "spoils" the lift. Consequently, the wing stalls. On most airplanes this critical angle of attack is about 15° to 20° (Fig. 3–3). Stalls will be thoroughly explained in the chapter on Proficiency Flight Maneuvers.

As noted earlier, when the propeller rotates, it "bites" into the air, thereby providing the force to pull (or push) the airplane forward. This forward motion causes the airplane to act on the air to produce lift. The propeller blades, just like a wing, are curved on one side and straight on the other side. Hence, as the propeller is rotated by the engine, forces similar to those of the wing create "lift" in a forward direction. This is called *thrust.*

Up to this point the discussion has related only to the "lifting" force. Before an understanding of how an airplane flies is complete, other forces must be discussed.

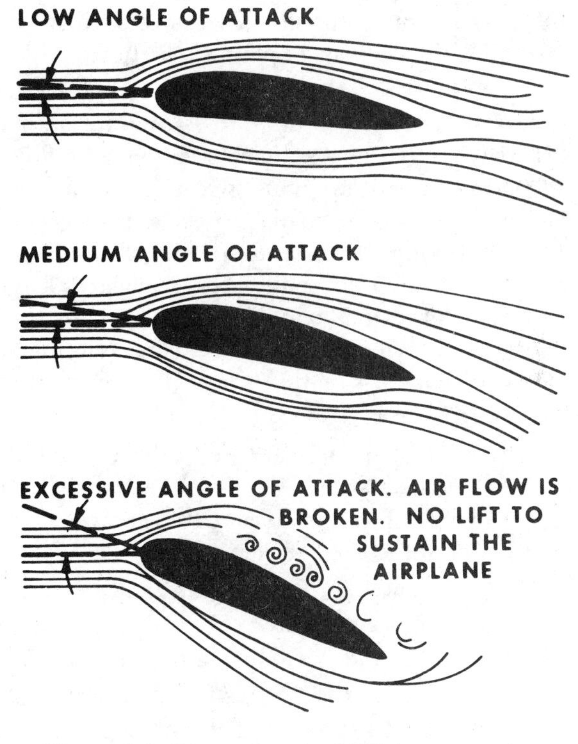

Figure 3–3 Increasing Angle of Attack to the Stall Point

Forces Acting on the Airplane

While the airplane is propelled through the air and sufficient lift is developed to sustain it in flight, there are certain other forces acting at the same time (Fig. 3–4).

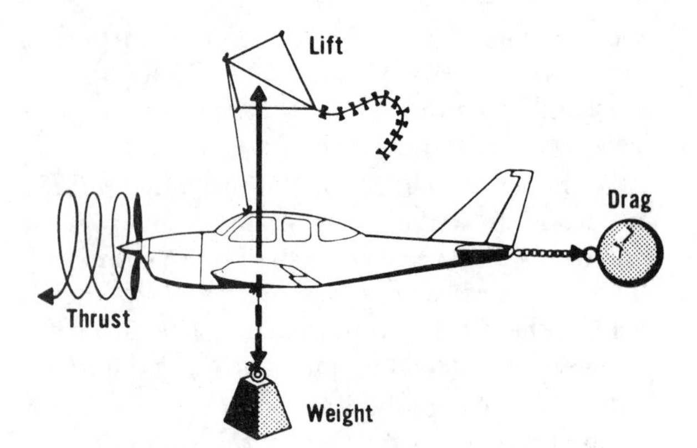

Figure 3–4 Four Forces Acting on the Airplane

Every particle of matter, including an airplane, is attracted downward toward the center of the earth by *gravitational* force. The amount of this force on the airplane is measured in terms of weight. To keep the airplane flying, *lift* must overcome the weight or gravitational force. The development of lift and thrust was explained earlier.

Another force that constantly acts on the airplane is called *drag.* It is the resistance created by the air particles striking and flowing around the airplane when it is moving through the air. Airplane designers are constantly trying to streamline wings, fuselages, and other parts to reduce the rearward force of drag as much as possible. The part of drag caused by form resistance and skin friction is termed *parasite drag* since it contributes nothing to the lift force.

A second part of the total drag force is caused by the wing's lift. As the wing deflects air downward to produce lift, the total lift force is not exactly vertical, but is tilted slightly rearward. This means that it causes some rearward drag force. This drag is called *induced drag*, and is the price paid to produce lift. The larger the angle of attack, the more the lift force on the wing tilts toward the rear and the larger the induced drag becomes. To give the airplane forward motion, the thrust must overcome drag.

In a steady flight condition (no change in speed or flightpath), the always present forces that oppose each other are also equal to each other. That is, lift equals weight, and thrust equals drag.

Another force which frequently acts on the airplane is *centrifugal* force. However, this force occurs only when the airplane is turning or changing the direction (horizontally or vertically) of the flightpath. Newton's law of energy states that "a body at rest tends to remain at rest, and a body in motion tends to remain moving at the same speed and in the same direction." Thus, to make an airplane turn from straight flight, a sideward-inward force must act on it (Fig. 3–5). The tendency of the airplane to keep moving in a straight line and outward from a turn is the result of inertia and it results in centrifugal force. Therefore, some impeding force is needed to overcome this centrifugal force so the airplane can move in the desired direction. The lift of the wings provides this counteracting force when the airplane's wings are banked in the desired direction. This is further discussed in this chapter in the section on Turning Flight.

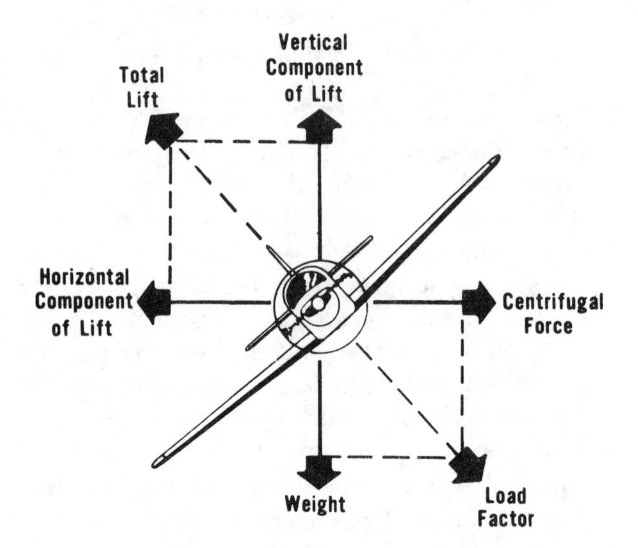

Figure 3–5 Forces Acting on the Airplane in a Turn

Since the airplane is in a banked attitude during a properly executed turn, the pilot will feel the centrifugal force by increased seat pressure, rather than the feeling of being forced to the side as is experienced in a rapidly turning automobile. The amount of force (G force) felt by seat pressure depends on the rate of turn. The pilot will, however, be forced to the side of the airplane (as in an automobile) if a turn is improperly made or the airplane is made to slip or skid.

One other force which will affect the airplane during certain conditions of flight, and which will be frequently referred to in the discussions on various flight maneuvers, is *torque effect* or left turning tendency. It is probably one of the least understood forces that affect an airplane.

Torque effect is the force which causes the airplane to have a tendency to swerve (yaw) to the left, and is created by the engine and propeller. There are four factors which contribute to this yawing tendency; (1) torque reaction of the engine and propeller, (2) the propeller's gyroscopic effect, (3) the corkscrewing effect of the propeller slipstream, and (4) the asymmetrical loading of the propeller. It is important that pilots understand why these factors contribute to torque effect.

One of Newton's laws states, "for every action there is an equal and opposite reaction." Hence, the rotation of the propeller, with a clockwise movement (as viewed from the cockpit), tends to roll or bank the airplane in a counterclockwise (to the left direction (Fig. 3–6).

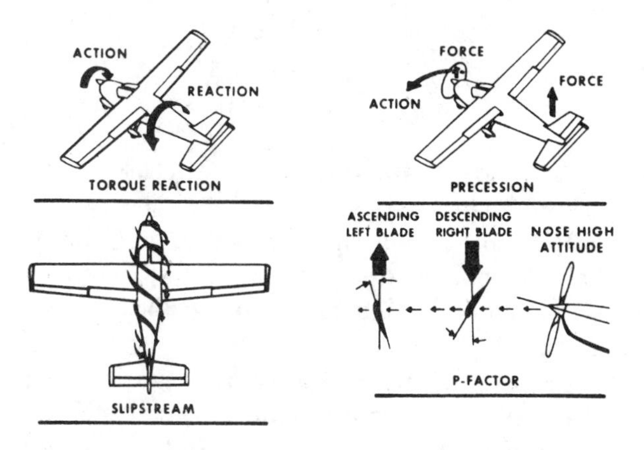

Figure 3–6 Left Turning Tendencies

This can be understood by visualizing a rubber-band-powered model airplane. Wind the rubber band in a manner that it will unwind and rotate the propeller in a clockwise direction. If the fuselage is released while the propeller is held the fuselage will rotate in a counterclockwise direction (looking from the rear). This effect of *torque reaction* is the same in a real propeller-driven airplane except that instead of the propeller being held by hand, its rotation is resisted by air.

This counter-rotational force causes the airplane to try to roll to the left. It will be noted in the case of a real airplane that the force is stronger when power is significantly advanced while the airplane is flying at very slow airspeed.

The second factor that causes the tendency of an airplane to yaw to the left is the gyroscopic properties of the propeller. Here, we are concerned with *gyroscopic precession* which is the resultant action or deflection of a spinning object when a force is applied to the outer rim of its rotational mass. When a force is applied to the object's axis, it is the same as applying the force to the outer rim. If the axis of a spinning gyroscope (propeller in this case) is tilted, the resulting force will be exerted 90° ahead in the direction of rotation and in the same direction as the applied force (Fig. 3–7). That force will be particularly noticeable during takeoff in a tailwheel type airplane if the tail is rapidly raised from a three-point to a level flight attitude. The abrupt change of attitude tilts the horizontal axis of the propeller, and the resulting precession produces a forward force on the right side (90° ahead in the direction of rotation), yawing the airplane's nose to the left. The amount of force created by this precession is directly related to the rate at which the propeller axis is tilted when the tail is raised.

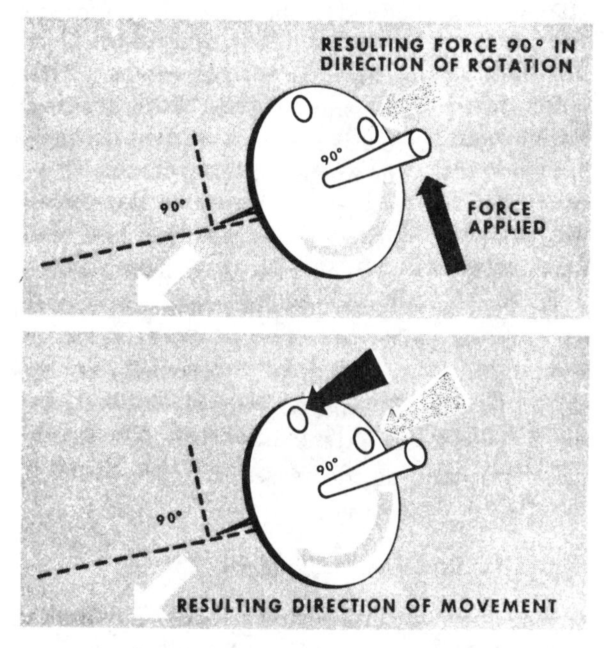

Figure 3–7 Gyroscopic Precession

The third factor that causes the airplane's left yawing tendency is the *corkscrewing* of the propeller slipstream, acting against the side of the fuselage and tail surfaces (Fig. 3–6). The high speed rotation of an airplane propeller results in a corkscrewing rotation to the slipstream as it moves rearward. At high propeller speeds and low forward speed, as in the initial part of a takeoff, the corkscrewing flow is compact and imposes considerable side forces on the airplane. As the airplane's forward speed increases, the corkscrew motion of the slipstream loosens or elongates, resulting in a straighter flow of air along the side of the fuselage toward the airplane's tail.

When this corkscrewing slipstream strikes the side of the fuselage and the vertical tail surface at airspeeds less than cruising, it produces a yawing moment which tends to revolve the airplane around its vertical axis. Since in most U.S. built airplanes propeller rotation is clockwise as viewed from the cockpit, the slipstream strikes the vertical tail surface on the left side, thus pushing the tail to the right and yawing the nose of the airplane to the left.

The fourth factor which causes the left yawing tendency is the asymmetrical loading of the propeller, frequently referred to as *P-factor* (Fig. 3–6). When an airplane is flying with a high angle of attack (with the propeller axis inclined), the bite of the downward-moving propeller blade is greater than the bite of the upward-moving blade. This is due to the downward-moving blade meeting the oncoming relative wind at a greater angle of attack than the upward-moving blade. Consequently, there is greater thrust on the downward-moving blade on the right side and this force causes the airplane to yaw to the left.

At low speeds the yawing tendency caused by P-factor is greater because the airplane is at a high angle of attack. Conversely, as the speed of the airplane is increased and the airplane's angle of attack is reduced, the asymmetrical loading decreases and the turning tendency is decreased.

Stability and Controllability

Besides being supported in flight by lift and propelled through the air by thrust, an airplane must be stable and controllable since it is free to revolve or move around three axes. These axes may be thought of as axles around which the airplane revolves, much like a wheel does. Each axis is perpendicular to the other two and all three intersect at the airplane's *center of gravity* (CG). The point around which the airplane's weight is evenly distributed or balanced is considered the CG of the airplane.

The axis which extends lengthwise through the fuselage from the nose to the tail is the *longitudinal axis*. The axis extending through the fuselage from wingtip to wingtip is the *lateral axis*. The axis which passes vertically through the fuselage at the center of gravity is the *vertical axis* (Fig. 3–8).

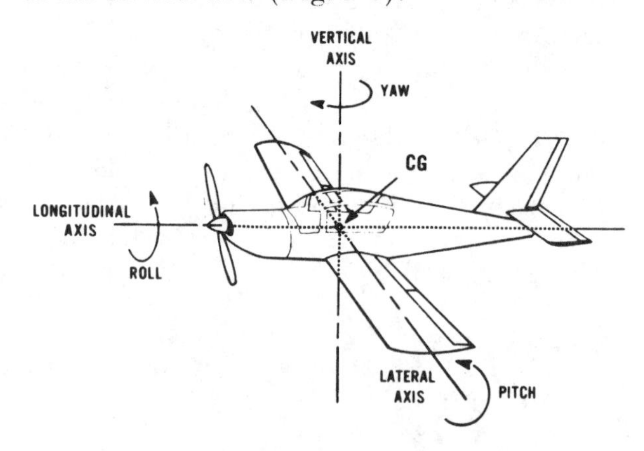

Figure 3–8 Axes of the Airplane

Rotation about the airplane's longitudinal axis is roll, rotation about its lateral axis is pitch, and rotation about its vertical axis is yaw.

Because of their ability to revolve about these axes, all airplanes must possess stability in varying degrees for safety and ease of operation. An unstable airplane would require that the pilot continually vary pressures on the flight controls and consequently would be difficult to control. The term "stability" means the ability of the airplane to return of its own accord to its original condition of flight, or the normal flight attitude, after it has been disturbed by some outside force. A ball in a round bowl is considered stable because after being pushed to one side it will roll back and forth until it finally comes to rest at the center of the bowl.

If an arrow having no feathered tail is shot from a bow, it usually will wobble or fall end-over-end as it travels, since there is no force produced to bring it back to its original point-first travel. The arrow is made stable, however, by adding pieces of feather near the rear of its shaft. Then, when the arrow is shot and begins to wobble, turn, or yaw, the air strikes the tail feathers at an angle and deflects the feathered end of the shaft to turn the arrow back to a straight path. This corrective action continues as long as the arrow has sufficient forward motion.

An airplane wing by itself is also unstable. It would flip over and continue to flip end-over-end as it flutters to the ground. Like the unstable arrow, the unstable wing needs some kind of "tail feathers" to balance it and keep it on a straight course. Like the stable arrow, airplanes have their "tail feathers" in the form of horizontal and vertical surfaces located at the rear of the fuselage. These surfaces are the horizontal stabilizer and the vertical stabilizer or fin.

If all the upward lift forces on the wing were concentrated in one place, there would be established a center of lift, which is usually called *center of pressure* (CP). In addition, if all the weight of the airplane were concentrated in one place, there would be a center of weight, or as it is termed, *center of gravity* (CG). Rarely, though, are the CP and CG located at the same point.

The locations of these centers in relation to each other have a significant effect on the stability of the airplane. If the center of the wing's lifting force (CP) is forward of the airplane's center of gravity (CG)—the airplane would always have a tendency to nose up and would have an inherent tendency to enter a stalled condition. Therefore, most airplanes are designed to have their CG located slightly forward of the CP, to create a nose-down tendency so the airplane will have a natural tendency to pitch downward away from a stalling condition (Fig. 3–9). This provides a safety feature in the characteristics of the airplane.

While the airplane is flying within its range of normal speeds, the airflow exerts a downward force on the horizontal stabilizer; thus,

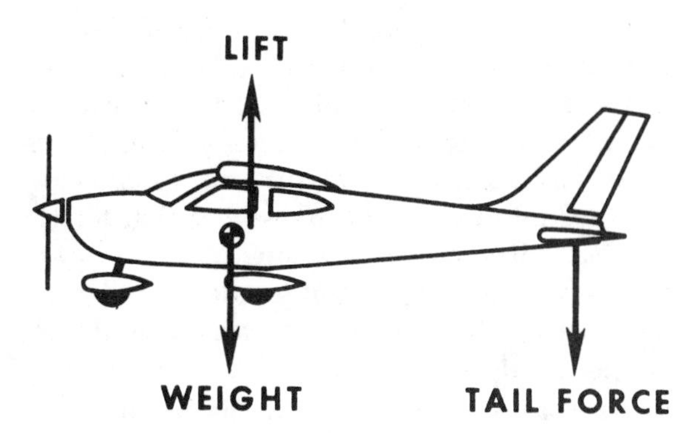

Figure 3–9 Balancing Forces of the Airplane

at normal cruise speed it partially offsets the inherent nose heaviness of the airplane. In addition, many airplanes have the line of thrust located lower than the CG. In this situation the propeller's thrust provides a nose-up pitching force to help overcome the inherent nose heaviness. With this balanced condition, the airplane characteristically will remain in level flight. However, when the power is reduced and the airspeed is decreased, the airflow exerts less downward force on the horizontal stabilizer. At the same time the nose-up force of thrust is also decreased (Fig. 3–10). Due to this unbalanced condition the airplane's nose will tend to lower and the airplane will enter a descent of its own accord.

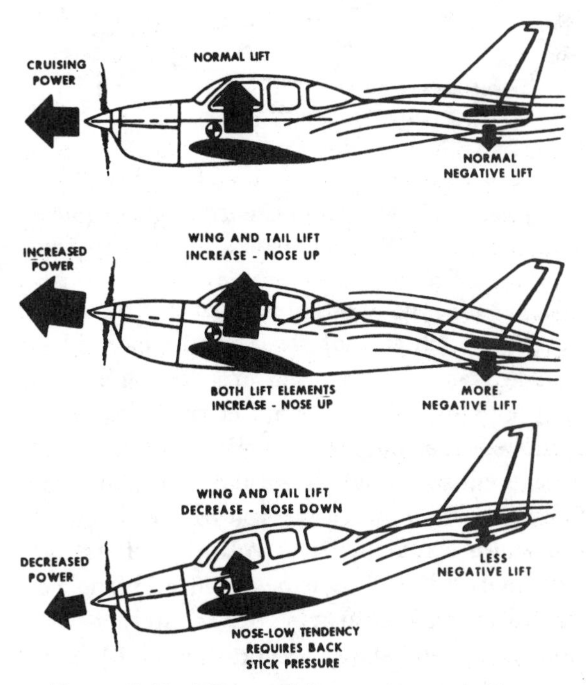

Figure 3–10 Effects of Power/Airspeed Changes

During the descent the airspeed will begin increasing. As a result, the downward force increases on the horizontal stabilizer, causing the nose to rise. This process will continue again and again if the airplane is dynamically stable (and if the pilot takes no action to stop it), but with each oscillation the nose-up and nose-down motion becomes less and less. Eventually, the airplane's descent attitude and airspeed will stabilize.

Like the feathered arrow, the most important factor producing directional stability is the weathervaning effect created by the fuselage and vertical fin of the airplane (Fig 3–11). It keeps the airplane headed into the relative wind. If the airplane yaws, or skids, the sudden rush of air against the surface of the fuselage and fin quickly forces the airplane back to its original direction of flight.

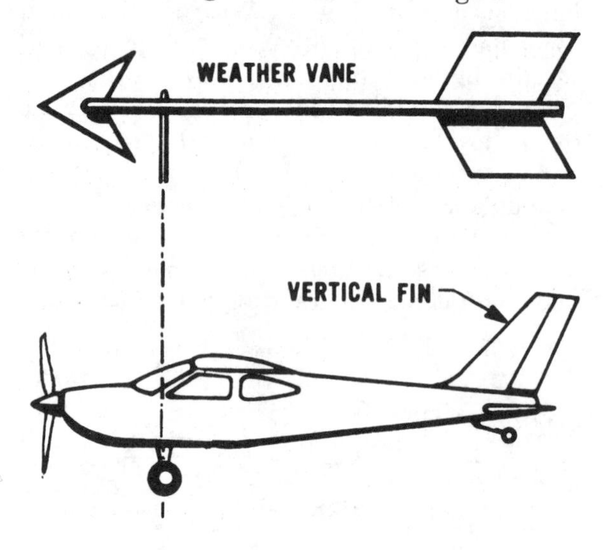

Figure 3–11 Weathervaning Tendency of the Airplane

Generally, in straight-and-level flight, the wings on each side of the airplane have identical angles of attack and are developing the same amount of lift. This laterally balanced condition normally keeps the airplane level. Occasionally, though, a gust of air will upset this balance by increasing the lift on one wing and cause the airplane to roll around its longitudinal axis. A well designed airplane has certain design features to counteract this momentary unbalanced condition and return the airplane to a wings-level attitude.

Most airplanes are designed so that the outer tips of the wings are higher than the wing roots attached to the fuselage. The upward angle thus formed by the wings is called *dihedral,* and is usually only a few degrees (Fig. 3–12).

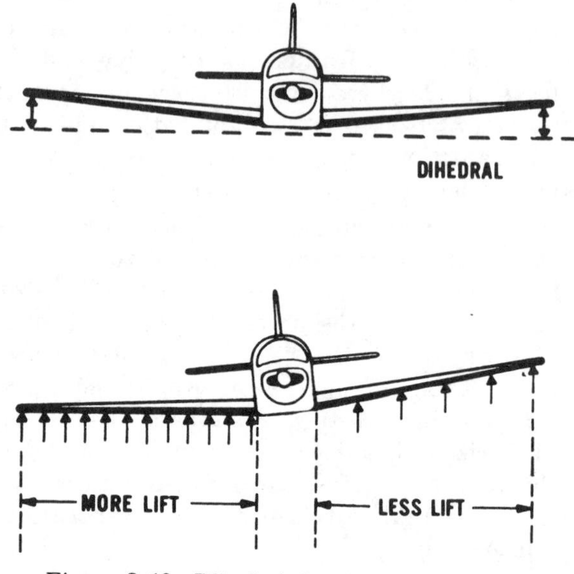

Figure 3–12 Dihedral for Lateral Stability

The rolling action of an airplane caused by gusts is constantly being corrected by the dihedral of the wings. If one wing gets lower than the other when the airplane is flying straight, it will have a different attitude in relation to the oncoming air. The result is that the lowered wing has a greater angle of attack and thus more lift than the raised wing and consequently will rise.

If this rising action causes the wing to go past the level attitude, the opposite wing will then have a greater angle of attack and more lift. A dynamically stable airplane will oscillate less and less and eventually will return to its original position as the oscillation dampens.

Although stability in an airplane is desirable, it must not be so strong that the pilot cannot overcome the inherent stability. The pilot must be able to control or maneuver the airplane at will about the airplane's three axes.

Roll, pitch, and yaw, the motions of an airplane about its longitudinal, lateral, and vertical axes, are controlled by the three control surfaces. This will be discussed in the chapter on The Effect and Use of Controls.

Turning Flight

An airplane, like any moving object, requires a sideward force to make it turn. In a normal turn, this force is supplied by banking the airplane so that lift is exerted inward as well as upward. The force of lift is thus separated into two components at right angles to each other (Fig. 3–5). The lift acting upward and opposing weight is called the *vertical lift component.* The lift acting horizontally and opposing inertia or centrifugal force is called the *horizontal lift component.* Thus the *horizontal lift component* is the sideward force that forces the airplane from straight flight and causes it to turn. The equal and opposite reaction to this sideward force is centrifugal force. If an airplane is not banked, no force is provided to make it turn unless the turn is skidded by rudder application. Likewise, if an airplane is banked, it will turn unless held on a constant heading with opposite rudder. Proper control technique assumes that an airplane is turned by banking, and that in a banking attitude it should be turning.

Banking an airplane in a level turn does not by itself produce a change in the *amount* of lift. However, the division of lift into horizontal and vertical components reduces the amount of lift supporting the weight of the airplane. Consequently, the reduced vertical component results in the loss of altitude unless the total lift is increased by (1) increasing the angle of attack of the wing, (2) increasing the airspeed, or (3) increasing the angle of attack and airspeed in combination. Assuming a level turn with no change in thrust, the angle of attack is increased by raising the nose until the vertical component of lift is equal to the weight. The greater the angle of bank, the weaker is the vertical lift component, and the greater is the angle of attack for the lift/weight balance necessary to maintain a level turn.

Climbing Flight

For an automobile to go uphill at the same speed that was being maintained on a level road, the driver must "step on the gas;" that is, power must be increased. This is because it takes more work to pull the car's weight up the hill and to maintain the same speed at which the car was moving along the level road. If the driver did not increase the power, the automobile might still climb up the incline, but it would gradually slow down to a speed slower than that at which it was moving on the level road.

Similarly, an airplane can climb at the cruise power setting with a sacrifice of speed, or it can, within certain limits, climb with added power and no sacrifice in speed. Thus, there is a definite relationship between power, attitude, and airspeed.

When transitioning from level flight to a climb, the forces acting on the airplane go through definite changes (Fig. 3–13). The first change, an increase in lift, occurs when back pressure is applied to the elevator control. This initial change is a result of the increase in the angle of attack which occurs when the airplane's pitch attitude is being raised. This results in a climbing attitude. When the inclined flightpath and the climb speed are established, the angle of attack and the corresponding lift again stabilize at approximately the original value.

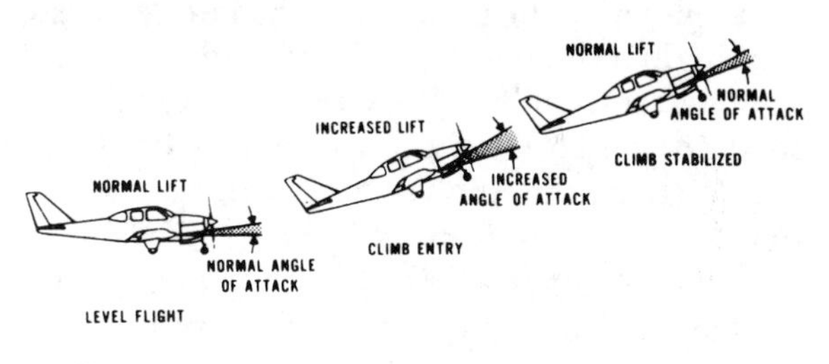

Figure 3–13 Angles of Attack During Climb

As the airspeed decreases to the climb speed, the downward force of the air striking the horizontal stabilizer becomes less, creating a longitudinally unbalanced condition which produces a tendency for the airplane to nose down. To overcome this tendency and maintain a constant climb attitude, additional back pressure must be applied to the elevator control.

The primary factor which affects an airplane's ability to climb is the amount of excess power available; that is, the power available above that which is required for straight-and-level flight. During the climb, lift operates perpendicularly to the flightpath, so that it is not directly opposing gravity to support the airplane's weight. With the flightpath inclined, the lift is partially acting rearward—creating what is termed induced drag. This adds to the total drag. Since weight is always acting perpendicular to the earth's surface, and drag is acting in a direction opposite to the airplane's flightpath during a climb, it is necessary for thrust to overcome both drag and gravity.

Descending Flight

When the power is reduced during straight-and-level flight, the thrust needed to balance the airplane's drag is no longer adequate. Due to the unbalanced condition, the drag causes a reduction in airspeed. This decrease in speed, in turn, results in a corresponding decrease in the wing's lift. The weight of the airplane now exceeds the force of lift so the resulting flightpath is downward as well as forward. Since the flightpath is inclined downward, the force of gravity is providing the forward thrust. In effect, the airplane is actually going "down hill."

As in entering a climb, the forces acting on an airplane again go through definite changes when transitioning from level cruising flight to a descent. When forward pressure is applied to the elevator control or the airplane's pitch attitude is allowed to lower, the wing's angle of attack is decreased, the lift is reduced, and the flightpath starts downward. This change in the flightpath is the result of the lift becoming less than the weight of the airplane as the angle of attack is reduced. This unbalance of lift and weight causes the airplane to descend with respect to the horizontal path of level flight. The initial reduction of lift which starts the airplane downward is momentary. When the flightpath stabilizes, the angle of attack again approaches the original value, and lift and weight stabilize.

Just as in climbing flight, the downward force on the horizontal stabilizer becomes less as the airspeed decreases. This produces an unbalanced condition which results in a tendency for the airplane to nose down. In order to maintain the desired descent attitude, it is usually necessary to apply slight back pressure on the elevator control.

As the descent is started, the airspeed may gradually increase, due to a component of weight now acting forward along the flightpath. The overall effect is that of increased thrust, which in turn would cause the airspeed to increase if the power were allowed to remain the same as that used for level cruise flight. For descent at the same airspeed as flown in level cruise flight, the power must be reduced as the descent begins.

The component of weight acting forward along the flightpath increases as the descent attitude is steepened and, conversely, will decrease as the descent attitude is shallowed. Therefore, the amount of power reduction for a descent at cruising speed will be determined by the rate of descent desired.

CHAPTER 4

The Effect and Use of Controls

This chapter briefly discusses the devices with which the pilot operates the airplane in the air and on the ground, and how those devices are to be used effectively.

To maneuver an airplane, the pilot must control its movement around its lateral, longitudinal, and vertical axes. This is accomplished by the use of the flight controls—elevators, ailerons, and rudder—which can be deflected from their neutral position into the flow of air as the airplane moves forward through the air. During flight, the flight controls have a natural "live pressure" due to the force of the airflow around them.

With this in mind, the pilot should think not of moving the flight controls, but of exerting force on them against this live pressure or resistance.

Elevators

The elevators control the movements of the airplane about its *lateral* axis. They form the rear part of the horizontal stabilizer, and are free to be moved up and down by the pilot, and are connected to a control stick or wheel in the cockpit by means of cables or rods. Applying forward pressure on the control causes the elevator surfaces to move downward. The flow of air striking the deflected elevator surfaces exerts an upward force, pushing the airplane's tail upward and the nose downward. Conversely, exerting back pressure on the control causes the elevator surfaces to move up, exerting a downward force to push the tail downward and the nose upward (Fig. 4–1).

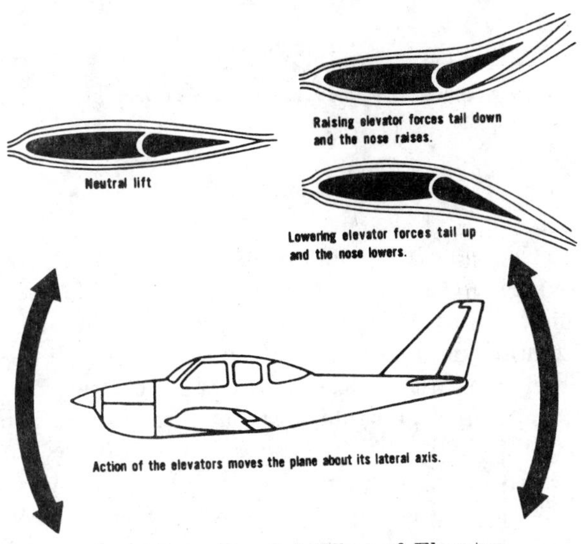

Figure 4–1 Use and Effect of Elevator

In effect, the elevators are the angle-of-attack control. When back pressure is applied on the control, the tail lowers and the nose rises, thus increasing the wing's angle of attack and lift.

Some airplanes have a movable horizontal surface called a "stabilator," which serves the same purpose as the horizontal stabilizer and elevators combined. When the cockpit control is moved, the complete stabilator is moved to raise or lower its leading edge, thus changing its angle of attack and amount of lift. In turn, this changes the wing's angle of attack and amount of lift.

Ailerons

The ailerons control the airplane's movement about its *longitudinal* axis. There are two ailerons, one at the trailing edge of each wing, near the wingtips. They are movable surfaces hinged to the wing's rear spar and are linked together by cables or rods so that when one aileron is deflected down, the opposite aileron moves up (Fig. 4–2).

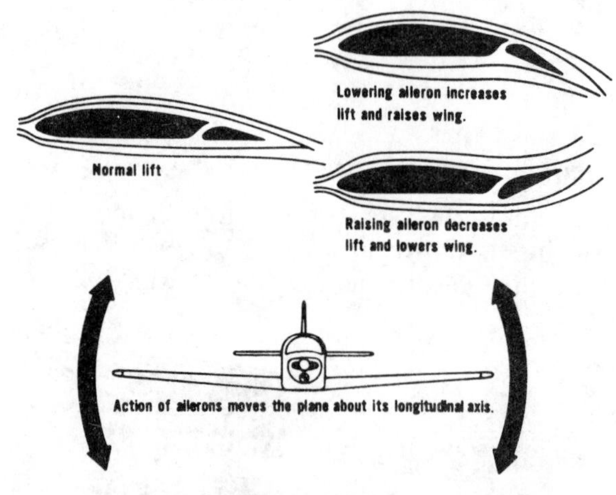

Figure 4–2 Use and Effect of Ailerons

Contrary to popular belief, the lift on the wings is the force that turns the airplane in flight—*not* the rudder. To obtain the horizontal component of lift required to pull the airplane in the desired direction of turn, the wings must be banked in that direction. When the pilot applies pressure to the left on the control stick or turns the control wheel toward the left, the right aileron surface deflects downward and the left aileron deflects upward. The force exerted by the airflow on the deflected surfaces raises the right wing and lowers the left wing (Fig. 4–3). This happens because the downward deflection of the right aileron changes the wing camber and increases the angle of attack and lift on that wing. Simultaneously, the left aileron moves upward and changes the effective camber, resulting in a decreased angle of attack, and less lift. Thus, decreased lift on the left wing and increased lift on the right wing causes the airplane to roll and bank to the left.

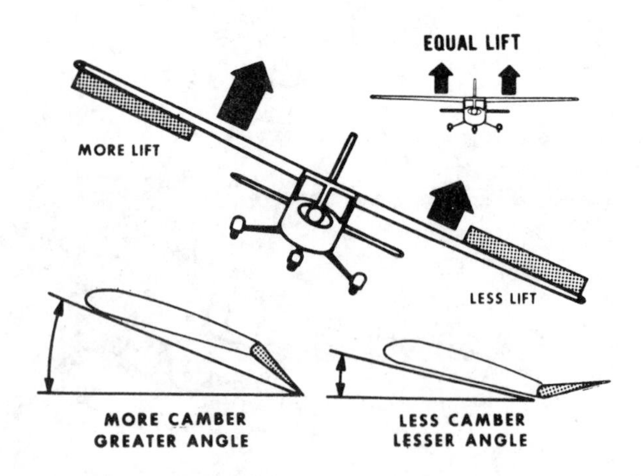

Figure 4–3 Forces Exerted by Ailerons

Since the downward deflected aileron produces more lift, it also produces more drag, while the opposite aileron has less lift and less drag. This added drag attempts to pull or veer the airplane's nose in the direction of the raised wing; that is, it tries to turn the airplane in the direction opposite to that desired (Fig. 4–4). This undesired veering is referred to as *adverse yaw*.

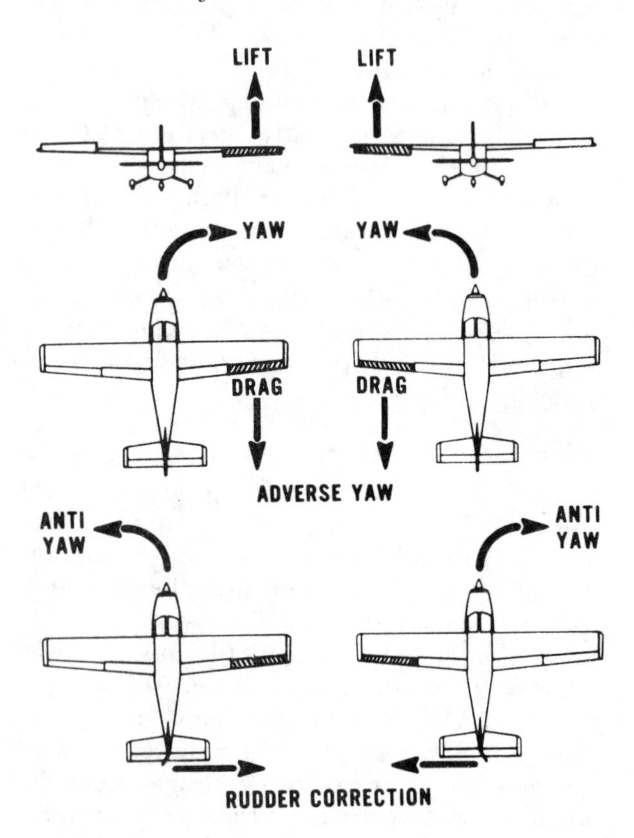

Figure 4–4 Cause and Effect of Adverse Yaw

To demonstrate this in flight, an attempt can be made to turn to the right without using the rudder pedals. As right aileron pressure is applied, the airplane rolls into a right bank and tries to turn to the right. But the adverse yaw, or the drag on the downward deflected left aileron, pulls the airplane's nose to the left. The airplane banks, but it turns hesitantly and sideslips. This is undesirable and corrective action should be taken by applying right rudder pressure.

When right rudder pressure is applied simultaneously with right aileron pressure, it keeps the airplane from yawing opposite to the desired direction of turn. In fact, the rudder must be used because the ailerons were used. Therefore, neither of those controls should be used separately when making normal turns.

To minimize this undesirable effect (adverse yaw), many airplanes are designed with differential-type ailerons or Frise-type ailerons.

Differential-Type Ailerons

Though not entirely eliminating adverse yaw, the "differential-type" aileron system raises one aileron a greater distance than the other aileron is lowered for a given movement of the control stick or wheel. In this case, since the raised aileron has as much or more surface area exposed to the airflow (thus increased drag) than the lowered aileron, the adverse yaw is greatly reduced.

Frise-Type Ailerons

The design of the aileron surface itself has also been improved by the "Frise-type" aileron. With this type of aileron, when pressure on the control stick or wheel is applied to one side, raising one of the ailerons, the *leading edge* of that aileron (which has an offset hinge) projects down into the airflow and creates drag. This helps equalize the drag created by the lowered aileron on the opposite wing and thus reduces adverse yaw (Fig. 4–5).

The Frise-type aileron also forms a slot so that the air flows smoothly over the lowered aileron. This helps to make the aileron more effective at high angles of attack. However, despite these improvements, some rudder action is still needed whenever ailerons are applied.

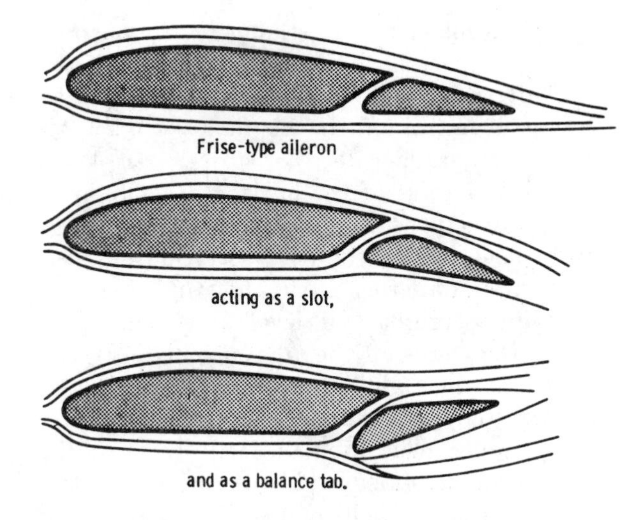

Figure 4–5 Frise-Type Ailerons

Rudder

The rudder controls movement of the airplane about its *vertical* axis. This is the motion called *yaw*. Like the other primary control surfaces, the rudder is a movable surface hinged to a fixed surface—in this case to the vertical stabilizer, or fin. Movement of the rudder is controlled by two rudder pedals—left and right. Its action is very much like that of the elevators, except that it moves in a different plane; the rudder deflects from side to side instead of up and down. When the rudder is deflected to one side, it protrudes into the airflow, causing a horizontal force to be exerted in the opposite direction (Fig. 4–6). This pushes the tail of the airplane in that direction and yaws the nose in the desired direction. When rudder is used for steering during ground taxiing, the propeller slipstream provides the force to yaw or turn the airplane in the desired direction.

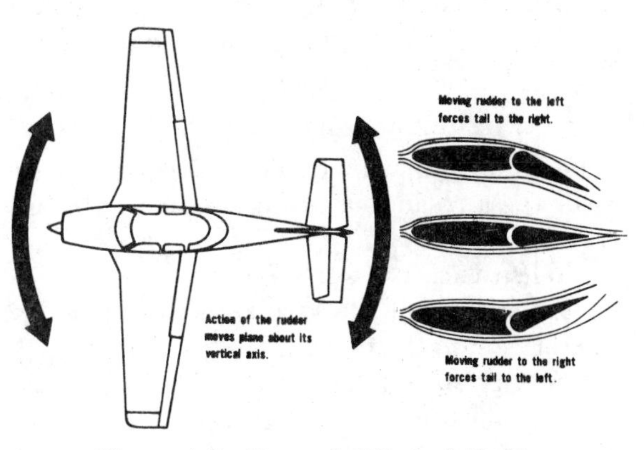

Figure 4–6 Use and Effect of Rudder

As mentioned earlier, the *primary* purpose of the rudder in flight is to counteract the effect of adverse yaw and to help provide directional control of the airplane (Fig. 4–4). In flight the rudder does *not* turn the airplane; instead, the force of the horizontal component of wing lift turns the airplane when the wings are banked. As in the demonstration of turning by use of ailerons alone, this can be verified by flying straight and level and then, after taking the hands off the aileron control, trying to turn to the right by applying right rudder pressure only. At first it may seem to work pretty well. The airplane will turn to the right but it will also skid to the left. (A skid in a turn is caused by insufficient angle of bank for a given radius of turn.) Since the airplane possesses inherent stability, it will tend to stop the skid by banking itself to the right.

If the pilot were to now neutralize the rudder, only a shallow banking turn would result. However, inasmuch as the purpose of this demonstration is to make a turn using only the rudder, continue to hold right rudder pressure. Since the airplane is slightly banked to the right, the rudder will force the nose of the airplane downward to the right. The reason for this is that yawing is the only movement the rudder can produce. As a result, the nose yaws downward, the airspeed increases, and the airplane starts losing altitude. At the same time, the airplane, being stable, attempts to stop the increased skidding by banking more steeply. The more steeply it banks, the more the nose is yawed downward by the right rudder action. The net result of holding rudder alone is a descending spiral unless back elevator pressure is applied. Thus, it can be seen that rudder alone cannot produce a coordinated turn. *Blending aileron, rudder, and elevator pressure does that.*

Use of Flight Controls

The following will always be true, regardless of the airplane's attitude in relation to the earth:

1. When back pressure is applied to the elevator control, the airplane's nose rises in relation to the pilot.
2. When forward pressure is applied to the elevator control, the airplane's nose lowers in relation to the pilot.
3. When right pressure is applied to the aileron control, the airplane's right wing lowers in relation to the pilot.
4. When left pressure is applied to the aileron control, the airplane's left wing lowers in relation to the pilot.
5. When pressure is applied to the right rudder pedal, the airplane's nose moves to the right in relation to the pilot.
6. When pressure is applied to the left rudder pedal, the airplane's nose moves to the left in relation to the pilot.

The preceding explanations should prevent the beginning pilot from thinking in terms of "up" or "down" in respect to the earth, which is only a relative state to the pilot. It will also make understanding of the functions of the controls much easier, particularly when performing steep banked turns and the more advanced maneuvers. Consequently, the pilot must be able to properly determine the control application required to place the airplane in any attitude or flight condition that is desired.

Coordinated use of ALL controls is extremely important in any turn. Applying aileron pressure is necessary to place the airplane in the desired angle of bank, while simultaneous application of rudder pressure is required to counteract the resultant adverse yaw. During a turn, the angle of attack must be increased by application of elevator pressure because more lift is required than when in straight-and-level flight. The steeper the turn, the more back elevator pressure is needed.

After the bank and turn are established, the airplane may continue in a constant turn when all pressure on the ailerons and rudder is released, or it may require some opposite aileron to prevent the bank angle from increasing. It may also require continued aileron pressure in the direction of turn to prevent returning to the wings-level attitude. This will be discussed in more detail in the chapter on Basic Flight Maneuvers.

After learning how the airplane will react when the flight controls are used, the pilot must learn how to use them properly. Rough and erratic usage of all or any one of the controls will cause the airplane to react accordingly; therefore, the pilot must form the habit of applying pressures smoothly and evenly.

The amount of force the airflow exerts on a control surface is governed by the airspeed and the degree that the surface is moved out of its neutral or streamlined position. Since the airspeed will not be the same in all maneuvers, the actual amount the control surfaces are moved is of little importance; but it is important that the pilot maneuver the airplane by applying sufficient control *pressures* to obtain a desired result, regardless of how far the control surfaces are actually moved.

The pilot's feet should rest comfortably against the rudder pedals. Both heels should support the weight of the feet on the cockpit floor with the ball of each foot touching the individual rudder pedals. The legs and feet should not be tense; they must be relaxed just as when driving an automobile.

When using the rudder pedals, pressure should be applied smoothly and evenly by pressing with the ball of one foot just as when using the brakes of an automobile. Since the rudder pedals are interconnected and act in opposite directions, when pressure is applied to one pedal, pressure on the other must be relaxed proportionally. When the rudder pedal must be moved significantly, heavy pressure changes should be made by applying the pressure with the ball of the foot while the heels slide along the cockpit floor. Remember, the ball of each foot must rest comfortably on the rudder pedals so that even slight pressure changes can be felt.

During flight, it is the *pressure* the pilot exerts on the control stick or wheel and rudder pedals that causes the airplane to move about its axes. When a control surface is moved out of its streamlined position (even slightly), the air flowing past it will exert a force against it and will try to return it to its streamlined position. It is this force that the pilot feels as pressure on the control stick or wheel and the rudder pedals.

Trim Devices

The secondary flight controls are the *trim tabs*, *balance tabs*, *or servo tabs*. They are used for trimming and balancing the airplane in flight and to reduce the force required of the pilot in actuating the primary flight control surfaces. These tabs are really small airfoils attached to, or recessed into, the trailing edge of the primary control surfaces (Fig. 4–7).

Trim Tabs. When an airplane's flight conditions (attitude, airspeed, loading, etc.) and configuration are changed, the control pressures required to maintain the new flight conditions are affected by the resulting changes in aerodynamic forces. The placement of fuel, passengers, baggage, or cargo sometimes results in the airplane being wing-heavy, tail-heavy, or nose-heavy. To counteract such unbalanced conditions the pilot will find it necessary to continually exert pressure on the control stick or wheel, or on the rudder pedals. Over a period of time this becomes annoying and fatiguing. To relieve the pilot of this tiring effort, most airplanes are equipped with trim tabs with which to trim the airplane for balanced flight.

A trim tab is a small, adjustable hinged surface, located on the trailing edge of the aileron, rudder, or elevator control surface. It is used to maintain balance in straight-and-level flight and during other prolonged flight conditions without the pilot having to hold pressure on the controls. This is accomplished by deflecting the tab in the direction *opposite* to that in which the primary control surface must be held. The force of the airflow striking the tab causes the main control surface to be deflected to a position that will correct the unbalanced condition of the airplane.

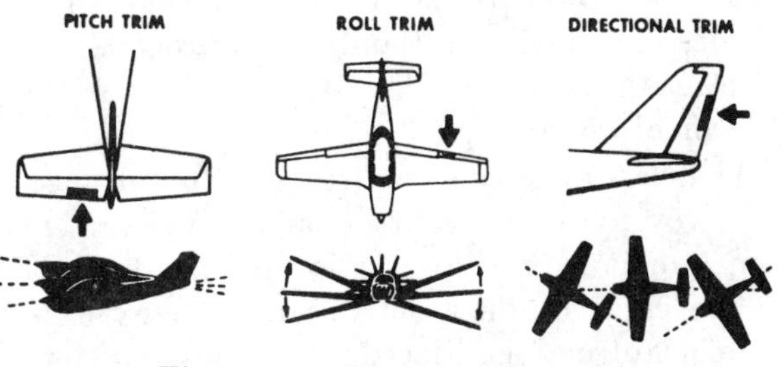

Figure 4–7 Effects of Trim Tabs

In many airplanes the trim tabs may be adjusted by controls in the cockpit, while in some of the older types they may be adjustable only on the ground. Those which can be controlled from the cockpit provide a trim control wheel or electric switch. To apply a trim force, the trim wheel or switch must be moved in the desired direction. The position in which the trim tab is set can usually be determined by reference to a trim indicator.

Balance Tabs. Balancing tabs look like trim tabs and are hinged in approximately the same places as trim tabs would be. The essential difference between the two is that the balancing tab is coupled to the control surface by a rod, so that when the primary control surface is moved in any direction the tab automatically is moved in the opposite direction. In this manner the airflow striking the tab counterbalances some of the air pressure against the primary control surface and enables the pilot to more readily move and hold it in position. These tabs also act as trim tabs as described above.

Servo Tabs. Servo tabs are very similar in operation and appearance to the trim tabs previously discussed. Servo tabs, sometimes referred to as flight tabs, are used primarily on the large airplanes. They aid the pilot in moving the control surface and in holding it in the desired position. Only the servo tab moves in response to movement of the pilot's flight control, and the force of the airflow on the servo tab then moves the primary control surface.

Flaps Control

The wing flaps are movable panels on the inboard trailing edges of the wings. They are hinged so that they may be extended downward into the flow of air beneath the wings to increase lift and drag. Their purpose is to permit a slower airspeed and a steeper angle of descent during a landing approach. In some cases they are also used to shorten the takeoff distance.

The flap operating control may be an electrical or hydraulic control on the instrument panel, or it may be a lever located on the floor to the right of the pilot's seat. In any case, the control may be placed in various positions by the pilot: UP, which raises flaps if they are in an extended position; NEUTRAL, which allows the flaps to remain in whatever intermediate position they may be at that time; and DOWN, which lowers the flaps if they are in the retracted or intermediate position (Fig. 4–8). In addition to the flap operating control, there is usually an indicator which shows the actual position of the flaps. On most general aviation airplanes the extent of travel of the flaps is approximately 30°–40°.

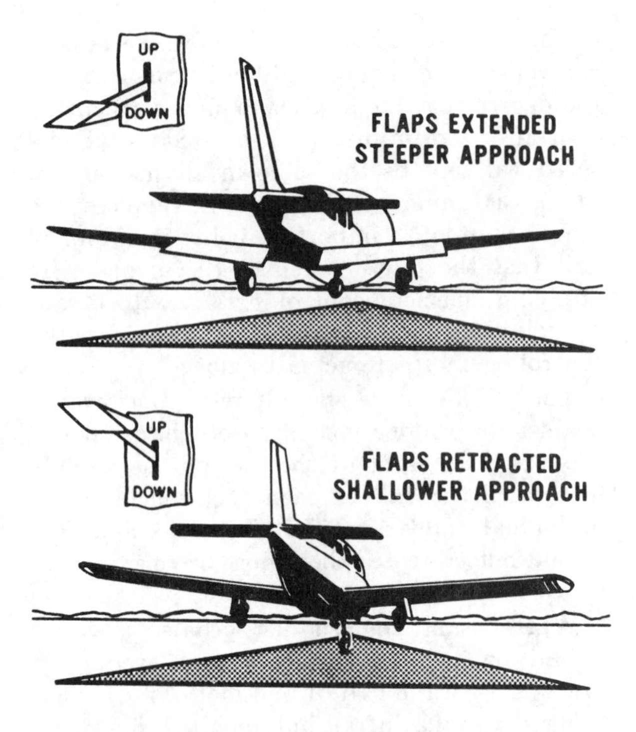

Figure 4–8 Use of Wing Flaps

Extending and retracting the flaps has a very noticeable effect on the airplane's performance. With a constant power setting while maintaining level flight, the airspeed will be lower with flaps extended because of the drag they create. If power is adjusted to maintain a constant airspeed while in level flight, the airplane's pitch attitude will usually be lower with flaps extended.

When the flaps are extended, the airspeed should be at or below the airplane's maximum flap extended speed (V_{fe}), because if they are extended above this airspeed, the force exerted by the airflow may result in damage to the flaps. If the airspeed limitations are exceeded unintentionally with the flaps extended, they should be retracted immediately regardless of airspeed.

It is extremely important that the pilot form the habit of *positively identifying* the flap control before attempting to use it to raise or lower the flaps. This will prevent inadvertently operating the landing gear control and retracting the gear instead of the flaps, particularly when on or near the ground.

A complete discussion of the use and effect of flaps is provided in the chapter describing Landing Approaches and Landings.

Retractable Landing Gear Control

Another control in airplanes so equipped, is the control for operating the retractable landing gear. Since the purpose of the landing gear is to support the airplane while it is on the ground, it is merely excess weight and drag during flight. Though the airplane cannot be relieved of this weight while in flight, the landing gear can be retracted into the airplane structure and out of the airflow, thus eliminating the unnecessary drag (Fig. 4–9).

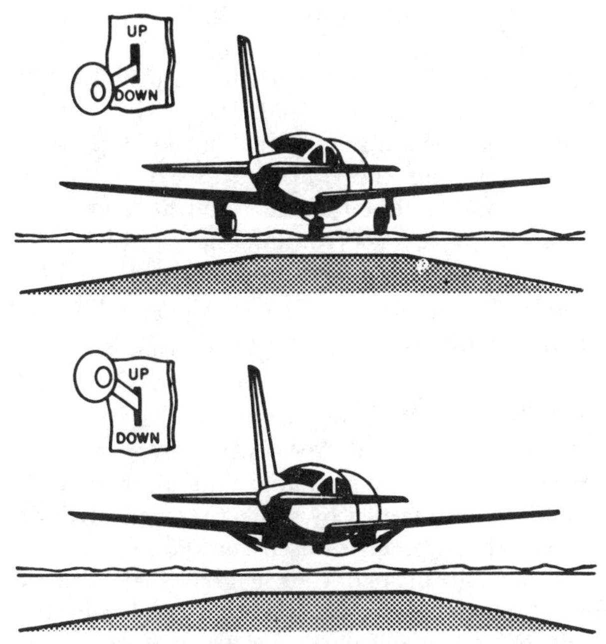

Figure 4–9 Use of Landing Gear Control

The control for operating the landing gear is a switch or lever, often in the shape of a wheel (to differentiate from the flap control which has an airfoil shape), on the instrument panel. When the control is moved to the DOWN position, the gear will extend; when the control is moved to the UP position the gear will retract. In addition to this operating control, an indicator or warning light is located on the instrument panel to show the position of the gear.

The landing gear should be operated only when the airspeed is at or below the airplane's maximum landing gear operating speed (V_{LO}). Operation at a higher airspeed may cause damage to the operating mechanism. When the gear is down and locked, the airplane should not be operated in excess of the airplane's maximum landing gear extended speed (V_{LE}).

It is extremely important that the pilot form the habit of *positively identifying* the landing gear control before attempting to use it to raise or lower the gear. Otherwise the pilot may inadvertently use the flap control and operate the flaps instead of the gear.

Throttle

The throttle is the control with which the pilot controls the engine's power output. In tandem-seating airplanes it is located on the left side of the cockpit along with other primary engine controls, and in side-by-side seating airplanes it is centrally located on the lower portion of the instrument panel. On some airplanes all the engine controls (throttle, mixture, and propeller) are installed on a separate unit called "engine control quadrant." By means of linkage the throttle is connected to the carburetor (or fuel control unit in fuel-injection engines) to regulate the amount of fuel-air mixture supplied to the engine, thereby controlling the power developed (Fig. 4–10).

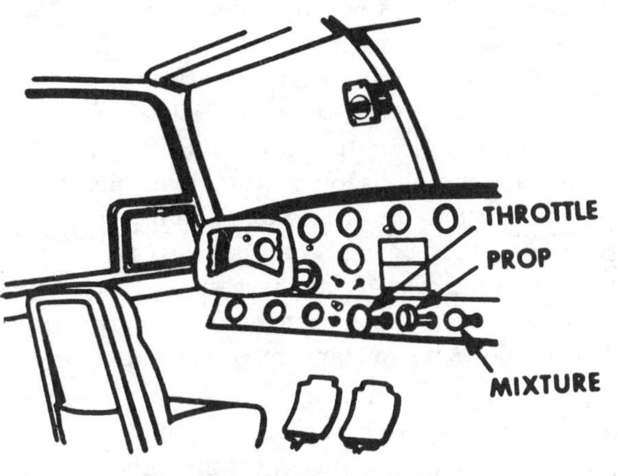

Figure 4–10 Engine Controls

The power output (and the engine RPM in the case of fixed-pitch propeller airplanes) is increased by pushing the throttle forward and decreased by pulling it aft. Unlike an automobile accelerator or gas pedal, the airplane's throttle has no spring return and will remain in any position to which it is set by the pilot. Undesired "creeping" can be prevented by means of a friction lock. Movement of the throttle should always be smooth, though positive, to ensure adequate and correct engine response when a change in power is needed. Rough or abrupt throttle usage may result in a delayed engine response or a complete loss of power, as well as in the possibility of exceeding the engine's limitations. Initial throttle movement should be made slowly, then followed by an increase in the rate of throttle movement as the engine responds. There is no situation that will be improved by rough or abrupt throttle usage, even when necessity requires quick changes in power.

Mixture Control

The fuel-air ratio of the combustible mixture delivered to the engine is controlled by the mixture control. This control, usually located adjacent to the throttle, is most often identified by a red knob (an indication to use it with caution). Like the throttle, it has no spring return and will remain in any position selected by the pilot. It also has a friction lock to prevent the control from "creeping" away from the desired setting.

The ratio of fuel-to-air is the most critical single factor affecting the power output of an engine. If the fuel-air mixture is too lean (too little fuel for the amount of air—in terms of weight), rough engine operation, sudden "cutting out," or an appreciable loss of power may occur. Lean mixtures must be avoided when an engine is operating near maximum output (such as takeoffs, climbs, and go arounds). At power settings in excess of 75% of the rated power, an excessively lean mixture will cause detonation, serious overheating, loss of power, and damage to the engine.

If the fuel-air mixture is too rich (too much fuel for the amount of air—in terms of weight), rough engine operation and an appreciable loss of power may also occur.

Carburetors and fuel control units normally are calibrated for sea-level operation, which means that the correct mixture of fuel and air will be obtained at sea level with the mixture control in the "full rich" position. As altitude increases, the air density decreases; that is, a cubic foot of air will not weigh as much as it would at a lower altitude. Consequently, as the flight altitude increases, the weight of air entering the cylinders will decrease, although the volume will remain the same. The amount of fuel entering the cylinders is more dependent on the *volume* of air than on the weight of air. Therefore, as the flight altitude increases, the amount of fuel will remain approximately the same for any given throttle setting if the position of the mixture control remains unchanged. Since the same amount (weight) of fuel but a lesser amount (weight) of air is entering the cylinders, the fuel-air mixture becomes richer as altitude increases (Fig. 4–10).

Moving the mixture control full forward to the RICH position provides the richest fuel mixture; this setting is used for all operations on the ground and at high power settings. Movement of the control aft toward the LEAN position progressively leans the mixture. The amount of leaning required varies with altitude. Moving the control full aft to IDLE CUT-OFF shuts off all fuel flow at the carburetor or the fuel control unit.

The manufacturer's recommendations for leaning the fuel mixture in the particular airplane should always be followed.

Propeller Control

On airplanes equipped with a controllable propeller the pilot's manual control of the propeller is maintained by a control lever adjacent to the throttle. It is used for the selection of the desired propeller or engine speed. Any rotational speed (revolutions per minute) within the prop governor's operating range can be selected by adjusting the lever fore and aft between the full DECREASE RPM position to the full INCREASE RPM position (Fig. 4–10).

The propeller control lever is connected to a propeller governor unit which, in turn, adjusts the propeller blades to establish the selected RPM settings. As the propeller control lever is moved forward, propeller and engine speed (RPM) will increase; if the control is moved aft, RPM will decrease.

In a constant-speed propeller system the governor automatically maintains the selected engine speed by varying the pitch of the propeller blades to compensate for varying engine loads, regardless of the airplane's attitude. A setting introduced into the governor by the pilot determines the RPM to be maintained and the governor then controls the pitch required to maintain the RPM.

CHAPTER 5

Ground Operations

This chapter discusses the basic procedures and techniques essential to the safe operation of the airplane on the ground prior to and after a flight. This includes the major points of ensuring that the airplane is in an airworthy condition, starting and stopping the engine, and taxiing the airplane to and from the parking area and the runway.

At many airports there usually is considerable activity in and around the parking area—aircraft may be operating at high power settings, taxiing in and out of the area, taking off and landing on nearby runways, and fuel trucks and people may be moving among the aircraft. Consequently, constant vigilance must be exercised at all times while performing ground operations.

The propeller is the most dangerous part of the airplane, since under certain light conditions, it is difficult to see a revolving propeller. This may give the illusion that it is not there. As a result, the files of FAA Aviation Safety Offices contain many cases that read: "Victim walked into a rotating propeller."

All pilots should form the habit of walking along the inner edge of the terminal ramp or parking area to the point at which the assigned airplane is located. When the parking spot is reached, the airplane should be approached from the rear. (This same procedure should be followed, in reverse order, whenever leaving the airplane.) While approaching the airplane, the pilot should note the presence of obstructions and articles such as fire extinguishers, fueling or maintenance equipment, or chocks that could be a hazard when taxiing the airplane.

The accomplishment of a safe, pleasant flight begins with a careful visual inspection before the pilot enters the airplane. In addition, a planned routine of starting, warm-up, taxiing, and before-takeoff checks will assure that the airplane is operating properly while there still is an opportunity to correct any discrepancy which may appear. When well organized, these checks can be made quickly, and soon will become matters of habit; the appearance, sound, and even the odor in the airplane will become familiar, and anything unfamiliar will alert the pilot that something is not as it has been or should be.

The use of an appropriate written checklist to inspect and start the airplane, as well as for all other ground checks and procedures, is highly recommended.

Checklists are guides for use in ensuring that all necessary items are checked in a logical sequence. The beginning pilot should not get the idea that the list is merely a crutch for poor memory—even the most experienced professional pilots never attempt to fly without an appropriate checklist. The habit of using a written checklist for the airplane being used should be so instilled in pilots that they will follow this practice throughout their flying activities (Fig. 5–1).

Figure 5–1 Use a Written Checklist

Visual Inspection of the Airplane

To a pilot, the airworthiness of the airplane is both a legal obligation and a direct responsibility. Careful personal attention to preflight procedures is the mark of a safe pilot and will be repaid not only in safety, but in lower airplane maintenance costs.

As the pilot approaches the airplane, the external visual inspection should be started by looking for hazardous obstructions in the parking area, and for dripping oil and fuel leaks under the airplane. Upon reaching the airplane all tiedowns, control locks, and chocks should be removed and the general appearance of the airplane checked for signs of damage such as dents, cracks, or scratches. Then, the preflight inspection should be performed in accordance with the printed checklist provided by the airplane manufacturer (Fig. 5–2).

Figure 5–2 Perform a Thorough Visual Inspection

Water and dirt contamination in the airplane's fuel system is potentially dangerous; therefore, the pilot should take certain actions to prevent contamination or eliminate contamination that may have occurred. If the airplane's weight and balance will not be adversely affected for the next flight the fuel tanks should be completely filled after each flight, or at least after the last flight of the day. The reason is that when air in the fuel tanks cools, the moisture it contains condenses into water, and contaminates the fuel. The more fuel there is in the tanks, the less moist air the tanks will contain and the less condensation and contamination will occur.

The pilot should *always* assume that the fuel in the airplane may be contaminated with water, and take the necessary steps to eliminate it during the preflight inspection. A substantial amount of fuel should be drained from the fuel strainer (gascolator) quick drain and, if possible, from each fuel tank sump into a transparent container to check for dirt and water. Water will be noticeable since it will sink to the bottom of the sample. Water (being heavier than gasoline), seeks the lowest levels in the fuel system—that is where the fuel drains are located. If water is found in the first sample, drain further samples until no trace appears.

A second preventive measure is to avoid refueling from cans and drums, which may introduce fuel contamination by dirt or other impurities.

Since each make and model of airplane has different features to inspect, it is impractical to provide an appropriate checklist here. Nonetheless, the following are some of the major items that should be given particular attention during a preflight inspection:

1. Check landing gear control DOWN (if retractable gear).
2. Turn master switch ON; check the fuel quantity gauges.
3. Check master switch and ignition switch OFF.
4. Visually check fuel supply in tanks; secure the tank caps.
5. Drain fuel tank sumps; check for contamination.
6. Drain fuel system sump (gascolator); check for contamination.
7. Check that fuel system vents are open.
8. Check oil level (ensure that dipstick is properly seated).
9. Check for obvious fuel or oil leaks.
10. Check cowling and inspection covers for security.
11. Check propeller and spinner for defects or nicks.
12. Check tires for cuts, wear, and proper inflation.
13. Check nose gear and landing gear shock struts for proper inflation. Check wheel wells (if retractable gear).
14. Check hydraulic lines and landing gear struts for leaks.
15. Inspect tailwheel spring, steering arms, steering chains, and tire inflation (if tailwheel type).
16. Remove pitot tube cover, if installed, and inspect pitot tube for clear opening.
17. Inspect static air source for clean opening.
18. Ensure that wings and control surfaces are free of mud, snow, ice, or frost.
19. Remove control surface lock, if installed.
20. Check for damage and operational interference of control surfaces or hinges.
21. Check landing flaps for signs of operational interference.
22. Check windshield and cabin windows for cleanliness.
23. Check carburetor air intake for obstructions.
24. Check baggage for proper storage and security.
25. Close and secure the baggage compartment door.

Cockpit Management

After entering the airplane, the pilot should first ensure that all necessary equipment, documents, and navigation charts are aboard. All too often a flight is begun before it is realized that certain essentials have been left behind. A check should be made for the Airworthiness Certificate, Registration Certificate, operating limitations, weight and balance data, and, if equipped with radio transmitter, the airplane's FCC Radio Station License. Of course, the pilot must also have valid FAA pilot and medical certificates and an FCC radiotelephone operator's permit.

If a cross-country flight is contemplated, a navigation computer and appropriate navigation charts, as well as a pencil and note pad should be aboard. The latter two articles are useful, even on local flights, for jotting down pertinent weather information and ATC clearances in certain terminal areas.

Regardless of what materials are to be used, they should be neatly arranged and organized in a manner that makes them readily available for use by the pilot. The cockpit or cabin should be checked for loose articles which might be tossed about if turbulence is encountered. All pilots should form the habit of "good housekeeping"; in the long run, it will pay off (as it does for professional pilots) in safe and efficient flying.

On each flight the pilot should be seated in the same position. If the seat is adjustable, it should be moved so that the pilot's knees are slightly bent and the balls of the feet placed on the rudder pedals. This will allow full movement of the pedals whenever necessary.

The pilot must be able to see inside and outside references without straining. Poor vision not only causes apprehension and confusion, but actually presents a hindrance to the control of the airplane. If the seat is not adjustable, cushions should be used to provide proper seating, but in their use, comfort and ease of control must not be sacrificed.

When the pilot is comfortably seated, the seatbelt should immediately be fastened and adjusted to a comfortably snug fit. This should be accomplished even though the engine is only to be run up momentarily.

If the seat is adjustable, it is important to ensure that the seat is locked in position. Many accidents have occurred as the result of acceleration or deceleration during takeoffs or landings when the seat suddenly moved too close or so far away from the controls that the pilot was unable to maintain control of the airplane.

Starting the Engine

The actual procedures for starting the engine will not be discussed here since there are as many different methods required as there are different engines, systems, starters, and propellers. The before-starting and starting checklist provided for the particular airplane being used should always be followed. There are, however, certain precautions pointed out here that apply to all airplanes.

Too many careless pilots start the engine with the tail of the airplane pointed toward an open hangar door, toward parked automobiles, or toward a group of bystanders. This is not only discourteous, thoughtless, and in violation of Federal Aviation Regulations, but often results in personal injury and serious damage to the property of others.

When ready to start the engine, the pilot should look around in all directions to be sure that nothing is or will be in the vicinity of the propeller, and that nearby persons and aircraft will not be struck by the propeller blast or the debris it might pick up from the ground.

If an electric starter is used, the pilot should always call "all clear," and wait for a response from persons who may be nearby before turning the ignition switch ON or activating the starter. While activating the starter, one hand should be kept on the throttle, to be ready to advance the throttle if the engine falters while starting or to prevent excessive RPM just after starting. A low power setting is recommended until the engine temperatures and oil pressure starts increasing.

As soon as the engine is operating smoothly, the oil pressure should be checked. If it does not rise to the manufacturer's specified value in about 30 seconds in summer or 60 seconds in winter, the engine is not receiving proper lubrication and should be shut down immediately to prevent internal damage.

Even though most airplanes are equipped with electric starters, every pilot should be familiar with the procedures and *dangers* involved in starting an engine by turning the propeller by hand (hand propping). Due to the associated hazards, this method of starting should be used only when absolutely necessary and when proper precautions have been taken. There have been many fatalities, serious injuries, and substantial property damage caused by the rotating propeller blades when the airplane suddenly moved forward, uncontrolled under its own power after hand starting.

It is recommended that an engine *never* be "hand propped" unless a qualified person thoroughly familiar with the operation of all the controls is seated at the controls and the brakes set. As an additional precaution, chocks should be placed in front of the main wheels. If this is not feasible, the airplane's tail should be securely tied down. NEVER ALLOW A PERSON WHO IS UNFAMILIAR WITH AIRPLANE CONTROLS TO HANDLE THE CONTROLS WHEN THE ENGINE IS STARTED BY AN OUTSIDE SOURCE.

When hand propping is necessary, the ground surface near the propeller should be firm and free of debris. Loose gravel, slippery grass, mud, or grease might cause the person "propping" the airplane to slip or fall into the rotating propeller as the engine starts.

First the ignition switch should be checked to be sure it is OFF. Then the blade to be swung should be rotated so that it is slightly above the horizontal position. The person doing the hand propping should face the blade squarely and stand close but not too close to the propeller blade. If standing too far away, it would be necessary to lean forward in an unbalanced position to reach the blade. This may cause the person to fall forward into the revolving blades when the engine starts.

After the throttle is set to the start position and the ignition switch turned ON, the propeller is swung by forcing the blade downward rapidly, pushing with the palms of both hands. If the blade is gripped tightly with the fingers, the person's body may be drawn into the propeller blades should the engine misfire and rotate in the opposite direction. As the blade is pushed down, the "hand-propper" should step backward away from the propeller. If the engine does not start, the propeller should not be repositioned for another attempt until it is *certain* the ignition switch is turned OFF.

When removing the wheel chocks after the engine starts remember that the propeller is almost invisible. There have been cases of serious injuries and fatalities because the person reached into the whirling propeller to remove the chocks. Before they are removed, the throttle should be set to idling and the chocks approached from the rear of the propeller—never from the front, or the side.

As stated previously, the procedure for starting should always be in accordance with the manufacturer's recommendations or checklist. Nonetheless, the following are some of the items that are essential in any starting procedure:

1. Remove control lock.
2. Check flight controls for freedom of movement.
3. Set wheel brakes.
4. Set carburetor heat control to COLD position.
5. Set mixture control to FULL RICH position.
6. Set propeller control (if equipped) to full INCREASE position.
7. Turn unnecessary electrical units OFF.
8. Set fuel selector to ON for desired tank.
9. Set throttle ¼″ to 1″ open.
10. Prime engine then lock the primer (if equipped).
11. Turn ignition switch to BOTH ON.
12. Turn master (battery) switch ON.
13. Turn rotating beacon ON.
14. Call out CLEAR.
15. Activate starter switch.
16. Adjust throttle for warmup (800–1000 RPM).
17. Check oil pressure.

Taxiing—General

Taxiing is the controlled movement of the airplane under its own power while on the ground. Since an airplane must be moved under its own power between the parking area and the runway, the pilot must thoroughly understand taxiing procedures and be proficient in maintaining positive control of the airplane's direction and speed of movement on the ground. In addition, the pilot must be alert and visually check the location and movements of everything else along the taxi path.

An awareness of other aircraft which are taking off, landing, or taxiing, and consideration for the right-of-way of others is essential to safety. To really observe the entire area, the pilot's eyes must cover almost a complete circle. While taxiing, the pilot must be sure the airplane's wings will clear all obstructions and other aircraft. If at any time there is doubt about the clearance of the wingtips, the pilot should stop the airplane and have someone check the amount of clearance from the object. If no help is available, the engine should be shut down. It may be necessary to have the airplane physically moved by a ground crew.

It is difficult to set any rule for a safe taxiing speed. What is safe under some conditions may be hazardous under others. The primary requirement of safe taxiing is safe positive control; the ability to stop or turn where and when desired. Normally, the speed should be at the rate where movement of the airplane is dependent on the throttle; that is, slow enough so when the throttle is closed the airplane can be stopped promptly.

Except while taxiing very slowly, it is best to slow down before attempting a turn. Otherwise, the turn may have too great a radius or could result in an uncontrollable swerve or a ground loop. The latter is most likely to occur when turning from a downwind heading to an upwind heading, due to the airplane's tendency to "weathervane" when the airplane is headed crosswind. This tendency is explained in more detail in the chapter "Landing Approaches and Landings." When turning from an upwind heading to a downwind heading, however, this precaution is not as important, since the weathervane tendency will usually cause a deceleration in the rate of turn.

Very sharp turns or attempting to turn at too great a speed must be avoided as both tend to exert excessive strains on the airplane, and such turns are difficult to control, once started. Many airplanes, particularly those with a short distance between the wheels, have a tendency to tip over and drag a wing tip on the ground, causing serious damage.

Since airplane flight control surfaces are designed for maximum effectiveness in flight, they have limited use during the airplane's movement on the ground. At normal taxi speeds, air pressures will usually not be felt on the control surface as they will during flight.

When taxiing at normal taxi speeds in a no-wind condition, the aileron and elevator control surfaces have little or no effect on directional control of the airplane and therefore should not be considered steering devices. Remember, the flight control's effect on the airplane is the result of pressure exerted by the airflow. Very little air pressure is created by the airflow from an idling propeller or from the slow forward movement of the airplane at a low taxi speed. Since the ailerons and elevator are not helpful in a no-wind condition, they should be held in neutral positions. Their use when taxiing in windy conditions will be discussed later.

The rudder pedals are the primary directional controls while taxiing. Steering with the pedals may be accomplished through the forces of airflow or propeller slipstream acting on the rudder surface, or through a mechanical linkage to the steerable nosewheel or tailwheel.

Initially, the pilot should taxi with the heels of the feet resting on the cockpit floor and the balls of the feet on the bottom of the rudder pedals (Fig. 5–3). The feet should be slid up onto the brake pedals only when it is necessary to depress the brakes. Only after considerable experience is gained, should the feet be positioned with the arches placed on the rudder pedals and the toes near but not quite touching the brake portion of the pedals. This permits the simultaneous application of rudder and brake whenever needed. The brakes are used primarily to stop the airplane at a desired point, to slow the airplane, or as an aid in making a sharp controlled turn. Whenever used, they must be applied smoothly, evenly, and cautiously at all times.

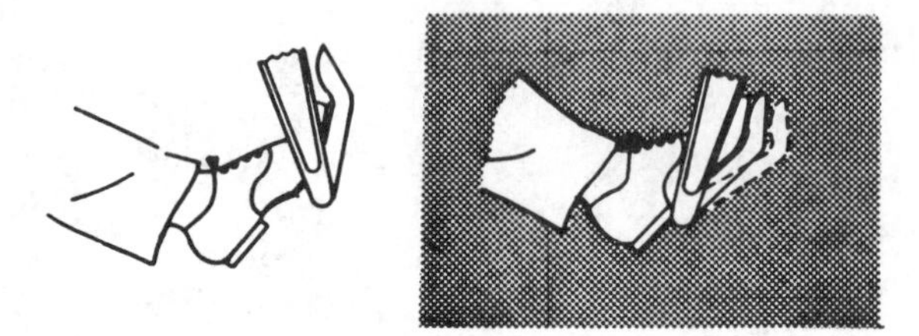

Figure 5–3 Position Feet Properly

More engine power may be required to start the airplane moving forward, or to start or stop a turn, than is required to keep it moving in any given direction. Therefore, there may be times when it is necessary to use a comparatively large amount of power. When used, the throttle should immediately be retarded once the airplane begins moving, to prevent accelerating too rapidly.

When first starting to taxi, the brakes should be tested immediately for proper operation. This is done by first applying power to start the airplane moving slowly forward, then retarding the throttle and simultaneously applying pressure smoothly to both brakes. If braking action is unsatisfactory, the engine should be shut down immediately.

To turn the airplane on the ground, the pilot should apply rudder in the desired direction of turn and use whatever power or brake that is necessary to control the taxi speed. The rudder should be held in the direction of the turn until just short of the point where the turn is to be stopped, then the rudder pressure released or slight opposite pressure applied as needed.

While taxiing, the pilot will have to anticipate the movements of the airplane and adjust rudder pressure accordingly. Since the airplane will continue to turn slightly even as the rudder pressure is being released, the stopping of the turn must be anticipated and the rudder pedals neutralized before the desired heading is reached. In some cases, it may be necessary to apply opposite rudder to stop the turn, depending on the taxi speed.

Usually when operating on a soft or muddy field, the taxi speed or power must be maintained slightly above that required under normal field conditions. Otherwise, the drag of the surface may cause the airplane to stop before adequate power can be applied to keep it in motion. This may require the use of full power in getting underway again, causing mud or stones to be picked up by the propeller, resulting in damage. When an airplane is allowed to stop under these conditions, it may be impossible to start moving again with its own power. On soft surfaces, such as turf, sand, or snow, the use of additional power will result in more blast on the airplane's tail, and consequently better rudder control.

The presence of moderate to strong headwinds and/or a strong propeller slipstream makes the use of the elevator necessary to maintain control of the pitch attitude while taxiing. This becomes apparent when considering the lifting action that may be created on the horizontal tail surfaces by either of those two factors. The elevator control in nosewheel type airplanes should be held in the neutral position, while in tailwheel type airplanes it should be held in the aft position to hold the tail down (Fig. 5-4).

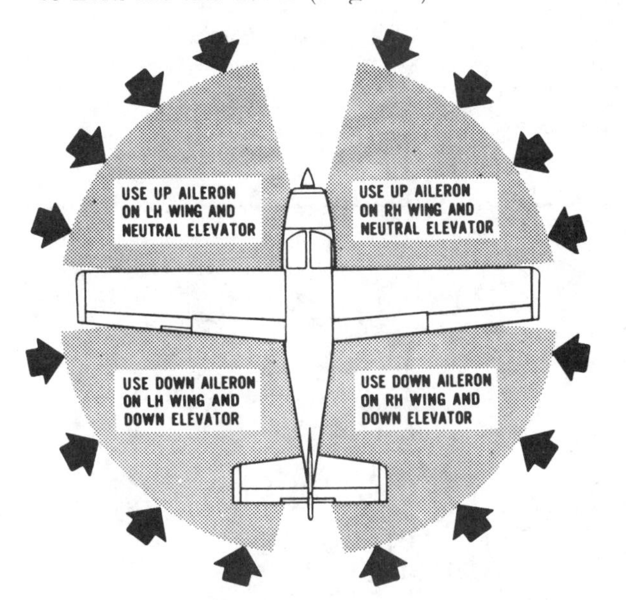

Figure 5-4 Position Controls Properly

Downwind taxiing will usually require less engine power after the initial ground roll is begun, since the wind will be pushing the airplane forward. To avoid overheating the brakes when taxiing downwind, keep engine power to a minimum. Rather than continuously riding the brakes to control speed, it is better to apply brakes only occasionally.

When taxiing in a crosswind, the wing on the upwind side usually will tend to be lifted by the wind unless the aileron control is held in that direction (upwind aileron UP). Moving the aileron into the UP position reduces the effect of wind striking that wing, thus reducing the lifting action. This control movement will also cause the opposite aileron to be placed in the down position, thus creating drag and possibly some lift on the downwind wing, further reducing the tendency of the upwind wing to rise.

When taxiing with a quartering tailwind, the elevator should be held in the down or neutral position, and the upwind aileron down. Since the wind is striking the airplane from behind, these control positions reduce the tendency of the wind to get under the tail and the wing and to nose the airplane over.

The application of these crosswind taxi corrections also helps to minimize the weathervaning tendency and ultimately result in making the airplane easier to steer.

In addition to learning taxiing techniques, all pilots should learn the standard hand signals used by ramp attendants for the direction of pilots operating airplanes in the parking area. These are shown in Fig. 5-5.

Taxiing—Tailwheel Type Airplanes

An airplane with a tailwheel landing gear arrangement has a tendency to "weathervane" or turn into the wind while it is being taxied. This tendency is much greater than in airplanes equipped with nosewheels. The reason for this is explained in the chapter on Landing Approaches and Landings.

The tendency of the tailwheel-type airplane to weathervane is greatest while taxiing directly crosswind; consequently, directional control is somewhat difficult. Without brakes, it is almost impossible to keep the airplane

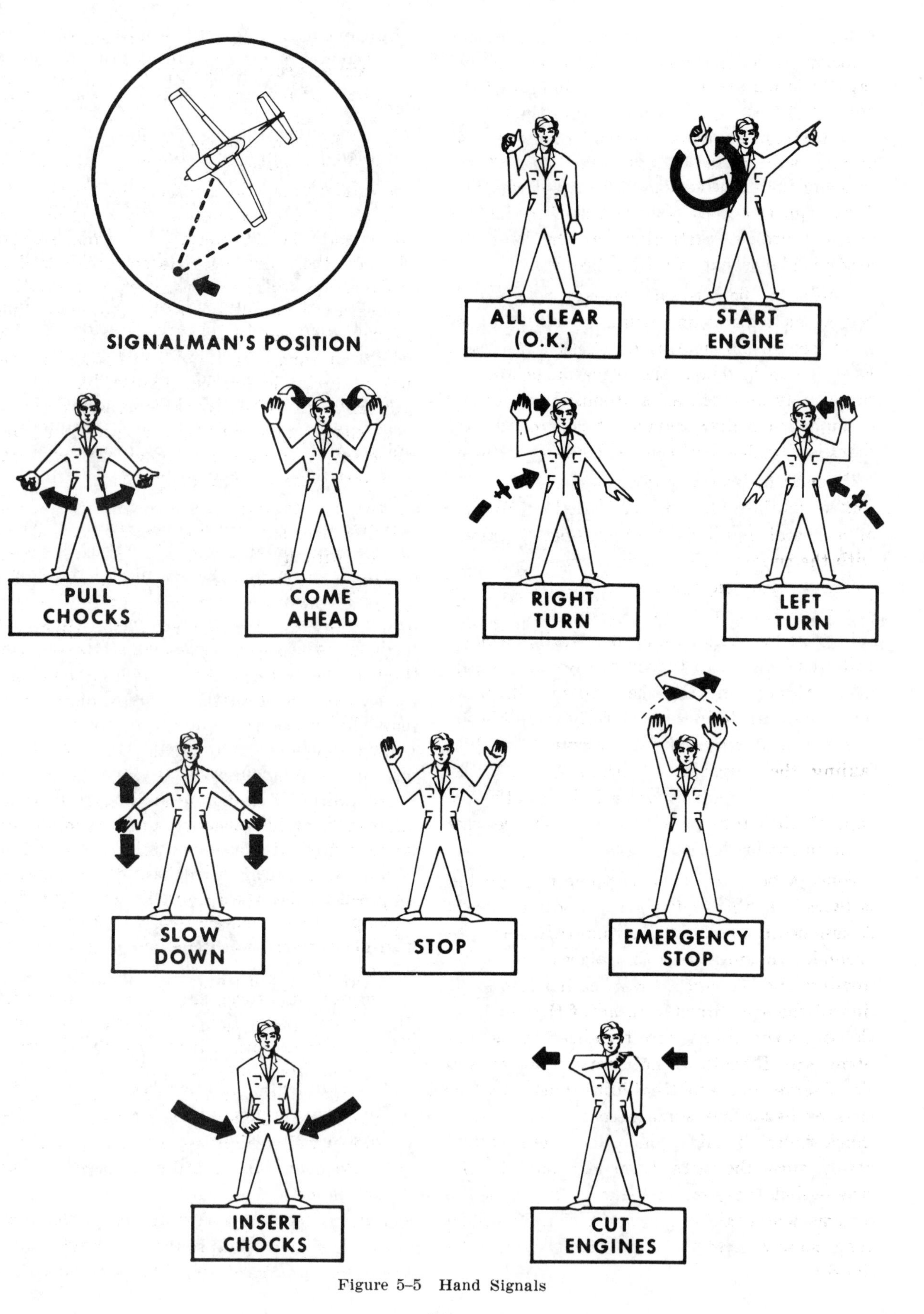

Figure 5-5 Hand Signals

from turning into any wind of considerable velocity since the airplane's rudder control capability mav be inadequate to counteract the crosswind.

In taxiing downwind the tendency to weathervane is increased, due to the tailwind decreasing the effectiveness of the flight controls. This requires a more positive use of the rudder and the brakes, particularly if the wind velocity is above that of a light breeze.

Unless the field is soft, or very rough, it is best when taxiing downwind to hold the elevator control in neutral or slightly forward. Even on soft fields, the elevator should be raised only as much as is absolutely necessary to maintain a safe margin of control in case there is a tendency of the airplane to nose over.

On most tailwheel type airplanes, directional control while taxiing is facilitated by the use of a steerable tailwheel which operates along with the rudder. The tailwheel steering mechanism remains engaged when the tailwheel is operated through an arc of 16° to 18° each side of neutral and then automatically becomes full swiveling when turned to a greater angle. The airplane may thus be pivoted within its own length, if desired, yet is fully steerable for slight turns while taxiing forward. While taxiing, the steerable tailwheel should be used for making normal turns and the pilot's feet kept off the brake pedals to avoid unnecessary wear on the brakes.

Since a tailwheel type airplane rests on the tailwheel as well as the main landing wheels, it assumes a nose high attitude when on the ground. In most cases this places the engine cowling high enough to restrict the pilot's vision of the area directly ahead of the airplane. Consequently, objects directly ahead of the airplane are difficult, if not impossible, to see. To observe and avoid colliding with any objects or hazardous surface conditions such as chuck holes or mire, the pilot should alternately turn the nose from one side to the other—that is zigzag, or make a series of short S-turns while taxiing forward. This must be done slowly, smoothly, positively, and cautiously.

Taxiing—Nosewheel Type Airplanes

Taxiing is somewhat simplied in an airplane equipped with a nosewheel. The good visibility and steerable nosewheel provide excellent ground handling characteristics. On this type airplane, the nosewheel usually is connected to the rudder pedals by a mechanical linkage and is steerable approximately 18° on either side of neutral. If the pilot desires to make a turn sharper than can be obtained with rudder pedal alone, brake pressure may be applied to decrease the radius of turn. However, since most nosewheels are not free swiveling, the radius of turn which can be obtained by the use of brake is limited. This may make it virtually impossible to pivot the airplane on one wheel. Consequently, the pilot will find it necessary to adequately plan the taxi path and anticipate all necessary turns.

Although taxiing with a steerable nosewheel requires less special pilot technique than required with a tailwheel, the habitual observance of all safety precautions is necessary. In spite of the good visibility and ground maneuvering provided by nosewheel airplanes, it is generally poor technique to use excessively high power settings, to taxi at high speeds, or to control the direction and taxi speed with only the brakes.

In starting to taxi a nosewheel airplane, the airplane should always be allowed to roll forward slowly to allow the nosewheel to become aligned straight ahead to avoid turning into an adjacent airplane or nearby obstruction. This also prevents side stress that would be imposed on the wheel assembly and strut if an attempt were made to force the airplane straight forward while the nosewheel is turned to one side.

Normally all turns should be started using the rudder pedal steering of the nosewheel. To tighten the turn after full pedal deflection is reached, brake may be applied as needed on the inside of the turn. When stopping the airplane, it is advisable to always stop with the nosewheel straight to relieve any strain on the nose gear and to make it easier to start moving ahead.

During crosswind taxiing, even the nosewheel type airplane has some tendency to weathervane. However, the weathervaning tendency is less than in tailwheel type airplanes because the main wheels are located further aft and the nosewheel's ground friction helps to resist the tendency. The nosewheel linkage from the rudder pedals provides adequate steering control for safe and efficient ground handling and normally only rudder pressure is necessary to correct for a crosswind.

Caution is particularly required when taxiing nosewheel equipped airplanes of the high-wing type, in strong quartering tailwinds. Because of the characteristic short coupling of the landing gear arrangement and the susceptibility of the tail and upwind wing to being lifted by the wind, a quartering tailwind can cause the high-wing airplane to flip over on its back. To prevent this when taxiing with a strong quartering tailwind, the elevator should be held in the down position (elevator control forward) and the aileron on the upwind side should be held in the down position (aileron control in the direction opposite that from which the wind is blowing). Furthermore, sudden bursts of power and sudden braking should be avoided whenever taxiing in these conditions.

Pretakeoff Check

The "pretakeoff check" is the systematic procedure for making a last-minute check of the engine, controls, systems, instruments, and radio prior to flight. Normally, it is performed after taxiing to a position near the takeoff end of the runway. Taxiing to that position usually allows sufficient time for the engine to warm up to at least minimum operating temperatures and ensures adequate lubrication of the internal moving parts of the engine before being operated at high power settings.

Some pilots devote too little attention to engine warmup. Most engines require that the oil temperature reach a certain degree before proper operation can be depended upon.

Today's air cooled engines generally are closely cowled and equipped with pressure baffles which direct the flow of air during flight to the engine in sufficient quantities for cooling purposes. On the ground, however, much less air is forced around these baffles and through the cowling, and any prolonged engine operation may cause overheating long before any indication of rising temperature is given by the oil temperature gauge.

Most modern high performance airplane engines are designed to be operated with very short warmup periods, the initial sluggishness of the oil being taken care of in the engine design. When operating these engines, a cylinder head temperature gauge usually is available to determine adequate operating temperatures. In the absence of such instrument, the takeoff may be made when the throttle can be advanced to full power without the engine faltering. Even in extremely cold weather the cylinder heads may overheat before any indication of such a condition shows on the oil temperature gauge. This is more likely to happen in very cold weather when the oil heats slower than in warm weather.

Before starting the pretakeoff check, the airplane should be positioned out of the way of other aircraft. It is recommended that the airplane be headed as nearly as possible into the wind to obtain more accurate operating indications and to minimize engine overheating when the engine is run up. After the airplane is properly positioned for the runup, it should be allowed to roll forward slightly so that the nosewheel or tailwheel will be aligned fore and aft.

During the engine runup the surface under the airplane should be firm, (a smooth turf or paved surface if possible) and free of debris. Otherwise the propeller will pick up pebbles, dirt, mud, sand, or other loose particles and hurl them backward, not only damaging the tail of the airplane, but often inflicting injury to the propeller itself. Inspection of the leading edge of almost any propeller which has been in use for any period of time will show the results of neglecting this precaution.

While performing the engine runup, the pilot must divide attention inside and outside the airplane. If the parking brake slips, or if application of the toe brakes is inadequate for the amount of power applied, the airplane could move forward unnoticed if attention is fixed inside the airplane.

Since each airplane has different features and equipment, the checklist provided by the airplane manufacturer must be used for the before takeoff check. The following, however, are some of the major items that should be checked or set before moving onto the takeoff runway:

1. Set parking brake or hold brake ON.
2. Check wing flaps for operation and set for takeoff.
3. Check flight controls for free and proper operation.
4. Set trim tabs for takeoff.
5. Adjust altimeter to reported altimeter setting or field elevation.
6. Set heading indicator to correspond with compass heading.
7. Set fuel selector to fullest tank.
8. Set propeller control (if equipped) to FULL INCREASE.
9. Set mixture control to RICH.
10. Set carburetor heat control to COLD.
11. Check engine temperature.
12. Adjust throttle to runup RPM.
13. Check each magneto for operation.
14. Set magneto switch to BOTH ON.
15. Check propeller governor (if equipped) for operation.
16. Set propeller control (if equipped) to FULL INCREASE.
17. Check engine instruments for normal indications.
18. Check engine idle speed.
19. Set flight instruments.
20. Check seat locked and seatbelt fastened.
21. Obtain takeoff clearance (if required).
22. Check cabin door locked.
23. Check runway and final approach for aircraft.
24. Check clock for time.
25. Recall takeoff "V-speeds" (critical performance speeds).

After Landing Check

During the after-landing roll, the airplane should be gradually slowed to normal taxi speed before turning off the landing runway. Any significant degree of turn at faster speeds could result in ground looping and subsequent damage to the airplane.

To give full attention to controlling the airplane during the landing roll, the after-landing check should be performed *only after* the airplane is brought to a complete stop clear of the active runway. There have been many cases of the pilot mistakenly grasping the wrong handle and retracting the landing gear, instead of the flaps, due to improper division of attention while the airplane was moving.

Because of different features and equipment in various airplanes, the after landing checklist provided by the manufacturer should be used. Some of the items may include:

1. Hold brakes ON.
2. Identify landing flap control and retract flaps.
3. Open engine cowl flaps (if equipped).
4. Recheck and set propeller control (if equipped) to FULL INCREASE.
5. Set trim tabs for takeoff.

Engine Shutdown Check

A flight is never complete until the engine is shut down and the airplane secured. Unless parking in a designated, supervised area, the pilot should select a location and heading which will prevent the propeller or jet blast of other airplanes from striking the airplane broadside. Whenever possible the airplane should be parked headed into the existing or forecast wind. After stopping on the desired heading, the airplane should be allowed to roll straight ahead enough to straighten the nosewheel or tailwheel. Finally, the pilot should always use the procedures in the manufacturer's checklist for shutting down the engine and securing the airplane. Some of the important items include:

1. Set the parking brakes ON.
2. Set throttle to IDLE or 1000 RPM.
3. Turn ignition switch OFF then ON.
4. Set propeller control (if equipped) to FULL INCREASE.
5. Turn electrical units and radios OFF.
6. Set mixture control to IDLE CUT-OFF.
7. Turn ignition switch to OFF when engine stops.
8. Turn master electrical switch to OFF.
9. Install control lock.
10. Place chocks at wheels.
11. Tie down airplane.

CHAPTER 6

Basic Flight Maneuvers

This chapter discusses and explains the fundamental flight maneuvers upon which all flying tasks and techniques are based. In learning to fly, as in any learning process, fundamentals must be mastered before the more advanced phases can be learned.

Maneuvering of the airplane is generally divided into four flight fundamentals; (1) straight-and-level, (2) turns, (3) climbs, and (4) descents. All controlled flight consists of either one, or a combination of more than one, of these basic maneuvers. Proper control of an airplane's attitude is the result of the pilot knowing when and how much to change the attitude, and then smoothly changing the attitude the required amount, or maintaining a constant attitude. When flying by reference to objects outside the airplane, the effects of the pilot's control application on the airplane's flight attitude can be seen by observing the relationship of the position of some portion of the airplane to the outside references.

At first, control of the airplane is a matter of *consciously* fixing the relationship of specific reference points on the airplane to the horizon. As basic flight skills are developed through experience and training, the pilot will acquire a continuous awareness of these relationships *without* conscious effort. The reference points will be used almost subconsciously in varying degrees to determine the attitude of the airplane during all maneuvers.

In establishing the reference points, the airplane should be placed approximately in the desired flight attitude, and then the specific points selected. No two pilots see this relationship exactly the same. The apparent position of the reference points will depend on each pilot's seat height and lateral position, and/or the pilot's eye level and line of sight.

Integrated Flight Instruction

In introducing the basic flight maneuvers, it is recommended that the "Integrated Flight Instruction" method be used. This means that each flight maneuver should be performed by using both outside visual references and the flight instruments.

When pilots use this technique, they achieve a more precise and competent overall piloting ability. That is, it results in less difficulty in holding desired altitudes, controlling airspeed during takeoffs, climbs, descents, and landing approaches, and in maintaining headings in the traffic pattern, as well as on cross-country flights.

The use of integrated flight instruction does not, and is not intended to, prepare pilots for flight in instrument weather conditions. It does, however, provide an excellent foundation for the future attainment of an instrument pilot rating, and will result in the pilot becoming a more accurate, competent, and safe pilot. Although integrated flight instruction should be used for all flight maneuvers, its use is specifically discussed here in only the Basic Flight Maneuvers.

A sharp lookout for other aircraft must be maintained at all times, particularly when using instrument references, to avoid the possibility of collision with other aircraft. Frequently, other aircraft are unnoticed until they suddenly appear within the limited area of the pilot's vision. Consequently, it is imperative that the pilot not only divide attention between controlling the airplane by outside visual references and flight instruments, but also be observant of other aircraft.

Attitude Flying

Airplane control is composed of three components: (1) pitch control, (2) bank control, and (3) power control (Fig. 6–1). Pitch control is the control of the airplane about its lateral axis by applying elevator pressure to raise or lower the nose, usually in relation to the horizon. Bank control is the control of the airplane about its longitudinal axis by use of the ailerons to attain the desired angle of bank in relation to the horizon. Power control is the control of power or thrust by use of the throttle to establish or maintain desired airspeeds in coordination with the attitude changes.

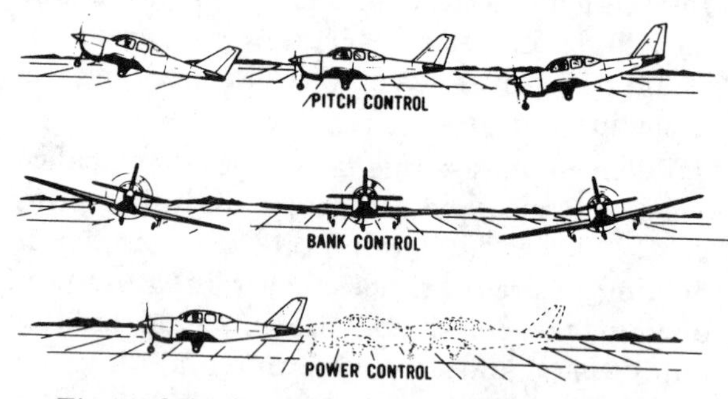

Figure 6–1 Components of Airplane Control

The attitude indicator (artificial horizon), heading indicator, altimeter, vertical speed indicator, and airspeed indicator, are the instruments used as references for control of the airplane. The attitude indicator shows directly both the pitch and bank attitude of the airplane; the heading indicator shows directly the airplane's direction of flight, and indirectly, the bank attitude; the altimeter indicates the airplane's altitude and, indirectly, the need for a pitch change; the vertical speed indicator shows the rate of climb or descent; and the airspeed indicator shows the results of power and/or pitch changes by the airplane's speed. The outside visual references used in controlling the airplane include the airplane's nose and wingtips to show both the airplane's pitch attitude and flight direction; the wings and frame of the windshield to show the angle of bank.

The objectives in these basic maneuvers are to learn the proper use of the controls for maneuvering the airplane, to attain the proper attitude in relation to the horizon by use of inside and outside references, and to emphasize the importance of dividing attention and constantly checking all reference points.

Straight-and-level Flight

Straight-and-level flight is just what the name implies—flight in which a constant heading and altitude are maintained. It is accomplished by making immediate corrections for deviations in direction and altitude from unintentional slight turns, descents, and climbs.

The pitch attitude for *level flight* (constant altitude) is usually obtained by selecting some portion of the airplane's nose as a reference point, and then keeping that point in a fixed position relative to the horizon. That position should be cross-checked occasionally against the altimeter to determine whether or not the pitch attitude is correct. If altitude is being gained or lost, the pitch attitude should be readjusted in relation to the horizon and then the altimeter rechecked to determine if altitude is now being maintained. The application of forward or back elevator pressure is used to control this attitude.

The pitch information obtained from the attitude indicator also will show the position of the nose relative to the horizon and will indicate whether elevator pressure is necessary to change the pitch attitude to return to level flight (Fig. 6–2).

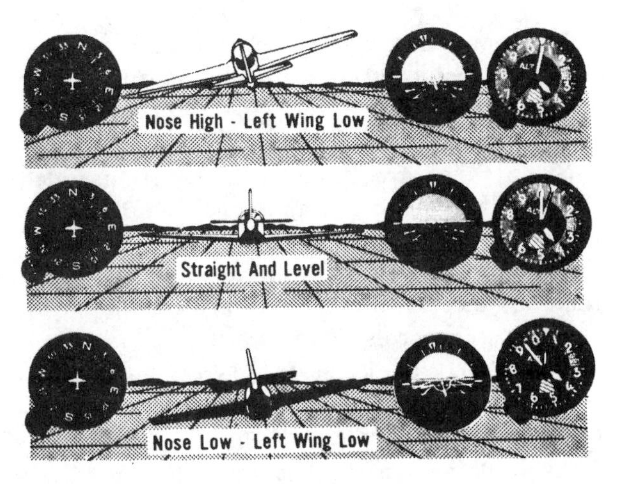

Figure 6–2 Outside and Instrument References for Level Flight

In all normal maneuvers the term "increase the pitch attitude" implies raising the nose in relation to the horizon; the term "decreasing the pitch" means lowering the nose.

To achieve *straight flight* (constant heading), the pilot selects two or more outside visual reference points directly ahead of the airplane (such as fields, towns, lakes, or distant clouds, to form points along an imaginary line) and keeps the airplane's nose headed along that line. Roads and section lines on the ground also offer excellent references—straight flight can be maintained by flying parallel or perpendicular to them. While using these references, an occasional check of the heading indicator should be made to determine that the airplane is actually maintaining flight in a constant direction.

Straight flight (laterally level flight) may also be accomplished by visually checking the relationship of the airplane's wingtips with the horizon. Both wingtips should be equidistant above or below the horizon (depending on whether the airplane is a high-wing or low-wing type), and any necessary adjustments should be made with the ailerons, noting the relationship of control pressure and the airplane's attitude.

Continually observing the wingtips has advantages other than being a positive check for leveling the wings. It also helps divert the pilot's attention from the airplane's nose, prevents a fixed stare, and automatically expands the radius of visual scanning (Fig. 6–3). In straight-and-level flight the wingtips can be used for both estimating the airplane's laterally level attitude or bank, and to a lesser degree, its pitch attitude.

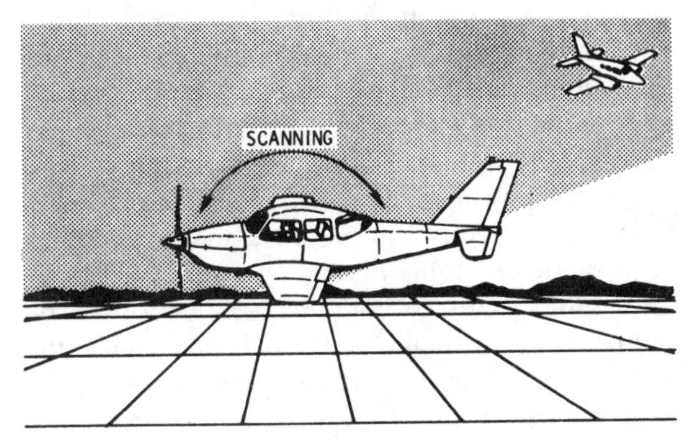

Figure 6–3 Visual Scanning While Checking Airplane's Attitude

Any time the wings are banked, even though very slightly, the airplane will turn. Thus, close attention should be given to the attitude indicator to detect small indications of bank, and to the heading indicator to note any change of direction.

When the wings are approximately level, straight flight could be maintained by simply exerting the necessary forces on the rudder in the desired direction. However, the practice of using rudder alone is not correct and may make precise control of the airplane difficult. Straight-and-level flight requires almost no application of control pressure if the airplane is properly trimmed and the air is smooth. For that reason, the pilot must not form the habit of constantly moving the controls unnecessarily.

When practicing this fundamental flight maneuver, the pilot should trim the airplane so it will fly straight and level without assistance. This is called "hands-off flight." The trim controls, when correctly used, are aids to smooth and precise flying. Improper trim technique usually results in flying that is physically tiring, particularly in prolonged straight-and-level flight. By using the trim tabs to relieve all control pressures, the pilot will find that it is much easier to hold a given altitude and heading. The airplane should be trimmed by first applying control pressure to establish the desired attitude, and then adjusting the trim so that the airplane will maintain that attitude without control pressure in "hands-off flight."

For all practical purposes, the airspeed will remain constant in straight-and-level flight with a constant power setting. Practice of intentional airspeed changes by increasing or decreasing the power, will provide an excellent means of developing proficiency in maintaining straight-and-level flight at various speeds. Significant changes in airspeed will, of course, require considerable changes in pitch attitude and pitch trim to maintain altitude. Pronounced changes in pitch attitude and trim will also be necessary as the flaps and landing gear are operated.

Turns

A turn is a basic flight maneuver used to change or return to the desired heading. It involves close coordination of all three flight controls—aileron, rudder, and elevator. Since turns are a part of most other flight maneuvers, it is important that the pilot thoroughly understand factors involved and learn to perform them well.

For purposes of this discussion, turns are divided into three classes: shallow, medium, and steep (Fig. 6–4).

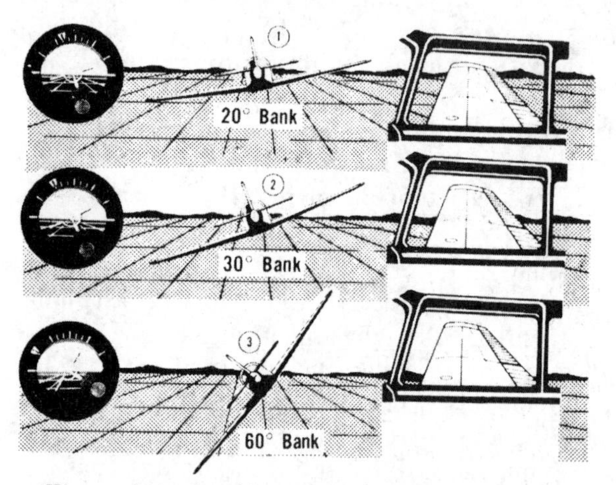

Figure 6–4 Outside and Instrument References During Turns

Shallow turns are those in which the bank is so shallow (less than approximately 20°) that the inherent stability of the airplane is acting to level the wings unless some control force is used to maintain the bank.

Medium turns are those resulting from a degree of bank (approximately 20° to 45°) at which the airplane tends to hold a constant bank without control force on the ailerons.

Steep turns are those resulting from a degree of bank (more than approximately 45°) at which the "overbanking tendency" of an airplane overcomes stability, and the bank tends to increase unless pressure is applied to the aileron controls to prevent it.

The actual bank angle (degrees of bank) required for a given rate of turn is a function of airspeed and airplane design.

As explained in the chapter Effect and Use of Controls, turns are made in an airplane by banking the wings, so that by changing the direction of the wing's lift toward one side or the other, the airplane will be pulled in that direction. This is done by applying coordinated aileron and rudder pressure to bank the airplane in the direction of the desired turn (Fig. 6–5).

Figure 6–5 Unbalanced Lift Results in Banking

It will be recalled that when an airplane is flying straight and level, the total lift is acting perpendicular to the wings and to the earth. As the airplane is banked into a turn, the lift then becomes the resultant of two components. One, the vertical component, continues to act perpendicular to the earth and opposes gravity. The other, the horizontal component, acts parallel to the earth's surface and opposes centrifugal force caused by the turn. These two lift components act at right angles to each other, causing the resultant lifting force to act perpendicular to the banked wings of the airplane. It is this lifting force that actually turns the airplane—*not the rudder* as misinformed pilots might believe. The fallacy that the rudder makes the airplane turn was demonstrated in the earlier explanation of the Effect and Use of Controls.

When applying aileron to bank the airplane, the depressed or lowered aileron (on the rising wing) produces a greater drag than the raised aileron (on the lowering wing). This increased aileron drag tends to yaw the airplane toward the rising wing, or opposite to the desired direction of turn, while the banking action is taking effect. Figure 4–4 illustrates this effect. To counteract the yawing tendency, rudder pressure must be applied simultaneously in the desired direction of turn. This produces a coordinated turn.

The adverse yaw produced by aileron drag has been a major consideration in design of the controls, and in most modern airplanes has been significantly reduced. This is accomplished by providing greater up travel than down travel of the ailerons, or by using Frise-type, or slotted ailerons.

After the bank has been established, in a theoretically perfect medium turn in smooth air, all pressure on the aileron control may be relaxed. The airplane will remain at the bank selected with no further tendency to yaw since there is no longer a deflection of the ailerons. As a result, pressure may also be relaxed on the rudder pedals, and the rudder allowed to streamline itself with the direction of the air passing it. If pressure is maintained on the rudder after the turn is established, the airplane will tend to skid to the outside of the turn. If a definite effort is made to center the rudder rather than let it streamline itself to the turn, it is probable that some opposite rudder pressure will be exerted inadvertently. This would tend to force the airplane to yaw opposite its original turning path. As a result the airplane would tend to slip to the inside of the turn. The ball in the turn indicator will be displaced off-center whenever the airplane is skidding or slipping sidewards (Fig. 6–6).

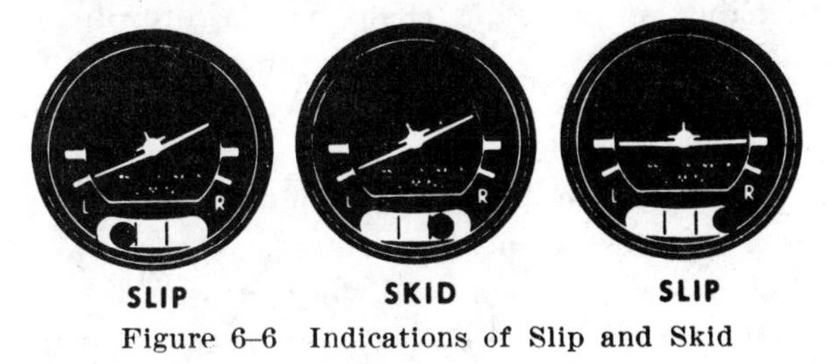

Figure 6–6 Indications of Slip and Skid

In all turns in which a constant altitude is to be maintained, it is necessary to increase the angle of attack by applying back elevator pressure. This is required because lift that is equal to the weight of the airplane plus the centrifugal force caused by the turn must be obtained from the wing to maintain altitude. Producing sufficient lift for the turn could be done either by increasing the power and airspeed or by increasing the angle of attack. The force of lift must be further increased as the turn steepens and the load factor and the centrifugal force build up, but must be slowly decreased as the airplane is being rolled back to level flight when completing the turn.

To stop the turn, the wings must be returned to laterally level flight by the use of the ailerons, and the resulting adverse yaw (now acting in the same direction as the turn), must be overcome by the coordinated application of rudder. The yaw effect will often be more apparent when rolling out of a turn than rolling into a turn, due to the higher angle of attack and wing loading, and the slower airspeed which exists when the rollout is started. A greater deflection of the ailerons is necessary to obtain the same control effect as when rolling into the turn from level flight. Due to the slower airspeed, this displacement of the ailerons requires less pressure on the aileron control, but it also imposes a greater drag on the "down" aileron since the wing is at a greater angle of attack. The greater drag on that aileron now tends to make the airplane yaw toward the direction of the turn rather than away from it; therefore, a greater rudder pressure must be used to counteract it.

To understand the relationship between airspeed, bank, and radius of turn, it must be recalled that the rate of turn at any given airspeed depends on the amount of the sideward force causing the turn; that is, the horizontal lift component. The horizontal lift component varies in proportion to the amount of bank. Thus, the rate of turn at a given airspeed increases as the angle of bank is increased. On the other hand, when a turn is made at a *higher* airspeed at a given bank angle, the inertia is greater and the centrifugal force created by the turn becomes greater, causing the turning rate to become slower. It can be seen, then, that at a given angle of bank, a higher airspeed will make the radius of the turn larger because the airplane will be turning at a slower rate.

With an airplane possessing positive lateral stability, a wing which has lowered as the result of a gust of air will usually tend to return to its original position. However, this inherent tendency should not be depended on because other factors may overcome it. The pilot should take appropriate control action to level the wing whenever necessary to prevent the airplane from entering an unintentional spiral.

Figure 6–7 illustrates the partial explanation of the airplane's overbanking tendency during a steep turn. As the radius of the turn becomes smaller, a significant difference develops between the speed of the inside wing and the speed of the outside wing. The wing on the outside of the turn travels a longer circuit than the inside wing, yet both complete their respective circuits in the same length of time. Therefore, the outside wing must travel faster than the inside wing and as a result it develops more lift. This creates a slight differential between the lift of the inside and outside wings and tends to further increase the bank.

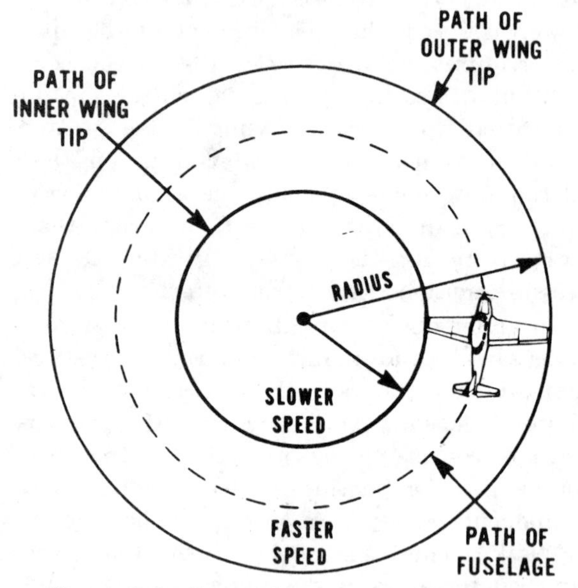

Figure 6–7 Cause of Overbanking Tendency

When changing from a shallow bank to a medium bank, the airspeed of the wing on the outside of the turn increases in relation to the inside wing as the radius of turn decreases, but the force created exactly balances the force of the inherent lateral stability of the airplane, so that at a given speed, no aileron pressure is required to maintain that bank. However, as the radius decreases further when the bank progresses from a medium bank to a steep bank, the lift differential overbalances the lateral stability, and counteractive pressure on the ailerons is necessary to keep the bank from steepening.

To establish the desired angle of bank, the pilot should use outside visual reference points as well as specific indications on the attitude indicator (Fig. 6–4).

The best outside reference for establishing the degree of bank is the angle formed by the raised wing of low-wing airplanes (the lowered wing of high-wing airplanes) and the horizon, or the angle made by the top of the engine cowling and the horizon. Since on most light planes the engine cowling is fairly flat, its horizontal angle to the horizon will give some indication of the approximate degree of bank. Also, information obtained from the attitude indicator will show the angle of the wings in relation to the horizon.

The pilot's posture while seated in the airplane is very important in all maneuvers, particularly during turns, since it will affect the alignment of outside visual references. At the beginning the student may lean to the side when rolling into the turn in an attempt to remain upright in relation to the ground rather than "ride" with the airplane. This tendency must be corrected at the outset, if the student is to learn to properly use visual references.

This tendency is also characteristic in airplanes which have side-by-side seats because the pilot is seated to one side of the longitudinal axis about which the airplane rolls. This makes the nose *appear* to rise when making a correct left turn and to descend in correct right turns.

As in all maneuvering, the pilot should form the habit of ensuring that the area toward which a turn is to be made is clear of other aircraft. This is especially vital in high-wing airplanes since once the bank is started, the lowered wing may conceal other aircraft in that direction. If nearby aircraft are not detected before lowering the wing, it would be too late to avoid a collision.

After the area has been cleared, the turn should be started by gradually and simultaneously applying pressure to both the aileron and rudder in the desired direction. This pressure will move the control surfaces out of their streamlined position and cause the airplane to bank and turn. The rate at which the airplane rolls into the turn is governed by the rapidity and amount of pressure applied.

When entering turns, pressure should be applied on the ailerons and the rudder progressively; that is, applied smoothly, and gradually built up, then gradually relaxed just before the desired degree of bank is attained.

The rotation of the airplane about its longitudinal axis when entering a turn should be accomplished in such manner that a pitch attitude also may be maintained which will produce flight at a constant altitude. Because of the decrease in the vertical lift component and the increase in load factor in the turn, the angle of attack required to maintain altitude will be slightly higher than in straight-and-level flight, if no power adjustment is made. Consequently, sufficient back pressure must be applied to the elevator control. The steeper the bank, of course, the greater the increase in required angle of attack; therefore, more back pressure is needed.

As soon as the airplane rolls from the wings-level attitude, the nose should also start to move along the horizon, increasing its rate of travel proportionately as the bank is increased. Any variation from this will be indicative of the particular control that is being misused. The following variations provide excellent guides:

1. If the nose starts to move before the bank starts, rudder is being applied too soon.
2. If the bank starts before the nose starts turning, or the nose moves in the opposite direction, the rudder is being used too late.
3. If the nose moves up or down when entering a bank, excessive or insufficient back elevator pressure is being applied.

As the desired angle of bank is established, aileron and rudder pressures should be released. This will stop the bank from increasing, for the aileron control surfaces will be neutral in their streamlined position. The back elevator pressure should *not* be released but should be held constant or sometimes increased to maintain a constant altitude. Throughout the turn, the pilot should cross-check the references and occasionally include the altimeter to determine whether or not the pitch attitude is correct. If gaining or losing altitude, the pitch attitude should be adjusted in relation to the horizon and then the altimeter and vertical speed indicator rechecked to determine if altitude is now being maintained.

During all turns, the ailerons and rudder are used to correct minor variations just as they are in straight-and-level flight. However, during very steep turns, considerably more back elevator pressure is required to maintain altitude than in shallow and medium turns, and additional power may be needed to maintain a safe airspeed. Frequently, there is a tendency for the airplane's nose to lower, resulting in a loss of altitude.

To recover from an unintentional nose-low attitude during a steep turn, the pilot should first reduce the angle of bank with coordinated aileron and rudder pressure. Then back elevator pressure should be used to raise the airplane's nose to the desired pitch attitude. After accomplishing this, the desired angle of bank can be reestablished. Attempting to raise the nose *first* by increasing back elevator pressure will usually cause a tight descending spiral, and could lead to overstressing the airplane.

The rollout from a turn is similar to the roll-in except that control pressures are used in the opposite direction. Aileron and rudder pressure are applied in the direction of the rollout or toward the high wing. As the angle of bank decreases, the elevator pressure should be released smoothly as necessary to maintain altitude, since the airplane is no longer banking and the effects of centrifugal force and the vertical component of lift are becoming less.

Since the airplane will continue turning as long as there is any bank, the rollout must be started before reaching the desired heading. The amount required to lead the heading will depend on the rate of turn and the rate at which the rollout will be made. As the wings become level, the control pressures should be gradually and smoothly released so that the controls are neutralized as the airplane assumes straight-and-level flight. As the rollout is being completed, attention should be given to outside visual references as well as the attitude indicator and heading indicator to determine that the wings are being leveled precisely and the turn stopped.

To make a precision 90° turn using outside references, the airplane should be aligned with a road or section line on the ground and then a turn made to where the airplane will be perpendicular to the road or line. In the absence of any ground reference, a point may be selected on the horizon directly off the wingtip, then the airplane turned until it is headed to that point. The selection of the reference should be made while visually scanning the area for other aircraft, obstacles, or nearby clouds.

Climbs

Climbs and climbing turns are basic flight maneuvers in which the pitch attitude and power result in a gain in altitude. A straight climb is one in which the airplane gains altitude while traveling straight ahead. Climbing turns are those in which the airplane gains altitude while turning.

As used in this chapter, a normal climb is one made at a pitch attitude and airspeed which, when constantly maintained, will give the best altitude gain with the engine controls set to the climb power recommended by the manufacturer.

As with the other maneuvers, climbs should be performed by using both flight instruments and outside visual references (Fig. 6–8). The normal climb speed recommended by the airplane manufacturer should be used. This is usually very close to the airplane's best rate-of-climb airspeed, but may be slightly higher to provide better engine cooling or increased flight visibility.

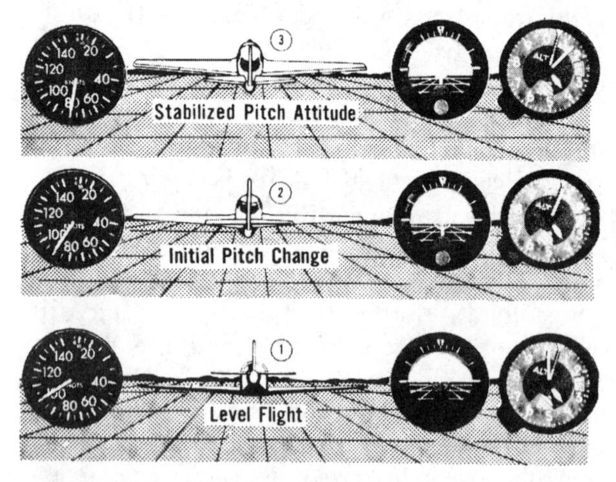

Figure 6–8 Outside and Instrument References for Climbing

As a climb is started, the airspeed will gradually diminish. This reduction in airspeed is gradual, rather than immediate, because of the initial momentum of the airplane. The thrust required to maintain straight-and-level flight at a given airspeed is not sufficient to maintain the same airspeed in a climb; climbing flight takes more power than straight-and-level flight. Consequently, the engine power controls must be advanced to a higher power setting.

The effects of torque at the climb power setting are a primary factor in climbs. Since the climb airspeed is lower than cruising speed, the airplane's angle of attack is relatively high. With these conditions, torque and asymmetrical loading of the propeller will cause the airplane to have a tendency to roll and yaw to the left. To counteract this, right-rudder pressure must be used. During the early practice of climbs and climbing turns, this may make coordination of the controls feel awkward, but after a little practice the correction for torque effects will become instinctive.

Trim is also a very important consideration during a climb. After the climbing attitude, power setting, and airspeed have been established, the airplane should be trimmed to relieve all pressures from the controls. If further adjustments are made in the pitch attitude, power, or airspeed, the airplane must be retrimmed.

As the airplane gains altitude during a climb, the manifold pressure gauge (if the airplane is so equipped) will indicate a loss in pressure (power), because the same volume of air going into the engine's induction system gradually decreases in density as altitude increases. Thus, the total pressure in the manifold (and consequently power) decreases. This will occur at the rate of approximately 1″ of manifold pressure for each 1,000-foot gain in altitude. During prolonged climbs, then, the throttle must be continually advanced if a constant power is to be maintained.

When performing a climb, the power should be advanced to the climb power settings recommended by the airplane manufacturer. If the airplane is appropriately equipped, this will include engine RPM and manifold pressure. Normally, the flaps and landing gear (if retractable) should be in the up position to reduce drag, although practice may also be accomplished with the gear down to simulate a climb immediately after takeoff.

To enter the climb, simultaneously advance the throttle and apply back pressure on the elevator. As the power is increased to the climb setting, the airplane's nose will tend to rise toward the climbing attitude. While the pitch attitude increases and as the airspeed decreases, progressively more right-rudder pressure must be used to compensate for torque effects and to maintain direction. Usually right-rudder trim will be required to relieve this pressure.

When the climb is established, back elevator pressure must be maintained to keep the pitch attitude constant. It will be noted that as the airspeed decreases, the elevators will try to return to their neutral or streamlined position and the airplane's nose will tend to lower. Nose-up elevator trim should be used to compensate for this so that the pitch attitude can be maintained without holding back pressure on the elevator control. Throughout the climb, since the power is fixed at the climb power setting, the airspeed must be controlled by the use of elevator.

The pilot should cross-check the airspeed indicator and the position of the airplane's nose in relation to the horizon, as well as the pitch attitude shown on the attitude indicator, to determine if the pitch attitude is correct. At the same time a constant heading should be held with the wings level if a straight climb is being performed, or a constant angle of bank if in a climbing turn.

To return to straight-and-level flight from a climbing attitude, it is necessary to start the level-off approximately 50 feet below the desired altitude. While approaching that altitude, the wings should be leveled and the nose lowered to the level flight attitude. The nose must be lowered gradually, however, because a loss of altitude will result if the pitch attitude is decreased to the cruising level flight position without allowing the airspeed to increase proportionately. As the nose is lowered and the wings leveled, the airplane should be retrimmed (Fig. 6–9).

After the airplane is in a level attitude, climb power should be retained temporarily so that the airplane will accelerate to desired cruise speed. When the airspeed reaches the desired cruise speed, the throttle setting and the propeller control (if so equipped) should then be reduced to appropriate cruise power settings, the mixture control adjusted to a lean position, and the airplane trimmed for "hands-off" flight.

In developing skills in the performance of *climbing turns*, the following factors should be considered:

1. With a constant power setting, the same pitch attitude and airspeed cannot be maintained in a bank as in a straight climb due to the decrease in effective lift and airspeed during a turn.

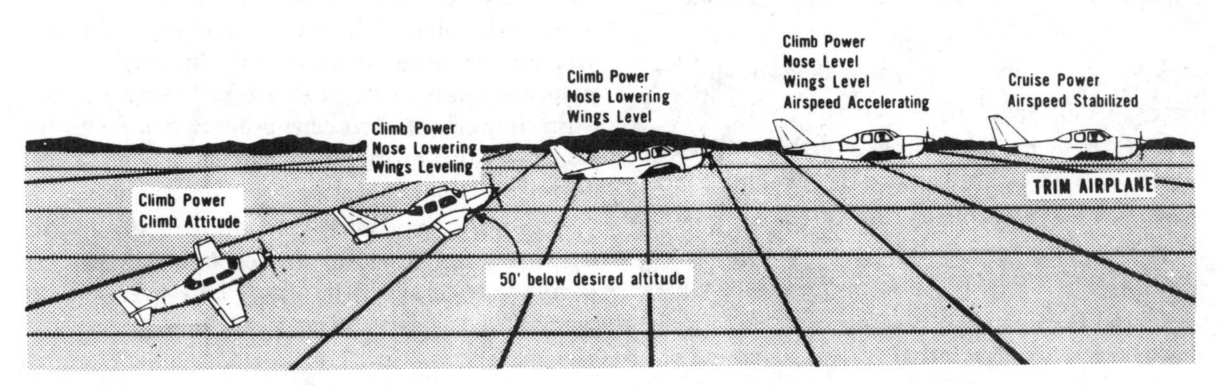

Figure 6–9 Leveling-Off From a Climb

2. The degree of bank should be neither too steep nor too shallow. Too steep a bank intensifies the effect mentioned in 1 above. If too shallow, the angle of bank may be difficult to maintain because of the inherent stability of the airplane.
3. A constant airspeed, a constant rate of turn, and a constant angle of bank must be stressed. The coordination of all controls is likewise a primary factor to be stressed and developed.
4. The airplane will have a greater tendency towards nose-heaviness than in a normal straight climb, due to the decrease in effective lift as is the case in all turns.
5. As in all maneuvers, attention should be diverted from the airplane's nose and divided among all references equally.
6. Proficiency in turns to the right as well as to the left should be developed.

All of the factors that affect the airplane during level (constant altitude) turns will affect it during climbing turns or any other turning maneuver. It will be noted that because of the low airspeed, aileron drag (adverse yaw) will have a more prominent effect than it did in straight-and-level flight and more rudder pressure will have to be blended with aileron pressure to keep the airplane in coordinated flight during changes in bank angle. Additional elevator back pressure and trim will also have to be used to compensate for centrifugal force, for loss of vertical lift, and to keep the pitch attitude constant.

During climbing turns the loss of vertical lift becomes greater as the angle of bank is increased, so shallow turns must be used to maintain an efficient rate of climb. If a medium- or steep-banked turn is used, the airplane will not climb so rapidly.

There are two ways to establish a climbing turn. Either establish a straight climb and then turn or establish the pitch and bank attitudes simultaneously from straight-and-level flight. The second method is usually preferred because the pilot can more effectively check the area for other aircraft while the climb is being established.

Descents

A descent, or glide, is a basic maneuver in which the airplane is losing altitude in a controlled descent with little or no engine power; forward motion is maintained by gravity pulling the airplane along an inclined path, and the descent rate is controlled by the pilot balancing the forces of gravity and lift.

Although power-off descents (glides) are directly related to the practice of power-off accuracy landings, as will be seen in later discussions, they have a specific operational purpose in normal landing approaches, and forced landings after engine failure. Therefore, it is necessary that they be performed more subconsciously than other maneuvers because most of the time during their execution, the pilot will be giving full attention to details other than the mechanics of performing the maneuvers. Since glides are usually performed relatively close to the ground, accuracy of their execution and the formation of proper technique and habits are of special importance.

Because the application of the controls is somewhat different in power-off descents than during power-on descents, gliding maneuvers require the perfection of a technique somewhat different from that required for ordinary power-on maneuvers. This control difference is caused primarily by two factors—the absence of the usual propeller slipstream, and the difference in the relative effectiveness of the various control surfaces at slow speeds.

The glide ratio of an airplane is the distance the airplane will, with power off, travel forward in relation to the altitude it loses. For instance, if an airplane travels 10,000 feet forward while descending 1,000 feet, its glide ratio is said to be 10 to 1. Technically, it is practically impossible to know exactly what the glide distance of the airplane will be, because so many things affect it; however, it is very important that the pilot has a fair idea of how far the airplane will glide under certain conditions.

The glide ratio of the airplane is affected by all four fundamental forces that act on the airplane (weight, lift, drag, and thrust). If all factors affecting the airplane are constant, the glide ratio will be constant. Therefore, in order to judge the gliding distance of the airplane, the pilot must keep all of these forces constant. Although the effect of wind will not be covered in this section, it is a very prominent force acting on the gliding distance of the airplane in relation to its movement over the ground. It is sufficient to say here that the stronger the headwind, the less the gliding distance.

In light general aviation type airplanes, the weight of the airplane can be considered to remain constant for any given loading condition, since the fuel burn-off has insignificant effect. However, the heavier the airplane the higher the airspeed must be to obtain the same glide ratio.

The lift of the airplane will remain constant for any one airspeed, since the angle of attack of the wing remains the same regardless of variations of the flightpath from the horizontal. Therefore, if a constant airspeed is maintained, the lift will be constant.

Under different conditions of flight, the drag factors may be varied through the operation of the landing gear and/or flaps. When the landing gear is lowered or the flaps are lowered, the drag is greater and the airspeed will decrease unless the pitch attitude is lowered. As the pitch attitude is lowered, the glidepath steepens and reduces the distance traveled. With the power off, a windmilling propeller also creates considerable drag, thereby retarding the airplane's forward movement.

Although the propeller thrust of the airplane is normally dependent on the power output of the engine, the throttle is in the closed position during a glide, so the thrust is constant. Since power is not used during a glide or power-off approach, the pitch attitude must be adjusted as necessary to maintain a constant airspeed.

The best speed for the glide is one which will give a minimum rate of descent and provide a safe margin of flying speed above a stall. Changes in the gliding airspeed will result in proportionate changes in glide ratio, for as the airspeed of the glide is reduced or increased from the optimum glide speed, the glide ratio is also changed. When below the optimum speed, the faster altitude will be lost. For this reason the pilot should *never* try to stretch a glide by reducing the airspeed below the airplane's recommended glide speed.

The most efficient gliding *speed* will vary with the gross weight of the airplane, the configuration of landing gear and flaps, and the windmilling of the propeller. Within certain limits, the wind component would also affect the speed required for the most efficient glide angle. In airplanes for which the manufacturer does not provide the optimum glide speed, and with which the pilot is not familiar, it can be determined by experimentation. This can be accomplished by establishing a power-off glide and noting the airspeed and vertical speed, and then gradually reducing the airspeed until the vertical speed reaches its minimum and starts to increase. The airspeed at that moment is the best glide speed *in still air.* When the correct pitch attitude and airspeed have been determined, the pilot should commit to memory the position of reference points in relation to the horizon, and the tone of the sound made by the air passing over the airplane structure.

As in straight-and-level flight, turns, and climbs, the pilot should perform descents by reference to both flight instruments and outside visual references (Fig. 6–10).

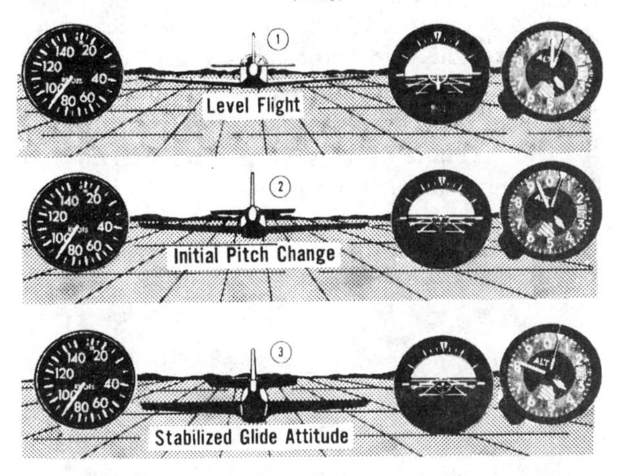

Figure 6–10 Outside and Instrument References for Descending

To enter the glide, the pilot should close the throttle and advance the propeller control (if so equipped) to low pitch (high RPM). A constant altitude should be held with back pressure on the elevator control until the airspeed decreases to the recommended glide speed, then the pitch attitude should be allowed to decrease to maintain that gliding speed. When the speed has stabilized, the airplane should be retrimmed for "hands-off flight."

When the approximate gliding pitch attitude is established, the airspeed indicator should be checked. If the airspeed is higher than the recommended speed, the pitch attitude is too low, and if the airspeed is less than recommended, the pitch attitude is too high; therefore, the pitch attitude should be readjusted accordingly. After the adjustment has been made, it is important to retrim the airplane so that it will maintain this attitude without the need to hold pressure on the elevator control.

When the proper glide has been established, flaps may be used, but then the pitch attitude will have to be changed accordingly to maintain the desired glide speed. Again the pitch attitude should be adjusted first, then the airspeed checked after it has had time to decrease. It is best to *always establish the proper flight attitudes by checking the visual reference first*, then use the flight instruments as a secondary check. It is good practice to always retrim the airplane after each pitch adjustment.

In order to maintain the most efficient glide in a turn, more altitude must be sacrificed than in a straight glide since this is the only way speed can be maintained without power. Turning in a glide decreases the glide performance of the airplane to an even greater extent than does a normal turn with power.

Skidding to the outside in a gliding turn, particularly when close to the ground, is even worse than skidding in a climbing turn. While the results are the same, the proximity of the ground makes the former more likely to end disastrously.

The level off from a glide must be started before reaching the desired altitude because of the airplane's downward inertia (Fig. 6–11). The amount of lead depends upon the rate of descent and the pilot's control technique. With too little lead, there will be a tendency to descend below the selected altitude. For example, assuming a 500-FPM rate of descent, the altitude must be led by 100–150 feet to level off at an airspeed higher than the glide speed At the lead point, power should be added to the appropriate level flight cruise setting so the desired airspeed will be attained at the desired altitude. Since the nose will tend to rise as the airspeed increases, the pilot should smoothly control the pitch attitude to attain the level flight attitude so that the level off is completed at the desired altitude.

When recovery is being made from a gliding turn, the pressure on the elevator control which was applied during the turn must be decreased or the nose will tend to rise too rapidly, making it difficult to attain the desired cruise speed and altitude. This error will require considerable attention and conscious control adjustment.

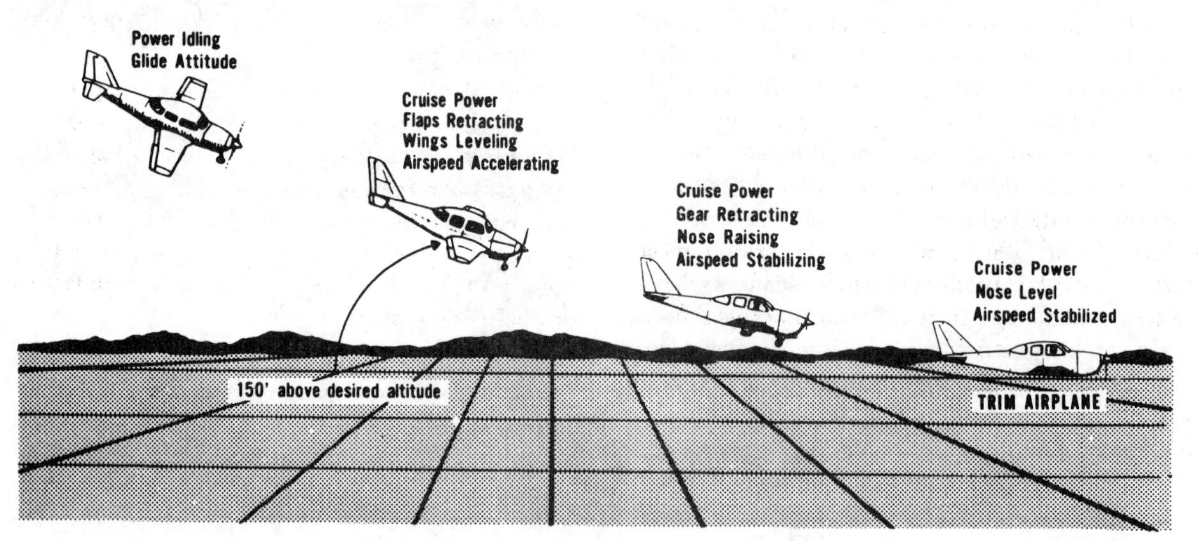

Figure 6–11 Leveling-Off From a Descent

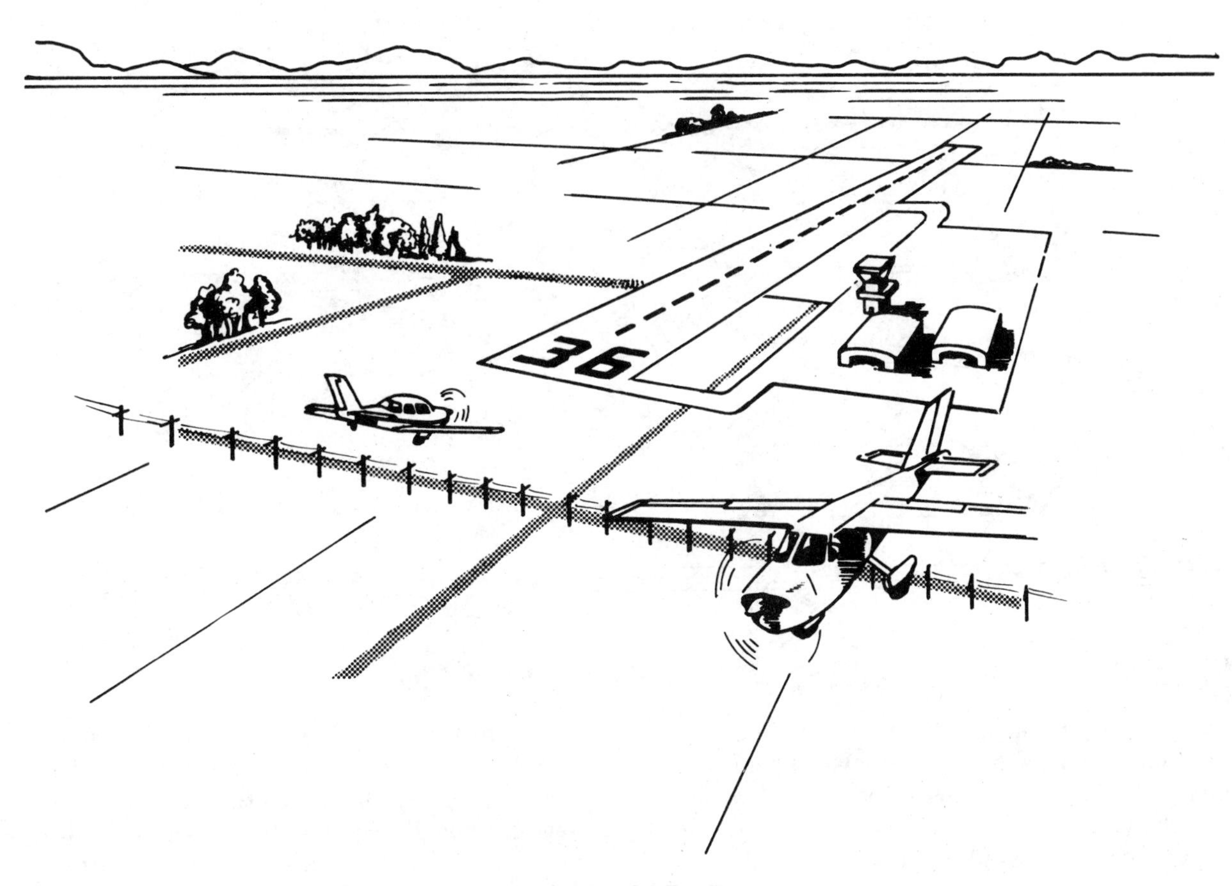

CHAPTER 7

Airport Traffic Patterns and Operations

This chapter explains the methods used for safely adjusting the flow of air traffic at and near airports, and discusses the major traffic services and landing approach aids that are available to the pilot at the busier terminal areas. Just as roads and streets are needed in order to utilize automobiles, airports or airstrips are needed to utilize airplanes. Every flight begins and ends at an airport or other suitable landing field. For that reason, it is essential that the pilot learn the traffic rules, traffic control procedures, traffic advisory services, and traffic pattern layouts that may be in use at various airports.

When an automobile is being driven on congested city streets, it can be brought to a stop so as to give way to conflicting traffic. An airplane, however, can only be slowed down. Even then, it may be traveling 60 to 180 miles per hour. Consequently, specific traffic patterns and traffic control procedures have been established at designated airports. The traffic patterns provide specific routes for takeoffs, departures, arrivals, and landings. The exact nature of each airport traffic pattern is dependent on the runway in use, wind conditions, obstructions, and other factors.

Control towers and radar facilities provide a means of adjusting the flow of arriving and departing aircraft, and render assistance to the pilot in busy terminal areas. Airport lighting and runway marking systems are used frequently to alert the pilot to abnormal conditions and hazards, so arrivals and departures can be made safely.

Airports vary in complexity from small grass or sod strips to major terminals having a complex of many paved runways and taxiways (Fig. 7-1). Regardless of the type of airport, the pilot must know and abide by the rules and general operating procedures applicable to the airport being used. These rules and procedures are based not only on logic or common sense, but also on courtesy, and their

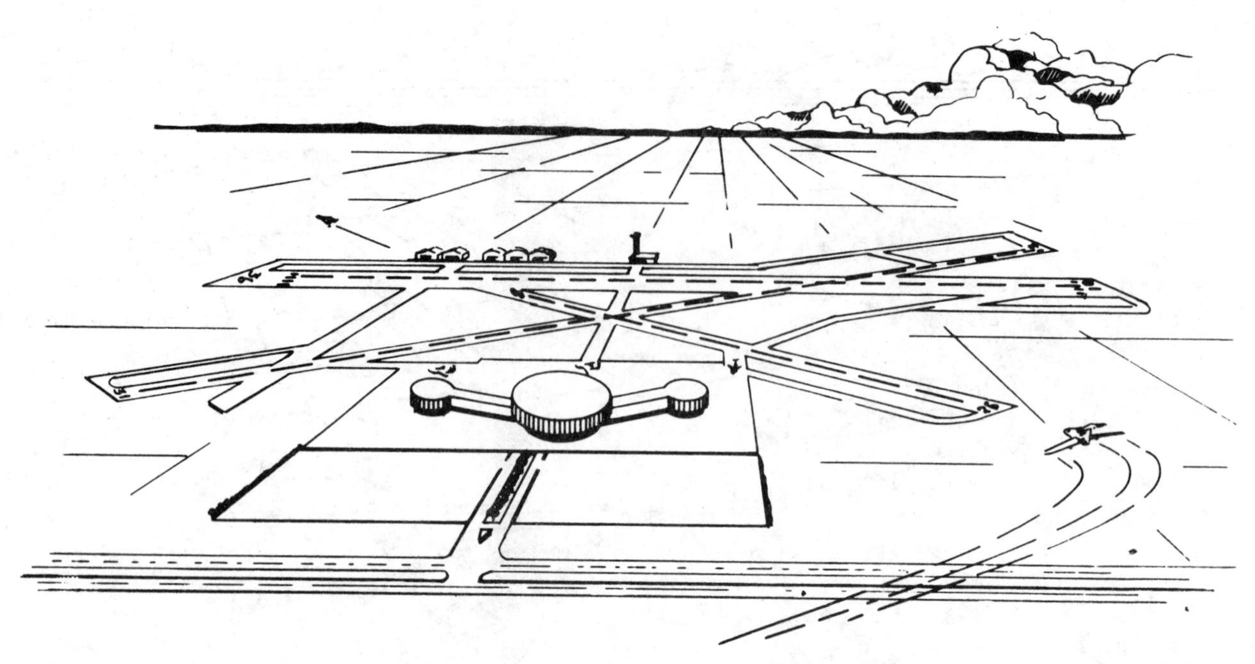

Figure 7–1 Major Airport Terminal Area

objective is to keep air traffic moving with maximum safety and efficiency. The use of any traffic pattern, service, or procedure does not, however, alter the *responsibility of each pilot* to see and avoid other aircraft.

Airport Traffic Patterns

To assure that air traffic flows into and out of an airport in an orderly manner, an airport traffic pattern is established appropriate to the local conditions, including the direction and placement of the pattern, the altitude at which it is to be flown, and the procedures for entering and leaving the pattern. Unless the airport displays approved visual markings indicating that turns should be made to the right, the pilot should make all turns in the pattern to the left.

When operating at an airport with a control tower, the pilot receives, by radio, a clearance to approach or depart as well as pertinent information about the traffic pattern. If there is no control tower, it is the pilot's responsibility to determine the direction of the traffic pattern, to comply with the appropriate traffic rules, and to display common courtesy toward other pilots operating in the area.

The pilot is not expected to have intimate knowledge of all traffic patterns at all airports, but if familiar with the *basic* rectangular pattern, it will be easy to make proper approaches and departures from most airports, regardless of whether they have control towers. At tower-controlled airports, the tower operator may instruct pilots to enter the traffic pattern at any point or to make a straight-in approach without flying the usual rectangular pattern. Many other deviations are possible if the tower operator and the pilot work together in an effort to keep traffic moving smoothly. It must be recognized that jets or large, heavy aircraft will frequently be flying wider and/or higher patterns than lighter aircraft and in many cases will make a straight-in approach for landing.

Compliance with the basic rectangular traffic pattern reduces the possibility of conflicts at airports where air traffic is not being controlled by an FAA control tower. While accident statistics for each year show an improvement over previous years, it is known that the majority of midair collisions occur in the vicinity of nontower airports, under visual flight rules (VFR) weather conditions. It is imperative then, that the pilot form the habit of exercising constant vigilance in the vicinity of airports even though the air traffic appears to be light.

The basic rectangular traffic pattern is illustrated in Fig. 7–2. The traffic pattern altitude is usually 1,000 feet above the elevation of the airport surface. The use of a common altitude at a given airport is the key factor in minimizing the risk of collisions at nontower airports. At all airports, the direction of traffic flow (in accordance with FAR Part 91) is always to the left, *unless* right turns are indicated by approved light signals or visual markings on the airport, or the control tower specifically directs otherwise.

It is recommended that while operating in the traffic pattern at nontower airports the pilot maintain an airspeed that conforms with the limits established by FAR 91 for tower-controlled airports: no more than 156 knots (180 MPH) for reciprocating engine aircraft or 200 knots (230 MPH) for turbine-powered airplanes. In any case, the speed should be adjusted, when practicable, so that it is compatible with the speed of other aircraft in the pattern.

The basic rectangular traffic pattern consists of four "legs" positioned in relation to the runway in use. In the following discussion, reference is made to a "90° turn" from one leg to the other since the ground track of each leg is perpendicular to the preceding one. The actual change in the airplane's heading during those turns will be more or less than 90° depending on the amount of correction necessary to counteract wind drift.

The *upwind leg* of the rectangular pattern is a straight course aligned with, and leading from, the takeoff runway. This leg begins at the point the airplane leaves the ground and continues until the 90° turn onto the crosswind leg is started.

On the upwind leg after takeoff, the pilot should continue climbing straight ahead until reaching a point beyond the departure end of the runway and within 300 feet of traffic pattern altitude. If leaving the pattern, the pilot should continue straight ahead, or depart by making a 45° left turn (right turn for a right-hand pattern).

The *crosswind leg* is the part of the rectangular pattern that is horizontally perpendicular to the extended centerline of the

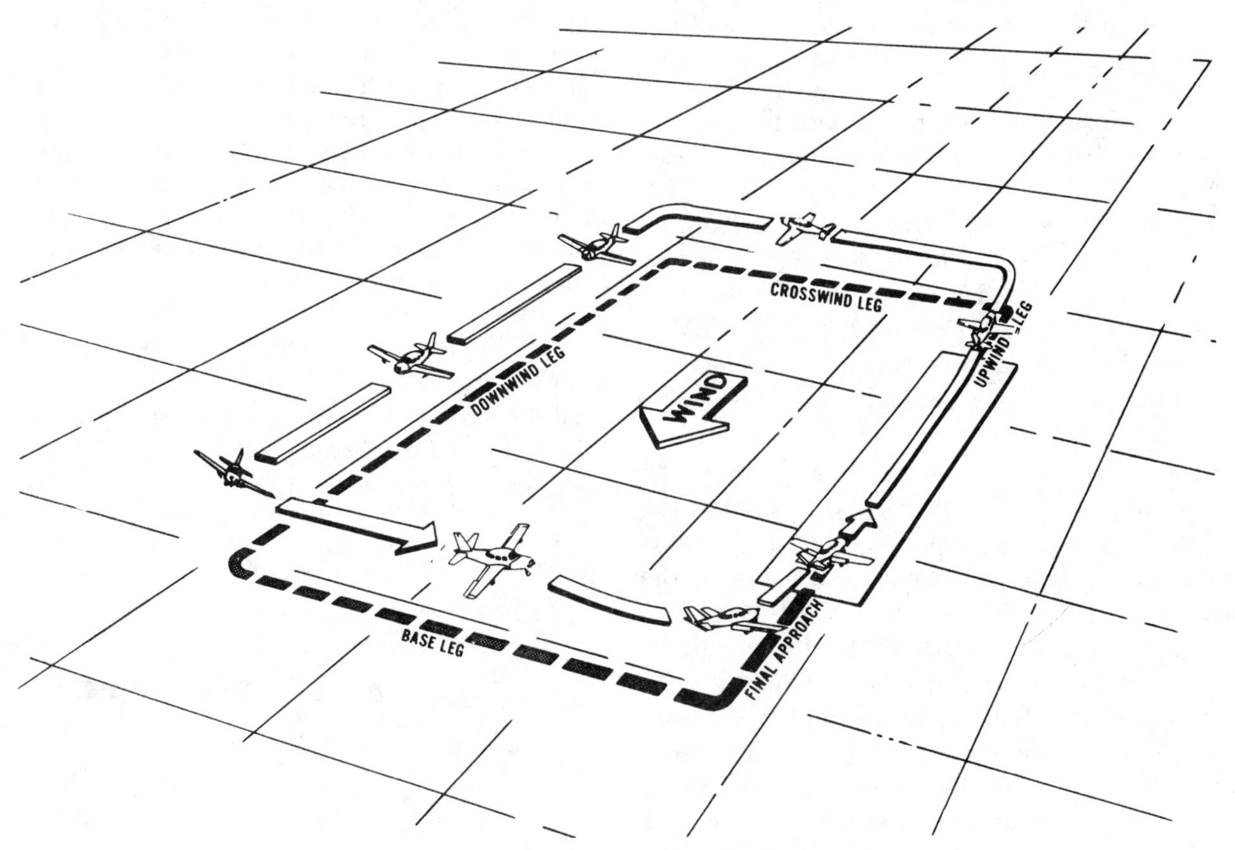

Figure 7-2 Basic Rectangular Traffic Pattern

takeoff runway and is entered by making a 90° turn from the upwind leg. On the crosswind leg the airplane proceeds to the downwind leg position.

Since in most cases the takeoff is made into the wind, the wind now will be approximately perpendicular to the airplane's flightpath. As a result, the airplane will have to be crabbed or headed slightly into the wind while on the crosswind leg to maintain a ground track that is perpendicular to the runway centerline extension. This factor will be further explained in a later chapter covering wind drift and maneuvering by reference to ground objects.

After reaching the prescribed altitude for the traffic pattern and when in the proper position to enter the downwind leg, a level medium bank 90° turn should be made into the downwind leg.

The *downwind leg* is a course flown parallel to the landing runway, but in a direction opposite to the intended landing direction. This leg should be approximately ½ to 1 mile out from the landing runway, and at the specified traffic pattern altitude. During this leg, the prelanding check should be completed and the landing gear extended if retractable. The downwind leg continues past a point abeam of the approach end of the runway to where a medium bank 90° turn is made onto the base leg.

The *base leg* is the transitional part of the traffic pattern between the downwind leg and the final approach leg. Depending on the wind condition, it is established at a sufficient distance from the approach end of the landing runway to permit a gradual descent to the intended touchdown point. The ground track of the airplane while on the base leg should be perpendicular to the extended centerline of the landing runway, although the longitudinal axis of the airplane may not be aligned with the ground track when it is necessary to crab into the wind to counteract drift. (This will be discussed in the chapter on Landing Approaches and Landings.) While on the base leg the pilot must ensure, before turning onto the final approach, that there is no danger of colliding with another aircraft that may be already on the final approach.

As stipulated in Federal Aviation Regulations, aircraft while on final approach to land, or while landing, have the right-of-way over other aircraft in flight or operating on the surface. When two or more aircraft are approaching an airport for the purpose of landing, the aircraft at the lower altitude has the right-of-way, but it shall not take advantage of this rule to cut in front of another which is on final approach to land, or to overtake that aircraft.

The *final approach leg* is a descending flightpath starting from the completion of the base-to-final turn and extending to the point of touchdown. This is probably the most important leg of the entire pattern, for here the pilot's judgment and technique must be keenest to accurately control the airspeed and descent angle while approaching the intended touchdown point. The various aspects are thoroughly explained in the chapter on Landing Approaches and Landings.

To enter the traffic pattern at an airport without a control tower, inbound pilots are expected to observe other aircraft already in the pattern and to conform to the traffic pattern in use. If no other aircraft are in the pattern, then traffic indicators on the ground and wind indicators must be checked to determine which runway and traffic pattern direction should be used (Fig. 7–3). Many airports have L-shaped traffic pattern indicators displayed with a segmented circle adjacent to the runway. The short member of the L shows the direction in which the traffic pattern turns should be made when using the runway parallel to the long member. For example, in Fig. 7–4, the airplane should fly a right-hand pattern. These indicators should, of course, be checked while at a distance well away from any pattern

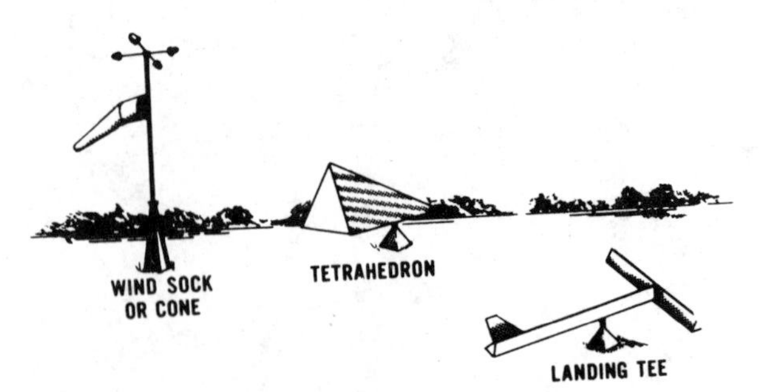

Figure 7–3 Wind and Landing Direction Indicators

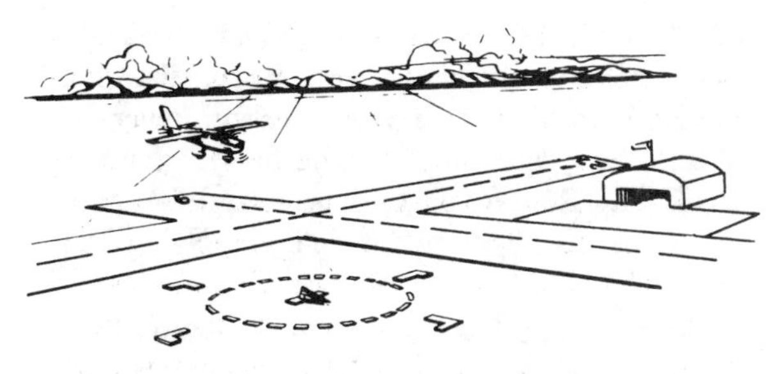

Figure 7–4 Left and Right Traffic Pattern Indicators

that might be in use, or while at a safe height well *above* generally used pattern altitudes. When the proper traffic pattern direction has been determined, the pilot should then proceed to a point well clear of the pattern before descending to the pattern altitude.

Generally, when approaching an airport for landing, the traffic pattern should be entered at a 45° angle to the downwind leg, headed toward a point abeam of the midpoint of the runway to be used for landing. Arriving airplanes should always be at the proper traffic pattern altitude before entering the pattern, and should stay clear of the traffic flow until established on the entry leg. Entries into traffic patterns while descending create specific collision hazards and must be avoided at all times.

The entry leg should be of sufficient length to provide a clear view of the entire traffic pattern, and to allow the pilot adequate time for planning the intended path in the pattern and the landing approach.

Non-controlled Airport Traffic

In addition to flying a basic rectangular traffic pattern, pilots operating at nontower-controlled airports are urged to use the communications radio to announce their positions and intentions to a ground radio station located at those airports or, if none is functioning, to broadcast "in the blind" on an appropriate radio frequency. This alerts other pilots to the presence of your airplane and helps in avoiding midair collisions.

FAA has over 180 Flight Service Stations (FSS) which provide, on a designated radio frequency, advisory information concerning the airport at which they are located. These advisories, when requested, will include the speed and direction of the surface wind and other pertinent airport conditions—as well as the favored runway under the existing wind condition.

In addition, the FSS will advise the pilot if there is observed or reported traffic in the traffic pattern, or in the vicinity, so the pilot can approach or depart the airport in such manner as to avoid disrupting or endangering other aircraft. These FSSs are listed, along with their assigned frequencies, in the Airport/Facility Directory published by the U.S. Department of Commerce.

When there is no FAA facility on the airport, a radio service called UNICOM can be very useful to the pilot. The presence of a UNICOM facility at an airport is indicated in the Airport/Facility Directory and on sectional aeronautical charts; there are approximately 4,000 airports with UNICOM in the United States. This is an informal voluntary advisory service provided by the airport operator for the convenience of pilots. The UNICOM operator will relay information about known traffic in the area, and about the airport conditions.

UNICOM provides convenient air/ground communication, but it should be remembered that the person providing the information may or may not be an experienced observer of air traffic. It may be a veteran pilot, and again it may be someone who has no flying experience at all.

As standard operating practice, all traffic inbound to an uncontrolled airport should continuously monitor the appropriate radio frequency as indicated on the aeronautical chart or in the Airport/Facility Directory. To avoid radio interference with other air traffic that may be using UNICOM at nearby airports, the arriving pilot should delay the initial call until about 5 miles from the airport, and then listen before making any transmission.

Departing pilots should monitor the proper frequency, broadcasting their position and intentions before taxiing onto the runway for takeoff. To minimize congestion on the communication frequencies, all radio transmissions should be brief and concise as possible.

When two-way radio communications are conducted, the pilot must have a Restricted Radiotelephone Operator Permit issued by the Federal Communications Commission, as required of everyone using a radio transmitter.

Radio Communications

When initiating radio communications with a ground facility, the following should be stated:

1. Identification of the ground station being called.
2. Identification of the airplane, by make and registration number.

Ground station call signs comprise the name of the location or airport, followed by the appropriate indication of the type of station:

OAKLAND TOWER (airport traffic control tower);

MIAMI GROUND (ground control position in tower);

KENNEDY APPROACH (radar or non-radar approach control position);

ST. LOUIS DEPARTURE (radar or non-radar departure control position);

WASHINGTON RADIO (FAA Flight Service Station).

During the initial contact with a ground station, the complete airplane call sign should be used. This includes the name or make of the airplane followed by the complete registration number. For example, "NEW YORK RADIO, MOONEY THREE ONE ONE FIVE ECHO, OVER." Communication transmissions may be continued by using the airplane make and only the last three characters of the airplane's registration number when so initiated by the ground station.

When initiating the call to any FSS, the pilot should also indicate the frequency on which a reply is expected. Some ground stations transmit on more than one frequency. For example, the New York FSS transmits on several VORTAC frequencies in the area, one of which is Riverhead VORTAC. These VORTACs are shown on charts, each having a different name and frequency. If the pilot has the receiver tuned to Riverhead VORTAC and wishes to call the New York FSS, the call should be made to "RIVERHEAD RADIO." This automatically tells New York FSS that the pilot is expecting a reply on the Riverhead VORTAC frequency. If the call is made to "NEW YORK RADIO," it would be necessary to tell the FSS to "REPLY ON RIVERHEAD VORTAC."

When a reply from the ground station is not received on the initial call, the pilot should recheck the radio frequency to make sure it is correct, and check the altitude—the airplane may be too low for two-way VHF radio communications.

When a reply is received, the same format as for initial callup should be used except, after the airplane identification, state the message to be sent or acknowledge the message received. The acknowledgment is usually made by saying the word "ROGER." For example, "APACHE ONE TWO THREE XRAY, ROGER." Pilots are expected to comply with ATC clearances/instructions if they acknowledge the message by "ROGER."

The radio is similar to a "party line" telephone—that is, many people use the same frequency. To avoid interfering with someone's communication it is necessary to first *listen* on the appropriate frequency to be sure no one else is transmitting. When the frequency is clear and after giving thought to what is going to be said, the pilot should hold the microphone close to the mouth and speak in a normal tone of voice. As few words as possible should be used to give a clear understanding of the message. Standard phraseology is explained in the Airman's Information Manual.

Tower-Controlled Airport Traffic

Not all airports have control towers, but it is essential that the pilot become familiar with the requirements and procedures where a control tower is used. Today's airport congestion resulted in the "airport traffic area" concept of traffic control for those airports that have a control tower in operation. Where an airport traffic area exists, it extends to a radius of 5 statute miles from the center of the airport, and extends up to (but not including) 3,000 feet above the airport surface (Fig. 7-5). Airport traffic areas are not depicted on aeronautical charts, but the airports at which these areas exist are identifiable on the charts by a control tower radio frequency and the letters CT adjacent to the airport symbol. Complete radio data for each airport having radio communication facilities is found in the Airport/Facility Directory.

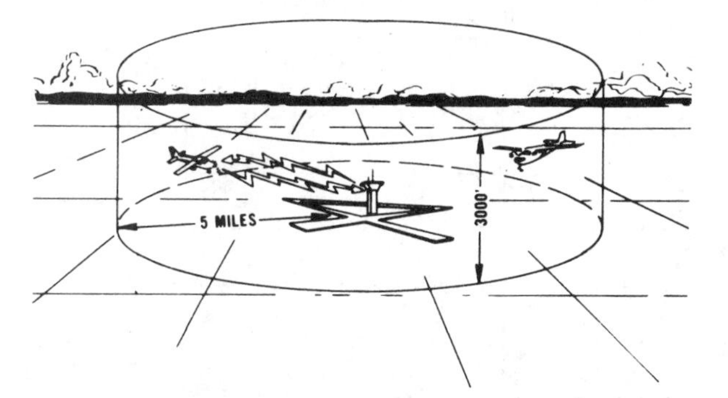

Figure 7–5 Airport Traffic Area

When operating at an airport where traffic control is being exercised by a federally operated control tower, pilots are required to maintain two-way radio communication with the tower while within the airport traffic area, unless the tower authorizes otherwise. Through this communication the pilot requests permission from the tower operator to taxi, take off, or land and receives a specific clearance along with pertinent information about the traffic pattern.

The tower controller will issue clearances (instructions) for aircraft to generally follow the desired flightpath (traffic pattern), as well as the proper taxi routes when operating on the ground. If not otherwise authorized or directed by the tower, pilots approaching to land must circle the airport to the left. However, an appropriate clearance must be received from the tower before the landing may be made.

Airport traffic control is based upon observed, or reported known traffic, and airport conditions. Tower controllers establish the sequence of arriving and departing aircraft by requiring the pilots to adjust their flightpaths and speeds as necessary to achieve proper spacing. These adjustments can only be based on observed air traffic, accurate position reports from pilots, and anticipated aircraft maneuvers.

On occasion it may be necessary for a pilot to maneuver the aircraft to maintain spacing with other traffic. The controller can anticipate minor maneuvering such as shallow "S" turns. The controller cannot, however, anticipate a major maneuver such as a 360° turn. If a pilot makes a 360° turn after obtaining a landing sequence, the result is usually a gap in the landing interval and more importantly, it causes a chain reaction which may result in a conflict with following air traffic, and interruption of the sequence established by the tower or approach controller. Should a pilot decide to make maneuvering turns to maintain spacing behind a preceding aircraft, it is necessary to advise the controller. Except when requested by the controller or in emergency situations, or when receiving radar service, a 360° turn should *never* be executed in the traffic pattern without first advising the controller.

Procedures in which light-gun signals are used apply only when the aircraft using the particular airport is not equipped with two-way radio, or when radio contact cannot be established or maintained (Figs. 7–6 and 7–7).

In addition to directing all incoming and outgoing air traffic by radiotelephone or light-gun signals, the tower also provides current weather data, altimeter setting, and any other information necessary to safe flight within the airport traffic area.

Figure 7–6 Controlling Airport Traffic With Light-Gun Signals

Color and type of signal	Aircraft on the ground	Aircraft in flight
Steady green---------	Cleared for takeoff-------	Cleared to land.
Flashing green-------	Cleared to taxi-----------	Return for landing (to be followed by steady green at proper time.
Steady red-----------	Stop----------------------	Give way to other aircraft and continue circling.
Flashing red---------	Taxi clear of landing area (runway) in use	Airport unsafe--do not land.
Flashing white------	Return to starting point on airport	
Alternating red and green	General warning signal-----exercise extreme caution	

Figure 7–7 Meanings of Light-Gun Signals

Tower operators are all licensed by the FAA, and the towers themselves are in most cases under the jurisdiction of the FAA, although some are independently run by the individual airport operator. If the tower is not Federally operated, communications are required if the aircraft is equipped with radio. If the aircraft's radio allows only reception from the tower, a listening watch should be maintained.

ATC Clearances

Air Traffic Control (ATC) clearances are issued solely for the purpose of preventing collision between aircraft. Whether operating VFR or IFR, a pilot must not deviate from the air traffic clearance without obtaining an amended clearance. If the pilot cannot comply with any provision of the clearance, ATC should be informed *immediately* giving a brief reason, and normally ATC will issue an amended clearance. In the event of emergency deviation from the provisions of an ATC clearance, the pilot must notify ATC as soon as possible and get an amended clearance.

Understanding and complying with air traffic clearances also speeds up traffic movement on an airport. Before starting to taxi, the pilot should always call for a taxi clearance from the tower. (At many airports, this is done on a special radio frequency assigned to Ground Control.) Then, when cleared to a particular runway, the pilot may taxi close to *but not onto* the specified runway. A clearance to move onto the runway, and to take off, must be obtained on the regular control tower frequency.

When giving takeoff or landing instructions, if a tower controller cautions the pilot about "wake turbulence," it is a warning that turbulence may exist behind another aircraft that has just made a takeoff or landing. After receiving such an advisory, the pilot should analyze the situation and determine the proper course of action. Even though a takeoff or landing clearance has been issued, it may be safer to wait or to change the intended operation in some way. In that case, the pilot should ask the controller for a revised clearance. "Cleared for takeoff" or "Cleared to land" means only that the runway is or will be clear for the takeoff or landing.

Sometimes clearances include the word "immediate," such as "Cleared for immediate takeoff." Such communications are to be interpreted as meaning that *if* the pilot takes off at once there will be adequate separation from other aircraft. *It is not an order to go.* The controller cannot be expected to do the pilot's thinking. If there is any reason to believe the takeoff or landing cannot proceed safely, it is the *pilot's responsibility* to decline the clearance. The controller's primary job is to aid in preventing collisions between aircraft, not to advise pilots on flight procedures.

The FAA desires to help the student pilot in acquiring sufficient practical experience in the operating environment. To receive additional assistance while operating in areas of concentrated air traffic, a student pilot need only identify oneself as a student pilot during the initial call to an FAA radio facility. For instance, "Dayton Tower, this is Fleetwing 1234, Student Pilot, over." This special identification will alert FAA air traffic control personnel and enable them to provide the student pilot with such extra assistance and consideration as may be needed. This procedure, however, is not mandatory.

Wake Turbulence Avoidance

Just as powered boats do on water, every airplane generates a wake while in flight. Initially, the disturbance was attributed to "prop wash." It is now known, however, that this disturbance is caused by a pair of counter rotating vortices trailing from the airplane's wingtips. The strong vortices from large aircraft pose particular problems to light aircraft. In fact, the wingtip vortices of these large aircraft can impose rolling moments that exceed the control capability of many light aircraft. The manner in which this hazardous phenomenon is created is explained in the chapter on Principles of Flight and Performance Characteristics.

This invisible turbulence can be extremely dangerous when it is encountered during takeoff and landing. Numerous airplanes have encountered wake turbulence of such severity that complete loss of control of the airplane resulted. The aircraft were, in some cases, at altitudes too low to recover. The most severe wake turbulence is produced by large and heavy commercial or military airplanes in the landing or takeoff configuration.

Under certain conditions, airport traffic controllers apply special procedures for separating light aircraft from heavy jet aircraft. The controllers also provide pilots with whom they are in communication, and who, in the tower's opinion, may be adversely affected by wake turbulence from a large airplane, the position and direction of flight of the large airplane followed by the phrase "CAUTION—WAKE TURBULENCE." Whether or not a warning has been given, however, THE PILOT IS EXPECTED TO ADJUST THE FLIGHTPATH AS NECESSARY TO AVOID WAKE ENCOUNTERS.

Vortices are generated whenever an airplane is developing lift, since trailing vortices are a byproduct of wing lift. Therefore, prior to takeoff or landing, pilots should always note the rotation or touchdown point of the preceding airplane (Fig. 7–8).

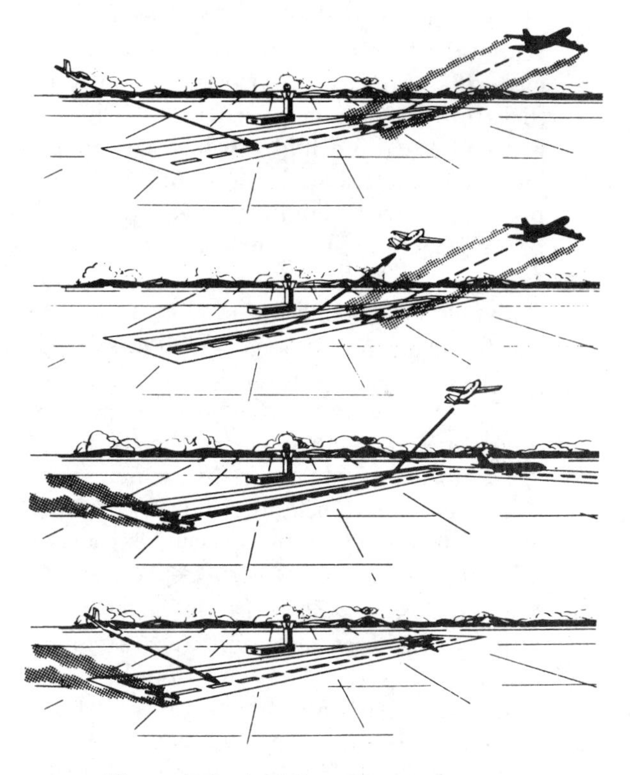

Figure 7–8 Avoiding Wingtip Vortices

Pilots of all aircraft should visualize the location of the vortex trail behind large aircraft and use proper vortex avoidance procedures to achieve safe operation. The following vortex avoidance procedures are recommended for the various situations:

1. Landing behind a large airplane—on the same runway. Stay at or above the large aircraft's approach flightpath—note its

touchdown point—then land beyond that point.

2. Landing behind a large airplane—when a parallel runway is closer than 2,500 feet, consider possible drift of the vortices to the other runway. Stay at or above the large airplane's final approach path—note its touchdown point—then land beyond that point.
3. Landing behind a la ge airplane—on a crossing runway: Cross above the large airplane's flightpath.
4. Landing behind a departing large airplane—on the same runway. Note the large airplane's rotation point—then land well prior to its rotation point.
5. Landing behind a departing large airplane—on a crossing runway. Note the large airplane's rotation point—if it is past the runway intersection—continue the approach but land prior to the intersection, avoiding flight below the large airplane's flightpath. Abandon the approach unless a landing is assured well before reaching the intersection.
6. Departing behind a large airplane: Note the large airplane's rotation point—then rotate prior to the large airplane's rotation point and continue to climb above and stay upwind of the large airplane's climbpath until turning clear of its wake. Avoid subsequent headings which will cross below and behind a large airplane. Be alert for any critical takeoff situation which could lead to a vortex encounter.
7. Intersection takeoffs—on the same runway: Be alert to adjacent large airplane operations particularly upwind of your runway. If an intersection takeoff clearance is received, avoid subsequent headings which will cross below a large airplane's path.
8. Departing or landing after a large airplane executing a low missed-approach or touch-and-go landing: Because vortices settle and move laterally near the ground, the vortex hazard may exist along the runway and in your flightpath after a large airplane has executed a low missed-approach or a touch-and-go landing, particularly in light quartering wind conditions. The pilot should assure that an interval of at least 2 minutes has elapsed before takoff or landing.

Visual Approach Slope Indicator (VASI)

There are many airports equipped with a Visual Approach Slope Indicator to assist pilots in the landing approach. The VASI provides a color-coded *visual* glidepath using a system of lights positioned alongside the runway, near the designated touchdown point. It ensures safety by providing a visual glidepath which clears all obstructions in the final approach area. Once the principles and color code of the lighting system are understood, flying the VASI is a simple matter of noting the light's colors and adjusting the airplane's rate of descent to stay on the visual glide slope.

The VASI is especially effective during approaches over water or featureless terrain where other sources of visual reference are lacking or misleading, and at night. It provides optimum descent guidance for landing and minimizes the possibility of undershooting or overshooting the designated touchdown area.

There are several types of VASIs in use including the 2-bar, 3-bar, and tri-color systems. The basic principle of the 2-bar and 3-bar VASI is that of color differentiation between red and white. Each light unit projects a beam of light having a white segment in the upper part of the beam and red segment in the lower part of the beam (Fig. 7–9).

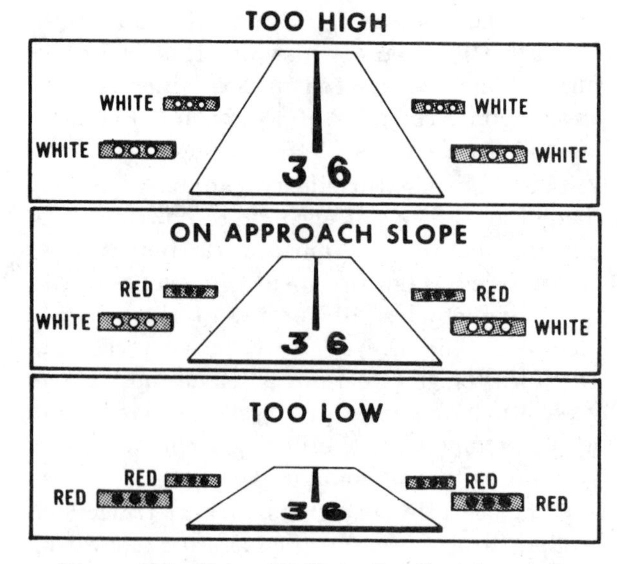

Figure 7–9 Using VASI on Landing Approach

When on the proper glidepath, the pilot will in effect, overshoot the downwind bars and undershoot the upwind bars (Fig. 7–8). Thus, the downwind bars will be seen as white and the upwind bars as red. From a position below the glidepath the pilot will see all the light bars as red, and from above the glidepath all the light bars will appear white. Passing through the glidepath from a low position is indicated to the pilot by a transition in color from red through pink to white. From a high position, passing through the path is indicated to the pilot by a transition in color from white through pink to red.

When the pilot is below the glidepath, the red bars tend to merge into one distinct red signal. A safe obstruction clearance may not exist when this distinct red signal is visible. The visual glidepath will separate into individual lights as the pilot approaches the runway threshold. At this point, the approach should be continued by reference to the runway touchdown zone.

Three-bar VASI installations provide two visual glidepaths. The lower glidepath is provided by the near and middle bars and is normally set at a 3 degree incline while the upper glidepath, provided by the middle and far bars, is normally ¼ degree higher. This higher glidepath is intended for use only by high cockpit aircraft to provide a sufficient threshold crossing height.

When using a 3-bar VASI it is not necessary to use all three bars. The near and middle bars constitute a two-bar VASI for using the lower glidepath. Also, the middle and far bars constitute a 2-bar VASI for using the upper glidepath. The Tri-color Approach Slope Indicator normally consists of a single light unit, projecting a three-color visual approach path into the final approach area of the runway upon which the system is installed. In these systems, a below glidepath indication is red, the above glidepath indication is amber and the on-path indication is green.

Automatic Terminal Information Service

Many of the tower-controlled airports provide Automatic Terminal Information Service (ATIS), which broadcasts on a specially assigned radio frequency a continuous recording of essential but routine noncontrol information such as current weather, communication requirements, and the runways and types of approaches in use.

There are three distinct advantages of this service: First, extensive use of ATIS by pilots greatly reduces the congestion on the regular tower and ground control radio frequencies, and reduces the routine workload on the controllers. This allows the controllers to devote more time to the specific control of arriving and departing aircraft; second, the ATIS broadcast provides more information than is contained in the normal tower or ground control instructions for taxi, takeoff, landing, weather, NOTAMs, etc. and third, the pilot can receive this information when cockpit duties are least pressing and can listen to as many repeats as desired.

The ATIS broadcast should be monitored prior to requesting taxi clearance or prior to requesting landing clearance. Arriving pilots should monitor the broadcast well in advance of entering the Airport Traffic Area. Each ATIS broadcast carries an identifying phonetic alphabet code word (Alpha, Bravo, Charlie, etc.). This code word is important. After receiving the ATIS broadcast, the pilot, on initial contact with ground control, tower, or approach control, should state that the information has been received and repeat the specific code word. Example—". . TULSA GROUND CONTROL, THIS IS CESSCRAFT SEVEN FOUR SIX FOUR CHARLIE, ON TERMINAL RAMP, READY TO TAXI. I HAVE INFORMATION *ECHO*. OVER. ."

When notified by the pilot that ATIS has been received, the control tower operator, in giving takeoff or landing instructions, will not repeat certain information contained in the ATIS broadcast. Monitoring the published ATIS radio frequency and using this information will assist the pilot in planning the flight into the airport traffic area with regard to runway, weather conditions, and the approach and departure routes of other air traffic. Airports at which this service is available are indicated on aeronautical charts and in the Airman's Information Manual.

Terminal Radar Service for VFR Aircraft

Another valuable service for avoiding conflict with other traffic is provided by radar traffic control facilities at many of the busy airport hubs. This service is available at the pilot's request through the approach control facility serving the area, on a workload permitting basis. The issuance of traffic information as observed on a radar display by the controller (Fig. 7–10), is based on the principle of assisting and advising a pilot that a particular radar target's (aircraft) position and track indicates it may intersect or pass in such close proximity to the pilot's intended flightpath that it warrants the pilot's attention. This is to alert the pilot to take appropriate evasive action should the need arise. This service is not intended to relieve the pilot of the responsibility for continued vigilance to see and avoid other aircraft. Information and the availability of this service is presented in the Airport/Facility Directory.

Figure 7–10 ATC Terminal Radar Service

Terminal Control Areas

The FAA has adopted additional air traffic rules for the control or segregation of all aircraft within airspace designated as "terminal control areas" (TCA) surrounding some of the nation's busiest airports. The major problem inherent in a terminal airspace operation

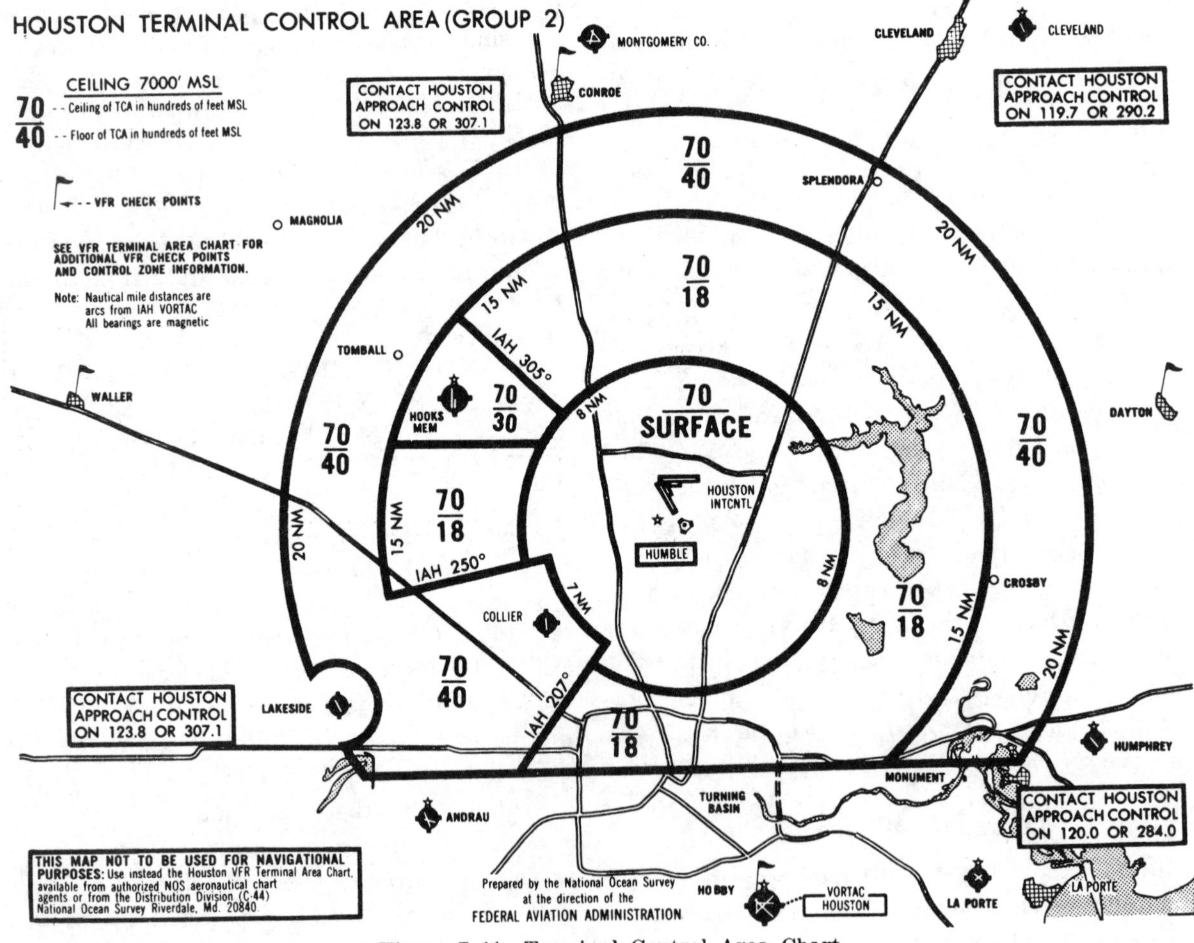

Figure 7–11 Terminal Control Area Chart

is the mixture of controlled and noncontrolled air traffic. The safest environment is, of course, one within which *all* aircraft are provided positive separation from each other.

To increase safety and to reduce the risk of midair collisions, a system has been developed to segregate air traffic operating to or from the airports within the designated terminal control area (TCA) from other traffic operating near that terminal area. *All* traffic operating within the terminal control area is separated and controlled by ATC. This requires that aircraft operating within this type designated airspace have certain communication and navigation equipment. Appropriate authorization from ATC must be received *prior* to operating within the terminal control area.

Horizontal and vertical boundaries of each terminal control area vary according to (1) the types of aircraft normally using the area and the nature of air operations in that area; (2) the facilities at the airports and the navigational aids available; (3) the air traffic capabilities to meet the needs of the terminal control concept. Although the configuration of each TCA is tailored to individual needs of the area, the general appearance is circular in nature with layers as in an upside-down wedding cake. The altitudes and distances outlining the TCA are shown on Aeronautical Sectional Charts and VFR Terminal Area Charts for individual terminals (Fig. 7–11). These are available for purchase at many airports.

The terminal control areas are classed as Group I or Group II depending on the number of aircraft operations and passengers they service. To operate within either group, the aircraft must be equipped with an operable VOR or TACAN receiver, two-way radio with capability for communicating with ATC and, with certain exceptions, an approved transponder. An encoding altimeter is required for Group I TCA. In addition, the pilot in command must have at least a Private Pilot Certificate to land or take off at an airport within Group I Terminal Control Area. A transponder is not required for helicopters or for an airplane if flying IFR to other than the primary airport of either Group I or Group II areas, or VFR to Group II areas. The specific requirements are stipulated in Federal Aviation Regulations, Part 91.

Runway and Taxiway Markings

In the interest of safety and efficiency of aircraft operations, the FAA has established for the guidance of pilots, standard runway and taxiway markings. Since most airports are marked in this manner it is important for the pilot to learn the meaning of the markings and how to use them.

Standard runway markings are painted in reflective white, while markings of taxiways and nontraffic areas such as blast pads and overruns are in reflective yellow.

All runways are marked with a runway direction number. The number is the whole number nearest the 10° increment of the magnetic azimuth of the centerline of the runway, measured clockwise from magnetic north. The last zero of the magnetic azimuth is not used, and single numbers are not preceded by a zero. For example, a runway which has a direction of 270° is designated as runway 27; a runway having a direction of 090° is called runway 9 (Fig. 7–12). To differentiate between two parallel runways, the runway direction number has a letter "L" or "R" following it.

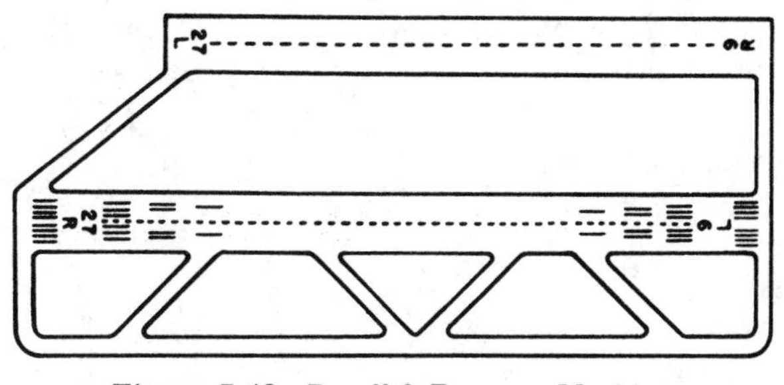

Figure 7–12 Parallel Runway Markings

When runways and/or taxiways have been permanently closed to aircraft traffic, all markings indicating a usable runway or taxiway are obliterated. Crosses are then painted near the ends, on each closed runway or taxiway. When a closed runway or taxiway is intersected by a usable runway or taxiway, the crosses are painted on the closed surface on each side of the usable surface. Naturally, the pilot should not use runways so marked, for takeoff or landing.

A line perpendicular to the runway centerline designates the beginning of the portion of a runway that is usable for landing. No landing should be made short of that line (Fig. 7–13).

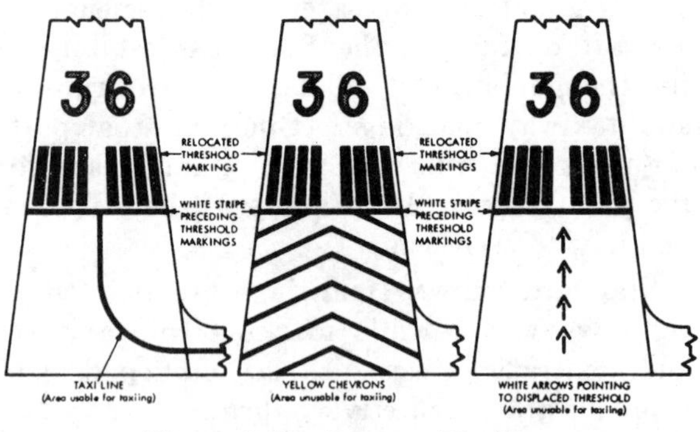

Figure 7–13 Runway Markings

An overrun or blast pad area at the approach end of a runway is marked by a series of equally spaced yellow chevrons to indicate the portion not usable for takeoff or landing. The apex of the chevrons points to and terminates at the threshold of the usable portion of the runway.

When it has been necessary to position the landing threshold up-runway from the end of the paving, it is known as a relocated or displaced threshold. There are two methods of marking this area (Fig. 7–13). When the paved area on the approach side of the relocated threshold is to be used only for taxiing, all markings indicating that the area was usable as a runway are obliterated. When the paved area on the approach side of the relocated threshold is not to be used for taxiing, this area is marked in the same manner as overrun or blast pad areas, as previously discussed. Where thresholds are displaced, the unusable portion is marked with a series of large white arrows. The arrows are placed along the centerline on the approach side of the relocated threshold and point to the designated landing area. In all cases a white stripe across the width of the full-strength runway precedes the threshold markings. Never land short of the stripe.

Runway shoulders which have been stabilized with materials that give the appearance of firm paving but are not intended for use by aircraft, are marked with a series of partial yellow chevrons.

The taxiway centerline is marked with a continuous yellow line. The edges of the taxiway are sometimes marked with two continuous lines 6 inches apart. Taxiway HOLDING LINES consist of two continuous lines and two dashed lines perpendicular to the centerline. Pilots should stop short of the holding line for runup or when instructed by ATC to "HOLD SHORT OF (runway, ILS critical area, etc.)." Aircraft exiting a runway are not clear of the runway until the aircraft has passed the runway holding line.

CHAPTER 8

Takeoffs and Departure Climbs

This chapter explains and describes the factors involved and the pilot techniques required for safely taking the airplane off the ground and departing the takeoff area under normal conditions, as well as in various situations where maximum takeoff performance of the airplane is absolutely essential.

The takeoff and departure climb involves the movement of the airplane from its starting position on the runway to the point where a positive climb to a safe maneuvering altitude has been well established. Since the takeoff requires both ground and in-flight operation, the pilot must be able to use the controls during their transition from ground functions to in-flight functions with maximum smoothness and coordination. Skill in blending these functions will improve the pilot's ability to control the airplane's direction of movement on and from the runway.

A thorough knowledge of takeoff and climb principles, both in theory and practice, will prove to be of extreme value throughout the pilot's career. It often may prevent an attempt to take off under critical conditions that would require performance beyond the capability of the airplane or the skill of the pilot. The takeoff itself, though relatively simple, often presents the most hazards of any part of a flight. Accident statistics show that takeoff accidents, although slightly less in frequency of occurrence, are much more tragic than landing accidents. Consequently, the importance of a thorough knowledge of the principal factors involved, and faultless technique and judgment cannot be overemphasized.

Takeoffs should always be made as nearly into the wind as practical. There are two reasons for this—first, the airplane's speed while on the ground is much less than if the takeoff were made downwind, thus reducing wear and stress on the landing gear; and second, a shorter ground roll and consequently much less space is required to develop the minimum lift necessary for takeoff and climb. Since the airplane depends on airspeed in order to fly, a headwind provides some of that airspeed, even with the airplane motionless, by reason of the wind flowing over the wings.

Although the takeoff and climb process is one continuous maneuver, it will be divided into three separate steps for purposes of explanation: (1) the takeoff roll, (2) the lift-off, and (3) the initial climb after becoming airborne (Fig. 8–1).

The *takeoff roll* is that portion of the takeoff procedure during which the airplane is accelerated from a standstill to an airspeed that provides sufficient lift for it to become airborne.

The *lift-off*, or rotation, is the act of becoming airborne as a result of the wings lifting the airplane off the ground or the pilot rotating the nose up, increasing the angle of attack to start a climb.

The *initial climb* begins when the airplane leaves the ground and a pitch attitude has been established to climb away from the takeoff area. Normally it is considered complete when the airplane has reached a safe maneuvering altitude, or an enroute climb has been established.

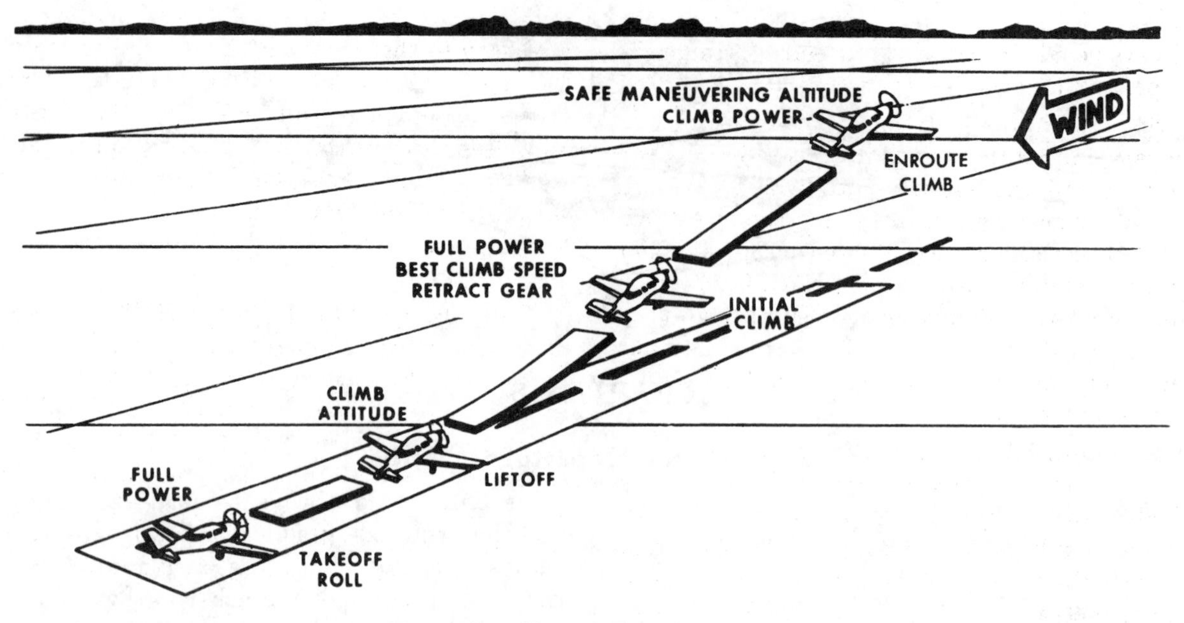

Figure 8–1 Normal Takeoff and Climb

The normal takeoff is one in which the airplane is headed directly into the wind or the wind is very light, and the takeoff surface is firm with no obstructions along the takeoff path, and is of sufficient length to permit the airplane to gradually accelerate to normal climbing speed.

Prior To Takeoff

Before taxiing onto the runway or takeoff area, the pilot should ensure that the engine is operating properly and that all controls, including flaps and trim tabs, are set for takeoff. In addition, the pilot must make certain that the takeoff path will be clear of other aircraft, vehicles, persons, or livestock, and that there will be adequate time to execute the takeoff before any aircraft in the traffic pattern turns onto the final approach. It is inadvisable to start a takeoff immediately behind another aircraft, particularly large, heavily-loaded transport airplanes, because of the wake turbulence they create.

While taxiing onto the runway, the pilot can select ground reference points that are aligned with the runway direction as aids to maintaining directional control during the takeoff. These may be distant trees, towers, buildings, mountain peaks, or the markings or lights along the runway.

Normal Takeoff Roll

After taxiing onto the runway, the airplane should be carefully aligned with the intended takeoff direction, and the tailwheel or nosewheel positioned straight, or centered. In airplanes equipped with a locking device, the tailwheel should be locked in the centered position. After releasing the brakes the throttle should be smoothly and continuously advanced to maximum allowable power. As the airplane starts to roll forward, the pilot should slide both feet down on the rudder pedals so that the toes or balls of the feet are on the rudder portions—not on the brake portions.

The use of maximum allowable power, even though it may appear that conditions do not require it, is recommended for every takeoff. Engine-wise there is little or no advantage in using reduced power. A takeoff at reduced power requires more time, and thus more revolutions of the engine; consequently, there is more wear on the pistons, rings, and cylinders. Airplane-wise, reduced power would be the equivalent of starting the takeoff at a point well down the runway. At some airports the runway length may be insufficient to permit this. Reduced power not only lengthens the takeoff roll but increases wear on the tires.

An abrupt application of power may cause the airplane to yaw sharply to the left because of the torque effects of the engine and propeller. With this in mind, the throttle should always be advanced smoothly and continuously to prevent any sudden swerving.

Smooth, gradual advancement of the throttle is very important, particularly in the case of high horsepower engines or tailwheel-type airplanes, since peculiarities in their takeoff characteristics are accentuated in proportion to the rapidity with which takeoff power is applied. The engine instruments should be monitored during the takeoff so as to note immediately any malfunction or indication of insufficient power.

In nosewheel-type airplanes, no pressures on the elevator control are necessary beyond those needed to steady it. Applying unnecessary pressure will only aggravate the takeoff and prevent the pilot from recognizing when pressure is actually needed to establish the takeoff attitude.

As speed is gained, the elevator control will tend to assume a neutral position if the airplane is correctly trimmed. At the same time, directional control should be maintained with smooth, prompt, positive rudder corrections (except in very large airplanes in which nosewheel steering is accomplished by a small hand-operated nosewheel steering mechanism) throughout the takeoff roll. The effects of torque or P-factor at the initial speeds tend to pull the nose to the left. The pilot must use whatever rudder pressure is needed to correct for these effects or for existing wind conditions to keep the nose of the airplane headed straight down the runway. The use of brakes for steering purposes is to be avoided, since they will cause slower acceleration of the airplane's speed, lengthen the takeoff distance, and possibly result in severe swerving.

With a tailwheel-type airplane of normal stability characteristics and the elevator trim set for takeoff, on application of maximum allowable power, the airplane will (when sufficient speed has been attained) normally assume approximately the correct takeoff pitch attitude of its own accord. That is, the tail will rise slightly. This attitude can then be maintained by applying slight back elevator pressure. If the elevator control is pushed forward during the takeoff roll to prematurely raise the tail, its effectiveness will rapidly build up as the speed increases, making it necessary to then apply back pressure to lower the tail to the proper takeoff attitude. This erratic change in attitude will delay the takeoff and lead to directional control problems. Rudder pressure must be used promptly and smoothly to counteract yawing forces so that the airplane continues straight down the runway.

While the speed of the takeoff roll increases, more and more pressure will be felt on the flight controls, particularly the elevators and rudder. Since the tail surfaces (except "T" tails) receive the full effect of the propeller slipstream, they become effective first. As the speed continues to increase, all of the flight controls will gradually become effective enough to maneuver the airplane about its three axes. It is at this point, in the taxi-to-flight transition, that the airplane is being flown more than taxied. As this occurs, progressively smaller rudder deflections are needed to maintain direction.

The feel of resistance to the movements of the controls, as well as the airplane's reaction to such movements, are the only real indicators of the degree of control attained. (Except on certain high performance airplanes, instruments are not reliable indicators in this regard.) This feel of resistance is not a measure of the airplane's speed but rather of its controllability. To determine the degree of controllability, the pilot must be conscious of the reaction of the airplane to the control pressures and immediately adjust the pressures as needed to control the airplane.

Normal Lift-Off

Since a good takeoff depends on the proper takeoff attitude, it is important to know how this attitude appears and how it is attained. The ideal takeoff attitude is that which requires only minimum pitch adjustments shortly after the airplane lifts off to attain the speed for the best rate of climb.

Each type of airplane has its own best pitch attitude for normal lift-off. However, varying conditions may make a difference in the required takeoff technique. A rough field, a smooth field, a hard surface runway, or a short or soft, muddy field, all call for a slightly different technique, as will smooth air in contrast to a strong, gusty wind. The different techniques for those other-than-normal conditions are discussed later in this chapter.

When all the flight controls become effective during the takeoff roll in a nosewheel type airplane, back elevator pressure should be gradually applied to raise the nosewheel slightly off the runway, thus establishing the takeoff or lift-off attitude. This is often referred to as "rotating." (In tailwheel-type airplanes, the tail should first be allowed to rise off the ground slightly to permit the airplane to accelerate more rapidly.) At this point, the position of the nose in relation to the horizon should be noted, then elevator pressure applied as necessary to hold this attitude. On both types of airplanes, of course, the wings must be kept level by applying aileron pressure as necessary.

The airplane may be allowed to fly off the ground while in this normal takeoff attitude. Forcing it into the air by applying excessive back pressure would only result in an excessively high pitch attitude and may delay the takeoff. As discussed earlier, excessive and rapid changes in pitch attitude result in proportionate changes in the effects of torque, thus making the airplane more difficult to control.

Although the airplane can be forced into the air, this is considered an unsafe practice and must be avoided under normal circumstances. If the airplane is forced to leave the ground by using too much back pressure before adequate flying speed is attained, the wing's angle of attack may be excessive, causing the airplane to settle back to the runway or even to stall. On the other hand, if sufficient back elevator pressure is *not* held to maintain the correct takeoff attitude after becoming airborne, or the nose is allowed to lower excessively, the airplane may also settle back to the runway. This would occur because the angle of attack is decreased and lift diminished to the degree where it will not support the airplane. It is important, then, to hold the attitude constant after rotation or lift-off.

Even as the airplane leaves the ground, the pilot must continue to be concerned with maintaining straight flight as well as holding the proper pitch attitude.

During takeoffs in a strong, gusty wind it is advisable that an extra margin of speed be obtained before the airplane is allowed to leave the ground, since a takeoff at the normal takeoff speed may result in lack of positive control, or a stall, when the airplane encounters a sudden lull in strong gusty wind, or other turbulent air currents. In this case the pilot should hold the airplane on the ground longer to attain more speed, then make a smooth, positive rotation to leave the ground.

Normal Initial Climb

Upon lift-off, the airplane should be flying at approximately the attitude which will allow it to accelerate to its best rate-of-climb airspeed. The best rate-of-climb speed (V_y) is that speed at which the airplane will gain the most altitude in the shortest period of time.

If the airplane has been properly trimmed, some back pressure may be required on the elevator control to hold this attitude until the proper climb speed is established. On the other hand, relaxation of any back pressure on the elevator control before this time may result in the airplane settling, even to the extent that it contacts the runway.

The airplane will pick up speed rapidly after it becomes airborne. However, only after it is certain the airplane will remain airborne and a definite climb is established, should the flaps and landing gear be retracted (if the airplane is so equipped).

It is recommended by engine manufacturers, as well as the FAA, that takeoff power be maintained until an altitude at least 500 feet above the surrounding terrain or obstacles is attained. The combination of best rate-of-climb (V_y) and maximum allowable power will give an additional margin of safety, in that sufficient altitude is attained in minimum time from which the airplane can be safely maneuvered in case of engine failure or other emergency. Also, in many airplanes, the use of maximum allowable power automatically gives a richer mixture for additional cooling of the engine during the climb-out.

Since the power on the initial climb is fixed at the takeoff power setting, the airspeed must be controlled by making slight pitch adjustments using the elevators. However, the pilot should not stare at the airspeed indicator when making these slight pitch changes, but should, instead, watch the attitude of the airplane in relation to the horizon. It is better to first make the necessary pitch change and hold the new attitude momentarily, and then glance at the airspeed indicator as a check to see if the new attitude is correct. Due to inertia, the airplane will not accelerate or decelerate immediately as the pitch is changed. It takes a little time for the airspeed to change. If the pitch attitude has been over- or under-corrected, the airspeed indicator will show (belatedly) a speed that is more or less than that desired. When this occurs, the cross-checking and appropriate pitch-changing process must be repeated until the desired climbing attitude is established.

When the correct pitch attitude has been attained, it should be held constant while cross-checking it against the horizon and other outside visual references. The airspeed indicator should be used only as a check to determine if the attitude is correct.

After the recommended climbing airspeed has been well established, and a safe maneuvering altitude has been reached, the power should be adjusted to the recommended climb setting and the airplane trimmed to relieve the control pressures. This will make it much easier to hold a constant attitude and airspeed.

During initial climb, it is important that the takeoff path remain aligned with the runway to avoid the hazards of drifting into obstructions, or the path of another aircraft which may be taking off from a parallel runway.

If necessary to take off immediately behind another airplane, the possibility of wake turbulence must be anticipated, especially if the wind condition is calm or straight down the runway. If turbulence is encountered, resulting in sudden deviations in flight attitudes, firm control pressures should be applied to make a shallow turn to fly out of the wake turbulence. When in smoother air the airplane can then be realigned with the original flightpath. If a crosswind is present, the turn should be made into the wind, since the wake turbulence will be blown downwind or away from your flightpath.

Crosswind Takeoff and Climb

While it is usually preferable to take off directly into the wind whenever possible or practical, there will be many instances when circumstances or wisdom will indicate otherwise. Consequently, the pilot must be familiar with the principles and techniques involved in crosswind takeoffs as well as those for normal takeoffs. A crosswind will affect the airplane during takeoff much as it does in taxiing. With this in mind, it can be seen that the technique for crosswind correction during takeoffs closely parallels the crosswind correction techniques used in taxiing, previously explained in this handbook. Additional discussions of crosswind factors are contained in Chapter 9.

Crosswind Takeoff Roll

The technique used during the initial takeoff roll in a crosswind is generally the same as used in a normal takeoff, except that the aileron control must be held INTO the crosswind. This raises the aileron on the upwind wing to impose a downward force on the wing to counteract the lifting force of the crosswind, and prevents that wing from rising.

As the airplane is taxied into takeoff position, it is essential that the windsock and other wind direction indicators be checked so that the presence of a crosswind may be recognized and anticipated. If a crosswind is indicated, FULL aileron should be held into the wind as the takeoff roll is started. This control position should be maintained while the airplane is accelerating until the ailerons start becoming sufficiently effective for maneuvering the airplane about its longitudinal axis.

While keeping the wings level with the aileron control, the takeoff path must be held straight with the rudder (Fig. 8–2). Normally, this will require applying downwind rudder pressure, since on the ground the airplane (especially the tailwheel-type) will tend to weathervane into the wind. When takeoff power is applied, torque or P-factor which yaws the airplane to the left, may be sufficient to counteract the weathervaning tendency caused by a crosswind from the right. On the other hand, it may also aggravate the tendency to swerve left when the wind is from the left. In any case, whatever rudder pressure is re-

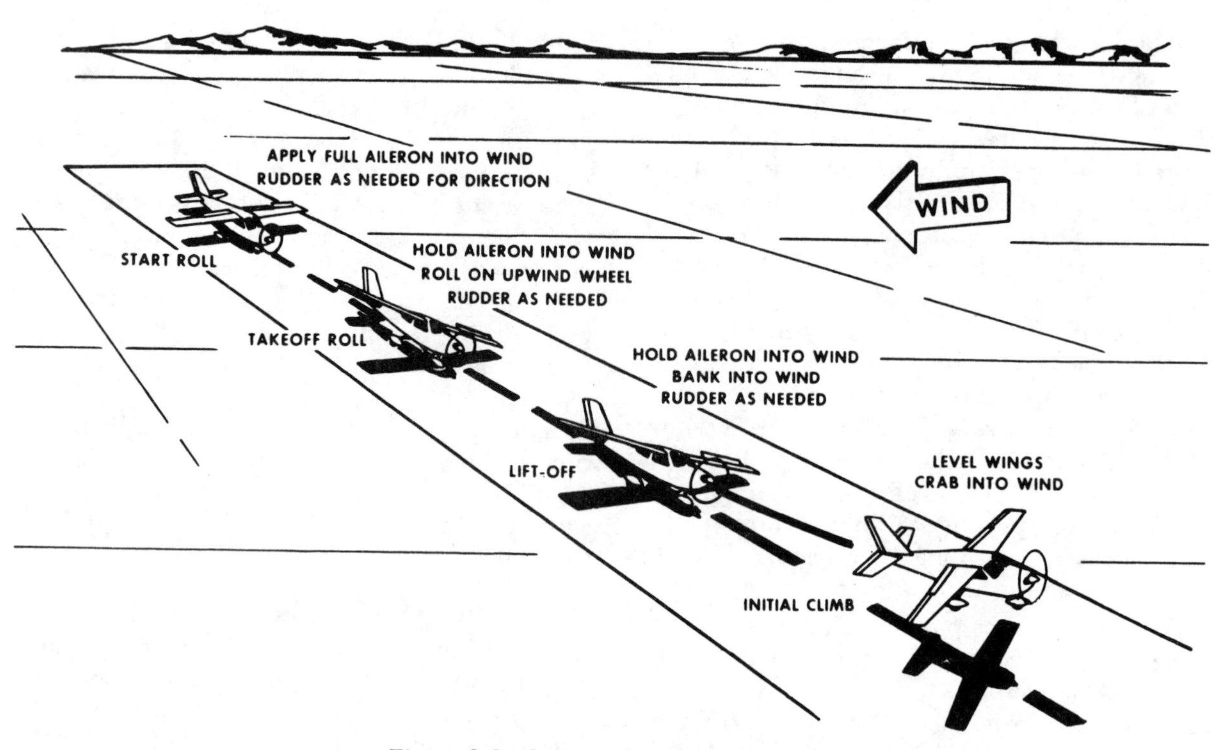

Figure 8–2 Crosswind Takeoff and Climb

quired to keep the airplane rolling straight down the runway should be applied.

As the forward speed of the airplane increases and the crosswind becomes more and more of a relative headwind, the mechanical holding of full aileron into the wind should be reduced. It is when increasing pressure is being felt on the aileron control that the ailerons are becoming more effective. As the aileron's effectiveness increases and the crosswind component of the relative wind becomes less effective, it will be necessary to reduce the aileron pressure gradually to keep the wings level. The crosswind component effect does not completely vanish, so some aileron pressure will have to be maintained throughout the takeoff roll to keep the crosswind from raising the upwind wing. The use of proper aileron pressure then, is simply to keep this from happening.

If the upwind wing rises, thus exposing more surface to the crosswind, a "skipping" action may result (Fig. 8–3). This is usually indicated by a series of very small bounces, caused by the airplane attempting to fly and then settling back onto the runway. During these bounces, the crosswind also tends to move the airplane sideways, and these bounces will develop into side skipping. This side skipping imposes severe side stresses on the landing gear and could result in structural failure.

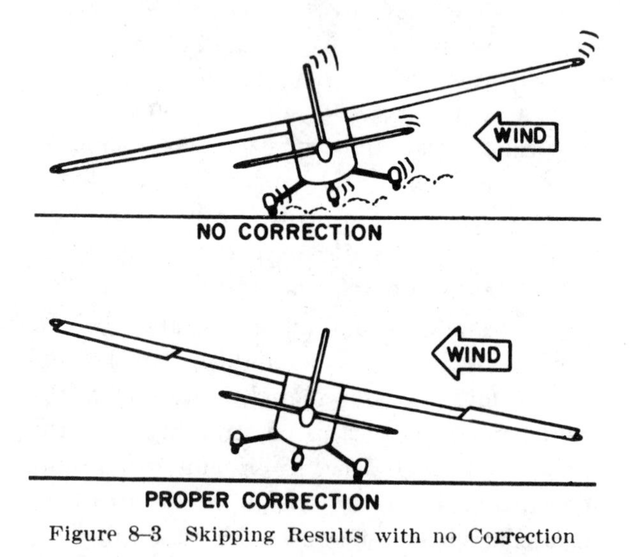

Figure 8–3 Skipping Results with no Correction

It is important then, during the crosswind takeoff roll, to hold sufficient aileron into the wind not only to keep the upwind wing from rising but to hold that wing down so that the airplane will, immediately after lift-off, be slipping into the wind enough to counteract drift.

Crosswind Lift-Off

As the nosewheel or tailwheel is being raised off the runway, the holding of aileron control into the wind may result in the downwind wing rising and the downwind main wheel lifting off the runway first, with the remainder of the takeoff roll being made on that one main wheel (Fig. 8–3). This is acceptable of course, and is preferable to side skipping.

If a significant crosswind exists, the main wheels should be held on the ground slightly longer than in a normal takeoff so that a smooth but very definite lift-off can be made. This procedure will allow the airplane to leave the ground under more positive control so that it will definitely remain airborne while the proper amount of drift correction is being established. More importantly, it will avoid imposing excessive side loads on the landing gear and prevent possible damage that would result from the airplane settling back to the runway while drifting.

As both main wheels leave the runway and ground friction no longer resists drifting, the airplane would be slowly carried sideways with the wind unless adequate drift correction is maintained by the pilot. It is important, then, to establish and maintain the proper amount of crosswind correction prior to lift-off; that is, aileron pressure toward the wind to keep the upwind wing from rising and rudder pressure as needed to prevent weathervaning.

Initial Crosswind Climb

If proper correction is being applied, as soon as the airplane is airborne it will be slipping into the wind sufficiently to counteract the drifting effect of the wind. This slipping should be continued until the airplane has climbed well above the ground. At that time, the airplane should be headed toward the wind to establish just enough "crab" to counteract the wind and then the wings rolled level. The climb while in this "crab," should be continued so as to follow a ground track aligned with the runway direction. The remainder of the climb technique is the same as used for a normal takeoff and climb.

Short Field Takeoff and Climb

Taking off and climbing from fields where the takeoff area is short or the available takeoff area is restricted by obstructions, requires that the pilot operate the airplane at the limit of its takeoff performance capabilities. To depart such an area safely, the pilot must exercise positive and precise control of airplane attitude and airspeed so that takeoff and climb performance results in the shortest ground roll and the steepest angle of climb.

The result achieved should be consistent with the flight performance section of the FAA approved Airplane Flight Manual or the Pilot's Operating Handbook. In all cases the power setting, flap setting, airspeed, and procedures prescribed by the airplane's manufacturer should be observed.

In order to accomplish maximum performance takeoff safely, the pilot must be well indoctrinated in the use and effectiveness of the best angle-of-climb speed (V_x) and best rate-of-climb speed (V_y) for the specific make and model of airplane being flown.

The speed for best angle-of-climb (V_x) is that which will result in the greatest gain in altitude for a given *distance* over the ground. It is usually slightly less than the speed for best rate-of-climb (V_y) which provides the greatest gain in altitude *per unit of time*. The specific speeds to be used for a given airplane are stated in the FAA approved Airplane Flight Manual or the Pilot's Operating Handbook. It will be found that in some airplanes, a deviation of 5 knots from the recommended speed will result in a significant reduction of climb performance. That being the case, precise control of airspeed has an important bearing on the safety of the operation.

Taking off from short fields requires that the takeoff be started from the very beginning of the takeoff area and the airplane accelerated as rapidly as possible. At the field threshold, the airplane is aligned with the intended takeoff path and maximum allowable power applied promptly while releasing the brakes. If the use of flaps is recommended by the airplane manufacturer, they should be extended the proper amount *before* starting the takeoff roll. This permits the pilot to give full attention to

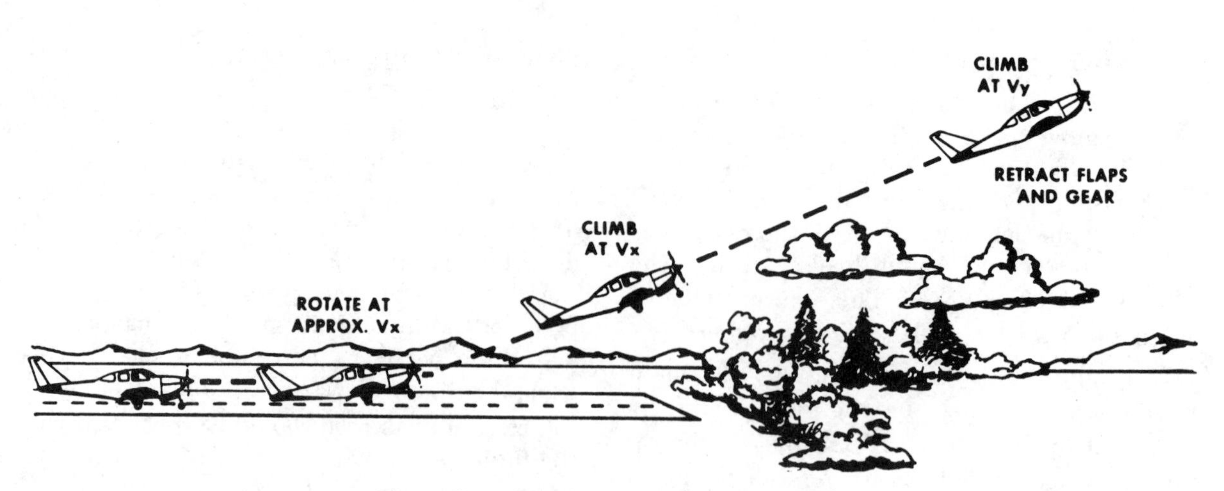

Figure 8–4 Short Field Takeoff and Climb

the proper technique and the airplane's performance throughout the takeoff. There is no significant advantage to extending flaps just prior to lift-off.

Takeoff power should be applied smoothly and continuously—there can be no hesitation—to accelerate the airplane as rapidly as possible. As the takeoff roll progresses, the airplane's pitch attitude and angle of attack should be adjusted to that which results in the minimum amount of drag and the quickest acceleration (Fig. 8–4). In nosewheel-type airplanes this will involve very little use of the elevator control, since the airplane is already in a low drag attitude. In tailwheel-type airplanes, the tail should be allowed to rise off the ground slightly, then held in this tail-low flight attitude until the proper lift-off or rotation airspeed is attained.

For the steepest climb-out and best obstacle clearance, the airplane should be allowed to roll with its full weight on the main wheels and accelerated to the lift-off speed. The airplane should be smoothly and firmly lifted off, or rotated, by applying back pressure on the elevator control as the best angle-of-climb speed (V_x) is attained. Since the airplane will accelerate more rapidly after lift-off, additional back pressure becomes necessary to hold a constant airspeed. After becoming airborne, a straight climb should be maintained at the best angle-of-climb speed (V_x) until the obstacles have been cleared or, if no obstacles are involved, until an altitude of at least 50 feet above the takeoff surface is attained. Thereafter the pitch attitude may be lowered slightly, and the climb continued at the best rate-of-climb speed (V_y) until reaching a safe maneuvering altitude. Unlike a short field approach where the power can be varied to maintain the desired approach airspeed, the short field takeoff requires the use of full takeoff power. Since the power setting is thereby fixed, airspeed must be controlled by adjusting the pitch attitude which in turn will also vary the climb angle. Remember that an attempt to pull the airplane off the ground prematurely, or to climb too steeply, may cause the airplane to settle back to the runway or into the obstacles.

On short field takeoffs, the flaps and landing gear should remain in takeoff position until well clear of obstacles (or as recommended by the manufacturer) and the best rate-of-climb speed (V_y) has been established. It is generally unwise for the pilot to be looking in the cockpit or reaching for flap and landing gear controls, until obstacle clearance is assured. When the best rate-of-climb speed has stabilized, retraction of the flaps may be started. It is usually advisable to raise the flaps in increments to avoid sudden loss of lift and settling of the airplane. After the flaps are fully retracted, the landing gear should be retracted (if so equipped) and the power reduced to the normal climb setting.

Soft Field Takeoff and Climb

Takeoffs and climbs from soft fields require the use of operational techniques for getting the airplane airborne as *quickly* as possible to eliminate the drag caused by tall grass, soft sand, mud, snow, etc., and may or may not require climbing over an obstacle. These same techniques are also useful on a rough field where it is advisable to get the airplane off the ground as soon as possible to avoid damaging the landing gear.

Soft surfaces or long wet grass usually retard the airplane's acceleration during the takeoff roll so much that adequate takeoff speed might not be attained if normal takeoff techniques were employed.

The correct takeoff procedure at fields with such restraining conditions is quite different from that appropriate for short fields with firm, smooth surfaces. To minimize the hazards associated with takeoffs from soft or rough fields, support of the airplane's weight must be transferred as rapidly as possible from the wheels to the wings as the takeoff roll proceeds. This is done by establishing and maintaining a relatively high angle of attack or nose-high pitch attitude as early as possible by use of the elevator control. Wing flaps may be lowered *prior* to starting the takeoff (if recommended by the manufacturer) to provide additional lift and transfer the airplane's weight from the wheels to the wings as early as possible.

The airplane should be taxied onto the takeoff surface at as fast a speed as possible, consistent with safety and surface conditions. Since stopping on a soft surface, such as mud or snow, might bog the airplane down, it should be kept in continuous motion with sufficient power while lining up for the takeoff roll.

As the airplane is aligned with the proposed takeoff path, takeoff power must be applied smoothly and as rapidly as the powerplant will accept it without faltering. As the nosewheel type airplane accelerates, enough back elevator pressure should be applied to establish a positive angle of attack and to reduce the weight supported by the nosewheel. In tailwheel-type airplanes, the tail should be kept low to maintain the inherent positive angle of attack and to avoid any tendency of the airplane to nose over as a result of soft spots, tall grass, or deep snow.

When the airplane is held at a nose-high attitude throughout the takeoff run the wings will, as speed increases and lift develops, progressively relieve the wheels of more and more of the airplane's weight, thereby minimizing the drag caused by surface irregularities or adhesion. If this attitude is accurately maintained, the airplane will virtually fly itself off the ground. It may even become airborne at an airspeed slower than a safe climb speed because of the action of "ground effect." This phenomenon produces an interim gain in lift during flight at very low altitude due to the effect the ground has on the flow pattern of the air passing along the wing. "Ground effect" is further explained in the chapter on Principles of Flight.

After becoming airborne, the nose should be lowered very gently with the wheels just clear of the surface to allow the airplane to accelerate to the best rate-of-climb speed (V_y), or best angle-of-climb speed (V_x) if obstacles must be cleared. Extreme care must be exercised immediately after the airplane becomes airborne and while it accelerates, to avoid settling back onto the surface. An attempt to climb prematurely or too steeply may cause the airplane to settle back to the surface as a result of losing the benefit of "ground effect." Therefore, it is recommended that no climb to an altitude higher than barely clear of the surface be attempted at an airspeed slower than the best angle-of-climb airspeed (V_x).

After a definite climb is established, and the airplane has accelerated to the best rate-of-climb speed (V_y), retract the landing gear and flaps, if so equipped.

In the event an obstacle must be cleared after a soft field takeoff, the climb-out must be performed at the best angle-of-climb airspeed (V_x) until the obstacle has been well cleared. After reaching this point the airspeed may then be accelerated to the best rate-of-climb (V_y) and the flaps and gear retracted. The power may then be reduced to the normal climb setting.

CHAPTER 9

Landing Approaches and Landings

This chapter discusses the factors that affect an airplane during the landing approach and the landing under normal and critical circumstances, and the pilot's techniques for positively controlling those factors. The pilot must be able to make the transition from in-flight control with accuracy, smoothness, and positiveness. So that the pilot may better understand the factors that will influence judgment and technique, the last part of the approach pattern and the actual landing will be divided into five phases—the base leg, the final approach, the roundout, the touchdown, and the after-landing roll.

The *base leg* is that portion of the airport traffic pattern along which the airplane proceeds from the downwind leg to the final approach leg and begins the descent to a landing. While on the base leg the pilot must accurately judge the distance in which the airplane must descend to the landing point and correct for wind drift so the ground track remains perpendicular to the extension of the centerline of the landing runway.

The *final approach* is the last part of the traffic pattern during which the airplane is aligned with the landing runway or area, and a straight line descent is made to the point of touchdown. The descent rate (descent angle) is governed by the airplane's height and distance from the intended touchdown point, and the airplane's speed over the ground.

The *roundout*, or flare as it is sometimes called, is that part of the final approach where the airplane makes a transition from the approach attitude to the touchdown or landing attitude.

The *touchdown* is the actual contact or touching of the main wheels of the airplane on the landing surface, as the full weight of the airplane is being transferred from the wings to the wheels.

The *after-landing roll*, or rollout, is the forward roll of the airplane on the landing surface after touchdown while the airplane's momentum decelerates to a normal taxi speed or a stop.

Normal Approach and Landing

This type of approach and landing involves the use of techniques for what is considered a "normal" situation; that is, when engine power is available, the wind is light or the final approach is made directly into the wind, the final approach path has no obstacles, and the landing surface is firm and of ample length to gradually bring the airplane to a stop. The selected landing point should be beyond the runway's approach threshold but well within the first one-third portion of the runway.

The factors involved and the techniques described for the normal approach and landing also have variable applications to the other-than-normal approaches and landings which are discussed later in this chapter. This being the case, the principles of normal operations are explained first and must be understood before proceeding to the more complex operations (Fig. 9–1).

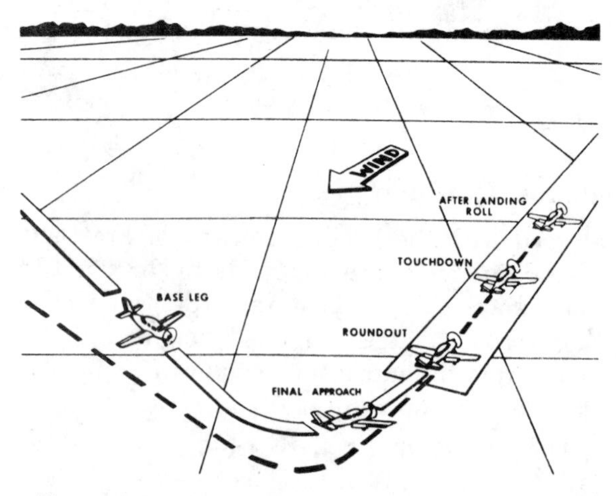

Figure 9–1 Segments of Approach and Landing

Base Leg

The placement of the base leg is one of the more important judgments to be made by the pilot in any landing approach. The pilot must accurately judge the altitude and distance from which a gradual descent will result in landing at the desired spot. The distance will depend on the altitude of the base leg, the effect of wind and amount of wing flaps used. When there is a strong wind on final approach or the flaps will be used to produce a steep angle of descent, the base leg must be positioned closer to the approach end of the runway than would be required with a light wind or no flaps. Normally, the landing gear should be extended and the before-landing check completed *prior* to reaching the base leg.

After turning onto the base leg, the pilot should start the descent with reduced power and an airspeed of approximately 1.4 V_{so}. (V_{so}—the stalling speed with power off, landing gear and flaps down.) For example, if V_{so} is 60 knots, the speed should be 1.4 times 60, or 84 knots. Landing flaps may be partially lowered if desired at this time. Full flaps are not recommended until the final approach is established and the landing assured. Drift correction should be established and maintained to follow a ground track perpendicular to the extension of the centerline of the runway on which the landing is to be made. Since the final approach and landing will normally be made into the wind, there will be somewhat of a crosswind during the base leg. This requires that the airplane be angled (crabbed) sufficiently into the wind to prevent drifting farther away from the intended landing spot.

The base leg should be continued to the point where a medium to shallow-banked turn will align the airplane's path directly with the centerline of the landing runway. This descending turn should be completed at a safe altitude which will be dependent upon the height of the terrain and any obstructions along the ground track. The turn to the final approach should also be sufficiently above the airport elevation to permit a final approach long enough for the pilot to accurately estimate the resultant point of touchdown, while maintaining the proper approach airspeed. This will require careful planning as to the starting point and the radius of the turn. Normally, it is recommended that the angle of bank not exceed a medium bank because the steeper the angle of bank, the higher the airspeed at which the airplane stalls. Since the base-to-final turn is made at a relatively low altitude, it is vitally important that a stall not occur at this point. If an extremely steep bank is needed to prevent overshooting the proper final approach path, it is advisable to discontinue the approach, go around, and plan to start the turn earlier on the next approach rather than risk a hazardous situation.

Final Approach

Immediately after the base-to-final approach turn is completed, the longitudinal axis of the airplane should be aligned with the centerline of the runway or landing surface, so that drift (if any) will be recognized immediately. On a normal approach, with no wind drift, the longitudinal axis should be kept aligned with the runway centerline throughout the approach and landing. (The proper way to correct for a crosswind will be explained under the section "Crosswind Approach and Landing." For now, only an approach and landing where the wind is light or straight down the runway will be discussed.)

After aligning the airplane with the runway centerline, the final flap setting should be completed and the pitch attitude adjusted as required for the desired rate of descent. Slight adjustments in pitch and power may be necessary to maintain the descent attitude and the desired approach airspeed. In the absence of the manufacturer's recommended airspeed, a speed equal to 1.3 V_{so} should be used; that is, if V_{so} is 60 knots, the speed should be 78 knots. When the pitch attitude and airspeed have been stabilized, the airplane should be retrimmed to relieve the pressures being held on the controls.

The descent angle should be controlled throughout the approach so that the airplane will land in the center of the first third of the runway. The descent angle is affected by all four fundamental forces that act on an airplane (lift, drag, and thrust, and weight). If all those forces are constant, the descent angle will be constant in a no-wind condition. The pilot can control those forces by adjusting: (1) the airspeed, (2) the attitude, (3) the power, and (4) the drag (flaps or forward slip. The wind also plays a prominent part in the gliding distance over the ground; naturally, the pilot has no control over the wind but may correct for its effect on the airplane's descent by appropriate pitch and power adjustments.

Considering the factors that affect the descent angle on the final approach, for all practical purposes at a given pitch attitude there is only one power setting for one airspeed, one flap setting, and one wind condition. A change in any one of these variables will require an appropriate coordinated change in the other controllable variables. For example, if the pitch attitude is raised too high without an increase of power, the airplane will settle very rapidly and touch down short of the desired spot (Fig. 9–2). For this reason, NEVER TRY TO STRETCH A GLIDE BY APPLYING BACK ELEVATOR PRESSURE ALONE to reach the desired landing spot. This will shorten the gliding distance if power is not added simultaneously. Therefore, the proper angle of descent and airspeed should be maintained by coordinating pitch attitude changes and power changes.

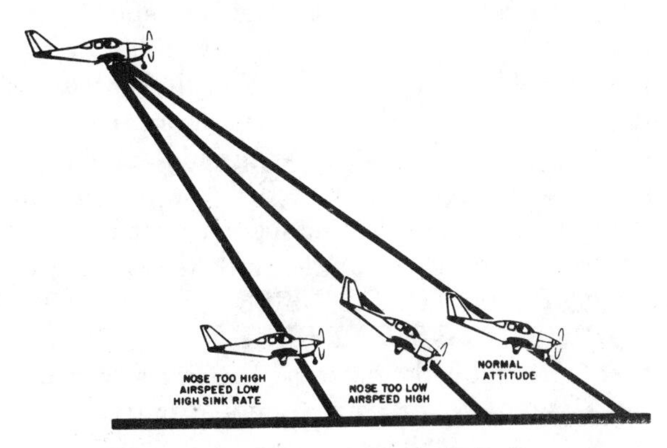

Figure 9–2 Effect of Pitch Attitude on Approach Angle

The objective of a good final approach is to descend at an angle and airspeed that will permit the airplane to reach the desired touchdown point at an airspeed which will result in a minimum of floating just before touchdown. To accomplish this it is essential that both the descent angle and the airspeed be accurately controlled. Since on a normal approach the power setting is not fixed as in a power-off approach, the power should be adjusted as necessary to control the airspeed, and the pitch attitude adjusted SIMULTANEOUSLY to control the descent angle or to attain the desired altitudes along the approach path. By lowering the nose and reducing power to keep approach airspeed constant, a descent at a higher rate can be made to correct for being too high in the approach. This is one reason for performing approaches with partial power; if the approach is too high, merely lower the nose and reduce the power. When the approach is too low, add power and raise the nose. On the other hand, if the approach is extremely high or low, it is advisable to reject the landing and execute a go-around. This procedure is explained later in this chapter.

The lift/drag factors may also be varied by the pilot to adjust the descent through the use of the landing flaps (Fig. 9–3, 9–4). When the flaps are lowered, the airspeed will decrease unless the power is increased or the pitch attitude lowered. After starting the final approach, the pilot must then estimate where the airplane will land through discerning judgment of the descent angle. If it appears that the airplane is going to overshoot or land slightly beyond the desired spot, more flaps may be used if not fully extended or the power reduced further, and the pitch attitude lowered. This will result in a steeper approach. If the spot is being undershot and a shallower approach is needed, the power and the pitch attitude should be increased to readjust the descent angle and the airspeed. NEVER RETRACT THE FLAPS TO CORRECT FOR UNDERSHOOTING since that will suddenly decrease the lift and cause the airplane to sink even more rapidly.

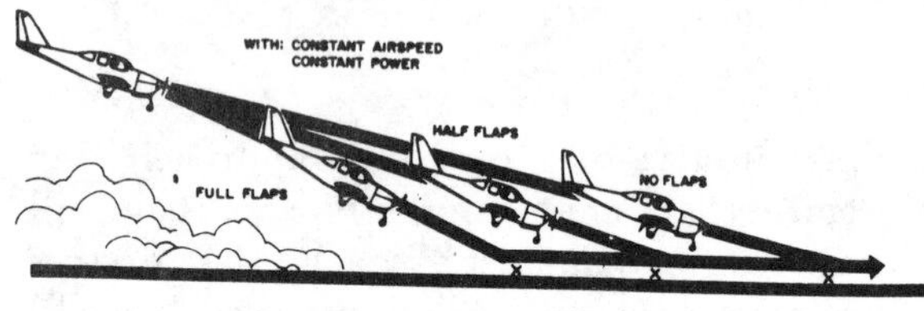

Figure 9–3 Effect of Flaps on Landing Point

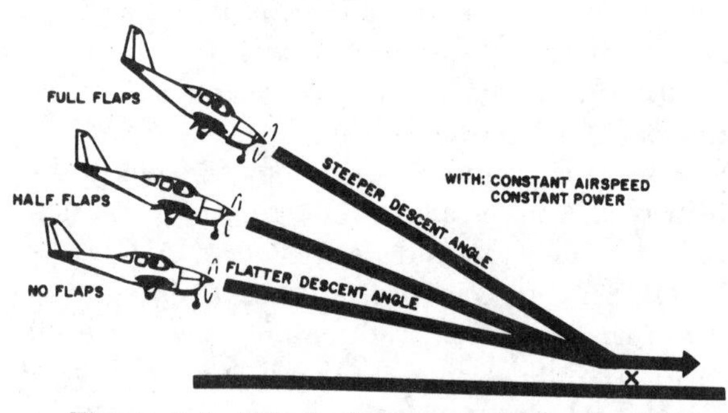

Figure 9–4 Effect of Flaps on Approach Angle

Generally, prior to the approach the airplane will have been trimmed to a state of balance and equilibrium in level flight at cruising power and airspeed. However, during the appoarch to a landing, the power is at a considerably lower than cruise setting and the airplane is flying at a relatively slow airspeed. It becomes apparent, then, that the airplane must be retrimmed on the final approach to compensate for the change in aerodynamic forces. With the reduced power and with a slower airspeed, the airflow produces less lift on the wings and less downward force on the horizontal stabiliber, resulting in a significant nose-down tendency. The elevator, then, must be trimmed more "nose-up." As pointed out in the go-around procedure in this chapter, and later in the elevator trim stall demonstration in Chapter 11, the pilot must be prepared to overpower this trim condition when the landing is rejected and a go-around is initiated.

It will be found that the roundout, touchdown, and landing roll are much easier to accomplish when they are preceded by a proper final approach with precise control of airspeed, attitude, power, and drag, resulting in a stabilized descent angle.

Estimating Height and Movement

During the approach, roundout, and touchdown, vision is of prime importance. To provide a wide scope of vision and to foster good judgment of height and movement, the pilot's head should assume a natural, straight-ahead position. The pilot's visual focus should not be fixed on any one side or any one spot ahead of the airplane, but should be changing slowly from a point just over the airplane's nose to the desired touchdown zone and back again, while maintaining a deliberate awareness of distance from either side of the runway within the pilot's peripheral field of vision.

Accurate estimation of distance is, besides being a matter of practice, dependent upon how clearly objects are seen; it requires that the vision be focused properly in order that the important objects stand out as clearly as possible.

Speed in particular, blurs objects at close range (Fig. 9–5). Most everyone has noted this in an automobile moving at high speed. Nearby objects seem to merge together in a blur, while objects farther away stand out clearly. The driver subconsciously focuses the eyes sufficiently far ahead of the automobile to see objects distinctly.

Figure 9–5 Focusing Too Close Blurs Vision

Similarly, the distance at which the pilot's vision is focused should be proportionate to the speed at which the airplane is traveling over the ground. Thus, as speed is reduced during the roundout, the distance ahead of the airplane at which it is possible to focus should be brought closer accordingly.

If the pilot attempts to focus on references that are too close or looks directly down, the references become blurred, and reactions will be either too abrupt or too late. In this case the pilot's tendency will be to overcontrol, roundout high, and make full-stall "drop-in" landings. When the pilot focuses too far ahead, accuracy in judging the closeness of the ground is lost and the consequent reactions will be slow since there will appear to be no necessity for any action, resulting in the airplane flying into the ground, nose first. The change of visual focus from a long distance to a short distance requires a definite time interval and, even though the time is brief, the airplane's speed during this interval is such that the airplane travels an appreciable distance, both forward and downward toward the ground.

If the focus is changed gradually, being brought progressively closer as speed is reduced, the time interval and the attendant pilot reactions will be reduced, and the whole landing process smoothed out.

Roundout (Flare)

The roundout is a slow, smooth transition from a normal approach attitude to a landing attitude. When the airplane, in a normal decent, approaches within what appears to be about 10 to 20 feet above the ground, the roundout or flare should be started, and once started should be a continuous process until the airplane touches down on the ground.

As the airplane reaches a height above the ground where a timely change can be made into the proper landing attitude, back elevator pressure should be gradually applied to slowly increase the pitch attitude and angle of attack. This will cause the airplane's nose to gradually rise toward the desired landing attitude. The angle of attack should be increased at a rate that will allow the airplane to continue settling slowly as forward speed decreases.

When the angle of attack is increased, the lift is momentarily increased, thereby decreasing the rate of descent (Fig. 9–6). Since power normally is reduced to idle during the roundout, the airspeed will also gradually decrease. This, in turn, causes lift to decrease again; it must be controlled by raising the nose and further increasing the angle of attack. During the roundout, then, the airspeed is being decreased to touchdown speed while the lift is being controlled so the airplane will settle gently onto the landing surface. The roundout should be executed at a rate that the proper landing attitude and the proper touchdown airspeed are attained simultaneously just as the wheels contact the landing surface.

The rate at which the roundout is executed depends on the airplane's height above the ground, the rate of descent, and the pitch attitude. A roundout started excessively high must be executed more slowly than one from a lower height to allow the airplane to descend to the ground while the proper landing attitude is being established. The rate of rounding out must also be proportionate to the rate of closure with the ground; that is, when the airplane appears to be descending very slowly, the increase in pitch attitude must be made at a correspondingly slow rate.

The pitch attitude of the airplane in a full-flap approach is considerably lower than in a no-flap approach. Therefore, to attain the proper landing attitude before touching down, the nose must travel through a greater pitch change when flaps are fully extended. Since the roundout is usually started at approximately the same height above the ground regardless of the degree of flaps used, the pitch attitude must be increased at a faster rate when full flaps are used. However, the roundout should still be executed at a rate proportionate to the airplane's downward motion.

Once the actual process of rounding out is started, the elevator control should not be pushed forward. If too much back pressure has been exerted, this pressure should be either slightly relaxed or held constant, depending on the degree of the error. In some cases, it may be necessary to advance the throttle slightly to prevent an excessive rate of sink, or a stall, all of which would result in a hard drop-in landing.

It is recommended, therefore, that the pilot form the habit of keeping one hand on the throttle throughout the approach and landing, should a sudden and unexpected hazardous situation require an immediate application of power.

Well before starting the roundout, it is imperative that the pilot *recheck* the landing gear for the "down-and-locked" indications, and place the propeller control in high RPM position, if the airplane is so equipped.

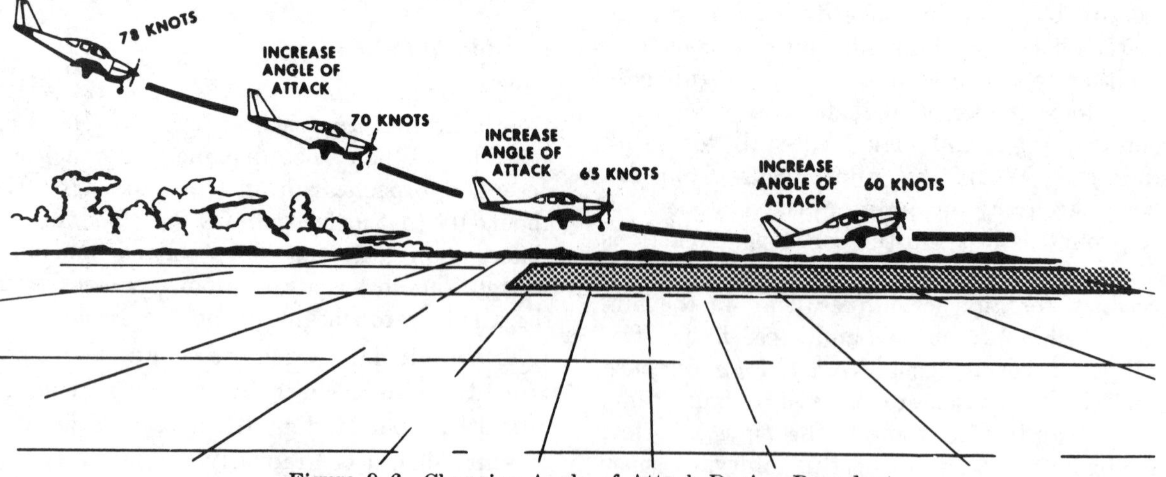

Figure 9–6 Changing Angle of Attack During Roundout

Touchdown

The touchdown is the gentle settling of the airplane onto the landing surface. The roundout and touchdown should be made with the engine idling, and the airplane at minimum controllable airspeed, so that the airplane will touch down on the main gear at approximately stalling speed. As the airplane settles, the proper landing attitude must be attained by application of whatever back elevator pressure is necessary.

Some pilots may try to force or fly the airplane onto the ground without establishing the proper landing attitude. It is paradoxical that the way to make an ideal landing is to try to hold the airplane's wheels a few inches off the ground as long as possible with the elevators. In most cases, when the wheels are within about two or three feet of the ground, the airplane will still be settling too fast for a gentle touchdown; therefore, this descent must be retarded by further back pressure on the elevators. Since the airplane is already close to its stalling speed and is settling, this added back pressure will only slow up the settling instead of stopping it. At the same time, though, it will result in the airplane touching the ground in the proper landing attitude.

Nosewheel-type airplanes should contact the ground in a tail-low attitude, with the main wheels touching down first so that little or no weight is on the nosewheel (Fig. 9–7). In tailwheel-type airplanes, the roundout and touchdown should be so timed that the wheels of the main landing gear and tailwheel touch down simultaneously (3-point landing). This requires fine timing, technique, and judgment of distance and altitude (Fig. 9–8).

In nosewheel-type airplanes, after the main wheels make initial contact with the ground, back pressure on the elevator control should be held to maintain a positive angle of attack for aerodynamic braking and to hold the nosewheel off the ground until the airplane decelerates. As the airplane's momentum decreases, back pressure may be gradually relaxed to allow the nosewheel to gently settle onto the runway. This will permit prompt steering with the nosewheel, if it is of the steerable type. At the same time it will cause a low angle of attack and negative lift on the wings to prevent floating or skipping, and will allow the full weight of the airplane to rest on the wheels for better braking action. The airplane should never be "flown on" the runway with excess speed.

When the wheels of tailwheel-type airplanes make contact with the ground, the elevator control should be carefully eased fully back to hold the tail down and the tailwheel on the ground. This provides more positive directional control of the airplane equipped with a steerable tailwheel, and prevents any tendency for the airplane to nose over. If the tailwheel is not on the ground, easing back on the elevator control may cause the airplane to become airborne again because the change in attitude will increase the angle of attack and produce enough lift for the airplane to fly.

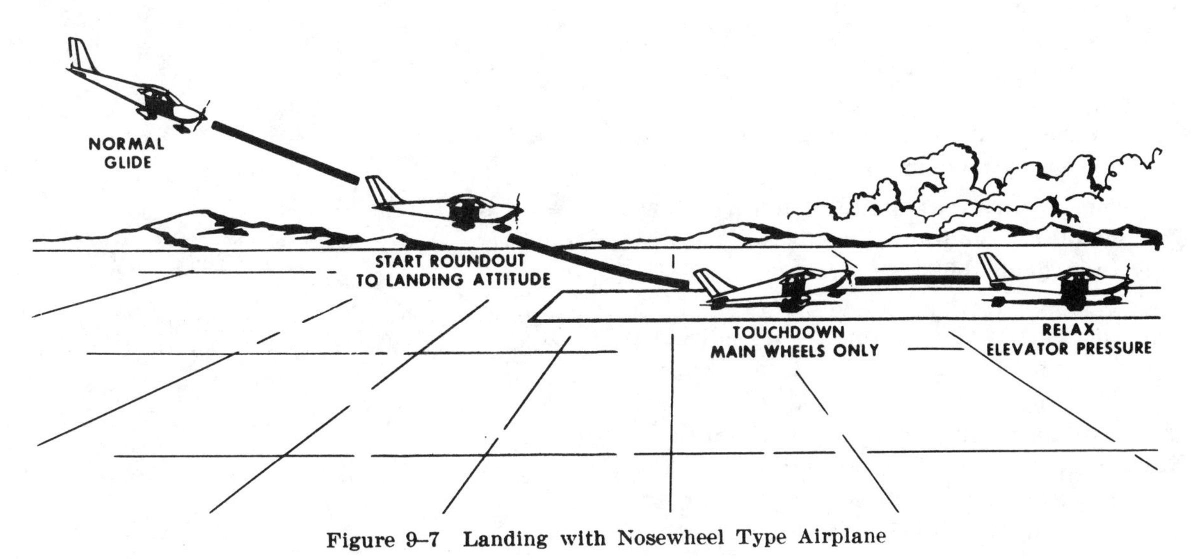

Figure 9–7 Landing with Nosewheel Type Airplane

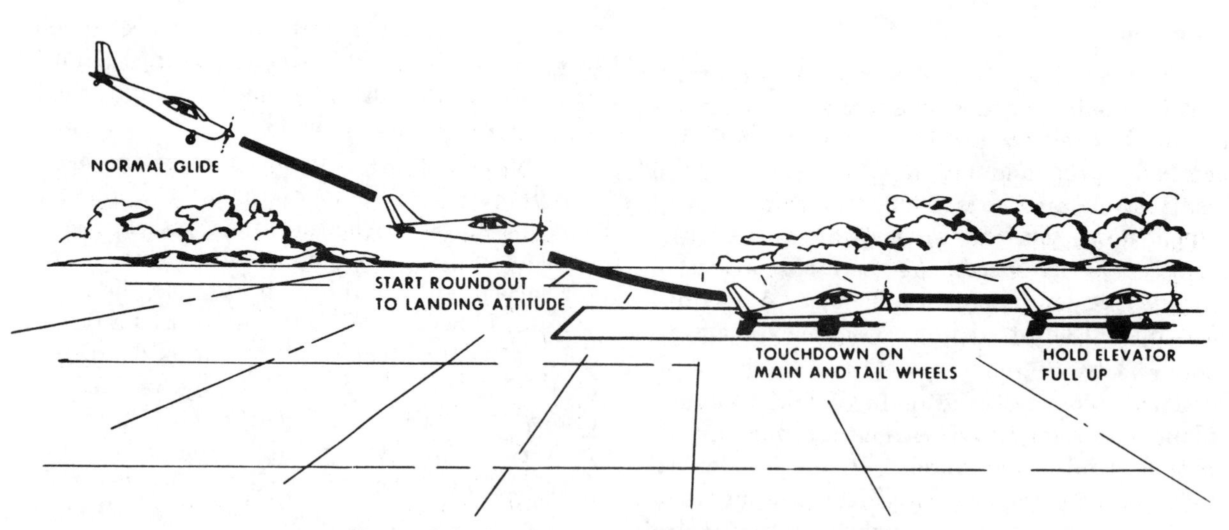

Figure 9–8 Landing with Tailwheel Type Airplane

It is extremely important that in either type of airplane, the touchdown occur with the airplane's longitudinal axis exactly parallel to the direction in which the airplane is moving along the runway. Failure to accomplish this not only imposes severe sideloads on the landing gear, but imparts groundlooping (swerving) tendencies, particularly to tailwheel-type airplanes. To avoid these side stresses or a groundloop, then, the pilot must never allow the airplane to touch down while in a crab or while dirfting.

Slips

A slip is a descent with one wing lowered and the airplane's longitudinal axis at an angle to the flightpath. It may be used for either of two purposes, or both of them combined. A slip may be used to steepen the approach path without increasing the airspeed, as would be the case if a dive were used. In can also be used to make the airplane move sideways through the air to counteract the drift which results from a crosswind (Fig. 9–9).

Formerly, slips were used as a normal means of controlling landing descents to short or obstructed fields, but they are now primarily used in the performance of crosswind landings and emergency landings. With the installation of wing flaps on modern airplanes, the use of slips to steepen or control the angle of descent is no longer a common procedure. However, the pilot still needs skill in performance of forward slips to correct for possible errors in judgment of the landing approach.

The primary purpose of "forward slips" is to dissipate altitude without increasing the airplane's speed, particularly in airplanes not equipped with flaps. There are many circumstances requiring the use of forward slips, such as in a landing approach over obstacles and in making forced landings, when it is always wise to allow an extra margin of altitude for safety in the original estimate of the approach. In the latter case, if the inaccuracy of the approach is confirmed by excess altitude when nearing the boundary of the selected field, the excess can be dissipated by slipping.

The use of slips has definite limitations. Some pilots may try to lose altitude by violent slipping rather than by smoothly maneuvering and exercising good judgment and using only a slight or moderate slip. In emergency

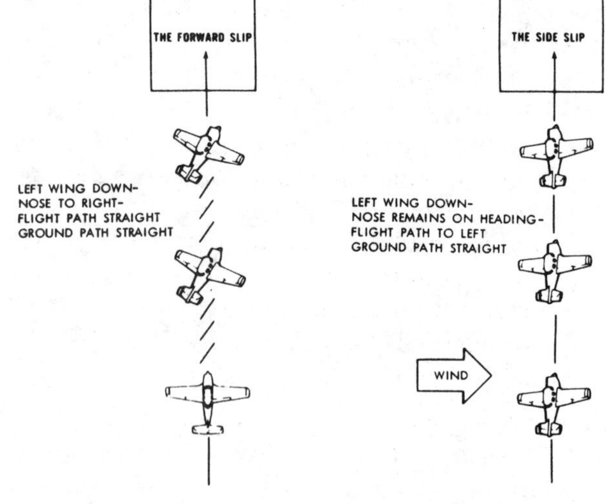

Figure 9–9 Forward Slip and Side Slip

landings, this erratic practice invariably will lead to trouble since enough excess speed may result to prevent touching down anywhere near the proper point, and very often will result in overshooting the entire field.

The "forward slip" is a slip in which the airplane's direction of motion continues the same as before the slip was begun (Fig. 9–9). If there is any crosswind, the slip will be much more effective if made toward the wind. Slipping should usually be done with the engine idling. There is little logic in slipping to lose altitude if the power is being used.

Assuming that the airplane is originally in straight flight, the wing on the side toward which the slip is to be made should be lowered by use of the ailerons. Simultaneously, the airplane's nose must be yawed in the opposite direction by applying opposite rudder so that the airplane's longitudinal axis is at an angle to its original flightpath. The degree to which the nose is yawed in the opposite direction from the bank should be such that the original ground track is maintained. The nose should also be raised as necessary to prevent the airspeed from increasing.

If a slip is used during the last portion of a final approach, the longitudinal axis of the airplane must be aligned with the runway just prior to touchdown so that the airplane will touch down headed in the direction in which it is moving over the runway. This requires timely action to discontinue the slip and align the airplane's longitudinal axis with its direction of travel over the ground at the instant of touchdown. Failure to accomplish this imposes severe sideloads on the landing gear and imparts violent groundlooping tendencies.

Discontinuing the slip is accomplished by leveling the wings and *simultaneously* releasing the rudder pressure while readjusting the pitch attitude to the normal glide attitude. If the pressure on the rudder is released abruptly the nose will swing too quickly into line and the airplane will tend to acquire excess speed.

Because of the location of the pitot tube and static vents, airspeed indicators in some airplanes may have considerable error when the airplane is in a slip. The pilot must be aware of this possibility and recognize a properly performed slip by the attitude of the airplane, the sound of the airflow, and the feel of the flight controls.

> NOTE: Forward slips with wing flaps extended should not be done in airplanes wherein the manufacturer's operating instructions prohibit such operation.

A sideslip, as distinguished from a forward slip (Fig. 9–9), is one during which the airplane's longitudinal axis remains parallel to the original flight path, but in which the flightpath changes direction according to the steepness of the bank. The sideslip is important in counteracting wind drift during crosswind landings and is discussed in a later section.

Go-Arounds (Rejected Landings)

Occasionally it may be advisable for safety reasons to discontinue the landing approach and make another approach under more favorable conditions. Extremely low base-to-final turns, overshot or low final approaches, the unexpected appearance of hazards on the runway, wake turbulence from a preceding airplane, or overtaking another airplane on the approach are hazardous conditions that would demand a go-around.

Although the need to discontinue a landing may arise at any point in the landing process, the most critical go-around will usually be one started when very close to the ground. Nevertheless, it is safer to make a go-around than to touch down while drifting or while in a crab, or to make a hard drop-in landing from a high roundout or bounced landing.

Regardless of the height above the ground at which it is begun, a safe go-around may be accomplished if an early decision is made, a sound plan is followed, and the procedure is performed properly. The earlier a dangerous situation is recognized and the sooner the landing is rejected and the go-around started, the safer the procedure will be. The pilot should never wait until the last moment to make the decision.

When the decision is made to discontinue an approach and perform a go-around, takeoff power should be applied *immediately* and the airplane's pitch attitude changed so as to slow or stop the descent. After the descent has been stopped, the landing flaps may be partially retracted or placed in the takeoff position, as recommended by the manufacturer.

Caution must be used, however, in retracting the flaps. Depending on the airplane's altitude and airspeed, it may be wise to retract the flaps intermittently in small increments to allow time for the airplane to accelerate progressively as they are being raised. A sudden and complete retraction of the flaps at a very low airspeed could cause a loss of lift resulting the airplane settling into the ground.

Unless otherwise specified in the airplane's operating manual, it is generally recommended that the flaps be retracted (at least partially) *before* retracting the landing gear—for two reasons. First, on most airplanes full flaps produce more drag than the landing gear; and second, in case the airplane should inadvertently touch down as the go-around is initiated, it is most desirable to have the landing gear in the down-and-locked position.

When takeoff power is applied, it will usually be necessary to hold considerable pressure on the controls to maintain straight flight and a safe climb attitude. Since the airplane has been trimmed for the approach (a low power and airspeed condition), the nose will tend to rise sharply and veer to the left unless firm control pressures are applied. Forward elevator pressure must be applied to hold the nose in a safe climbing attitude; right rudder pressure must be increased to counteract torque, or P-factor, and to keep the nose straight. The airplane must be held in the proper flight attitude regardless of the amount of control pressure that is required. Frequently, this pressure is quite strong.

While holding the airplane straight and in a safe climbing attitude, the pilot should retrim the airplane to relieve at least the heavy control pressures. Since the airspeed will build up rapidly with the application of takeoff power, and the controls will become more effective, this initial trim is to relieve the heavy pressures until a more precise trim can be made for the lighter pressures.

It is advisable to retract the landing gear only after the initial or rough trim has been accomplished and when it is certain the airplane will remain airborne. During the initial part of an extremely low go-around, the airplane may "mush" onto the runway and bounce. This situation is not particularly dangerous if the airplane is kept straight, and a constant, safe pitch attitude maintained. The airplane will be approaching safe flying speed rapidly and the advanced power will cushion any secondary touchdown.

If the pitch attitude is increased excessively in an effort to prevent the airplane from mushing onto the runway, it may cause the airplane to stall. This would be especially likely if no trim correction is made and the flaps remain fully extended. *Never attempt to retract the landing gear until after a rough trim is accomplished and a positive rate of climb is established or the descent stopped.*

After a positive rate of climb is established and the landing gear is retracted, the airplane should be allowed to accelerate to the best rate of climb speed (V_y) before the final flap retraction is accomplished.

From this point on, the procedure is identical with that for a normal climb after takeoff.

After-Landing Roll

The landing process must never be considered complete until the airplane decelerates to the normal taxi speed during the landing roll or has been brought to a complete stop when clear of the landing area. Many accidents have occurred as a result of pilots abandoning their vigilance and positive control after getting the airplane on the ground.

The pilot must be alert for directional control difficulties immediately upon and after touchdown due to the ground friction on the wheels. The friction creates a pivot point on which a moment arm can act. This is especially true in tailwheel-type airplanes because unlike nosewheel-type airplanes, the center of gravity (CG) is *behind* the main wheels (Fig. 9–10). Any difference between the direction the tailwheel-type airplane is traveling and the direction it is headed will produce a moment about the pivot point of the wheels and the airplane will tend to swerve. Nosewheel-type airplanes make the task of directional control much earlier because the center of gravity, being *ahead* of the main landing wheels, presents a moment arm which tends to straighten the airplane's path during the touchdown and

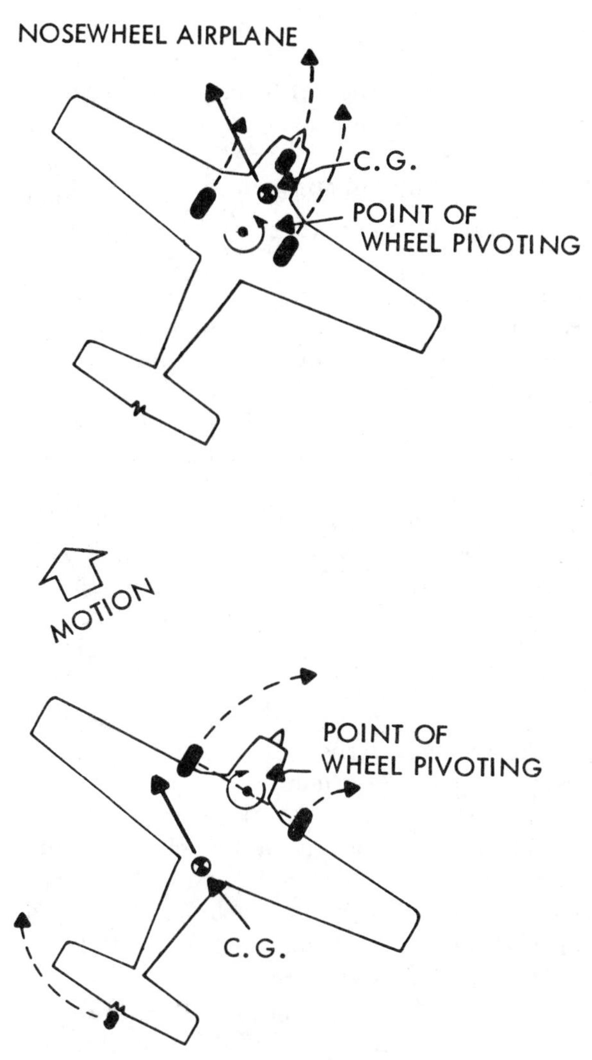

Figure 9–10 Effect of CG on Directional Control

after landing roll. However, this should not lull the pilot into a false sense of security.

Loss of directional control may lead to an aggravated, uncontrolled, tight turn on the ground, or a *"ground loop."* The combination of centrifugal force acting on the CG and ground friction of the main wheels resisting it during the ground loop may cause the airplane to tip or lean enough for the outside wingtip to contact the ground, and may even impose a sideward force which could collapse the landing gear. Tailwheel-type airplanes are most susceptible to ground loops late in the after-landing roll because rudder effectiveness decreases with the decreasing flow of air along the rudder surface as the airplane slows.

The rudder serves the same purpose on the ground as it does in the air—it controls the yawing of the airplane. The effectiveness of the rudder, however, is dependent on the airflow which, of course, depends on the speed of the airplane. As the speed decreases and the nosewheel or tailwheel has been lowered to the ground, the steerable nose or tailwheel provides more positive directional control.

The brakes of an airplane serve the same primary purpose as do the brakes of an automobile—that is, to reduce speed on the ground. In airplanes, however, they may also be used as an aid in directional control when more positive control is required than could be obtained with rudder, nosewheel steering, or tailwheel steering alone.

To use brakes, the pilot should slide the toes or feet up from the rudder pedals to the brake pedals. If rudder pressure is being held at the time braking action is needed, that pressure should not be released as the feet or toes are being slid up to the brake pedals, because control may be lost before brakes can be applied.

During the ground roll, the airplane's direction of movement may be changed by *carefully* applying pressure on one brake or uneven pressures on each brake in the desired direction. Caution must be exercised, however, when applying brakes to avoid overcontrolling.

The ailerons, too, serve the same purpose on the ground as they do in the air—they change the lift and drag components of the wings. During the after-landing roll they should be used to keep the wings level in much the same way they were used in flight. If a wing starts to rise, aileron control should be applied toward that wing to lower it. The amount required will depend on speed because as the forward speed of the airplane decreases, the ailerons will become less effective. Techniques for using ailerons in crosswind conditions are explained further in the section on crosswind landings.

After a nosewheel-type airplane is on the ground, back pressure on the elevator control may be gradually relaxed to place normal weight on the nosewheel to aid in better steering.

With a tailwheel-type airplane, the elevator control should be held back as far as possible and as firmly as possible, until the airplane stops. This provides more positive control with tailwheel steering, tends to shorten the after-landing roll, and prevents bouncing and skipping.

If available runway permits, the speed of the airplane should be allowed to dissipate in a normal manner by the friction and drag of the wheels on the ground. Brakes may be used if needed to help slow the airplane.

After the airplane has been slowed sufficiently and has been turned onto a taxiway or clear of the landing area, it should be brought to a complete stop. Only *after* this is done should the pilot retract the flaps and "clean up" the airplane. Too many accidents have occurred as a result of the pilot unintentionally operating the landing gear control and retracting the gear instead of the flap control when the airplane was still rolling. The habit of positively identifying either of these controls before actuating them, should be formed from the very beginning of flight training and continued in all future flying activities.

Crosswind Approach and Landing

Many runways or landing areas are such that landings must be made while the wind is blowing across rather than parallel to the landing direction; therefore, all pilots should be prepared to cope with these situations when they arise. The same basic principles and factors involved in a normal approach and landing apply to a crosswind approach and landing. Therefore, only the additional techniques required for correcting for wind drift are discussed here.

Crosswind landings are a little more difficult to perform than are crosswind takeoffs, mainly due to different problems involved in maintaining accurate control of the airplane while its speed is decreasing rather than increasing as on takeoff.

There are two usual methods of accomplishing a crosswind approach and landing—the crab method, and the wing-low method. Although the crab method may be easier for the pilot to maintain during final approach, it requires a high degree of judgment and timing in removing the crab immediately prior to touchdown. The wing-low method is recommended in most cases although a combination of both methods may be used.

Crosswind Final Approach

The crab method is executed by establishing a heading (crab) toward the wind with the wings level so that the airplane's ground track remains aligned with the centerline of the runway. This crab angle is maintained until just prior to touchdown, when the longitudinal axis of the airplane must be quickly aligned with the runway to avoid sideward contact of the wheels with the runway. If a long final approach is being flown, the pilot may use the crab method until just before the roundout is started and then smoothly changing to the wing-low method for the remainder of the landing.

The wing-low method will compensate for a crosswind from any angle, but more important, it enables the pilot to simultaneously keep the airplane's ground track and the longitudinal axis aligned with the runway centerline throughout the final approach, roundout, touchdown, and after-landing roll. This prevents the airplane from touching down in a sideward motion and imposing damaging side loads on the landing gear.

To use the wing-low method, the pilot aligns the airplane's heading with the centerline of the runway, notes the rate and direction of drift, then promptly applies drift correction by lowering the upwind wing (Fig. 9–11). The amount the wing must be lowered depends on the rate of drift. When the wing is lowered, the airplane will tend to turn in that direction. It is necessary, then, to simultaneously apply sufficient opposite rudder pressure to prevent the turn and keep the airplane's longitudinal axis aligned with the runway. In other words, the drift is controlled with aileron, and the heading with rudder. The airplane will now be side slipping into the wind just enough that both the resultant flightpath and the ground track are aligned with the runway. If the crosswind diminishes, this crosswind correction must be reduced accordingly or the airplane will begin slipping away from the desired path.

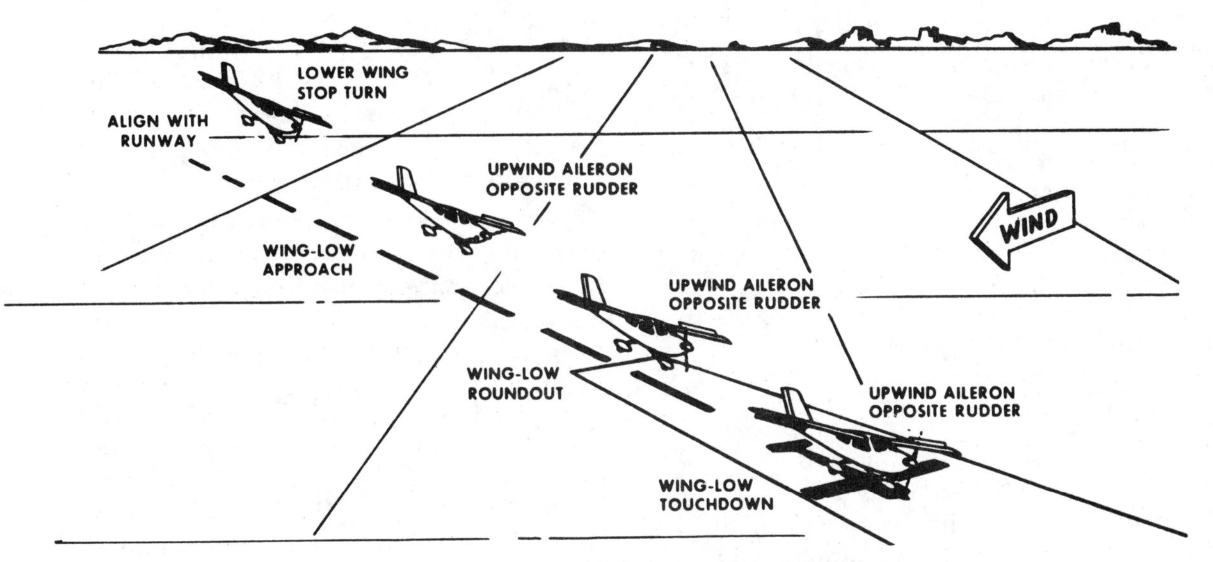

Figure 9–11 Crosswind Approach and Landing

To correct for very strong crosswind, the slip into the wind must be increased by lowering the upwind wing a considerable amount. As a consequence, this would result in a greater tendency of the airplane to turn. Since turning is not desired, considerable opposite rudder must be applied to keep the airplane's longitudinal axis aligned with the runway. In some airplanes, there may not be sufficient rudder travel available to compensate for the strong turning tendency caused by the steep bank. If the required bank is so steep that full opposite rudder will not prevent a turn, the wind is too strong to safely land the airplane on that particular runway with those wind conditions. Since the airplane's capability would be exceeded, it is imperative that the landing be made on a more favorable runway either at that airport or at an alternate airport.

Flaps can and should be used during most approaches since they tend to have a stabilizing effect on the airplane. However, the degree to which flaps should be extended will vary with the airplane's handling characteristics, as well as the wind velocity. Full flaps may be used so long as the crosswind component is not in excess of the airplane's capability or unless the manufacturer recommends otherwise.

Crosswind Roundout (Flare)

Generally, the roundout can be made as in a normal landing approach but the application of a crosswind correction must be continued as necessary to prevent drifting (Fig. 9–12).

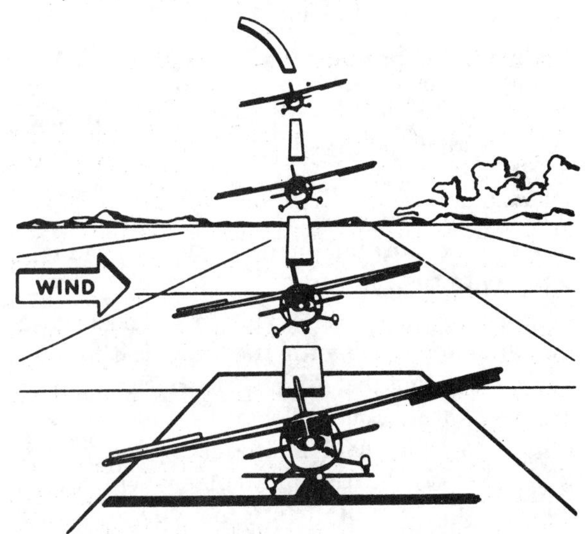

Figure 9–12 Crosswind Roundout and Touchdown

Since the airspeed decreases as the roundout progresses, the flight controls gradually become less effective; as a result, the crosswind correction being held would become inadequate. When using the wing-low method then, it is necessary to gradually increase the deflection of the rudder and ailerons to maintain the proper amount of drift correction.

Do not level the wings; keep the upwind wing down throughout the roundout. If the wings are leveled, the airplane will begin drifting and the touchdown will occur while drifting. Remember, the primary objective is to land the airplane without subjecting it to any side loads which result from touching down while drifting and to prevent ground looping while the landing is being accomplished.

Crosswind Touchdown

If the crab method of drift correction has been used throughout the final approach and roundout, the crab must be removed the instant before touchdown by applying rudder to align the airplane's longitudinal axis with its direction of movement. This requires timely and accurate action. Failure to accomplish this results in severe sideloads being imposed on the landing gear and imparts ground looping tendencies.

If the wing-low method is used, the crosswind correction (aileron into the wind and opposite rudder) should be maintained throughout the roundout, and the touchdown made on the upwind main wheel (Fig. 9–12).

During gusty or high wind conditions, prompt adjustments must be made in the crosswind correction to assure that the airplane does not drift as the airplane touches down.

As the forward momentum decreases after initial contact, the weight of the airplane will cause the downwind main wheel to gradually settle onto the runway.

In those airplanes having nosewheel steering interconnected with the rudder, the nosewheel may not be aligned with the runway as the wheels touch down because opposite rudder is being held in the crosswind correction. This is the case in airplanes which have no centering cam built into the nose gear strut to keep the nosewheel straight until the strut is compressed. To prevent swerving in the direction the nosewheel is offset, the corrective rudder pressure must be promptly relaxed just as the nosewheel touches down.

Crosswind After-Landing Roll

Particularly during the after-landing roll, special attention must be given to maintaining directional control by use of rudder, or nosewheel/tailwheel steering, while keeping the upwind wing from rising by use of aileron.

When an airplane is airborne it moves with the air mass in which it is flying regardless of the airplane's heading and speed. However, when an airplane is on the ground it is unable to move with the air mass (crosswind) because of the resistance created by ground friction on the wheels.

Characteristically, an airplane has a greater profile or side area, behind the main landing gear than forward of it (Fig. 9–13). With the main wheels acting as a pivot point and the greater surface area exposed to the cross-

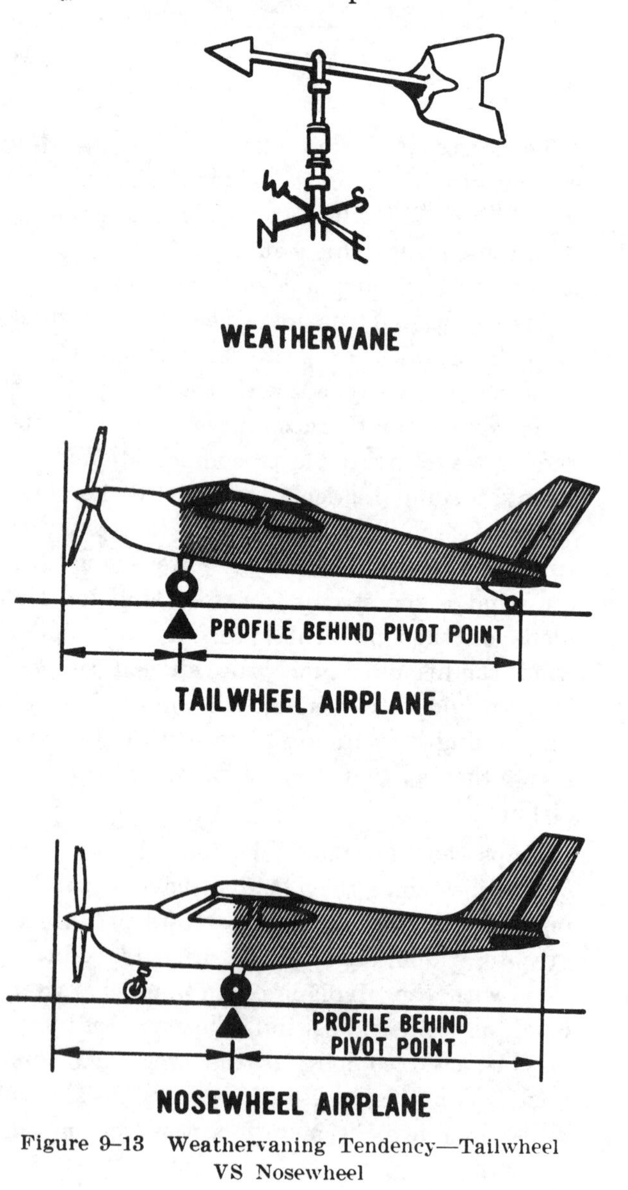

Figure 9–13 Weathervaning Tendency—Tailwheel VS Nosewheel

wind behind that pivot point, the airplane will tend to turn or "weathervane" into the wind.

Though it is characteristic of most airplanes, this weathervaning tendency is more prevalent in the tailwheel-type because the airplane's surface area behind the main landing gear is greater than in nosewheel-type airplanes.

Wind acting on an airplane during crosswind landings is the result of two factors—one is the natural wind which acts in the direction the air mass is traveling, while the other is induced by the movement of the airplane and acts parallel to the direction of movement. Consequently, a crosswind has a headwind component acting along the airplane's ground track and a crosswind component acting 90° to its track. The resultant or relative wind, then, is somewhere between the two components. As the airplane's forward speed decreases during the after-landing roll, the headwind component decreases and the relative wind has more of a crosswind component. The greater the crosswind component the more difficult it is to prevent weathervaning. The headwind component and the crosswind component can be determined by reference to Figure 9–14. For example: A relative wind at 20 knots at an angle of 60 degrees to the runway has a headwind component of 10 knots and a 90 degree crosswind component of 18 knots. Federal Aviation Regulations require that all airplanes, type-certificated since 1962, have safe ground handling characteristics with a *90 degree* crosswind component equal to 0.2 V_{so}. Thus, an airplane that stalls at 55 knots in the landing configuration, must have no uncontrollable ground looping (weathervaning) tendencies with a *90 degree* crosswind component of 11 knots (0.2 x 55). It is imperative that pilots determine the maximum crosswind component of each airplane then fly, and avoid operations in wind conditions that exceed the capability of the airplane.

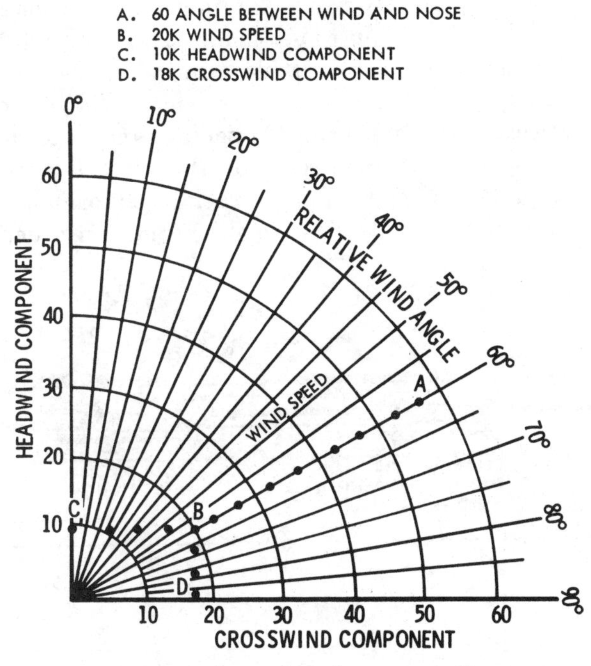

Figure 9–14 Crosswind Component Chart

While the airplane is decelerating during the after-landing roll, more and more aileron must be applied to keep the upwind wing from rising. Since the airplane is slowing down there is less airflow around the ailerons and they become less effective. At the same time the relative wind is becoming more of a crosswind and exerting a greater lifting force on the upwind wing. Consequently, when the airplane is coming to a stop the aileron control must be held fully toward the wind.

Turbulent-Air Approach and Landing

Power-on approaches at an airspeed slightly above the normal approach speed should be used for landing in significantly turbulent air. This provides for more positive control of the airplane when strong horizontal wind gusts, or up and down drafts, are experienced.

Like other power-on approaches (when the pilot can vary the amount of power), the angle of descent is controlled primarily by pitch adjustments, and the airspeed controlled primarily by changes in power. Nevertheless, a *coordinated combination* of both pitch and power adjustments is usually required. As in most other landing approaches, the proper approach attitude and airspeed require a minimum roundout or flare and should result in little or no floating during the landing.

To maintain good control, the approach in turbulent air with a gusty crosswind may require the use of partial wing flaps. With less than full flaps, the airplane will be in a higher nose-up attitude. Thus, it will require less of a pitch change to establish the landing attitude, and the touchdown will be at a higher airspeed to ensure more positive control. However, the speed should not be so excessive that the airplane will "float" past the desired landing area.

These landing approaches are usually performed at the normal approach speed plus one-half of the wind gust factor. If the normal speed is 70 knots and the wind gusts increase 15 knots, an airspeed of 77 knots is appropriate. In any case, the airspeed and the amount of flaps should be as the airplane manufacturer recommends.

An adequate amount of power should be used to maintain the proper airspeed throughout the approach, and the throttle retarded to idling position only after the main wheels contact the landing surface. Care must be exercised, however, in closing the throttle before the pilot is ready for touchdown, as in this situation the sudden or premature closing of the throttle may cause a sudden increase in the descent rate which could result in a hard landing.

Landings from power approaches in severe turbulence should be such that touchdown is made with the airplane in approximately level flight attitude. In nosewheel-type airplanes, the nose-up pitch attitude at touchdown should be only enough to prevent the nosewheel from contacting the surface before the main wheels have touched.

In tailwheel-type airplanes, touchdown should be made smoothly on the main wheels, with the tailwheel held well clear of the runway (Fig. 9–15). This is called a "wheel landing" and requires very careful timing and control usage to prevent bouncing. These wheel landings can be best accomplished by holding the airplane in level flight attitude until the main wheels touch, then immediately but smoothly retarding the throttle, and holding sufficient forward elevator pressure to hold the main wheels on the ground. The airplane should never be forced onto the ground by excessive forward pressure.

If the touchdown is made at too high a rate of descent as the main wheels strike the landing surface, the tail is forced down by its own weight. In turn, when the tail is forced down, the wing's angle of attack increases resulting in a sudden increase in lift and the airplane may become airborne again. Then as the airplane's speed continues to decrease, the tail may again lower onto the runway. If the tail is allowed to settle too quickly, the airplane may again become airborne. This process, often called "porpoising," usually intensifies even though the pilot tries to stop it. The best corrective action is to reject the landing and execute a go-around procedure.

Short Field Power Approach and Landing

This maximum performance operation requires the use of procedures and techniques for the approach and landing at fields which have a relatively short landing area or where an approach must be made over obstacles which limit the available landing area. As in short field takeoffs, it is one of the most critical of the maximum performance operations, since it requires that the pilot fly the airplane at one of its crucial performance capabilities while close to the ground in order

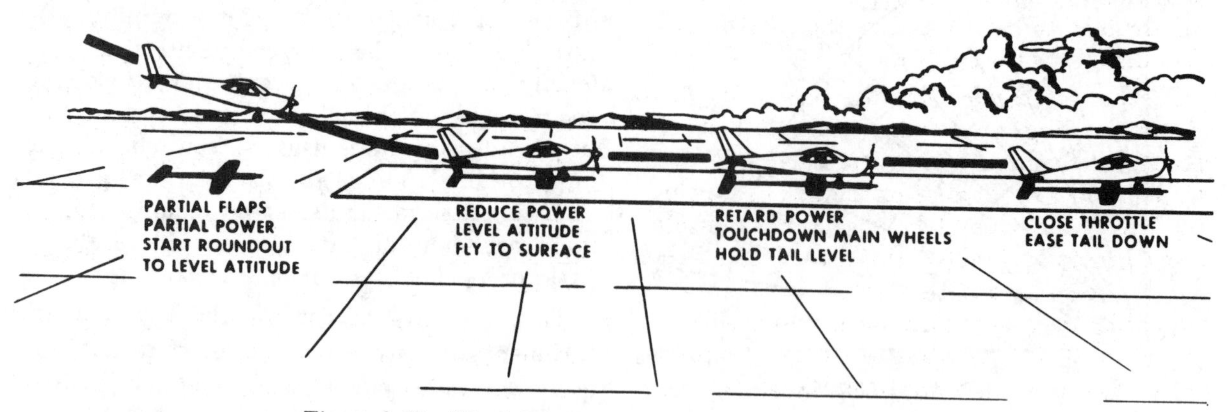

Figure 9–15 Wheel Landing with Tailwheel Type Airplane

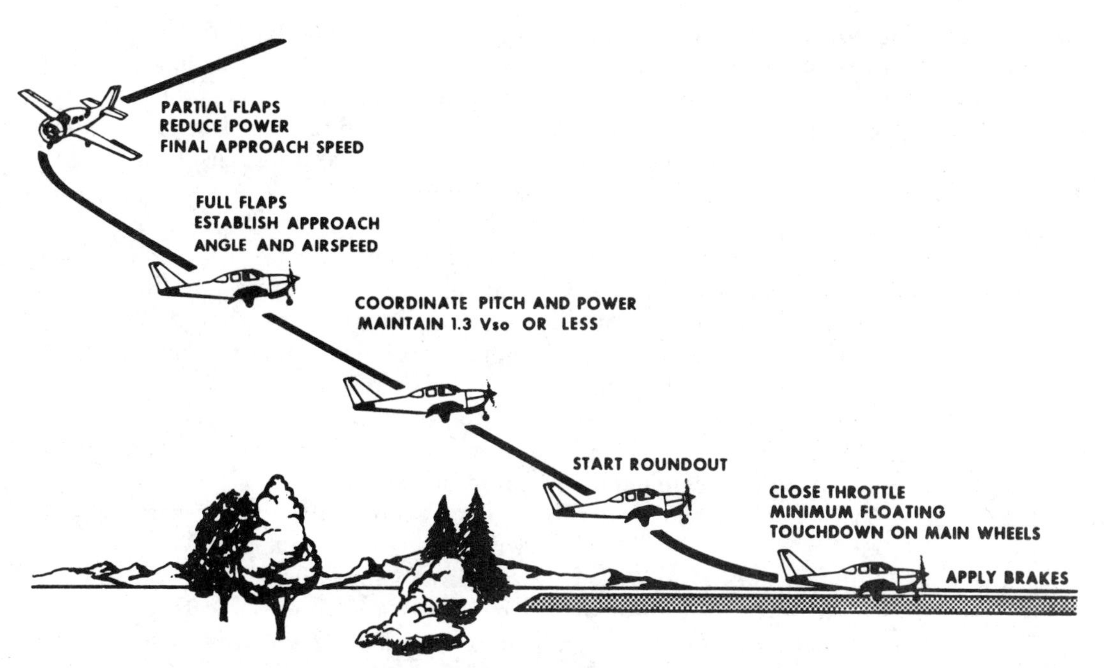

Figure 9–16 Short Field Approach and Landing

to safely land within confined areas. This low-speed type of power-on approach is closely related to the performance of "flight at minimum controllable airspeeds" described in the chapter on Proficiency Flight Maneuvers.

In the performance of power-off type approaches and landings, it is not always possible to predict the *exact* spot where the airplane will touch down—due to airspeed and rate or angle of descent variations caused by wind, up-and-down drafts, pitch changes, and floating after roundout. However, to land within a short field or a confined area. the pilot *must* have precise, positive control of the rate of descent and airspeed to produce an approach that will clear any obstacles, result in little or no floating during the roundout, and permit the airplane to be stopped in the shortest possible distance (Fig. 9–16).

The procedures for landing in a short field or for landing approaches over 50-foot obstacles as recommended in the FAA approved Airplane Flight Manual or the Pilot's Operating Handbook, should be used. These procedures generally involve the use of full flaps, and the final approach started from an altitude of at least 500 feet higher than the touchdown area. In the absence of the manufacturer's recommended approach speed, a speed of *not more* than 1.3 V_{so} should be used—that is, in an airplane which stalls at 60 knots with power off and flaps and landing gear extended, the approach speed should be no higher than 78 knots. In gusty air, no more than one-half the gust factor may be added. An excessive amount of airspeed could result in touchdown too far from the runway threshold or an after landing roll that exceeds the available landing area.

After the landing gear and full flaps have been extended, the pilot should simultaneously adjust the power and the pitch attitude to establish and maintain the proper descent angle and airspeed.

Since short field approaches are power-on approaches, the pitch attitude is adjusted as necessary to establish and maintain the desired rate or angle of descent, and power is adjusted to maintain the desired airspeed. However, a coordinated *combination of both pitch and power adjustments* is usually required. When this is done properly, very little change in the airplane's pitch attitude is necessary to make corrections in the angle of descent and only small power changes are needed to control the airspeed.

If it appears that the obstacle clearance is excessive and touchdown would occur well beyond the desired spot leaving insufficient room to stop, power may be reduced while lowering the pitch attitude to increase the rate of descent. If it appears that the descent angle will not ensure safe clearance of obstacles, power should be increased while simultaneously raising the pitch attitude to decrease the rate of descent. Care must be taken, however, to avoid an excessively low airspeed. If the speed is allowed to become too slow, an increase in pitch and application of full power may only result in a further rate of descent. This occurs when the angle of attack is so great and creating so much drag that the maximum available power is insufficient to overcome it. This is generally referred to as operating in the "region of reverse command" or operating on the "back side of the power curve."

Because the final approach over obstacles is made at a steep approach angle and close to the airplane's stalling speed, the initiation of the roundout or flare must be judged accurately to avoid flying into the ground, or stalling prematurely and sinking rapidly. A lack of floating during the flare, with sufficient control to touch down properly, is one verification that the approach speed was correct.

Touchdown should occur at the minimum controllable airspeed with the airplane in approximately the pitch attitude which will result in a power-off stall when the throttle is closed. Care must be exercised to avoid closing the throttle rapidly before the pilot is ready for touchdown, as closing the throttle may result in an immediate increase in the rate of descent and a hard landing.

Upon touchdown, nosewheel-type airplanes should be held in this positive pitch attitude as long as the elevators remain effective, and tailwheel-type airplanes should be firmly held in a three-point attitude. This will provide aerodynamic braking by the wings.

Immediately upon touchdown, and closing the throttle, the brakes should be applied evenly and firmly to minimize the after-landing roll. The airplane should be stopped within the shortest possible distance consistent with safety.

Soft Field Approach and Landing

Landing on fields that are rough or have soft surfaces, such as snow, sand, mud, or tall grass requires unique techniques. When landing on such surfaces, the pilot must control the airplane in a manner that the wings support the weight of the airplane as long as practical, to minimize drag and stresses imposed on the landing gear by the rough or soft surface.

The approach for the soft field landing is similar to the normal approach used for operating into long, firm landing areas. The major difference between the two is that during the soft field landing, the airplane is held 1 to 2 feet off the surface as long as possible to dissipate the forward speed sufficiently to allow the wheels to touch down gently at minimum speed.

The use of flaps during soft field landings will aid in touching down at minimum speed and is recommended whenever practical. In low wing airplanes, however, the flaps may suffer damage from mud, stones, or slush thrown up by the wheels. If flaps are used, it is generally inadvisable to retract them during the after-landing roll because the need for flap retraction usually is less important than the need for total concentration on maintaining full control of the airplane.

The final approach airspeed used for *short field* landings is equally appropriate to soft field landings, but there is no reason for a steep angle of descent unless obstacles are present in the approach path. Touchdown on a soft or rough field should be made at the lowest possible airspeed with the airplane in a nose-high pitch attitude.

In tailwheel-type airplanes, the tailwheel should touch down simultaneously with or just before the main wheels, and then should be held down by maintaining firm back elevator pressure throughout the landing roll. This will minimze any tendency for the airplane to nose over and will provide aerodynamic braking.

In nosewheel-type airplanes, after the main wheels touch the surface, the pilot should hold sufficient back elevator pressure to keep the nosewheel off the ground until it can no longer aerodynamically be held off the field surface. At this time the pilot should very gently lower the nosewheel to the surface. A slight addition of power during and immediately after touchdown usually will aid in easing the nosewheel down.

The use of brakes on a soft field is *not needed and should be avoided* as this may tend to impose a heavy load on the nose gear due to premature or hard contact with the landing surface, causing the nosewheel to dig in. It may also tend to cause a noseover in a tailwheel-type airplane. The soft or rough surface itself will provide sufficient reduction in the airplane's forward speed. Often it will be found that upon landing on a very soft field, the pilot will need to increase power to keep the airplane moving and from becoming stuck in the soft surface.

Emergency Approaches and Landings (Actual)

In spite of the remarkable reliability of present-day airplane engines, the pilot should always be prepared to cope with emergencies which may involve a *forced* landing caused by partial or complete engine failure.

A study by the National Transportation Safety Board reveals several factors that may interfere with a pilot's ability to act promptly and properly when faced with such an emergency:

1. Reluctance to Accept the Emergency Situation: A pilot whose mind is allowed to become paralyzed at the thought that the aircraft will be on the ground in a short time, regardless of what is done, is severly handicapped in the handling of the emergency. An unconscious desire to delay this dreaded moment may lead to such errors as failure to lower the nose to maintain flying speed, delay in the selection of the most suitable touchdown area within reach, and indecision in general.
2. Desire to Save the Aircraft: A pilot who has been conditioned to expect to find a relatively safe landing area whenever the instructor closed the throttle for a simulated forced landing may ignore all basic rules of airmanship to avoid a touchdown in terrain where aircraft damage is unavoidable. The desire to save the aircraft, regardless of the risks involved, may be influenced by the pilot's financial stake in the aircraft and the certainty that an undamaged aircraft implies no bodily harm. There are times when a pilot should be more interested in sacrificing the aircraft so that all occupants can safely walk away from it.
3. Undue Concern About Getting Hurt: Fear is a vital part of our self-preservation mechanism. When fear leads to panic we invite that which we want to avoid the most. The survival records favor those who maintain their composure and know how to apply the general concepts and techniques that have been developed throughout the years.

A competent pilot is constantly on the alert for suitable forced-landing fields. Naturally, the perfect forced-landing field is an established airport, or a hard-packed, long smooth field with no high obstacles on the approach end; however, these ideal conditions may not be readily available, so the best available field must be selected. Cultivated fields are usually satisfactory, and plowed fields are acceptable if the landing is made parallel to the furrows (Fig. 9–17). In any case, fields in which there are large boulders, ditches, or other features which present a hazard during the landing, should be avoided. A landing with the landing gear retracted may be advisable in soft or snow-covered fields to eliminate the possibly of the airplane nosing over, or damaging the gear, as a result of the wheels digging in.

Several factors must be considered in determining whether a field is of adequate length. When landing on a level field into a strong headwind, the distance required for a safe landing will naturally be much less than the distance required for landing with a tailwind. If it is impossible to land directly into the wind because maneuvering to an upwind approach would place the airplane at a dangerously low altitude, or a suitable field into the wind is not available, then the landing should be made crosswind or downwind. A large field that is crosswind, or even downwind, may be safer than a smaller field which is directly into the wind. Whenever possible the pilot should select a field that is wide enough to allow extending the base leg and delaying the turn onto the final approach to correct for any error in planning (Fig. 9–18).

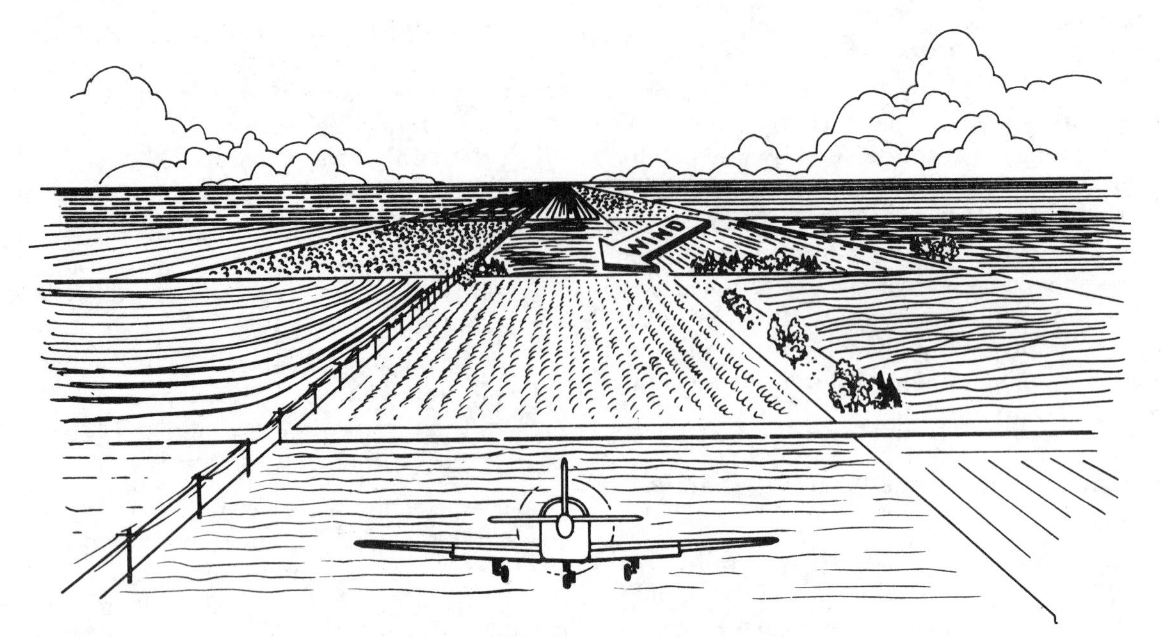

Figure 9–17 Land Parallel to Furrows

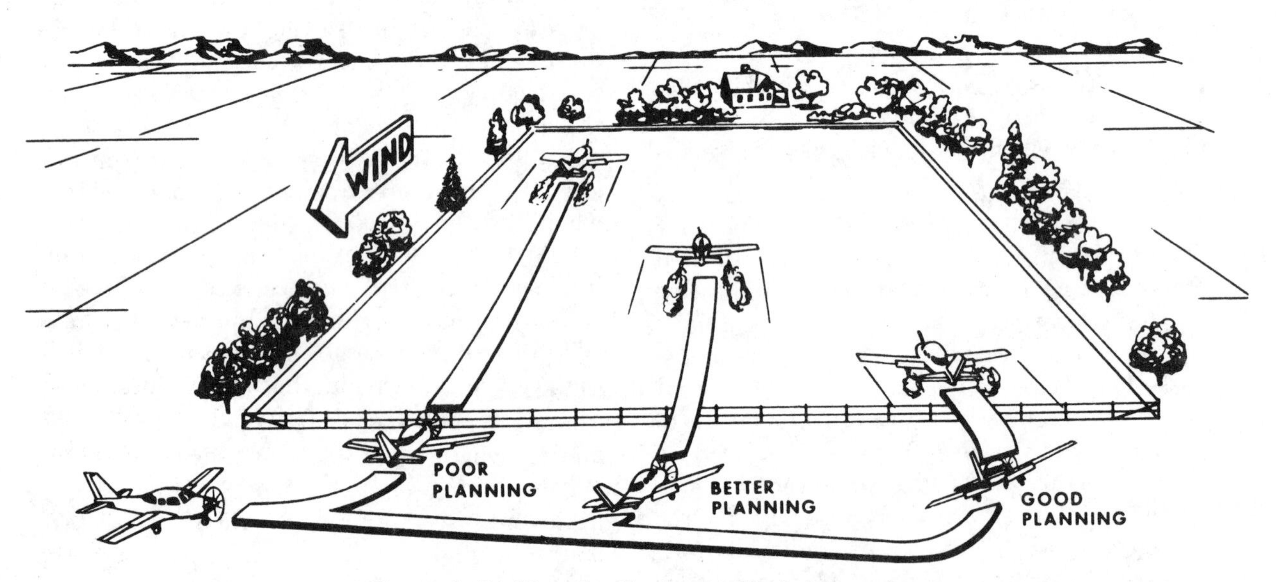

Figure 9–18 Plan the Turn onto Final Approach

The direction and speed of the wind are important factors during any landing, particularly in a forced landing, since the wind affects the airplane's gliding distance over the ground, the path over the ground during the approach, the groundspeed at which the airplane contacts the ground, and the distance the airplane rolls after the landing. All these effects should be considered during the selection of a field.

As a general rule, all landings should be made with the airplane headed into the wind, but this cannot be a hard and fast rule, since many other factors may make it inadvisable in the case of an actual forced landing (Fig. 9–19). Examples of such factors are:

1. Insufficient altitude may make it inadvisable or impossible to attempt to maneuver into the wind.
2. Ground obstacles may make landing into the wind impractical or inadvisable because they shorten the effective length of the available field.

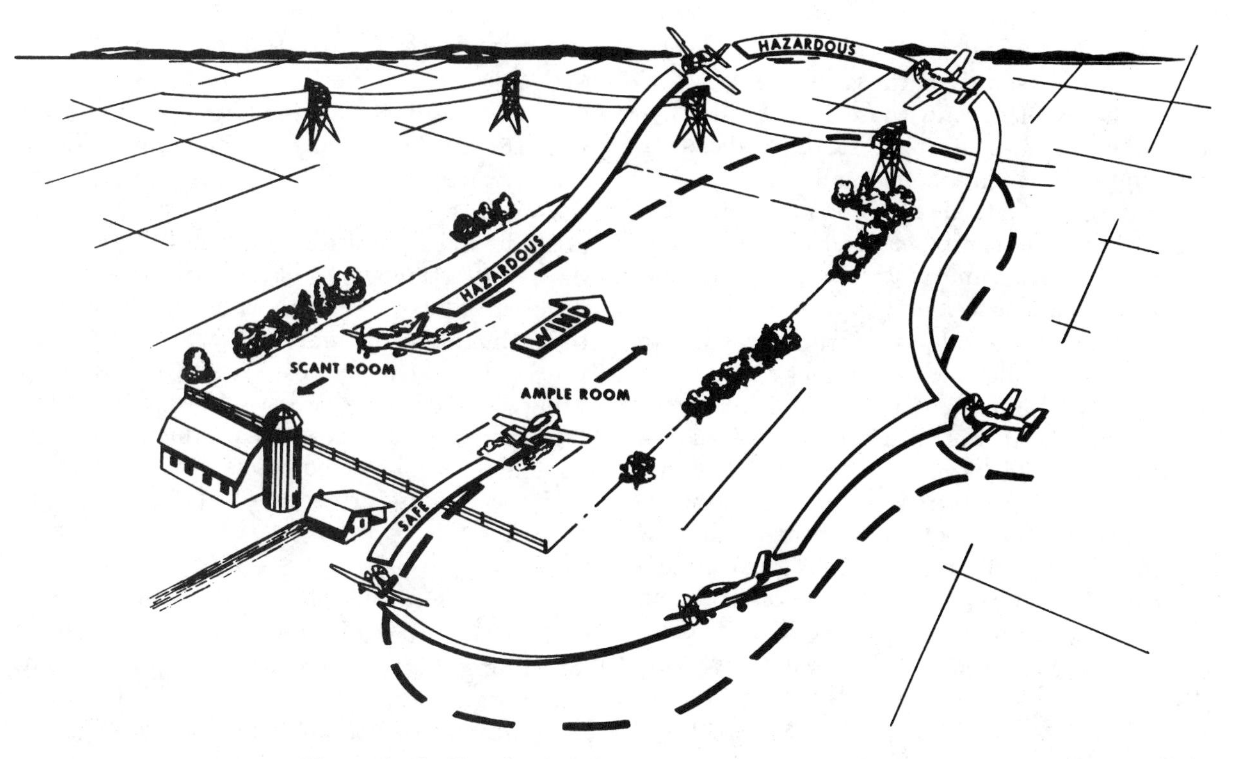

Figure 9–19 Use Good Judgment in Choosing Direction

3. Distance from a suitable field upwind from the present position may make it impossible to reach the field from the altitude at which the engine failure occurs.
4. The best available field may be downhill and at such an angle to the wind that a downwind landing uphill would be preferable and safer.

The altitude available is, in many ways, the controlling factor in the successful accomplishment of a forced landing. If an actual engine failure should occur immediately after takeoff and before a safe maneuvering altitude is attained, it is usually inadvisable to attempt to turn back to the field from which the takeoff was made (Fig. 9–20). Instead, it is generally safer to immediately establish the proper glide attitude, and select a field directly ahead or slightly to either side of the takeoff path.

The decision to continue straight ahead is often a difficult one to make unless the problems involved in attempting to turn back are seriously considered. In the first place, the takeoff was in all probability made into the wind. To get back to the takeoff field, a downwind turn must be made. This, of course, increases the groundspeed and rushes the pilot even more in the performance of procedures and in planning the landing approach. Secondly, the airplane will be losing considerable altitude during the turn and might still be in a bank when the ground is contacted, thus resulting in the airplane cartwheeling (which would be a catastrophe for the occupants as well as the airplane). Last, but certainly not least, after turning downwind the apparent

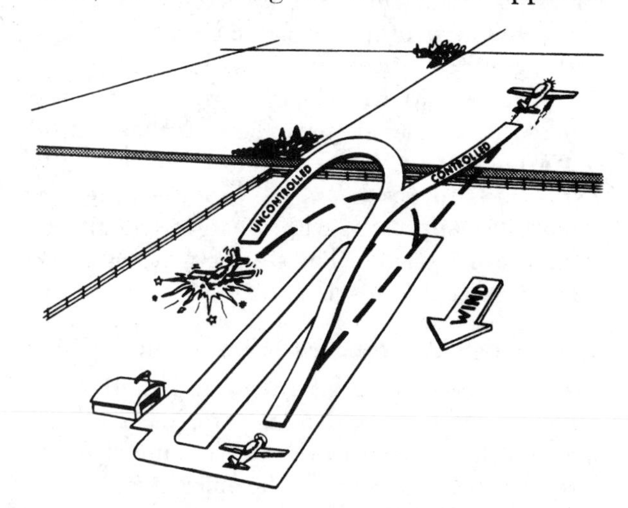

Figure 9–20 Turning Back May Cause Serious Problems

increase in groundspeed could mislead the pilot into attempting to prematurely slow down the airplane and cause it to stall. On the other hand, continuing straight ahead or making only a slight turn allows the pilot more time to establish a safe landing attitude, and the landing can be made as slowly as possible; but more importantly, the airplane can be landed while *under control.*

Concerning the subject of turning back to the runway following an engine failure on takeoff, each pilot should determine the minimum altitude at which an attempt of such a maneuver would be made in a particular aircraft. Experimentation at a safe altitude should give the pilot an approximation of height lost in a descending 180° turn at idle power. By adding a safety factor of about 25 percent the pilot should arrive at a practical "decision height." It speaks for itself that the ability to make a 180° turn does not necessarily mean that the departure runway can be reached in a power-off glide; this depends on the wind, the distance traveled during the climb, the height reached, and the glide distance of the airplane without power.

When a forced landing is imminent, wind direction and speed should always be considered, but the main object is to complete a safe landing in the largest and best field available. This involves getting the airplane on the ground in as near a normal landing attitude as possible without striking obstructions. If the pilot gets the airplane on the ground under control, it may sustain damage, but the occupants will probably get no worse than a shaking up.

During actual forced landings, it is recommended that the airplane be maneuvered to conform with a 360° overhead approach, a 180° side approach, a 90° approach, or a straight-in approach, with whatever modifications are necessary. These approaches are described later in this chapter.

Emergency Approaches (Simulated)

From time-to-time on dual flights, the instructor should give *simulated* forced landings by retarding the throttle and calling "Simulated Forced Landing." The objective of these simulated forced landings is to develop the pilot's accuracy, judgment, planning, technique, and confidence when little or no power is available.

A simulated forced landing may be given with the airplane in any configuration. When the instructor calls "simulated forced landing" the pilot should immediately establish a glide attitude and ensure that the landing gear and flaps are retracted (if so equipped). If the airspeed is above the proper glide speed, altitude should be maintained (while retracting the landing gear and flaps), and the airspeed allowed to dissipate to best glide speed. When the proper glide speed is attained, the nose should then be lowered to maintain that speed and the cockpit procedures performed during the glide.

A constant gliding speed should be maintained, because variations of gliding speed nullify all attempts at accuracy in judgment of gliding distance and the landing spot. The many variables such as altitude, obstructions, wind direction, landing direction, landing surface and gradient, and landing distance requirements of the airplane will determine the pattern and approach techniques to use.

Utilizing any combination of normal gliding maneuvers, from wings level to spirals, the pilot should eventually arrive at the normal "key" position at a normal traffic pattern altitude for the selected landing area. From this point on, the approach will be as nearly as possible a normal power-off approach (Fig. 9–21).

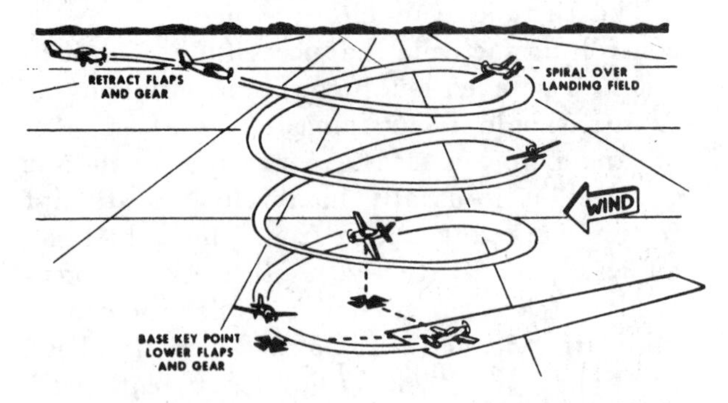

Figure 9–21 Remain Over Intended Landing Area

With the greater choice of fields afforded by higher altitudes, the inexperienced pilot may be inclined to delay making a decision, and with considerable altitude in which to maneuver, errors in maneuvering and estimation of glide distance may develop.

All pilots should learn to determine the wind direction and estimate its speed from the windsock at the airport, smoke from factories or houses, dust, brush fires, windmills, etc., and should constantly check against these while in flight.

Once a field has been selected, the student pilot should always be required to indicate it to the instructor. Normally, the student should be required to plan and fly a pattern for landing on the field first selected until the simulated forced landing is terminated by the instructor. This will give the instructor an opportunity to explain and correct any errors; it will also give the student an opportunity to see the results of the errors. However, if the student realizes during the approach that a poor field has been selected—one that would obviously result in disaster if a landing were to be made— and there is a more advantageous field within gliding distance, a change to the better field should be permitted. The hazards involved in these last-minute decisions, such as excessive maneuvering at very low altitudes, should be thoroughly explained by the instructor.

Slipping the airplane, using flaps, varying the position of the base leg, and varying the turn onto final approach should be stressed as ways of correcting for misjudgment of altitude and glide angle.

Eagerness to get down is one of the most common faults of inexperienced pilots during simulated forced landings. In giving way to this, they forget about speed and arrive at the edge of the field with too much speed to permit a safe landing. Too much speed may be just as dangerous as too little; it results in excessive floating and overshooting the desired landing spot. It should be impressed on the students that they cannot dive at a field and expect to land on it, particularly with today's sleek, modern airplanes.

During all simulated forced landings, the instructor should control the throttle, ensure that the engine is kept warm and cleared, and advance the throttle when the simulated forced landing approach is ended. When the throttle is reopened by the instructor after the termination of the approach, no doubt should exist in the student's mind as to who has control of the airplane. Either the instructor or the student should have complete control, since many near accidents have occurred from such misunderstandings.

Every simulated forced landing approach should be terminated as soon as it can be determined whether a safe landing could have been made. However, in no case should it be continued to a point where it creates an undue hazard or an annoyance to persons or property on the ground. NOTE: Regulations state that aircraft may not be operated closer than 500 feet to any person, vessel, vehicle, or structure.

In addition to flying the airplane from the point of simulated engine failure to where a reasonable safe landing could be made, the student should also be taught certain emergency cockpit procedures. The habit of performing these cockpit procedures should be developed to such an extent that, when an engine failure actually occurs, the student will check the critical items that would be necessary to get the engine operating again while selecting a field and planning an approach. Combining the two operations—accomplishing emergency procedures and planning and flying the approach—will be difficult for the student during the early training in forced landings.

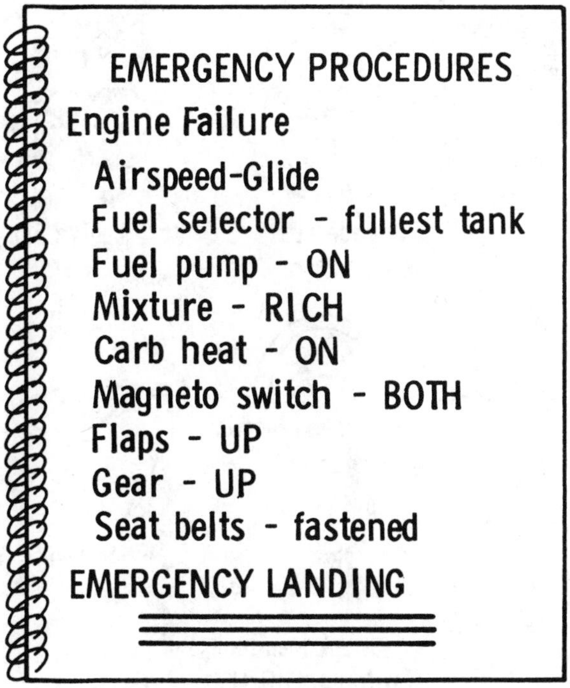

EMERGENCY PROCEDURES

Engine Failure

Airspeed-Glide
Fuel selector - fullest tank
Fuel pump - ON
Mixture - RICH
Carb heat - ON
Magneto switch - BOTH
Flaps - UP
Gear - UP
Seat belts - fastened

EMERGENCY LANDING

Figure 9–22 Check the Critical Items

There are definite steps and procedures to be followed in any simulated forced landing. Although these may differ somewhat from the procedures used in an actual emergency, they should be learned thoroughly by the student, and each step called out to the instructor. The use of a checklist is strongly recommended. Most airplane manufacturers provide a checklist of the appropriate items (Fig. 9–22).

Critical items to be checked should include the position of the fuel tank selector, the quantity of fuel in the tank selected, the fuel pressure gauge to see if the electric fuel pump is needed, the position of the mixture control, the position of the magneto switch, and the use of carburetor heat (Fig. 9–23). Many actual forced landings have been made and later found to be a result of the fuel selector valve being positioned to an empty tank while the other tank had plenty of fuel. It may be wise to change the position of the fuel selector valve even though the fuel gauge indicates fuel in all tanks, as fuel gauges have been known to be inaccurate. No doubt many actual forced landings could have been prevented if the pilots had developed the habit of checking these critical items during their flight training to the extent that it carried over into their later flying.

Instruction in emergency procedures should by no means be limited to simulated forced landings caused by power failures. Other emergencies associated with the operation of the airplane should be explained, demonstrated, and practiced if practicable. Among these emergencies are such occurrences as fire in flight, electrical or hydraulic systems malfunctions, unexpected severe weather conditions,

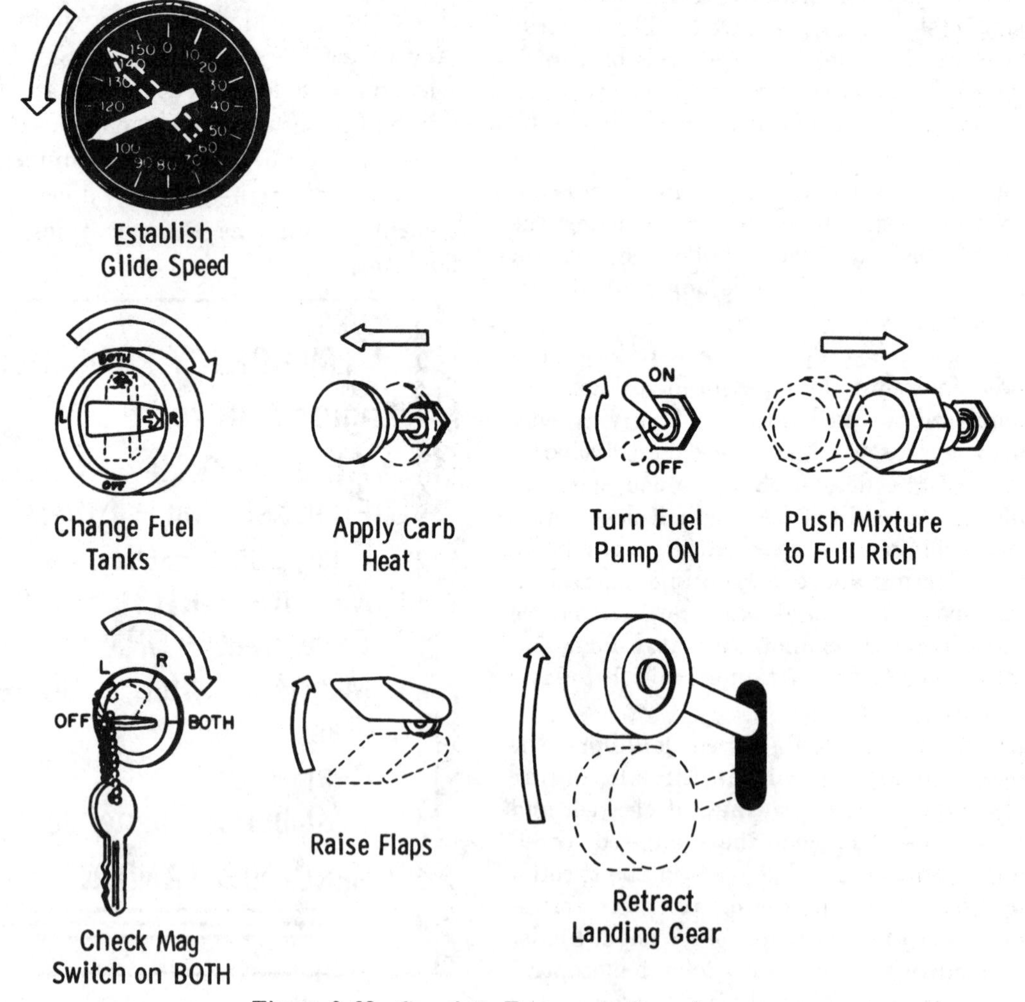

Figure 9–23 Complete Emergency Procedures

engine overheating, imminent fuel exhaustion, and the emergency operation of airplane systems and equipment.

Power-Off Accuracy Approaches

These are approaches and landings made by gliding with the engine idling, through a specific pattern to a touchdown beyond and within 200 feet of a designated line or mark on the runway. The objective is to instill in the pilot the judgment, technique, and procedures necessary for accurately flying the airplane, with no power, to a safe landing.

The ability to estimate the distance an airplane will glide to a landing is the real basis of all power-off accuracy approaches and landings. This will largely determine the amount of maneuvering that may be done from a given altitude. In addition to the ability to estimate distance, it requires the ability to maintain the proper glide while maneuvering the airplane (Fig. 9–24).

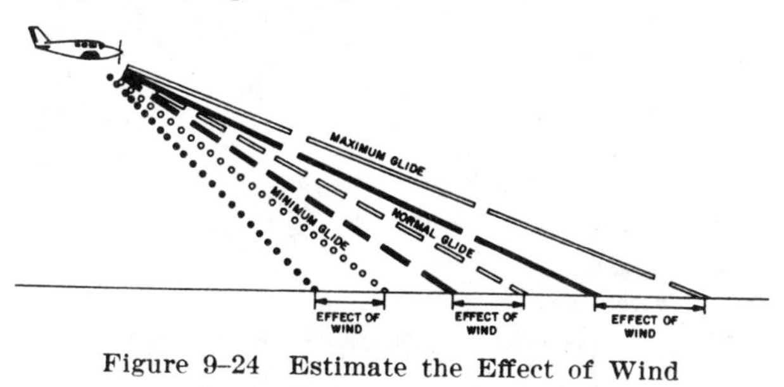

Figure 9–24 Estimate the Effect of Wind

With experience and practice, altitudes up to approximately a thousand feet can be estimated with fair accuracy, while above this level the accuracy in judgment of height above the ground decreases, since all features tend to merge. Therefore, the best aid in perfecting the ability to judge height above this altitude is through the indications of the altimeter and associating them with the general appearance of the earth.

The judgment of altitude in feet, hundreds of feet, or thousands of feet is not nearly so important though, as the ability to estimate gliding angle and its resultant distance. The pilot who knows the normal glide angle of the airplane can estimate with reasonable accuracy, the approximate spot along a given ground path at which the airplane will land, regardless of altitude. But, the pilot who also has the ability to accurately estimate altitude, can judge just how much maneuvering is possible during the glide, which is important to the choice of landing areas in an actual emergency.

The objective of a good final approach is to descend at an angle that will permit the airplane to reach the desired landing area, and at an airspeed that will result in a minimum of floating just before touchdown. To accomplish this it is essential that both the descent angle and the airspeed be accurately controlled.

Unlike a "normal approach" when the power setting is variable, on a "power-off approach" the power is fixed at the idle setting. Therefore, pitch attitude rather than power must be adjusted to control the airspeed. This, of course, also changes the glide or descent angle. By lowering the nose to keep the approach airspeed constant, the descent angle will steepen. If the approach is too high, lower the nose; when the approach is too low, raise the nose. However, if the pitch attitude is raised *too* high, the airplane will settle very rapidly due to a slow airspeed and insufficient lift. For this reason, NEVER TRY TO STRETCH A GLIDE to reach the desired landing spot (Fig. 9–25).

Uniform approach patterns such as the 90°, 180°, or 360° power-off approaches, are described on the following pages. Practice in these approaches provides the pilot with a basis on which to develop judgment in gliding distance and in planning an approach.

The basic procedure in these approaches involves closing the throttle at a given altitude, and gliding to a "key position." This position, like the pattern itself, must not be allowed to become the primary objective, as it is merely a convenient point in the air from which the pilot can best judge whether the glide will safely terminate at the desired spot. The selected "key position" should, of course, be one that is appropriate for the available altitude and the wind condition. From the key position, the pilot must constantly evaluate the situation.

It must be emphasized that, although accurate spot touchdowns are important, safe and properly-executed approaches and landings are vital. *The pilot must never sacrifice a good approach or landing just to land on the desired spot.*

Figure 9–25 Never Try to Stretch the Glide

90° Power-Off Approach

This approach is made from a base leg and requires only a 90° turn onto the final approach. The approach path may be varied by positioning the base leg closer to or farther out from the approach end of the runway according to wind conditions (Fig. 9–26).

The glide from the key position on the base leg through the 90° turn to the final approach is the final part of all accuracy landing maneuvers.

The 90° power-off approach is usually begun from a rectangular pattern at approximately 800 feet above the ground or at normal

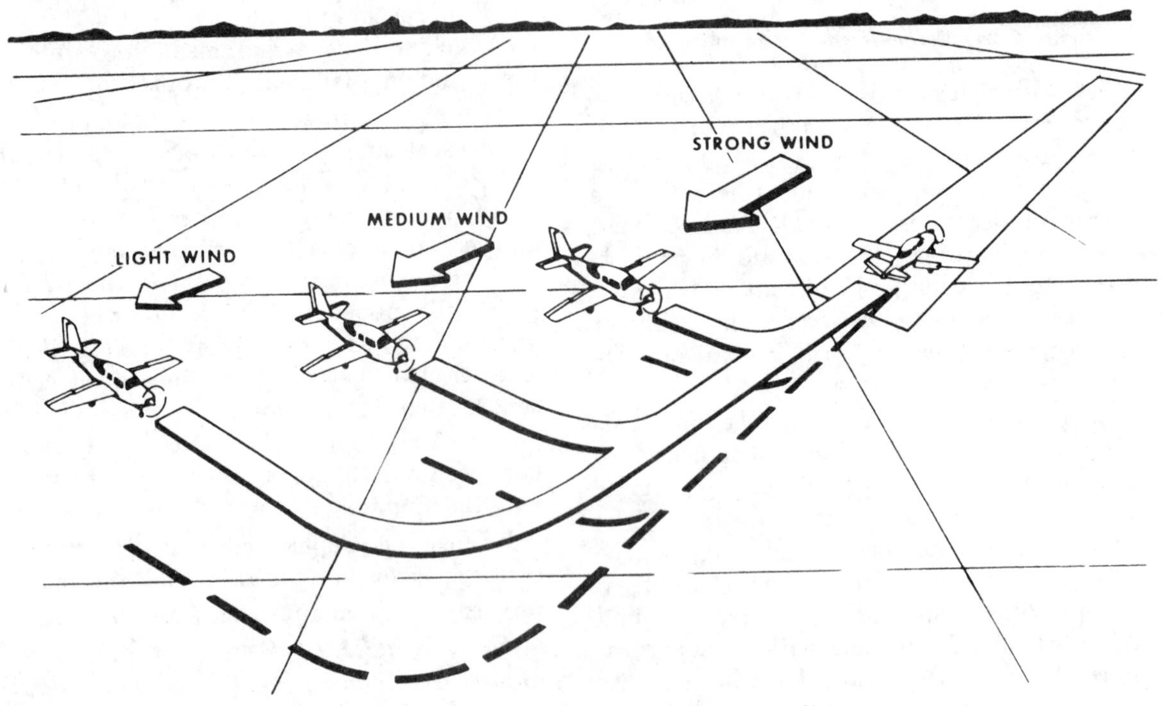

Figure 9–26 Plan the Base Leg for Wind Condition

traffic pattern altitude. The airplane should be flown onto a downwind leg at the same distance from the landing surface as in a normal traffic pattern. The prelanding procedure should be completed on the downwind leg, including extension of the landing gear if the airplane is equipped with retractable gear.

After a medium-banked turn onto the base leg is completed, the throttle should be retarded slightly and the airspeed allowed to decrease to the normal base-leg speed (Fig. 9–27). On the base leg, the airspeed, wind drift correction, and altitude should be maintained while proceeding to the "45° key position". At this position the intended landing spot will appear to be on a 45° angle from the airplane's nose.

The pilot can determine the strength and direction of the wind from the amount of crab necessary to hold the desired ground track on the base leg. This will help in planning the turn onto the final approach and in lowering the correct amount of flaps.

At the 45° key position, the throttle should be closed completely, the propeller control (if so equipped) advanced to the full increase RPM position, and altitude maintained until the airspeed decreases to the manufacturer's recommended glide speed. In the absence of a recommended speed, use 1.4 V_{so}. When this airspeed is attained, the nose should be lowered to maintain the gliding speed and the controls retrimmed.

The base-to-final turn should be planned and accomplished so that upon rolling out of the turn the airplane will be aligned with the runway centerline. When on final approach, the wing flaps are lowered and the pitch attitude adjusted as necessary to establish the proper descent angle and airspeed (1.3 V_{so}), then the controls retrimmed. Slight adjustments in pitch attitude or flaps setting may be necessary to control the glide angle and airspeed. However, NEVER TRY TO STRETCH THE GLIDE NOR RETRACT THE FLAPS to reach the desired landing spot. The final approach may be made with or without the use of slips.

After the final approach glide has been established, full attention must be given to making a good, safe landing rather than concentrating on the selected landing spot. The probability of landing on the spot was already determined by the base leg position and the flap setting. In any event it is better to execute a good landing two hundred feet from the spot than to make a poor landing precisely on the spot.

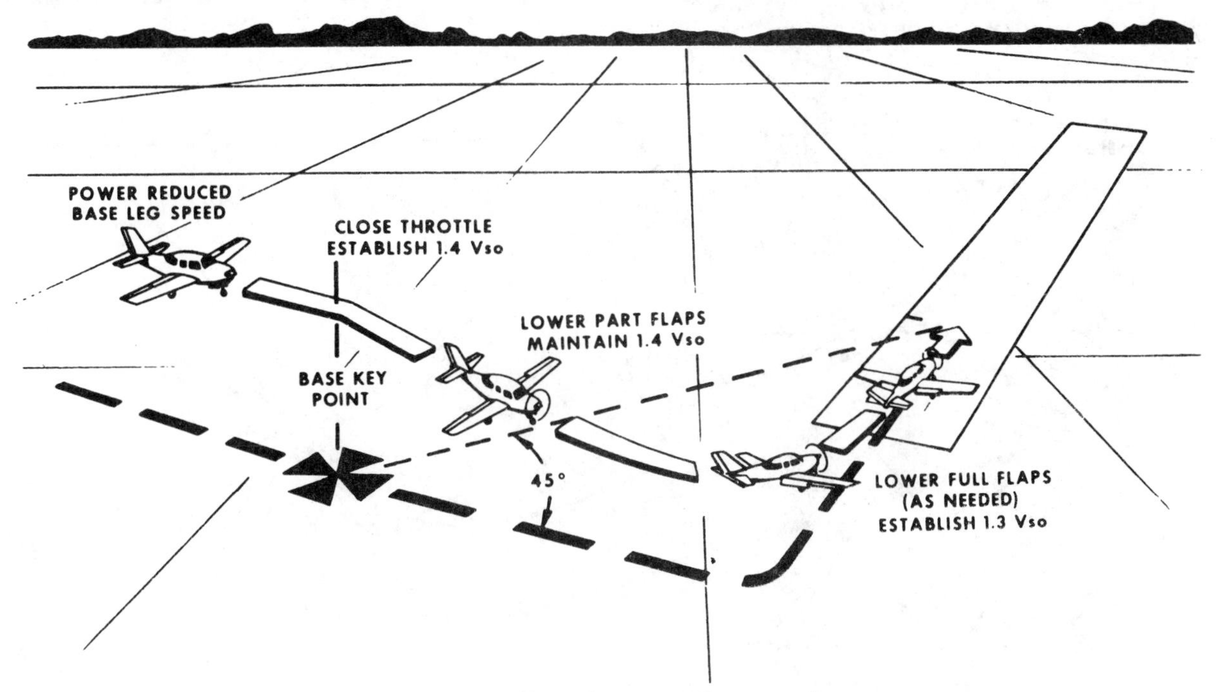

Figure 9–27 90° Power-Off Approach

180° Power-Off Approach

This approach is executed by gliding with the power off from a given point on a downwind leg to a preselected landing spot (Fig. 9–28). Actually it is an extension of the principles involved in the 90° power-off approach just described. Its objective, then, is to further develop judgment in estimating distances and glide ratios, in that the airplane must be flown without power from a higher altitude and through a 90° turn to reach the base leg position at a proper altitude for executing the 90° approach. Consequently, the 180° approach requires more planning and judgment than does the 90° approach.

In the execution of 180° power-off approaches, the airplane is flown on a downwind heading parallel to the landing runway and the landing gear extended (if retractable). The altitude from which this type of approach should be started will vary with the type of airplane, but it should usually not exceed 1,000 feet above the ground, except with large airplanes. Greater accuracy in judgment and maneuvering is required at higher altitudes.

When abreast of or opposite the desired landing spot, the throttle should be closed and altitude maintained while decelerating to the manfacturer's recommended glide speed, or 1.4 V_{so}. The point at which the throttle is closed is the "downwind key position."

The turn from the downwind leg to the base leg should be a uniform turn with a medium or slightly steeper bank. The degree of bank and amount of this initial turn will depend upon the glide angle of the airplane and the velocity of the wind. Again, the base leg should be positioned as needed for the altitude, or wind condition; that is, position the base leg to conserve or dissipate altitude so as to reach the desired landing spot.

The turn onto the base leg should be made at an altitude high enough and close enough to permit the airplane to glide to what would normally be the base key position in a 90° power-off approach.

Although the "key position" is important, it must not be overemphasized nor considered as a fixed point on the ground. Many inexperienced pilots may gain a conception of it

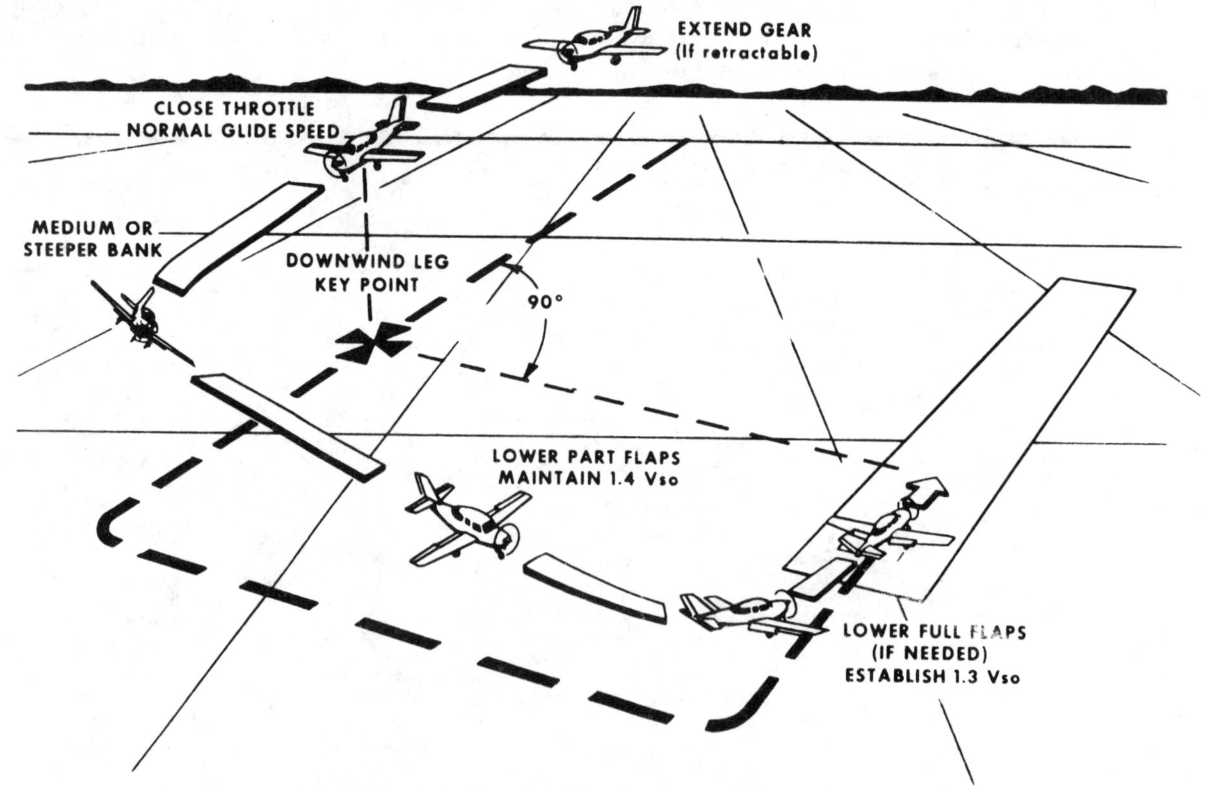

Figure 9–28 180° Power-Off Approach

as a particular landmark such as a tree, crossroad, or other visual reference, to be reached at a certain altitude. This will result in a mechanical conception and leave the pilot at a total loss any time such objects are not present. Both altitude and geographical location must be varied as much as is practical to eliminate any such conception. After reaching the base key position, the approach and landing are the same as in the 90° power-off approach.

360° Power-Off Approach

This advanced power-off approach is one in which the airplane glides through a 360° change of direction to the preselected landing spot. The entire pattern is designed to be circular but the turn may be shallowed, steepened, or discontinued at any point to adjust the accuracy of the flightpath.

The 360° approach is started from a position over the approach end of the landing runway or slightly to the side of it, with the airplane headed in the proposed landing direction and the landing gear and flaps retracted (Fig. 9–29).

It is usually initiated from approximately 2,000 feet or more above the ground—where the wind may vary significantly from that at lower altitudes. This must be taken into account when maneuvering the airplane to a point from which a 90° or 180° power-off approach can be completed.

After the throttle is closed over the intended point of landing, the proper glide speed should immediately be established, and a medium-banked turn made in the desired direction so as to arrive at the downwind key position opposite the intended landing spot. At or just beyond the downwind key position, the landing gear should be extended if the airplane is equipped with retractable gear. The altitude at the downwind key position should be approximately 1,000 to 1,200 feet above the ground.

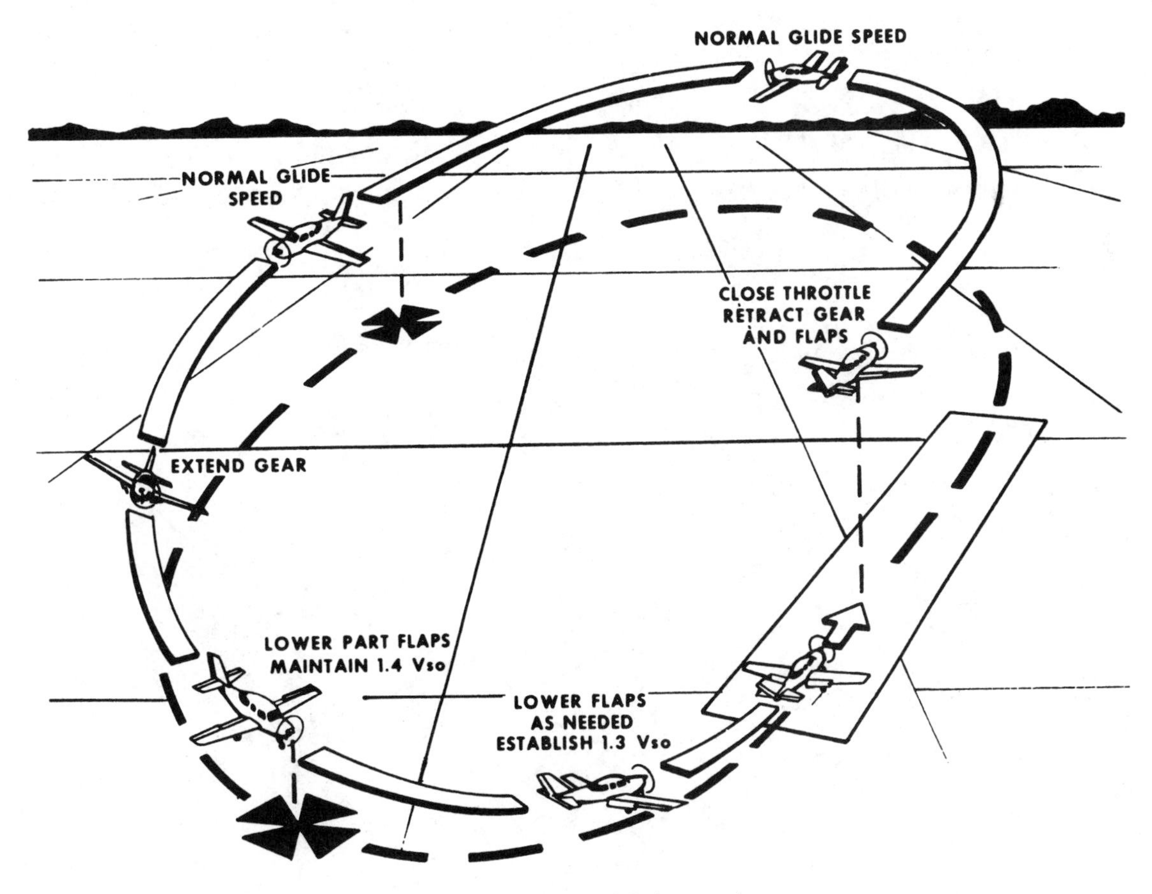

Figure 9–29 360° Power-Off Approach

After reaching that point, the turn should be continued to arrive at a base leg key position, at an altitude of about 800 feet above the terrain. Flaps may be used at this position, as necessary, but full flaps should not be used until established on the final approach.

The angle of bank can be varied as needed throughout the pattern to correct for wind conditions and to align the airplane with the final approach. The turn-to-final should be completed at a minimum altitude of 300 feet above the terrain.

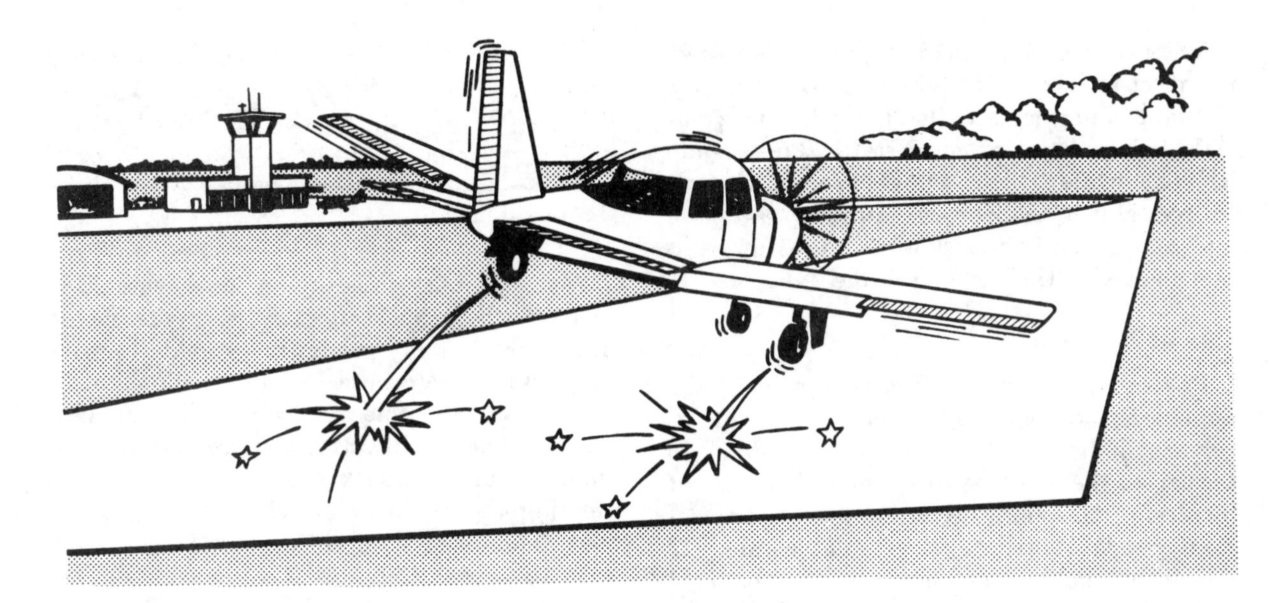

CHAPTER 10

Faulty Approaches and Landings

This chapter discusses the factors contributing to faulty approaches and landings and describes the appropriate actions for making recoveries. A thorough knowledge of these factors is invaluable to the pilot in preventing landing accidents.

The explanations of approaches and landings up to this point have been devoted mainly to normal situations, in which the landings were ideally executed. However, in addition to occasional errors in judgment during some part of the approach and landing, numerous variables such as traffic, wind shift, or wind gusts create situations requiring corrections or recoveries to assure a safe landing. Therefore, pilot skill in anticipation of, recognition of, and recovery from abnormal situations is equal in importance to normal approach and landing skills.

Low Final Approach

When the base leg is too low, insufficient power is used, landing flaps are extended prematurely, or the velocity of the wind is misjudged, sufficient altitude may be lost to cause the airplane to be well below the proper final approach path. In such a situation, the pilot would have to apply considerable power to fly the airplane (at an excessively low altitude) up to the runway threshold. The corrective action for such a situation has been discussed in the preceding chapter; that is, when it is realized the runway will not be reached unless appropriate action is taken, power must be applied immediately to maintain the airspeed while the pitch attitude is raised to increase lift and stop the descent. When the proper approach path has been intercepted, the correct approach attitude should be reestablished and the power reduced again. DO NOT increase the pitch attitude without increasing the power, since the airplane will decelerate rapidly and may approach the critical angle of attack and stall. If there is any doubt about the approach being safely completed, it is advisable to EXECUTE AN IMMEDIATE GO-AROUND.

Slow Final Approach

When the airplane is flown at too slow an airspeed on the final approach, the pilot's judgment of the rate of sink (descent) and the height of roundout may be defective. During an excessively slow approach, the wing is operating near the critical angle of attack and, depending on the pitch attitude changes and control usage, the airplane may stall or sink rapidly, contacting the ground with a hard impact.

Whenever a slow-speed approach is noted, the pilot should apply power to accelerate the airplane and increase the lift to reduce the sink rate and to prevent a stall. This should be done while still at a high enough altitude to reestablish the correct approach airspeed and attitude. If too slow and too low it is best to EXECUTE A GO-AROUND.

Use of Throttle

Power can be used effectively during the approach and roundout to compensate for errors in judgment. Power can be added to accelerate the airplane to increase lift without increasing the angle of attack; thus, the descent can be slowed to an acceptable rate. If the proper landing attitude has been attained and the airplane is only slightly high, the landing attitude should be held constant and sufficient power applied to help ease the airplane onto the ground. After the airplane has touched down, it will be necessary to close the throttle so that the additional thrust and lift will be removed and the airplane will stay on the ground.

High Roundout

Sometimes when the airplane appears to temporarily stop moving downward, the roundout has been made too rapidly and the airplane is flying level too high above the runway. Continuing the roundout would further reduce the airspeed, resulting in an increase in angle of attack to the critical angle. This would result in the airplane stalling and dropping hard onto the runway. To prevent this, the pitch attitude should be held constant until the airplane decelerates enough to again start descending. Then the roundout can be continued to establish the proper landing attitude. This technique should be used *only when there is adequate airspeed.* It may be necessary to add a slight amount of power to keep the airspeed from decreasing excessively and to avoid losing lift too rapidly.

When the proper landing attitude is attained, the airplane is approaching a stall because the airspeed is decreasing and the critical angle of attack is being approached, even though the pitch attitude is no longer being increaseed (Fig. 10–1).

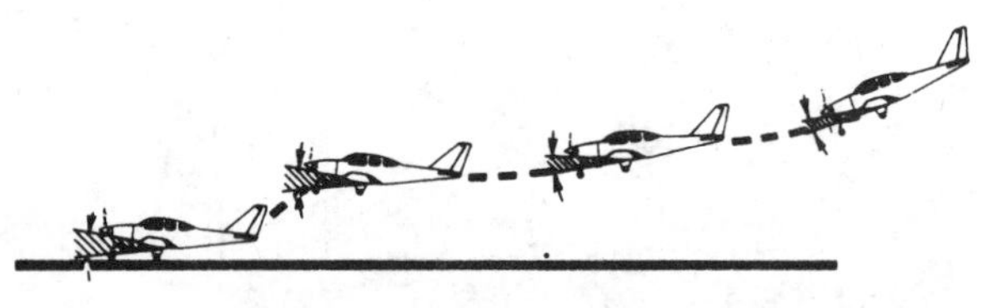

Figure 10–1 Rounding Out too High

Although back pressure on the elevator control may be relaxed slightly, the nose should not be lowered any perceptible amount to make the airplane descend when fairly close to the runway unless some power is added momentarily. The momentary decrease in lift that would result from lowering the nose and decreasing the angle of attack may be so great that the airplane might contact the ground with the nosewheel, which could collapse. It is recommended that a GO-AROUND be executed any time it appears that the nose should be lowered significantly.

Late or Rapid Roundout

Starting the roundout too late or pulling the elevator control back too rapidly to prevent the airplane from touching down prematurely can impose a heavy load factor on the wing and cause an accelerated stall.

Suddenly increasing the angle of attack and stalling the airplane during a roundout is a dangerous situation since it may cause the airplane to land extremely hard on the main landing gear, and then bounce back into the air. As the airplane contacts the ground, the tail will be forced down very rapidly by the back pressure on the elevator and by inertia acting downward on the tail.

Recovery from this situation requires prompt and positive application of power prior to occurrence of the stall. This may be followed by a normal landing if sufficient runway is available—otherwise the pilot should EXECUTE A GO-AROUND immediately.

Floating During Roundout

If the airspeed on final approach is excessive, it will usually result in the airplane "floating" (Fig. 10–2). Before touchdown can be made, the airplane may be well past the desired landing point and the available runway may be insufficient. When diving an airplane on final approach to land at the proper point, there will be an appreciable increase in airspeed. Consequently, the proper touchdown attitude cannot be established without producing an excessive angle of attack and lift. This will cause the airplane to gain altitude or "balloon."

Figure 10–2 Floating During Roundout

Any time the airplane floats, judgment of speed, height, and rate of sink must be especially keen. The pilot must smoothly and gradually adjust the pitch attitude as the airplane decelerates to touchdown speed and starts to settle, so the proper landing attitude is attained at the moment of touchdown. The slightest error in judgment and timing will result in either ballooning or bouncing.

The recovery from floating will depend on the amount of floating and the effect of a crosswind, as well as the amount of runway remaining. Since prolonged floating utilizes considerable runway length, it should be avoided especially on short runways or in strong crosswinds. If a landing cannot be made on the first third of the runway, or the airplane drifts sideways, the pilot should EXECUTE A GO-AROUND.

Ballooning During Roundout

If the pilot misjudges the rate of sink during a landing and thinks the airplane is descending faster than it should, there is a tendency to increase the pitch attitude and angle of attack too rapidly. This not only stops the descent, but actually starts the airplane climbing. This climbing during the roundout is known as "ballooning" (Fig. 10–3). Ballooning can be dangerous because the height above the ground is increasing and the airplane may be rapidly approaching a stalled condition. The altitude gained in each instance will depend on the airspeed or the rapidity with which the pitch attitude is increased.

When ballooning is slight, a constant landing attitude should be held and the airplane allowed to gradually decelerate and settle onto

Figure 10–3 Ballooning During Roundout

the runway. Depending on the severity of ballooning, the use of throttle may be helpful in cushioning the landing. By adding power, thrust can be increased to keep the airspeed from decelerating too rapidly and the wings from suddenly losing lift, but throttle must be closed immediately after touchdown. Also, it must be remembered that torque will have been created as power was applied; therefore, it will be necessary to use rudder pressure to keep the airplane straight as it settles onto the runway.

When ballooning is excessive, it is best to EXECUTE A GO-AROUND IMMEDIATELY; DO NOT ATTEMPT TO SALVAGE THE LANDING. Power must be applied before the airplane enters a stalled condition.

The pilot must be extremely cautious of ballooning when there is a crosswind present because the crosswind correction may be inadvertently released or it may become inadequate. Because of the lower airspeed after ballooning, the crosswind affects the airplane more. Consequently, the wing will have to be lowered even further to compensate for the increased drift. It is imperative that the pilot make certain that the appropriate wing is down and that directional control is maintained with opposite rudder. If there is any doubt, or the airplane starts to drift, EXECUTE A GO-AROUND.

Bouncing During Touchdown

When the airplane contacts the ground with a sharp impact as the result of an improper attitude or an excessive rate of sink, it tends to "bounce" back into the air. Though the airplane's tires and shock struts provide some springing action, the airplane does *not* bounce as does a rubber ball. Instead, it rebounds into the air because the wing's angle of attack was abruptly increased, producing a sudden addition of lift. (Fig. 10–4).

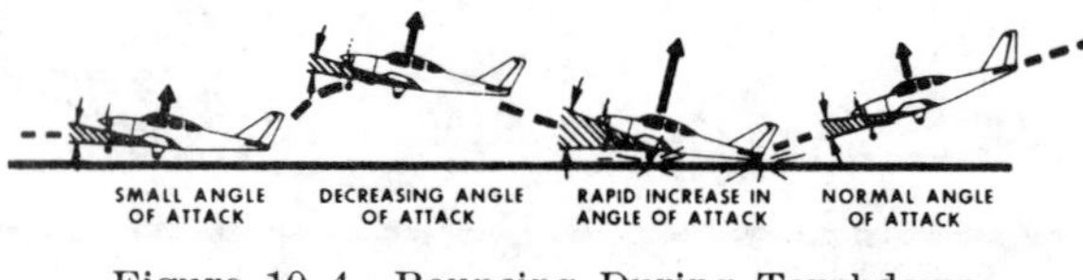

Figure 10–4 Bouncing During Touchdown

The abrupt change in angle of attack is the result of inertia instantly forcing the airplane's tail downward when the main wheels contact the ground sharply. The severity of the "bounce" depends on the airspeed at the moment of contact and the degree to which the angle of attack or pitch attitude was increased.

Since a bounce occurs when the airplane makes contact with the ground before the proper touchdown attitude is attained, it is almost invariably accompanied by the application of excessive back elevator pressure. This is usually the result of the pilot realizing too late that the airplane is not in the proper attitude and attempting to establish it just as the second touchdown occurs.

The corrective action for a bounce is the same as for ballooning and similarly depends on its severity. When it is very slight and there is no extreme change in the airplane's pitch attitude, a followup landing may be executed by applying sufficient power to cushion the subsequent touchdown, and smoothly adjusting the pitch to the proper touchdown attitude.

In the event a very slight bounce is encountered while landing with a crosswind, crosswind correction must be maintained while the next touchdown is made. Remember that since the subsequent touchdown will be made at a slower airspeed, the upwind wing will have to be lowered even further to compensate for drift.

When a bounce is severe, the safest procedure is to EXECUTE A GO-AROUND IMMEDIATELY. No attempt to salvage the landing should be made. Full power should be applied while simultaneously maintaining directional control, and lowering the nose to a safe climb attitude. The go-around procedure should be continued even though the airplane may descend and another bounce may be encountered. It would be extremely foolish to attempt a landing from a bad bounce since airspeed diminishes very rapidly in the nose-high attitude and a stall may occur before a subsequent touchdown could be made.

Extreme caution and alertness must be exercised any time a bounce occurs, but particularly when there is a crosswind. The crosswind correction will almost invariably be released by inexperienced pilots when the airplane bounces. When one main wheel of the airplane strikes the runway, the other wheel will touch down immediately afterwards, and the wings will become level. Then, with no crosswind correction as the airplane bounces, the wind will cause the airplane to roll with the wind, thus exposing even more surface to the crosswind and drifting the airplane more rapidly.

Hard Landing

When the airplane contacts the ground during landings, its vertical speed is instantly reduced to zero. Unless provision is made to slow this vertical speed and cushion the impact of touchdown, the force of contact with the ground may be so great as to cause structural damage to the airplane.

The purpose of pneumatic tires, rubber or oleo shock absorbers, and other such devices is, in part, to cushion the impact and to increase the time in which the airplane's vertical descent is stopped. The importance of this cushion may be understood from the computation that a 6 inch free fall on landing is equal, roughly, to a 340-foot per minute descent. Within a fraction of a second the airplane must be slowed from this rate of vertical descent to zero, without damage.

During this time, the landing gear together with some aid from the lift of the wings must supply whatever force is needed to counteract the force of the airplane's inertia and weight (Fig. 10–5).

Figure 10–5 Hard Landing and Bouncing

The lift decreases rapidly, however, as the airplane's forward speed is decreased, and the force on the landing gear increases as the shock struts and tires are compressed by the impact of touchdown. When the descent stops, the lift will be practically zero, leaving the landing gear alone to carry both the airplane's weight and inertia force. The load imposed at the instant of touchdown may easily be three or four times the actual weight of the airplane depending on the severity of contact.

Touchdown in a Drift or Crab

At times the pilot may correct for wind drift by crabbing on the final approach. If the roundout and touchdown are made while the airplane is drifting or in a crab, it will contact the ground while moving sideways. This will impose extreme side loads on the landing gear, and if severe enough, may cause structural failure.

The most effective method to prevent drift in primary training aircraft is the "wing-low method." This technique keeps the longitudinal axis of the airplane aligned with both the runway and the direction of motion throughout the approach and touchdown.

There are three factors that will cause the longitudinal axis and the direction of motion to be misaligned during touchdown; drifting, crabbing, or a combination of both.

If the pilot has not taken adequate corrective action to avoid drift during a crosswind landing, the main wheels' tire treads offer resistance to the airplane's sideward movement in respect to the ground. Consequently, any sidewise velocity of the airplane is abruptly decelerated, with the result that the inertia force is as shown in Fig. 10–6. This creates a moment around the main wheel when it contacts the ground, tending to overturn or tip the airplane. If the windward wingtip is raised by the action of this moment, all the weight and shock of landing will be borne by one main wheel. This could cause structural damage.

Not only are the same factors present that are attempting to raise a wing, but the crosswind is also acting on the fuselage surface

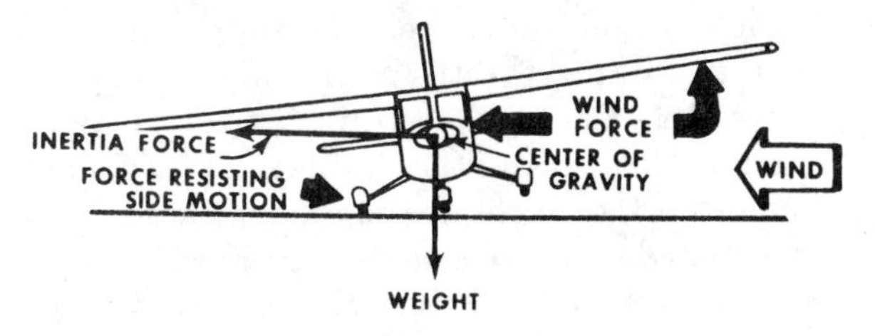

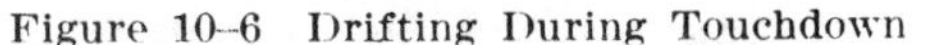
Figure 10–6 Drifting During Touchdown

behind the main wheels tending to yaw (weathervane) the airplane into the wind. This often results in a ground loop (Fig. 10–7).

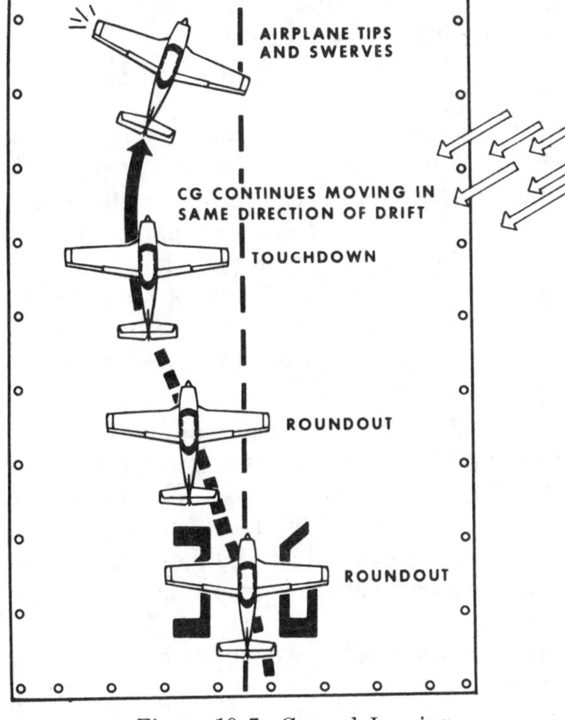

Figure 10–7 Ground Looping

Ground Loop

A ground loop is an uncontrolled turn during ground operation that may occur while taxiing or taking off, but especially during the after-landing roll. It is not always caused by drift or weathervaning although these things may cause the initial swerve. Careless use of the rudder, an uneven ground surface, or a soft spot that retards one main wheel of the airplane may also cause a swerve. In any case, the initial swerve tends to make the airplane ground loop, whether it be a tailwheel-type or nosewheel-type.

As explained in the chapter on Landing Approaches, due to the characteristics of an airplane equipped with a tailwheel, the forces that cause a ground loop increase as the swerve increases. The initial swerve develops centrifugal force and this, acting at the center of gravity (which is located behind the main wheels), swerve the airplane even more. If allowed to develop, the centrifugal force produced may become great enough to tip the airplane until one wing strikes the ground. (Fig. 10–7).

Airplanes having a nosewheel are somewhat less prone to ground loop. Since the center of gravity is located forward of the main landing gear on these airplanes, any time a swerve develops, centrifugal force acting on the center of gravity will tend to stop the swerving action.

If the airplane touches down while drifting or in a crab, the pilot should apply aileron toward the high wing and stop the swerve with the rudder. Brakes should be used to correct for turns or swerves only when the rudder is inadequate. The pilot must exercise caution when applying corrective brake action because it is very easy to over-control and aggravate the situation.

If brakes are used, sufficient brake should be applied on the low-wing wheel (outside of the turn) to stop the swerve. When the wings are approximately level, the new direction must be maintained until the airplane has slowed to taxi speed or has stopped.

Wing Rising After Touchdown

When landing in a crosswind there may be instances when a wing will rise during the after landing roll. This may occur whether or not there is a loss of directional control, depending on the amount of crosswind and the degree of corrective action.

Any time an airplane is rolling on the ground in a crosswind condition, the upwind wing is receiving a greater force from the wind than the downwind wing. This causes a lift differential. Also, the wind striking the fuselage on the upwind side may further raise the wing by tending to tip or roll the fuselage.

When the effects of these two factors are great enough, one wing may rise even though directional control is maintained. If no correction is applied, it is possible that a wing will rise sufficiently to cause the other one to strike the ground.

In the event a wing starts to rise during the landing roll, the pilot should immediately apply more aileron pressure toward the high wing and continue to maintain direction. The sooner the aileron control is applied, the more effective it will be. The further a wing is allowed to rise before taking corrective action, the more airplane surface is exposed to the force of the crosswind. This diminishes the effectiveness of the aileron.

Summarizing Corrective Actions

To summarize, there are four control measures that can be used to control the airplane on the ground. They are rudder, brakes, ailerons, and throttle.

During the after-landing roll, the pilot can maintain directional control by using varying degrees of rudder, brake, and/or throttle. Whether they are used in any sequence, or all simultaneously, depends on the nature of the landing problem. It is very important that control be maintained through the landing roll. If the airplane turns excessively, the pilot should not try to realign it with the runway immediately. Instead, it should be held straight until the airplane is under full control, then it should be gradually realigned with the runway. A number of accidents occur because the pilot over-controls while attempting to realign the airplane. Keep the wings level by maintaining direction and using the ailerons.

If the airplane has attained a landing attitude and is still well above the ground, a stall or rapid rate of sink must not be allowed to develop. Add power, or EXECUTE A GO-AROUND and plan another approach.

Any time the airplane approaches a stalling condition, whether after ballooning or contacting the ground and bouncing, apply full power, adjust the pitch attitude, and go around. It is unsafe to continue the landing.

During the after-landing roll, if a wing starts down, use aileron to raise it and, if necessary, use throttle to increase its effectiveness. Note that as the forward speed of the aircraft decreases, the ailerons become less effective.

Throttle increases both the thrust and lift of the airplane by increasing the relative wind over the wings and the slipstream over the tail surfaces. This causes these surfaces to be more sensitive and effective. During turns or swerves, adding throttle will pull the airplane forward and resist the turning of the airplane, but it will also increase torque. When large amounts of throttle are used, torque must be anticipated and corrected. As a safety rule remember, if used in time, the throttle will get the pilot out of almost any difficulty.

CHAPTER 11

Proficiency Flight Maneuvers

This chapter describes the flight training maneuvers and related factors that are useful in developing a high degree of pilot skill. Although most of these maneuvers are not performed as such in normal everyday flying, the elements and principles involved in each are applicable to performance of the customary pilot operations. They aid the pilot in analyzing the effect of wind and other forces acting on the airplane and in developing a fine control touch, coordination, and division of attention for accurate and safe maneuvering of the airplane. Some of the maneuvers, particularly those requiring the maximum performance of the airplane, have a distinct operational use. Therefore, the pilot should acquire a thorough understanding of the factors involved and the techniques recommended.

Maneuvering by Reference to Ground Objects

These maneuvers, referred to as "ground track" or "ground reference" maneuvers, are performed at a relatively low altitude while applying wind drift correction as needed to follow a predetermined track or path over the ground. They are designed to develop the ability to control the airplane, and recognize and correct for the effect of wind while dividing attention among other matters. This requires planning ahead of the airplane, maintaining orientation in relation to ground objects, flying appropriate headings to follow a desired ground track, and being cognizant of other air traffic in the immediate vicinity.

During these maneuvers, pilots should be alert for available forced landing fields. Due to the altitudes at which these maneuvers are performed, there is little time available to search for a suitable field for landing in the event the need arises.

Drift and Ground Track Control

Whenever any object is free from the ground, it is affected by the medium with which it is surrounded. This means that a free object will move in whatever direction and speed that the medium moves.

Such an example is a power boat crossing a river (Fig. 11–1). If the river were still, the boat could head directly to a point on the opposite shore and travel on a straight course

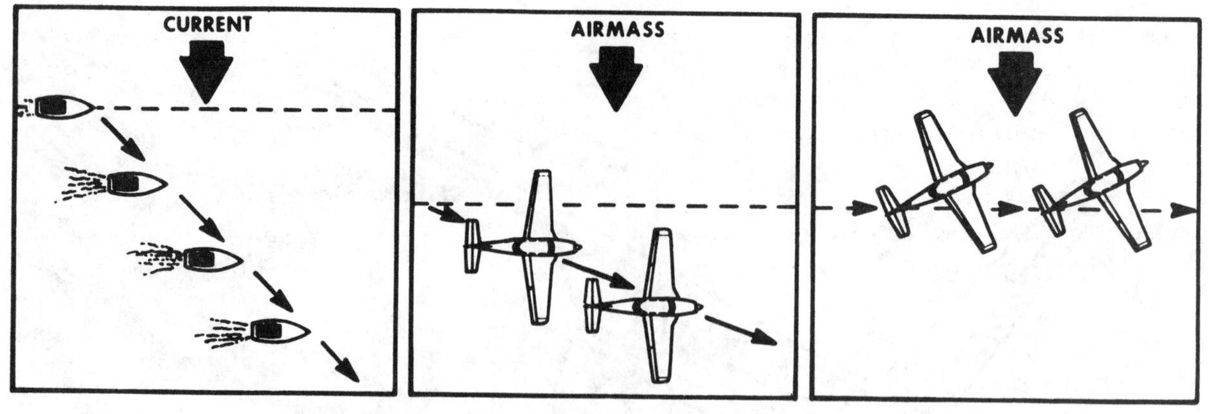

Figure 11–1 Drift and Drift Correction

to that point without drifting. However, if the river were flowing swiftly, the water current would have to be considered. That is, as the boat progresses forward through its own power, it must also move upstream at the same rate that the river is moving it downstream. This is accomplished by angling the boat upstream sufficiently to counteract the downstream flow. If this is done, the boat will follow the desired track across the river from the departure point directly to the intended destination point. Should the boat not be headed sufficiently upstream, it would drift with the current and run aground at some point downstream on the opposite bank.

As soon as an airplane becomes airborne, it is free of ground friction. Its path is then affected by the air mass in which it is flying. Therefore, the airplane (like the boat) will not always track along the ground in the exact direction that it is headed. When flying with the longitudinal axis of the airplane aligned with a road, it may be noted that the airplane gets closer to or farther from the road without any turn having been made. This would indicate that the air mass is moving sideward in relation to the airplane. Since the airplane is flying within this moving body of air (wind), it moves or drifts with the air in the same direction and speed, just like the boat moved with the river current (Fig. 11–1).

When flying straight and level and following a selected ground track, the preferred method of correcting for wind drift is to head (crab) the airplane sufficiently into the wind to cause the airplane to move forward into the wind at the same rate that the wind is moving it sideways. Depending on the wind velocity, this may require a large crab angle or one of only a few degrees. When the drift has been neutralized, the airplane will follow the desired ground track (Fig. 11–1).

To understand the need for drift correction during flight, consider a flight with a wind velocity of 30 knots from the left and 90° to the direction the airplane is headed. After one hour, the body of air in which the airplane is flying will have moved 30 nautical miles to the right. Since the airplane is moving with this body of air, it too will have drifted 30 nautical miles to the right. In relation to the air, the airplane moved forward, but in relation to the ground it moved forward as well as 30 nautical miles to the right.

There are times when the pilot needs to correct for drift while in a turn (Fig. 11–2). Throughout the turn the wind will be acting on the airplane from constantly changing angles. The time it takes for the airplane to progress through any part of a turn is governed by the relative wind angle and speed. This is due to the constantly changing groundspeed. When the airplane is headed into the wind the groundspeed is decreased; when headed downwind the groundspeed is increased. Through the crosswind portion of a turn the airplane must be crabbed sufficiently into the wind to counteract drift. To follow a desired circular ground track the crab angle must be varied in a timely manner because of the varying groundspeed as the turn progresses. Thus, the faster the groundspeed the faster the crab must be established; the slower the groundspeed the slower the crab angle may be established. It can be seen then that the steepest bank and fastest rate of turn should be made on the downwind portion of the turn and the shallowest bank and slowest rate of turn on the upwind portion.

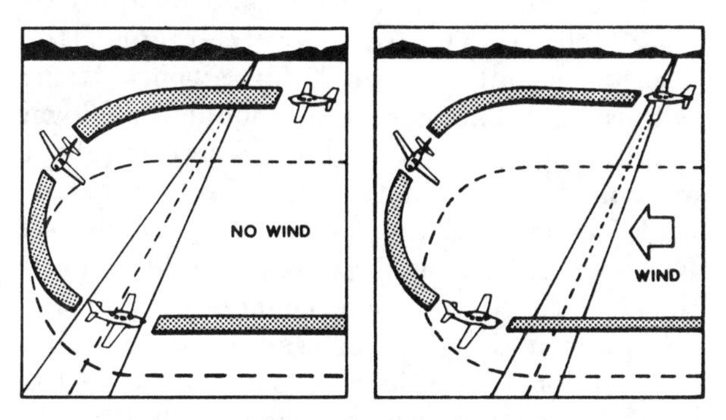

Figure 11–2 Effect of Wind During a Turn

The principles and techniques of varying the angle of bank to change the rate of turn and crab angle for controlling wind drift during a turn are the same for all ground track maneuvers involving changes in direction of flight.

When there is no wind, it would be simple to fly along a ground track with an arc of exactly 180° and a constant radius because the flightpath and ground track would be identical. This can be demonstrated by approaching a road at a 90° angle and, when directly over the road, rolling into a medium-banked turn, then maintaining the same angle of bank throughout the 180° of turn (Fig. 11–2).

To complete the turn, the rollout should be started at a point where the wings will become level as the airplane again reaches the road at a 90° angle and will be directly over the road just as the turn is completed. This would be possible only if there were absolutely no wind and if the angle of bank and the rate of turn remained constant throughout the entire maneuver.

If the turn were made with a constant angle of bank and a wind blowing directly across the road, it would result in a constant radius turn *through the air.* However, the wind effects would cause the *ground track* to be distorted from a constant radius turn or semicircular path. The greater the wind velocity, the greater would be the difference between the desired ground track and the flightpath. To counteract this drift, the flightpath can be controlled by the pilot in such a manner as to neutralize the effect of the wind, and cause the ground track to be a constant radius semicircle.

The effects of wind during turns (Fig. 11–3) can be demonstrated after selecting a road, railroad, or other ground reference which forms a straight line parallel to the wind, by flying into the wind directly over and along the line and then making a turn with a constant medium angle of bank for 360° of turn. The airplane will return to a point directly over the line but slightly downwind from the starting point, the amount depending on the wind velocity and the time required to complete the turn. The path over the ground will be an elongated circle, although in reference to the air it is a perfect circle. Straight flight during the upwind segment after completion of the turn is necessary to bring the airplane back to the starting position.

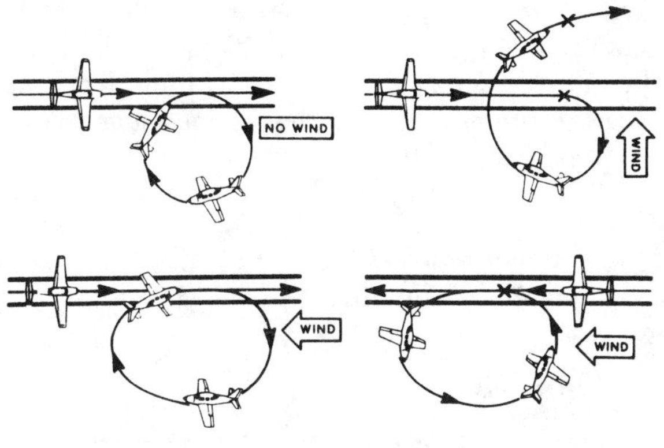

Figure 11–3 Effect of Wind During Turns

A similar 360° turn may be started at a specific point over the reference line, with the airplane headed directly downwind. In this demonstration the effect of wind during the constant banked turn will drift the airplane to a point where the line is reintercepted, but the 360° turn will be completed at a point downwind from the starting point.

Still another reference line which lies directly crosswind may be selected and the same procedure repeated, showing that if wind drift is not corrected the airplane will, at the completion of the 360° turn, be headed in the original direction but will have drifted away from the line a distance dependent on the amount of wind.

From these demonstrations, it can be seen where and why it is necessary to increase or decrease the angle of bank and the rate of turn to achieve a desired track over the ground. The principles and techniques involved can be practiced and evaluated by the performance of the ground track maneuvers discussed in this chapter.

Rectangular Course

The "rectangular course" is a practice maneuver in which the ground track of the airplane is equidistant from all sides of a selected rectangular area on the ground. While performing the maneuver, the altitude and airspeed should be held constant.

Like those of other ground track maneuvers, one of the objectives is to develop division of attention between the flightpath and ground references, while controlling the airplane and watching for other aircraft in the vicinity. Another objective is to develop recognition of drift toward or away from a line parallel to the intended ground track. This will be helpful in recognizing drift toward or from an airport runway during the various legs of the airport traffic pattern.

For this maneuver, a square or rectangular field (or an area bounded on four sides by section lines or roads), the sides of which are approximately a mile in length, should be selected well away from other air traffic (Fig. 11–4). The altitude flown should be approximately 600 to 1,000 feet above the ground (the altitude usually required for airport traffic patterns). The airplane should be flown parallel to and at a uniform distance (about one-fourth to one-half mile away) from the field boundaries, not above the boundaries. For best results, the flightpath should be positioned outside the field boundaries just far enough that they may be easily observed from either pilot seat by looking out the side of the airplane. If an attempt is made to fly directly above the edges of the field, the pilot will have no usable reference points to start and complete the turns. The closer the track of the airplane is to the field boundaries, the steeper the bank necessary at the turning points. Also, the pilot should be able to see the edges of the selected field while seated in a normal position and looking out the side of the airplane during either a left hand or right hand course—the distance of the ground track from the edges

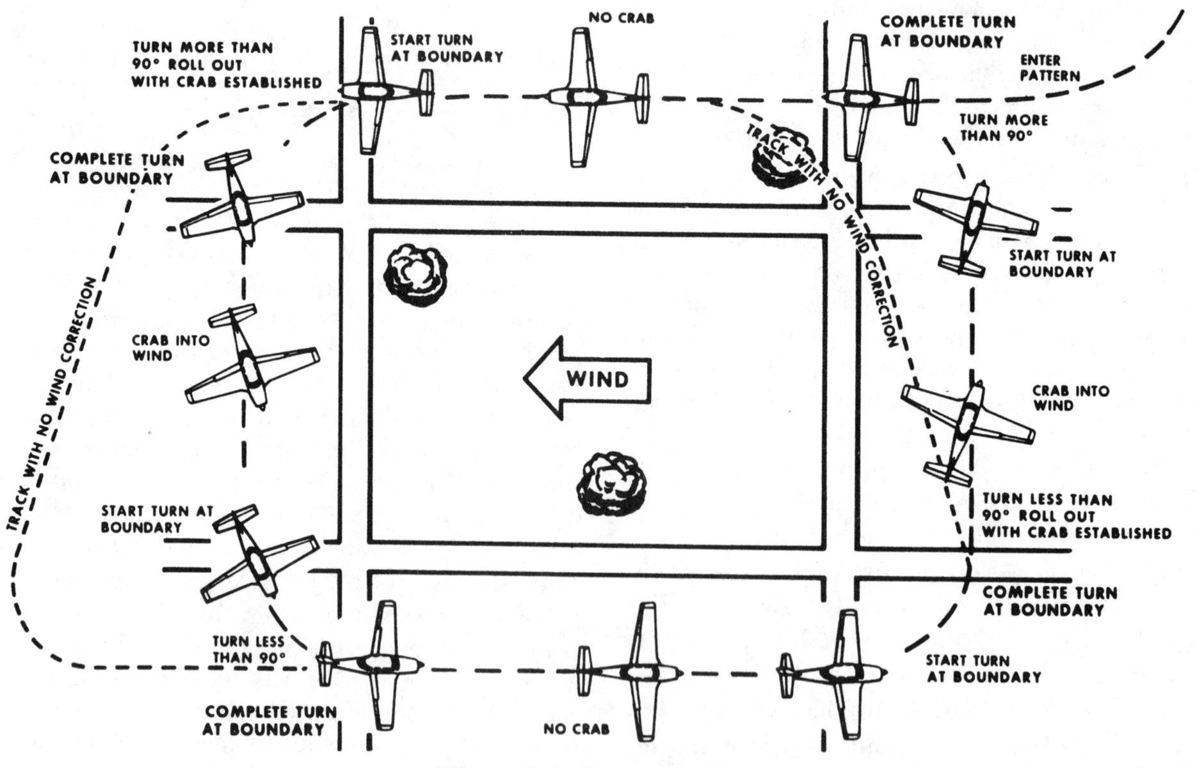

Figure 11–4 Rectangular Course

of the field should be the same regardless of whether the course is flown to the left or right. All turns should be started when the airplane is abeam the corners of the field boundaries and the bank normally should not exceed 45°. These should be the determining factors in establishing the distance from the boundaries for performing the maneuver.

Although the rectangular course may be entered from any direction, this discussion assumes entry on an upwind heading.

While the airplane is on the upwind leg, the next field boundary should be observed as it is being approached, to plan the turn onto the crosswind leg. Since the wind is a headwind on this leg, it is retarding the airplane's groundspeed and during the turn onto the crosswind leg will try to drift the airplane toward the field. For this reason, the roll-in to the turn must be slow and the bank relatively shallow to counteract this effect. As the turn progresses, the headwind component decreases, allowing the groundspeed to increase. Consequently, the bank angle and rate of turn must be increased gradually to assure that upon completion of the turn the crosswind ground track will continue the same distance from the edge of the field. Completion of the turn with the wings level should be accomplished at a point aligned with the upwind corner of the field.

Simultaneously, as the wings are rolled level, the proper drift correction must be established with the airplane crabbed into the wind. This requires that the turn be less than a 90° change in heading. If the turn has been made properly, the field boundary will again appear to be one-fourth to one-half mile away. While on the crosswind leg the crab angle should be adjusted as necessary to maintain a uniform distance from the field boundary.

As the next field boundary is being approached, the pilot should plan the turn onto the downwind leg. Since a crab angle is being held into the wind and away from the field while on the crosswind leg, this next turn will require a turn of more than 90°. Since the crosswind will become a tailwind, causing the groundspeed to increase during this turn, the bank initially must be medium and progressively increased as the turn proceeds. To complete the turn, the rollout must be timed so that the wings become level at a point aligned with the crosswind corner of the field just as the longitudinal axis of the airplane again becomes parallel to the field boundary. The distance from the field boundary should be the same as from the other sides of the field.

On the downwind leg the wind is a tailwind and results in an increased groundspeed. Consequently, the turn onto the next leg must be entered with a fairly fast rate of roll-in with relatively steep bank. As the turn progresses, the bank angle must be reduced gradually because the tailwind component is diminishing, resulting in a decreasing groundspeed.

During and after the turn onto this leg (the equivalent of the base leg in a traffic pattern) the wind will tend to drift the airplane away from the field boundary. To compensate for the drift, the amount of turn must be more than 90°.

Again, the rollout from this turn must be such that as the wings become level, the airplane is crabbed slightly toward the field and into the wind to correct for drift. The airplane should again be the same distance from the field boundary and at the same altitude, as on other legs. The base leg should be continued until the upwind leg boundary is being approached. Once more the pilot should anticipate drift and turning radius. Since drift correction was held on the base leg, it is necessary to turn less than 90° to align the airplane parallel to the upwind leg boundary. This turn should be started with a medium bank angle with a gradual reduction to a shallow bank as the turn progresses. The rollout should be timed to assure paralleling the boundary of the field as the wings become level.

Usually drift should not be encountered on the upwind or the downwind leg, but it may be difficult to find a situation where the wind is blowing exactly parallel to the field boundaries. This would make it necessary to crab slightly on all the legs. It is important to anticipate the turns to correct for groundspeed, drift, and turning radius. When the wind is behind the airplane, the turn must be faster and steeper; when it is ahead of the airplane, the turn must be slower and shallower. These same techniques apply while flying in airport traffic patterns.

S-Turns Across a Road

An "S-turn across a road" is a practice maneuver in which the airplane's ground track describes semicircles of equal radii on each side of a selected straight line on the ground (Fig. 11–5). The straight line may be a road, fence, railroad, or section line which lies perpendicular to the wind, and should be of sufficient length for making a series of turns. A constant altitude should be maintained throughout the maneuver. The altitude should be low enough to easily recognize drift but in no case lower than 500 feet above the highest obstruction.

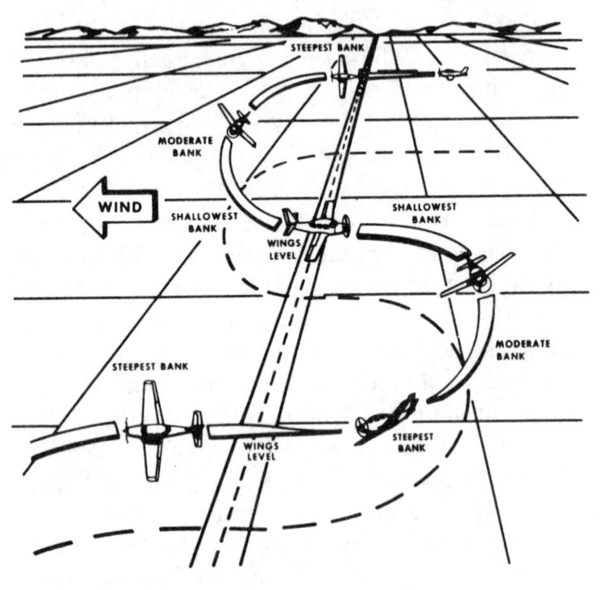

Figure 11–5 "S"-Turns Across a Road

The objectives are to develop the ability to compensate for drift during turns, orient the flightpath with ground references, and divide the pilot's attention. The maneuver consists of crossing the road at a 90° angle and immediately beginning a series of 180° turns of uniform radius in opposite directions, recrossing the road at a 90° angle just as each 180° turn is completed.

Since turns to effect a constant radius ground track require a changing roll rate and angle of bank to establish the crab needed to compensate for the wind, both will increase or decrease as groundspeed increases or decreases.

The bank must be steepest when beginning the turn on the downwind side of the road and must be shallowed gradually as the turn progresses from a downwind heading to an upwind heading. On the upwind side, the turn should be started with a relatively shallow bank and then gradually steepened as the airplane turns from an upwind heading to a downwind heading.

In this maneuver the airplane should be rolled from one bank directly into the opposite just as the reference line on the ground is crossed.

Before starting the maneuver, a straight ground reference line or road that lies 90° to the direction of the wind should be selected, then the area checked to ensure that no obstructions or other aircraft are in the immediate vicinity. The road should be approached from the upwind side, at no less than 500 feet AGL on a downwind heading. When directly over the road, the first turn should be started immediately. With the airplane headed downwind, the groundspeed is greatest and the rate of departure from the road will be rapid; so the roll into the steep bank must be fairly rapid to attain the proper crab angle. This prevents the airplane from flying too far from the road and from establishing a ground track of excessive radius.

During the latter portion of the first 90° of turn when the airplane's heading is changing from a downwind heading to a crosswind heading, the groundspeed becomes less and the rate of departure from the road decreases. The crab angle will be at the maximum when the airplane is headed directly crosswind.

After turning 90°, the airplane's heading becomes more and more an upwind heading, the groundspeed will decrease, and the rate of closure with the road will become slower. If a constant steep bank were maintained, the airplane would turn too quickly for the slower rate of closure, and would be headed perpendicular to the road prematurely. Because of the decreasing groundspeed and rate of closure while approaching the upwind heading, it will be necessary to gradually shallow the bank during the remaining 90° of the semicircle, so that the crab angle is removed completely and the wings become level as the 180° turn is completed at the moment the road is reached.

At the instant the road is being crossed again, a turn in the opposite direction should be started. Since the airplane is still flying into the headwind, the groundspeed is relatively slow. Therefore, the turn will have to be started with a shallow bank so as to avoid an excessive rate of turn which would establish the maximum crab angle too soon. The degree of bank should be that which is necessary to attain the proper crab so the ground track describes an arc the same size as the one established on the downwind side.

Since the airplane is turning from an upwind to a downwind heading, the groundspeed will increase and after turning 90°, the rate of closure with the road will increase rapidly. Consequently, the angle of bank and rate of turn must be progressively increased so that the airplane will have turned 180° at the time it reaches the road. Again, the rollout must be timed so the airplane is in straight-and-level flight directly over and perpendicular to the road.

Throughout the maneuver a constant altitude should be maintained, and the bank should be changing constantly to effect a true semicircular ground track.

Often there is a tendency to increase the bank too rapidly during the initial part of the turn on the upwind side, which will prevent the completion of the 180° turn before recrossing the road. This is apparent when the turn is not completed in time for the airplane to cross the road at a perpendicular angle. To avoid this error, the pilot must visualize the desired half circle ground track, and increase the bank during the early part of this turn. During the latter part of the turn, when approaching the road, the pilot must judge the closure rate properly and increase the bank accordingly, so as to cross the road perpendicular to it just as the rollout is completed.

Turns Around a Point

In this training maneuver, the airplane is flown in two or more complete circles of uniform radii or distance from a prominent ground reference point using a maximum bank of approximately 45° while maintaining a constant altitude. Its objective, as in other ground reference maneuvers, is to help the pilot develop the ability to subconsciously control the airplane while dividing attention between the flightpath and ground references and watching for other air traffic in the vicinity.

The factors and principles of drift correction that are involved in "S turns" and "eights" are also applicable in this maneuver. As in other ground track maneuvers, a constant radius around a point will, if any wind exists, require a constantly changing angle of bank and angles of crab. The closer the airplane is to a direct downwind heading where the groundspeed is greatest, the steeper the bank and the faster the rate of turn required to establish the proper crab; the more nearly it is to a direct upwind heading where the groundspeed is least, the shallower the bank and the slower the rate of turn required to establish the proper crab. It follows, then, that throughout the maneuver the bank and rate of turn must be gradually varied in proportion to the groundspeed.

The point selected for turns around a point should be prominent, easily distinguished by the pilot, and yet small enough to present a precise reference (Fig. 11-6). Isolated trees, crossroads, or other similar small landmarks

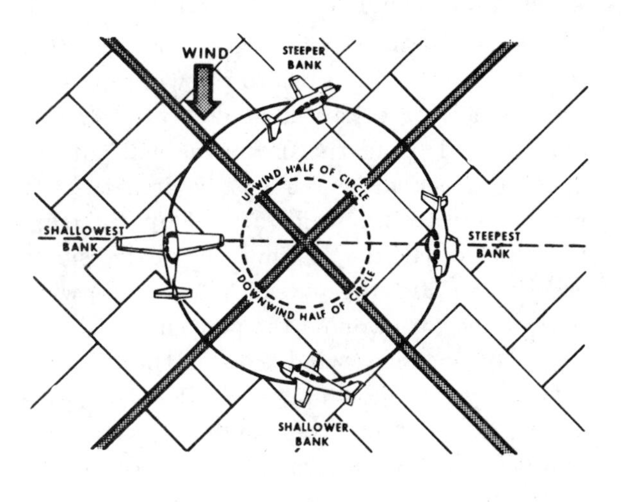

Figure 11-6 Turns Around a Point

are usually suitable. The point should, however, be in an area away from communities, livestock, or groups of people on the ground to prevent possible annoyance or hazard to others. Since the maneuver is performed at a relatively low altitude (not less than 500 feet), the area selected should also afford an opportunity for a safe emergency landing in the event it becomes necessary.

To enter turns around a point, the airplane should be flown on a downwind heading to one side of the selected point at a distance equal to the desired radius of turn. In a high-wing airplane, the distance from the point must permit the pilot to see the point throughout the maneuver even with the wing lowered in a bank. If the radius is too large, the lowered wing will block the pilot's view of the point.

When any significant wind exists, it will be necessary to roll into the initial bank at a rapid rate so that the steepest bank is attained abeam of the point when the airplane is headed directly downwind. It will be seen that by entering the maneuver while heading directly downwind, the steepest bank can be attained immediately. Thus, if a maximum bank of 45° is desired, the initial bank will be 45° if the airplane is at the correct distance from the point. Thereafter, the bank must be shallowed gradually until the point is reached where the airplane is headed directly upwind. At this point, the bank should be gradually steepened until the steepest bank is again attained when heading downwind at the initial point of entry.

Just as "S turns" and "eights" require that the airplane be crabbed into the wind in addition to varying the bank, so do "turns around a point." During the downwind half of the circle, the airplane's nose must be progressively crabbed toward the inside of the circle; during the upwind half the nose must be progressively crabbed toward the outside. The downwind half of the turn around the point may be compared to the downwind side of the "S turn across a road;" the upwind half of the turn around a point may be compared to the upwind side of the "S turn across a road."

As the pilot becomes experienced in performing turns around a point and has a good understanding of the effects of wind drift and varying of the bank angle and crab angle to required, entry into the maneuver may be from any point. When entering this maneuver at any point, the radius of the turn must be carefully selected, taking into account the wind velocity and groundspeed so that an excessive bank is not required later on to maintain the proper ground track.

Eights Along a Road

An "eight along a road" is a maneuver in which the ground track consists of two complete adjacent circles of equal radii on each side of a straight road or other reference line on the ground. The ground track resembles a figure "8" (Fig. 11–7). Like the other ground reference maneuvers, its objective is to develop division of attention while compensating for drift, maintaining orientation with ground references, and maintaining a constant altitude.

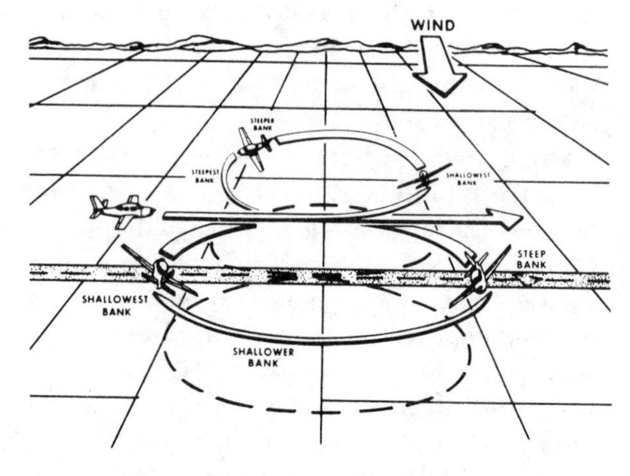

Figure 11–7 Eights Along a Road

Although "eights along a road" may be performed with the wind blowing parallel to the road or directly across the road, for simplification purposes, only the latter situation is explained since the principles involved in either case are common.

A reference line or road which is perpendicular to the wind should be selected and the airplane flown parallel to and directly above the road (not less than 500 feet above the ground). Since the wind is blowing across the flightpath, the airplane will require some crabbing to stay directly above the road during the initial straight and level portion. Before starting the maneuver, the area should be checked to ensure clearance of obstructions and avoidance of other aircraft.

Usually the first turn should be made toward a downwind heading starting with a medium bank. Since the airplane will be turning more and more directly downwind, the groundspeed will be gradually increasing and the rate of departing the road will tend to become faster. Thus, the bank and rate of turn must be increased to establish a crab to keep the airplane from exceeding the desired distance from the road when 180° of change in direction is completed. The steepest bank, then, must be attained when the airplane is headed directly downwind.

As the airplane completes 180° of change in direction, it will be flying parallel to and crabbing toward the road with the wind acting directly perpendicular to the ground track. At this point, the pilot should visualize the remaining 180° of ground track required to return to the same place over the road from which the maneuver started.

While the turn is continued toward an upwind heading, the wind will tend to keep the airplane from reaching the road, with a decrease in groundspeed and rate of closure. The rate of turn and crab angle, therefore, must be decreased proportionately so that the road will be reached just as the 360° turn is completed. To accomplish this, the bank must be decreased so that when headed directly upwind, it will be at the shallowest angle. In the last 90° of the turn, the bank may be varied to correct any previous errors in judging the turning rate and closure rate. The rollout should be timed so that the airplane will be straight and level over the starting point, with enough drift correction to hold it over the road.

After momentarily flying straight and level along the road, the airplane is then rolled into a medium bank turn in the opposite direction to begin the circle on the upwind side of the road. The wind will still be decreasing the groundspeed and trying to drift the airplane back toward the road; therefore the bank must be decreased slowly during the first 90° change in direction in order to reach the desired distance from the road and attain the proper crab angle when 180° change in direction has been completed.

As the remaining 180° of turn continues, the wind becomes more of a tailwind and increases the airplane's groundspeed. This causes the rate of closure to become faster; consequently, the angle of bank and rate of turn must be increased further to attain sufficient crab to keep the airplane from approaching the road too rapidly. The bank will be at its steepest angle when the airplane is headed directly downwind.

In the last 90° of the turn, the rate of turn should be reduced to bring the airplane over the starting point on the road. The rollout must be timed so the airplane will be straight and level, crabbing into the wind, and flying parallel to and over the road.

Eights Across a Road

This maneuver is a variation of "eights along a road" and involves the same principles and techniques. The primary difference is that at the completion of each loop of the figure eight the airplane should cross an intersection of roads, or a specific point on a straight road (Fig. 11–8). The loops should be across the road and the wind should be perpendicular to the road. Each time the road is crossed, the crossing angle should be the same and the wings of the airplane should be level. The "eights" also may be performed by rolling from one bank immediately to the other, directly over the road.

Eights Around Pylons

This training maneuver is an application of the same principles and techniques of correcting for wind drift as used in "turns around a point" and the same objectives as other ground track maneuvers. In this case, however, two points or "pylons" on the ground are used as references, and turns around each "pylon" are made in opposite directions to follow a ground

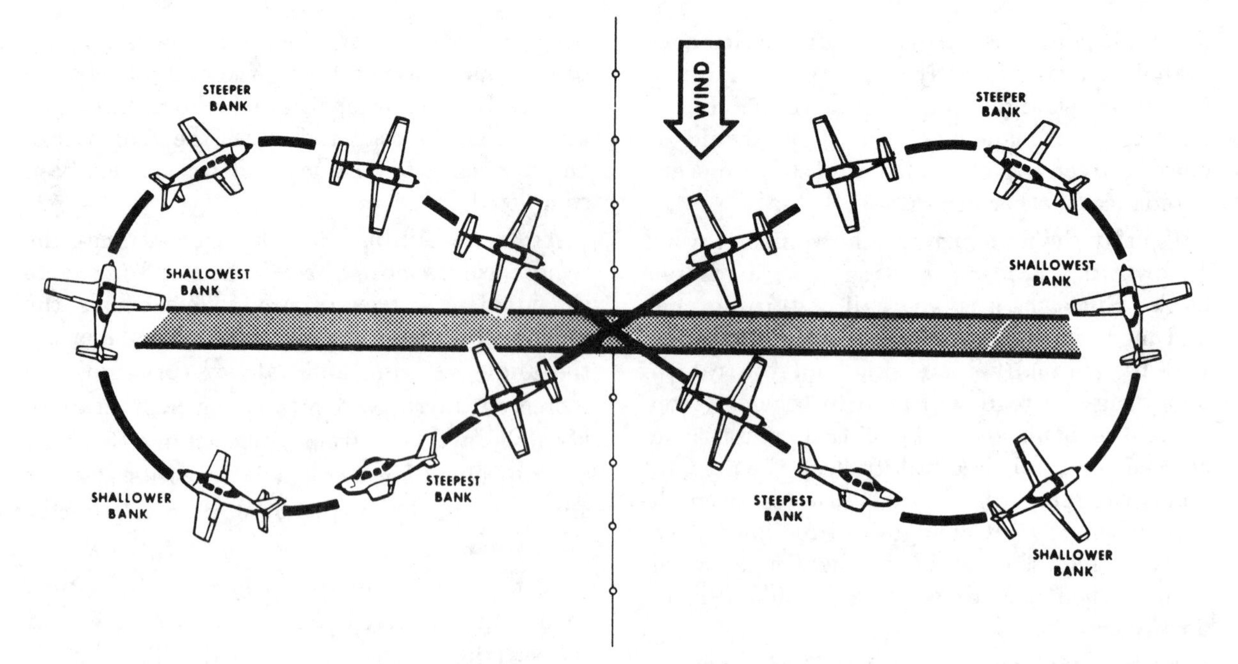

Figure 11–8 Eights Across a Road

track in the form of a figure "8" (Fig. 11–9). The pattern involves flying downwind between the pylons and upwind outside of the pylons. It may include a short period of straight-and-level flight while proceeding diagonally from one pylon to the other.

The pylons selected should be on a line 90° to the direction of the wind and should be in an area away from communities, livestock, or groups of people, to avoid possible annoyance or hazards to others. The area selected should be clear of hazardous obstructions and other air traffic. Throughout the maneuver a constant altitude of at least 500 feet above the ground should be maintained.

The "eight" should be started with the airplane on a downwind heading when passing between the pylons. The distance between the pylons and the wind velocity will determine the initial angle of bank required to maintain a constant radius from the pylons during each turn. The steepest banks will be necessary just after each turn entry and just before the rollout from each turn where the airplane is headed downwind and the groundspeed is greatest; the shallowest banks will be when the airplane is headed directly upwind and the groundspeed is least.

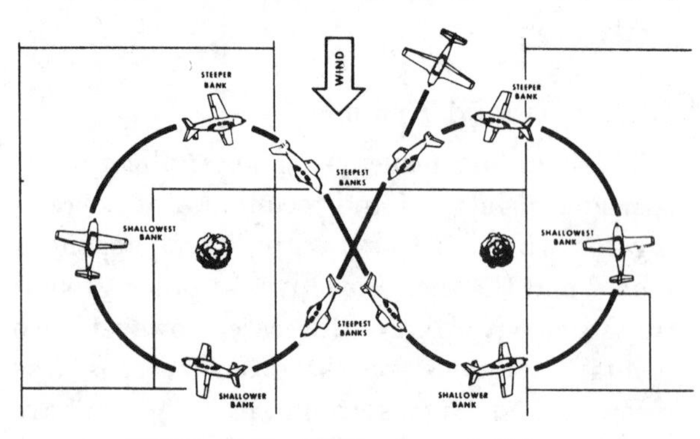

Figure 11–9 Eights Around Pylons

The rate of bank change will depend on the wind velocity—the same as it does in "S turns" and "turns about a point"—and the bank will be changing continuously during the turns. The adjustment of the bank angle should be gradual from the steepest bank to the shallowest bank as the airplane progressively heads into the wind, followed by a gradual increase until the steepest bank is again reached just prior to rollout. If the airplane is to proceed diagonally from one turn to the other, the rollout from each turn must be completed on the proper heading with sufficient crab angle to ensure that after brief straight-and-level flight, the airplane will arrive at the point where a turn of the same radius can be made around the other pylon. The straight-and-level flight segments must, therefore, be tangent to both circular patterns.

Eights-On-Pylons (Pylon Eights)

This training maneuver also involves flying the airplane in circular paths, alternately left and right, in the form of a figure "8" around two selected points or "pylons" on the ground. In this case, however, no attempt is made to maintain a uniform distance from the pylon. Instead, the airplane is flown at such an altitude and airspeed that a line parallel to the airplane's lateral axis, and extending from the pilot's eye *appears to pivot* on each of the pylons (Fig. 11–10). The altitude which is appropriate for the airplane being flown is called the "pivotal altitude" and is governed by the groundspeed. While not truly a ground *track* maneuver as were the preceding maneuvers, the objective is similar—to develop the ability to maneuver the airplane accurately while dividing one's attention between the flightpath and the selected points on the ground.

In explaining the performance of eights-on-pylons the term "wingtip" is frequently considered as being synonymous with the proper reference line, or pivot point on the airplane. This interpretation, however, is not always correct. High-wing, low-wing, swept-wing and taper-wing airplanes, as well as those with tandem or side-by-side seating, will all present different angles from the pilot's eye to the wingtip (Fig. 11–11). Therefore, in the correct performance of eights-on-pylons, as in other maneuvers requiring a lateral reference, the pilot should use a sighting reference line which, from eye level, parallels the lateral axis of the airplane.

The sighting point or line, while not necessarily on the wingtip itself, may be positioned in relation to the wingtip (ahead, behind, above, or below); but even then it will differ for each pilot, and from each seat in the airplane. This is especially true in tandem (fore and aft) seat airplanes. However, in side-by-side type airplanes, there will be very little variation in the sighting lines for different persons if those persons are seated so that the eyes of each are at approximately the same level.

An explanation of the pivotal altitude is also essential. There is a specific altitude at which, when the airplane turns at a given groundspeed, a projection of the sighting reference line to the selected point on the ground will appear to pivot on that point. Since different airplanes fly at different airspeeds, the groundspeed will be different (Fig. 11–12). Hence, each airplane will have its own pivotal altitude. The pivotal altitude does not vary with the angle of bank being used unless the bank is steep enough to affect the groundspeed.

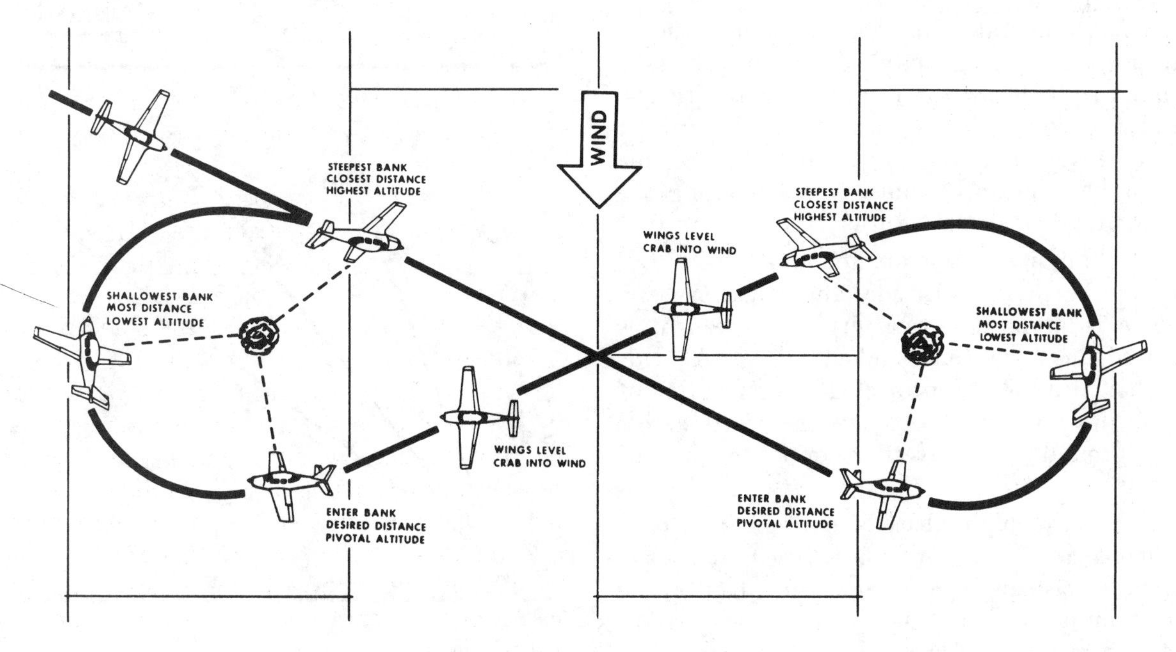

Figure 11–10 Eights-On-Pylons

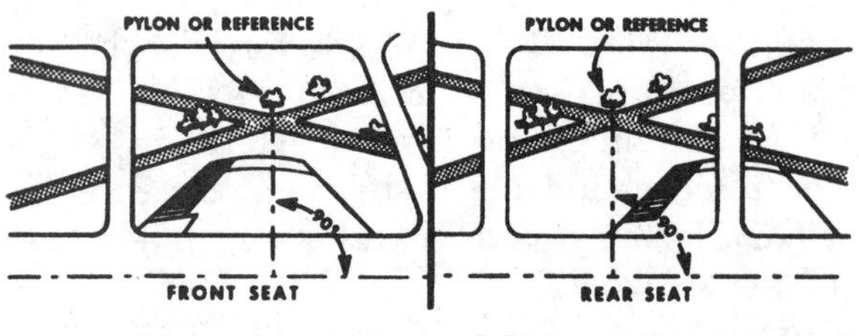

Figure 11–11 Line of Sight to Pylon

The angle of bank is affected by the distance from the pylon. At any altitude above that pivotal altitude, the projected reference line will appear to move rearward in a circular path in relation to the pylon. Conversely, when the airplane is below the pivotal altitude, the projected reference line will appear to move forward in a circular path.

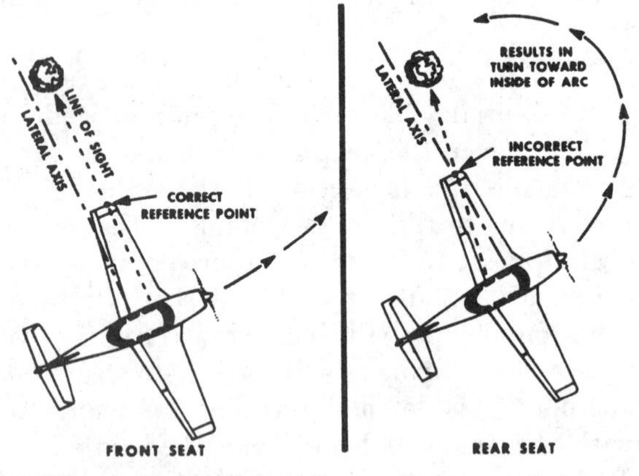

Figure 11–12 Select Proper Reference Point

To demonstrate this, the airplane is flown at normal cruising speed at an altitude estimated to be below the proper pivotal altitude, and then placed in a medium-banked turn. It will be seen that the projected reference line of sight appears to move forward along the ground as the airplane turns.

A climb is then made to an altitude well above the pivotal altitude, and when the airplane is again at normal cruising speed, it is placed in a medium-banked turn. At this higher altitude, the projected reference line of sight now appears to move backward across the ground in a direction opposite that of flight.

After the high altitude extreme has been demonstrated, the power is reduced, and a descent at cruising speed begun in a continuing medium bank around the pylon. The apparent backward travel of the projected reference line with respect to the pylon will slow down as altitude is lost, stop for an instant, then start to reverse itself, and would move forward if the descent were allowed to continue below the pivotal altitude.

The altitude at which the line of sight apparently ceased to move across the ground was the pivotal altitude. If the airplane descended below the pivotal altitude, power should be added to maintain airspeed while altitude is regained to the point at which the projected reference line moves neither backward nor forward but actually pivots on the pylon. In this way the pilot can determine the pivotal altitude of the airplane (Fig. 11–13).

The pivotal altitude is critical and will change with variations in groundspeed. Since the headings throughout the turns continually vary from directly downwind to directly upwind, the groundspeed will constantly change. This will result in the proper pivotal altitude varying slightly throughout the eight. There-

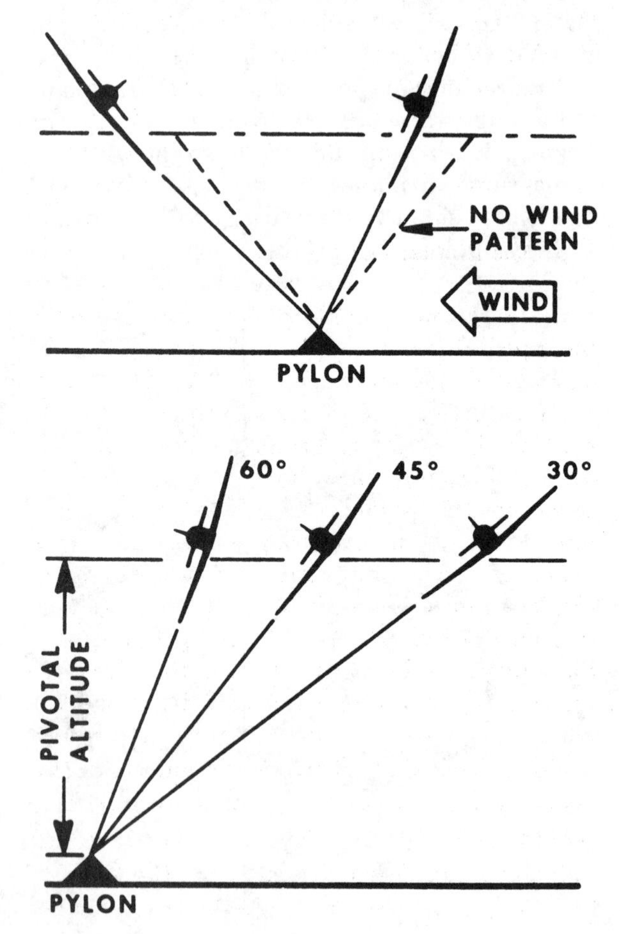

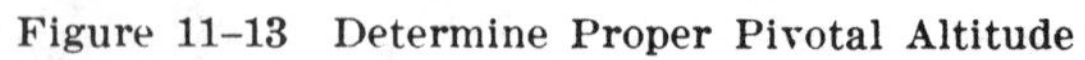

Figure 11–13 Determine Proper Pivotal Altitude

fore, adjustment must be made for this by climbing or descending as necessary to hold the reference line or point on the pylons. This change in altitude will be dependent on how much the wind affects the groundspeed.

Before beginning the maneuver, select two points on the ground along a line which lies 90° to the direction of the wind. The area in which the maneuver is to be performed should be checked for obstructions and other air traffic, and it should be located where a disturbance to groups of people, livestock, or communities will not result.

The selection of proper pylons is of importance to good eight-on-pylons. They should be sufficiently prominent to be readily seen by the pilot when completing the turn around one pylon and heading for the next, and should be adequately spaced to provide time for planning the turns and yet not cause unnecessary straight-and-level flight between the pylons. Approximately 3 to 5 seconds of straight-and-level flight should be sufficient for checking the area properly before entering the next turn.

For uniformity, the eight is usually begun by flying diagonally crosswind between the pylons to a point downwind from the first pylon so that the first turn can be made into the wind.

As the airplane approaches a position where the pylon appears to be just ahead of the wingtip, the turn should be started by lowering the upwind wing to place the pilot's line of sight reference on the pylon. As the turn is continued, the line of sight reference can be held on the pylon by gradually increasing the bank. The reference line should appear to pivot on the pylon. As the airplane heads into the wind, the groundspeed decreases; consequently the pivotal altitude is lower and the airplane must descend to hold the reference line on the pylon. As the turn progresses on the upwind side of the pylon, the wind becomes more of a crosswind and drifts the airplane closer to the pylon. Since a constant distance from the pylon is *not* required on this maneuver, no correction to counteract drifting should be applied. Therefore, with the airplane drifting closer to the pylon, the angle of bank must be increased to hold the reference line on the pylon.

If the reference line appears to move ahead of the pylon, the pilot should increase altitude. If the reference line appears to move behind the pylon, the pilot should decrease altitude. Varying rudder pressure to yaw the airplane and force the wing and reference line forward or backward to the pylon is a dangerous technique and must *not* be attempted.

As the airplane turns toward a downwind heading, the rollout from the turn should be started to allow the airplane to proceed diagonally to a point on the downwind side of the second pylon. The rollout must be completed in the proper crab angle to correct for wind drift, so that the airplane will arrive at a point downwind from the second pylon the same distance it was from the first pylon at the beginning of the maneuver.

Upon reaching that point, a turn is started in the opposite direction by lowering the upwind wing to again place the pilot's line of sight reference on the pylon. The turn is then continued just as in the turn around the first pylon but in the opposite direction.

The most common error in attempting to hold a pylon is incorrect use of the rudder. When the projection of the reference line moves forward with respect to the pylon, many pilots will tend to press the inside rudder to yaw the wing backward; when the reference line moves behind the pylon they will press the outside rudder to yaw the wing forward. The rudder is to be used only as a coordination control.

The eights-on-pylons is an advanced training maneuver that provides practice in developing coordination skills while the pilot's attention is directed at maintaining a pivotal position on a selected pylon.

Recognition of Stalls

A stall occurs when the smooth airflow over the airplane's wing is disrupted, and the lift degenerates rapidly (Fig. 11-14). This can occur AT ANY AIRSPEED, IN ANY ATTITUDE, WITH ANY POWER SETTING.

Figure 11-14 Loss of Lift Results in a Stall

The practice of stall recovery and the development of awareness of imminent stalls are of primary importance in pilot training. The objectives in performing intentional stalls are to familiarize the pilot with the conditions that produce stalls, to assist in recognizing an approach stall, and to develop the habit of taking prompt preventive or corrective action.

To become proficient, pilots must recognize the flight conditions that are conducive to stalls and know how to apply the necessary corrective action. They should learn to recognize an approaching stall by sight, sound, and feel. The following cues may be useful in recognizing the approaching stall:

1. Vision is useful in detecting a stall condition by noting the attitude of the airplane. This sense can be fully relied on *only* when the stall is the result of an unusual attitude of the airplane. However, since the airplane can also be stalled from a normal attitude, vision in this instance would be of little help in detecting the approaching stall.
2. Hearing is also helpful in sensing a stall condition, since the tone level and intensity of sounds incident to flight decrease as the airspeed decreases. In the case of fixed-pitch propeller airplanes in a power-on condition, a change in sound due to loss of RPM is particularly noticeable. The lessening of the noise made by the air flowing along the airplane structure as airspeed decreases is also quite noticeable, and when the stall is almost complete, vibration and its incident noises often increase greatly.
3. Kinethesia, or the sensing of changes in direction or speed of motion, is probably the most important and the best indicator to the trained and experienced pilot. If this sensitivity is properly developed, it will warn of a decrease in speed or the beginning of a settling or "mushing" of the airplane.
4. The feeling of control pressures is also very important. As speed is reduced, the "live" resistance to pressures on the controls becomes progressively less. Pressures exerted on the controls tend to become movements of the control surfaces, and the lag between those movements and the response of the airplane becomes greater, until in a complete stall all controls can be moved with almost no resistance, and with little immediate effect on the airplane.

Intentional stalls should be performed at an altitude that will provide adequate height above the ground for recovery and return to normal level flight. Though it depends on the degree to which a stall has progressed, most stalls require some loss of altitude during recovery. The longer it takes to sense the approaching stall, the more complete the stall is likely to become, and the greater the loss of altitude to be expected.

Several types of stall warning indicators have been developed that warn the pilot of an approaching stall. The use of such indicators is valuable and desirable, but the reason for practicing stalls is to learn to *recognize* stalls without the benefit of warning devices.

Fundamentals of Stall Recovery

During the practice of intentional stalls the real objective is not to *learn how to stall* an airplane but to *learn how to recognize an incipient stall and take prompt corrective action* (Fig. 11–15). Though the recovery actions must be taken in a coordinated manner, they are broken down into three steps here for explanation purposes.

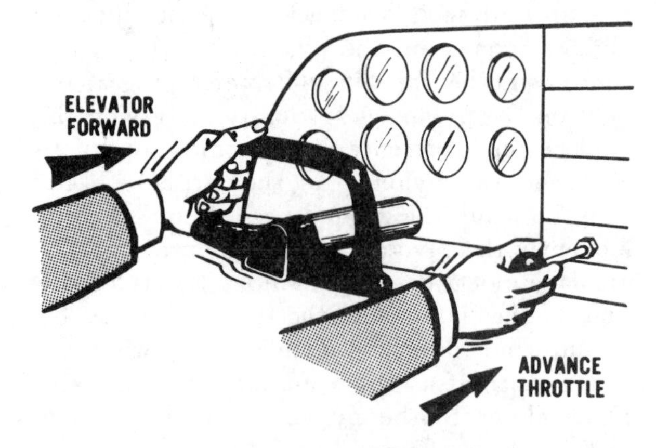

Figure 11–15 Stall Recovery Procedure

First, at the indication of a stall, the pitch attitude and angle of attack must be decreased positively and immediately. Since the basic cause of a stall is always an excessive angle of attack, the cause must first be eliminated by releasing the back elevator pressure that was necessary to attain that angle of attack or by moving the elevator control forward. This lowers the nose and returns the wing to an effective angle of attack. The amount of elevator control pressure or movement used depends on the design of the airplane, the severity of the stall, and the proximity of the ground. In some airplanes a moderate movement of the elevator control—perhaps slightly forward of neutral—is enough, while in others a forcible push to the full forward position may be required. However, an excessive negative load thrown on the wings by excessive forward movement of the elevator may impede, rather than hasten, the stall recovery. The object is to reduce the angle of attack but only enough to allow the wing to regain lift.

Second, the maximum allowable power should be applied to increase the airplane's speed and assist in reducing the wing's angle of attack. Generally, the throttle should be promptly but smoothly advanced to the maximum allowable position.

Although stall recoveries should be practiced without as well as with the use of power, in most actual stalls the application of more power, if available, is an integral part of the stall recovery. Usually, the greater the power applied, the less the loss of altitude.

Maximum allowable power applied at the instant of a stall will usually not cause overspeeding of an engine equipped with a fixed-pitch propeller, due to the heavy air load imposed on the propeller at slow airspeeds. It will be necessary, however, to reduce the power as airspeed is gained after the stall recovery so the airspeed will not become excessive. During practice stalls the tachometer indication should never be allowed to exceed the red line (maximum allowable RPM) marked on the instrument.

Third, straight-and-level flight should be regained with coordinated use of all controls.

Practice in both power-on and power-off stalls is important because it simulates stall conditions that could occur during normal flight maneuvers. For example, the power-on stalls are practiced to show what could happen if the airplane were climbing at an excessively nose-high attitude immediately after takeoff or during a climbing turn. The power-off turning stalls are practiced to show what could happen if the controls are improperly used during a turn from the base leg to the final approach. The power-off straight-ahead stall simulates the attitude and flight characteristics of a particular airplane during the final approach and landing.

Usually the first few practices should include only approaches to stalls, with recovery initiated as soon as the first buffeting or partial loss of control is noted. In this way the pilot can become familiar with the indications of an imminent stall without actually stalling the airplane. Recovery should be practiced first without the addition of power, and then with the addition of power to determine how effective power will be in executing a safe recovery.

Stall accidents usually result from an inadvertent stall at a low altitude in which a recovery was not accomplished prior to contact with the surface. As a preventive measure, stalls should be practiced at a safe altitude using a recovery technique that will result in a minimum loss of altitude. To recover with a minimum loss of altitude requires an application of power, reduction in the angle of attack (lowering the airplane's pitch attitude), and termination of the sink without entering another stall.

Use of Ailerons in Stall Recoveries

Different types of airplanes have different stall characteristics. Most modern airplanes are designed so that the wings will stall progressively outward from the wing roots to the wingtips. This is the result of designing the wings in a manner that the wingtips have less *angle of incidence* than do the wing roots. Such a design feature causes the tips of the wings to have a smaller angle of attack than the wing roots during flight (Fig. 11–16).

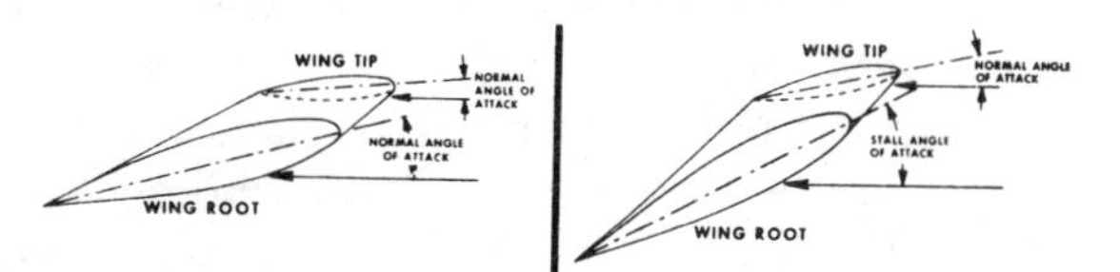

Figure 11–16 Angles of Attack at Wing Root and Wingtip

Since a stall is caused by exceeding the critical angle of attack, the wing roots of an airplane will exceed the critical angle before the wingtips and, therefore, the roots will stall first. The wings are designed in this manner so that aileron control will be available at high angles of attack (slow airspeed) and give the airplane more stable stalling characteristics (Fig. 11–17).

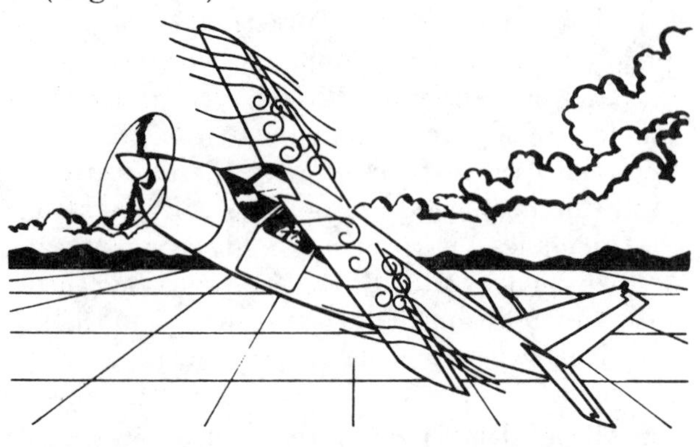

Figure 11–17 Wing Root Stalls Before Wingtip

When the airplane is approaching a completely stalled condition, the wingtips continue to provide some degree of lift and the ailerons still have some control effect. During recovery from a stall, the return of lift begins at the tips and progresses toward the roots. Thus, the ailerons can be used to level the wings.

Using the ailerons requires finesse to avoid an aggravated stall condition. For example, if the right wing dropped during the stall and *excessive* aileron control were applied to the left to raise the wing, the aileron deflected downward (right wing) would produce an even greater angle of attack (and drag), and possibly a more complete stall at the tip as the critical angle of attack is exceeded. The increase in drag created by the high angle of attack on that wing might cause the airplane to yaw in that direction. This adverse yaw could result in a spin unless directional control were maintained by rudder, and/or the aileron control sufficiently reduced.

Even though excessive aileron pressure may have been applied, a spin will not occur if directional (yaw) control is maintained by timely application of coordinated rudder pressure. Therefore, it is important that the rudder be used properly during both the entry and the recovery from a stall. Thus, the primary use of the rudder in stall recoveries is to counteract any tendency of the airplane to yaw, or slip. The correct recovery technique then, would be to decrease the pitch attitude by applying forward elevator pressure to break the stall, advancing the throttle to increase airspeed, and simultaneously maintaining direction with *coordinated* use of aileron and rudder.

Stall Characteristics

Because of engineering design variations, the stall characteristics for all airplanes cannot be specifically described; however, the similarities found in light general aviation training-type airplanes are noteworthy enough to be considered. It will be noted that the power-on and power-off stall warning indications will be different. The power-off stall will have less noticeable clues (buffeting, shaking) than the power-on stall. In the power-off stall the predominant clue can be the elevator control position (full up elevator against the stops) and high sink rate. When performing the power-on stall the buffeting will likely be the predominant clue that provides a positive indication of the stall. For the purpose of airplane certification, the stall warning may be furnished either through the inherent aerodynamic qualities of the airplane or by a stall warning device that will give a clear distinguishable indication of the stall. Most modern airplanes are equipped with a stall warning device.

The factors that affect the stalling characteristics of the airplane are: balance (load distribution), bank (wing loading), pitch attitude (critical angle of attack), coordination (control movement), drag (gear or flaps), and power. The pilot should learn the effects of each on the stall characteristics of the airplane being flown and what should be done to effect the proper correction. It should be reemphasized here that a stall can occur at *any* airspeed, in *any* attitude, or at *any* power setting, depending on the total number of factors affecting the particular airplane.

A number of factors may be induced as the result of other factors. For example, when the airplane is in a nose-high turning attitude, the angle of bank has a tendency to increase. This occurs because with the airspeed decreasing, the airplane begins flying in a smaller and smaller arc. Since the outer wing is moving in a larger radius and thus traveling faster than the inner wing, it has more lift and causes an overbanking tendency. At the same time, because of the decreasing airspeed and decreasing lift on both wings, the pitch attitude tends to lower. In addition, since the airspeed is decreasing while the power setting remains constant, the effect of torque becomes more prominent, causing the airplane to yaw.

During the practice of nose-high turning stalls, to compensate for these factors and to maintain a constant flight attitude until the stall occurs, aileron pressure must be continually adjusted to keep the bank attitude constant. At the same time, back elevator pressure must be continually increased to maintain the pitch attitude, as well as right rudder pressure increased to prevent adverse yaw from changing the turn rate. If the bank is allowed to become too steep, the vertical component of lift decreases and makes it even more difficult to maintain a constant-pitch attitude.

Whenever practicing turning stalls, a constant pitch and bank attitude should be maintained until the stall occurs. Whatever control pressures are necessary should be applied even though the controls appear to be crossed (aileron pressure in one direction, rudder pressure in the opposite direction). During the entry to a power-on turning stall to the right in particular, the controls will be crossed to some extent. This is due to right rudder pressure being used to overcome torque and left aileron pressure being used to prevent the bank from increasing.

Full Stalls—Power-Off

The practice of power-off stalls is usually performed with normal landing approach conditions in simulation of an accidental stall occurring during landing approaches. Airplanes equipped with flaps and/or retractable landing gear should, therefore, be in the landing configuration. Airspeed in excess of the normal approach speed should not be carried into a stall entry since it could result in an abnormally nose-high attitude. *Before executing these practice stalls the pilot must be sure the area is clear of other air traffic.*

After extending the landing gear, applying carburetor heat (if applicable), and retarding the throttle to idling (or normal approach power), the airplane should be held at a constant altitude in level flight until the airspeed decelerates to that of a normal approach, and then smoothly nosed down into the normal approach attitude to maintain that airspeed. Wing flaps should then be extended and pitch attitude adjusted to maintain the airspeed.

When the approach attitude and airspeed have stabilized, the airplane's nose should be smoothly raised to an attitude which will induce a stall (Fig. 11–18). Directional control should be maintained with the rudder, the wings held level by use of the ailerons, and a constant pitch attitude maintained with the elevator until the full stall occurs. The full stall will be evidenced by such clues as full up-elevator, high sink rate, uncontrollable nose-down pitching and possible buffeting.

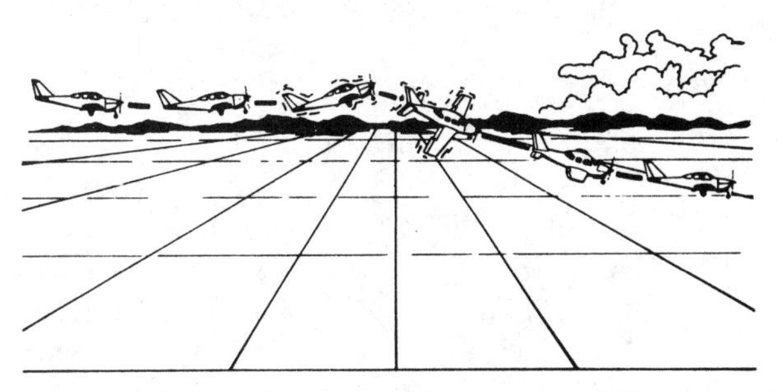

Figure 11–18 Full Stall Power Off

Recovery from the stall should be accomplished by reducing the angle of attack, releasing back elevator pressure, and advancing the throttle to maximum allowable power. Right rudder pressure is necessary to overcome the engine torque effects as power is advanced and the nose is being lowered.

The nose should be lowered as necessary to regain flying speed. Then the airplane should be returned to the normal straight-and-level flight attitude. When in normal level flight, the throttle should be returned to cruise power setting, and the flaps and landing gear retracted. After recovery is complete, a climb or go around procedure should be initiated as the situation dictates.

Recovery from power-off stalls should also be practiced from moderately banked turns to simulate an accidental stall during a turn from base leg to final approach. During the practice of these stalls, care should be taken that the turn continues at a uniform rate until the complete stall occurs. If the power-off turn is not properly coordinated while approaching the stall, wallowing may result when the stall occurs or, if the airplane is in a slip, the outer wing may stall first and whip downward abruptly. This does not affect the recovery procedure in any way; the stall must first be broken, the heading maintained, and the wings leveled by coordinated use of the controls. In the practice of turning stalls no attempt should be made to stall the airplane on a predetermined heading. However, to simulate a turn from base to final approach, the stall normally should be made to occur within a heading change of approximately 90°.

After the full stall occurs, the recovery should be made straight ahead with minimum loss of altitude, and is accomplished in accordance with the recovery procedure discussed earlier.

Practice recoveries from power-off type stalls should be accomplished both with, and without, the addition of power, and may be effected either just after the stall occurs, or after the nose has pitched down through the level flight attitude. Performance is unsatisfactory if a secondary stall occurs or if the pilot fails to take proper action to avoid excessive airspeed, excessive loss of altitude, or a spin.

Full Stalls—Power-On

Power-on stall recoveries are practiced from straight climbs, and climbing turns with 15° to 20° banks, to simulate an accidental stall occurring during takeoffs and departure climbs. Therefore, airplanes equipped with flaps and/or retractable landing gear normally should be in the takeoff configuration. However, power-on stalls should also be practiced with the airplane in a clean configuration (flaps and/or gear retracted) as in departure and normal climbs (Fig. 11–19).

After establishing the takeoff or departure configuration, the airplane should be slowed to the normal lift-off speed while clearing the area for other air traffic. When the desired speed is attained, the power should be set at takeoff power for the takeoff stall or the recommended climb power for the departure stall while establishing a climb attitude. The purpose of reducing the speed to lift-off speed before the throttle is advanced to the recommended setting is to avoid an excessively steep nose-up attitude for a long period before the airplane stalls.

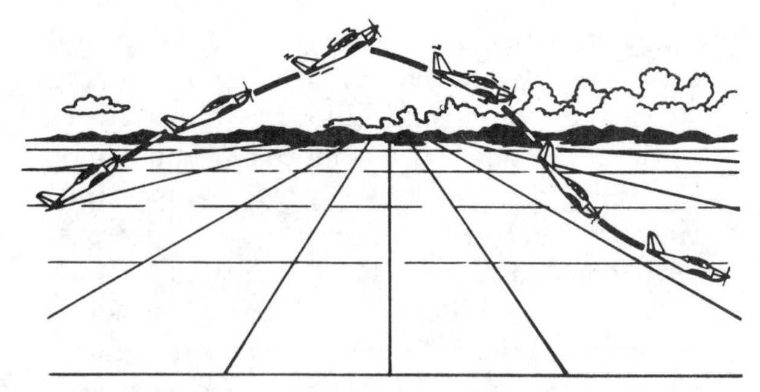

Figure 11–19 Full Stall Power On

After the climb attitude is established, the nose is then brought smoothly upward to an attitude obviously impossible for the airplane to maintain and is held at that attitude until the full stall occurs. In most airplanes it will be found that after attaining the stalling attitude, the elevator control must be moved progressively further back as the airspeed decreases until, at the full stall, it will have reached its limit and cannot be moved back farther.

Recovery from the stall should be accomplished by immediately reducing the angle of attack by positively releasing back elevator pressure and smoothly advancing the throttle to maximum allowable power. In this case, however, since the throttle is already at the climb power setting, the addition of power will be relatively slight.

The nose should be lowered as necessary to regain flying speed. Then the airplane should be returned to the normal straight-and-level flight attitude. When in normal level flight, the throttle should be returned to cruise power setting.

The pilot must recognize instantly when the stall has occurred and take prompt action to prevent a prolonged stalled condition. Performance is unsatisfactory if a secondary stall occurs, or if the pilot fails to take proper action to avoid excessive airspeed, excessive loss of altitude, or a spin.

Secondary Stall

This stall is called a secondary stall since it may occur after a recovery from a preceding primary stall. It is caused by attempting to hasten the completion of a stall recovery before the airplane has regained sufficient flying speed. When this stall occurs, the back elevator pressure should again be released just as in a normal stall recovery. When sufficient airspeed has been regained, the airplane can then be returned to straight-and-level flight.

This stall usually occurs when the pilot becomes too anxious to return to straight-and-level flight after a stall or spin recovery (Fig. 11–20).

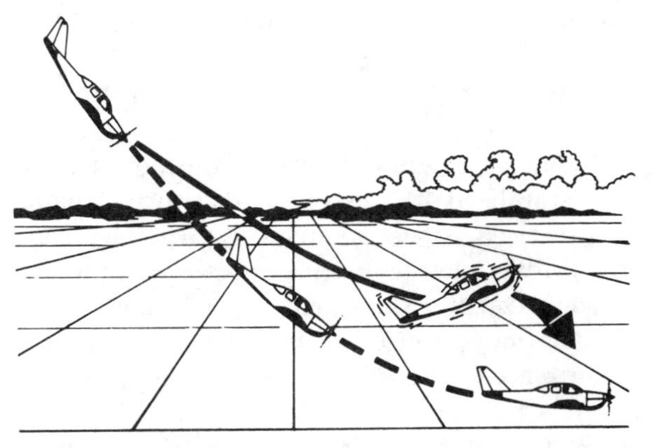

Figure 11–20 Recovering too Abruptly Causes Secondary Stall

Imminent Stalls—Power-On or Power-Off

An imminent stall is one in which the airplane is approaching a stall but is not allowed to completely stall. This stall maneuver is primarily for practice in retaining (or regaining) full control of the airplane immediately upon recognizing that it is almost in a full stall or that a full stall is likely to occur if timely preventive action is not taken (Fig. 11–21).

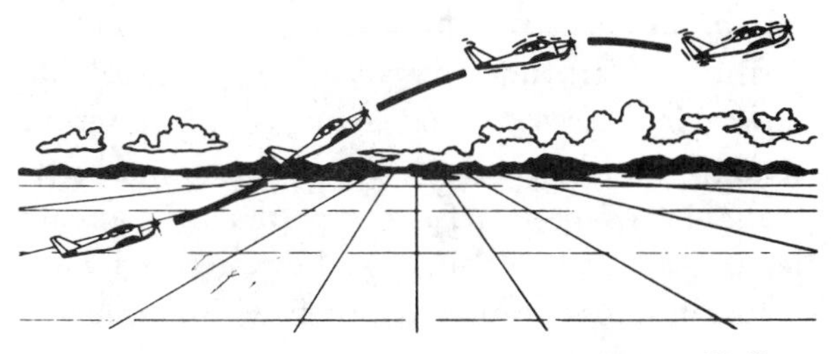

Figure 11–21 Imminent or Approach to a Stall

The practice of these stalls is of particular value in developing the pilot's sense of feel for executing maneuvers in which maximum airplane performance is required. These maneuvers, therefore, require flight with the airplane just on the verge of a stall, and recovery initiated before a full stall occurs. As in all maneuvers that involve significant changes in altitude or direction, the pilot must ensure that the area is clear of other air traffic before executing the maneuver.

These stalls may be entered and performed in the attitudes and with the same configuration as any of the basic full stalls or other maneuvers described in this chapter. However, instead of allowing a complete stall, when the *first* buffeting or decay of control effectiveness is noted, the angle of attack must be reduced immediately by releasing the elevator back pressure and applying whatever additional power is necessary. Since the airplane will not have been completely stalled, the pitch attitude needs to be decreased only to a point where minimum controllable airspeed is attained or until adequate control effectiveness is regained.

The pilot must promptly recognize the indications of an imminent stall and take timely, positive control action to prevent a full stall. Performance is unsatisfactory if a full stall occurs, if an excessively low pitch attitude is attained, or if the pilot fails to take timely action to avoid excessive airspeed, excessive loss of altitude, or a spin.

Maneuvering at Minimum Controllable Airspeed

This maneuver demonstrates the flight characteristics and degree of controllability of an airplane at its minimum flying speed. The ability to determine the characteristic control responses of any airplane is of great importance to pilots. They must develop this awareness in order to avoid stalls in any airplane they may fly at the slower airspeeds which are characteristic of takeoffs, climbs, and landing approaches.

Maintaining sufficient lift and adequate control of an airplane during maximum performance maneuvers depends upon a certain minimum airspeed. By definition, the term "flight at minimum controllable airspeed" means a speed at which any further increase in angle of attack or load factor, or reduction in power will cause an immediate stall. This critical airspeed will depend upon various circumstances, such as the gross weight and CG location of the airplane, maneuvering load imposed by turns and pullups, and the existing density altitude.

The objective of maneuvering at minimum controllable airspeed then, is to develop the pilot's sense of feel and ability to use the controls correctly, and to improve proficiency in performing maneuvers in which very low airspeeds are required.

Practice in flight at minimum control speeds should cover two distinct flight situations: (1) establishing and maintaining the airspeed appropriate for landing approaches and go-arounds in the airplane used; and (2) turning flight at the slowest airspeed at which the particular airplane is capable of continued controlled flight without stalling.

Maneuvering at minimum control speed should be performed using both instrument indications and outside visual reference. It is important that pilots form the habit of frequent reference to the flight instruments for airspeed, altitude, and attitude indications while flying at very slow speeds.

To begin the maneuver the throttle is gradually reduced from cruising position. While the airplane is losing airspeed, the position of the nose in relation to the horizon should be noted and should be raised as necessary to maintain altitude (Fig. 11–22). When the airspeed reaches the maximum allowable for landing gear operation, the landing gear (if equipped with retractable gear) should be extended and all gear-down checks performed. As the airspeed reaches the maximum allowable speed for flap operation, full flaps should be lowered and the pitch attitude adjusted to maintain altitude. Additional power will be required as the speed further decreases to maintain the airspeed just above a stall.

During these changing flight conditions it is important to retrim the airplane as often as necessary to compensate for changes in control pressures. If too much speed is lost, or too little power is used, further back pressure on the elevator control may result in a loss of altitude or a stall. When the desired pitch attitude and minimum control airspeed have been established, it is important to continually cross-check the attitude indicator, altimeter and airspeed indicator, as well as outside references to ensure that accurate control is being maintained.

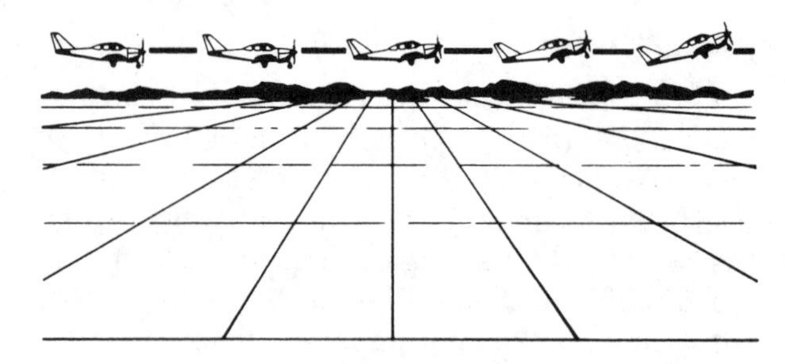

Figure 11–22 Minimum Airspeed at which Control is Maintained

When the attitude, airspeed, and power have been stabilized in straight flight, turns should be practiced to determine the airplane's controllability characteristics at this minimum speed. During the turns, power and pitch attitude may need to be increased to maintain the airspeed and altitude. If an excessively steep turn is made, the loss of vertical lift may result in a stall. A stall may also occur as a result of abrupt or rough control movements when flying at this critical airspeed.

Abruptly raising the flaps while at minimum controllable airspeed will also result in lift suddenly being lost, causing the airplane to lose altitude or perhaps stall.

Once flight at minimum controllable airspeed is set up properly for level flight, a descent or climb at minimum controllable airspeed can be established by adjusting the power to maintain the desired airspeed, and simultaneously adjusting the pitch attitude as necessary to establish the desired rate of descent or climb. A positive climb, however, may not be possible at altitude due to a lack of available power in excess of that required to maintain straight-and-level flight at the minimum controllable airspeed. In some airplanes an attempt to climb at such a slow airspeed may result in a loss of altitude, even with maximum power applied.

Accelerated Maneuver Stalls

Though the stalls just discussed normally occur at a specific airspeed, the pilot must understand thoroughly that all stalls result solely from attempts to fly at excessively high angles of attack. It will be recalled from previous discussions that during flight the angle of attack of an airplane wing is determined by a number of factors, the most important of which are: (1) the airspeed, (2) the gross weight of the airplane, and (3) the load factors imposed by maneuvering.

At the same gross weight, airplane configuration, and power setting, a given airplane will consistently stall at the same indicated airspeed *if no acceleration is involved.* The airplane will, however, stall at a higher indicated airspeed when excessive maneuvering loads are imposed by steep turns, pullups, or other abrupt changes in its flightpath. Stalls entered from such flight situations are called "accelerated maneuver stalls," a term which has no reference to the airspeeds involved.

Stalls which result from abrupt maneuvers tend to be more rapid, or severe, than the unaccelerated stalls, and because they occur at higher-than-normal airspeeds they may be unexpected by a pilot who has not experienced and understood them. Failure to take immediate steps toward recovery when an accelerated stall occurs may result in a complete loss of flight control, notably power-on spins.

It must be emphasized here that FAR Part 23, "Airworthiness Standards: Normal, Utility, and Acrobatic Airplanes," under which light planes are type-certificated, may *prohibit* the performance of these maneuvers (Fig. 11–23). For airplanes certificated in the Normal category, acrobatic maneuvers, including spins, *are not authorized.* Acrobatic flight means an intentional maneuver involving an abrupt change in an aircraft's attitude, an abnormal attitude, or abnormal acceleration not necessary for normal flight. For Utility category airplanes, limited acrobatics are authorized; however, information about the authorized maneuvers shown in the type certification flight tests must be available to the pilot, together with recommended entry speeds. *No other accelerated maneuver is authorized.*

UNITED STATES OF AMERICA
DEPARTMENT OF TRANSPORTATION—FEDERAL AVIATION ADMINISTRATION
STANDARD AIRWORTHINESS CERTIFICATE

1. NATIONALITY AND REGISTRATION MARKS	2. MANUFACTURER AND MODEL	3. AIRCRAFT SERIAL NUMBER	4. CATEGORY
N1917Z	SPEEDWING 440 B	00-000	Normal

5. AUTHORITY AND BASIS FOR ISSUANCE
This airworthiness certificate is issued pursuant to the Federal Aviation Act of 1958 and certifies that, as of the date of issuance, the aircraft to which issued has been inspected and found to conform to the type certificate therefor, to be in condition for safe operation, and has been shown to meet the requirements of the applicable comprehensive and detailed airworthiness code as provided by Annex 8 to the Convention on International Civil Aviation, except as noted herein.
Exceptions: None

6. TERMS AND CONDITIONS
Unless sooner surrendered, suspended, revoked, or a termination date is otherwise established by the Administrator, this airworthiness certificate is effective as long as the maintenance, preventive maintenance, and alterations are performed in accordance with Parts 21, 43, and 91 of the Federal Aviation Regulations, as appropriate, and the aircraft is registered in the United States.

DESIGNATION NUMBER: 8-3-97

UNITED STATES OF AMERICA
DEPARTMENT OF TRANSPORTATION—FEDERAL AVIATION ADMINISTRATION
STANDARD AIRWORTHINESS CERTIFICATE

1. NATIONALITY AND REGISTRATION MARKS	2. MANUFACTURER AND MODEL	3. AIRCRAFT SERIAL NUMBER	4. CATEGORY
N3488C	THRUSTER-A21	00-000	Utility

5. AUTHORITY AND BASIS FOR ISSUANCE
This airworthiness certificate is issued pursuant to the Federal Aviation Act of 1958 and certifies that, as of the date of issuance, the aircraft to which issued has been inspected and found to conform to the type certificate therefor, to be in condition for safe operation, and has been shown to meet the requirements of the applicable comprehensive and detailed airworthiness code as provided by Annex 8 to the Convention on International Civil Aviation, except as noted herein.
Exceptions: None

6. TERMS AND CONDITIONS
Unless sooner surrendered, suspended, revoked, or a termination date is otherwise established by the Administrator, this airworthiness certificate is effective as long as the maintenance, preventive maintenance, and alterations are performed in accordance with Parts 21, 43, and 91 of the Federal Aviation Regulations, as appropriate, and the aircraft is registered in the United States.

DATE OF ISSUANCE	FAA REPRESENTATIVE	DESIGNATION NUMBER
12-12-62	I. M. INSPECTOR	3-0-00

Any alteration, reproduction, or misuse of this certificate may be punishable by a fine not exceeding $1,000, or imprisonment not exceeding 3 years, or both. THIS CERTIFICATE MUST BE DISPLAYED IN THE AIRCRAFT IN ACCORDANCE WITH APPLICABLE FEDERAL AVIATION REGULATIONS.

FAA Form 8100-2 (7-67) FORMERLY FAA FORM 1362

Figure 11–23. Normal and Utility Airworthiness Certificate

Accelerated maneuver stalls, therefore, should *not* be performed in any airplane which is *prohibited* from such maneuvers by its type-certification restrictions. If they are permitted, they should be performed with a bank of approximately 45° and in no case at a speed greater than the airplane manufacturer's recommended airspeeds, or the design maneuvering speed specified for the airplane. The design maneuvering speed is the *maximum* speed at which the airplane can be stalled, or the controls deflected fully, without exceeding the airplane's limit load factor. At or below this speed the airplane will usually stall before the limit load factor can be exceeded. Those speeds must not be exceeded because of the extremely high structural loads which are imposed on the airplane, especially if there is turbulence. In most cases these stalls should be performed at no more than 1.2 times the normal stall speed.

The objective of demonstrating accelerated stalls is not to develop competency in setting up the stall, but rather to learn how they may occur and to develop the ability to recognize such stalls immediately, and to take prompt, effective recovery action. It is important that recoveries be made at the first indication of an imminent stall, or immediately after the stall has fully developed; a prolonged stall condition should never be allowed.

An airplane will stall during a coordinated steep turn exactly as it does from straight flight, except that the pitching and rolling actions tend to be more sudden. If the airplane is slipping toward the inside of the turn at the time the stall occurs, it tends to roll rapidly toward the outside of the turn as the nose pitches down because the outside wing stalls before the inside wing. If the airplane is skidding toward the outside of the turn, it will have a tendency to roll to the inside of the turn because the inside wing stalls first. If, however, the coordination of the turn at the time of the stall is accurate, the airplane's nose will pitch away from the pilot just as it does in a straight flight stall, since both wings stall simultaneously.

The accelerated stall demonstrations that follow are entered by establishing the desired flight attitude, then smoothly, firmly, and progressively increasing the angle of attack until a stall occurs. Because of the rapidly changing flight attitude, sudden stall entry, and possible loss of altitude, it is extremely vital that the area be clear of other aircraft and the entry altitude be adequate for safe recovery.

Excessive Back-Pressure (Accelerated) Stall

This demonstration stall, as in all stalls, is accomplished by exerting excessive back elevator pressure. Most frequently it would occur during improperly executed steep turns, stall and spin recoveries, and pullouts from steep dives. The objectives are to determine the stall characteristics of the airplane and develop the ability to instinctively recover when a stall occurs at other than normal stall speed or flight attitudes. A high-speed stall, although usually demonstrated in steep turns, may actually be encountered any time excessive back pressure is applied and/or the angle of attack is increased too rapidly. This stall should never be practiced with wing flaps in the extended position due to the lower "G" load limitations in that configuration.

From straight-and-level flight at maneuvering speed or less, the airplane should be rolled into a steep level flight turn and back pressure gradually applied on the elevator control. After the turn and bank are established, back pressure on the elevator control should be smoothly and steadily increased. The resulting centrifugal force will push the pilot's body down in the seat, increase the wing loading, and decrease the airspeed somewhat. After the airspeed reaches the design maneuvering speed or within 20 knots above the unaccelerated stall speed, back elevator pressure should be firmly increased until a definite stall occurs (Fig. 11–24). These speed restrictions must be observed to prevent exceeding the load limit of the airplane.

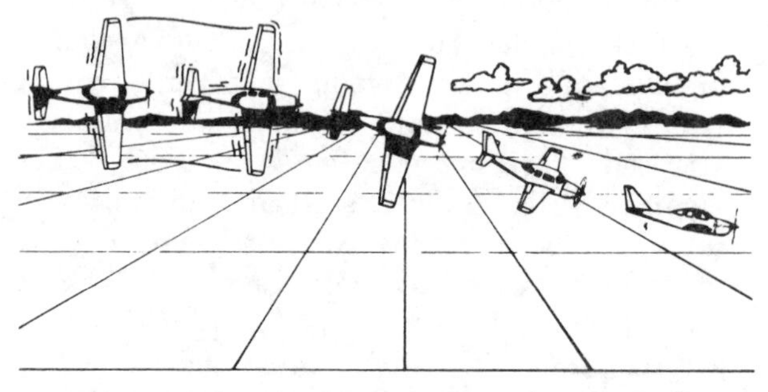

Figure 11–24 Accelerated (High Speed) Stall

When the airplane stalls, recovery should be made promptly, releasing sufficient back pressure and increasing power to reduce the angle of attack. If an uncoordinated turn is made, one wing may tend to drop suddenly, causing the airplane to roll in that direction. If this occurs, power must be added, the excessive back pressure released to break the stall *first*, then the airplane returned to straight-and-level flight with coordinated control pressure.

The pilot should recognize when the stall is imminent and take prompt action to prevent a completely stalled condition. It is imperative that a prolonged stall, excessive airspeed, excessive loss of altitude, or a spin be avoided.

Crossed Control Stall

The objective of this demonstration maneuver is to show the effect of improper control technique and to emphasize the importance of using coordinated control pressures whenever making turns. This type stall occurs with the controls "crossed"—that is, aileron pressure applied in one direction and rudder pressure in the opposite direction.

In addition, when excessive back elevator pressure is applied a "cross-control stall" may result (Fig. 11–25).

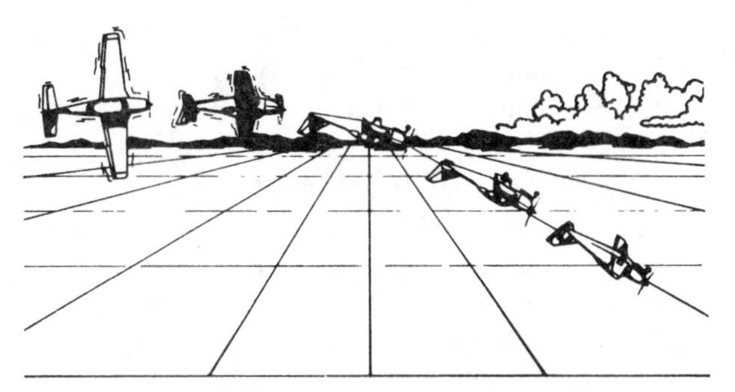

Figure 11-25 Crossed Controls Stall

This is a stall that is most apt to occur during a poorly planned and executed base-to-final approach turn and often is the result of overshooting the centerline of the runway during that turn. Normally, the proper action to correct for overshooting the runway is to increase the rate of turn by using coordinated aileron and rudder. At the relatively low altitude of a base-to-final approach turn, however, improperly trained pilots may be apprehensive of steepening the bank to increase the rate of turn. So, rather than steepening the bank, they hold the bank constant and attempt to increase the rate of turn by adding more rudder pressure in an effort to align it with the runway.

The addition of inside rudder pressure will cause the speed of the outer wing to increase and thus create greater lift on that wing. To keep that wing from rising and to maintain a constant angle of bank, opposite aileron pressure would need to be applied. The added inside rudder pressure will also cause the nose to lower in relation to the horizon. Consequently, additional back elevator pressure would be required to maintain a constant pitch attitude. The resulting condition, then, is a turn with rudder applied in one direction, aileron in the opposite direction, and excessive back elevator pressure—a pronounced cross-control condition.

Since the airplane is in a skidding turn during the crossed control condition, the wing on the outside of the turn speeds up and produces more lift than the inside wing; thus the airplane starts to increase its bank. The down aileron on the inside of the turn helps drag that wing back, slowing it up and decreasing its lift, which requires more aileron application. This further causes the airplane to roll. The roll may be so fast that it is possible the bank will be vertical or past vertical before it can be stopped.

For the demonstration of the maneuver it is important that it be entered at a safe altitude because of the possible extreme nose-down attitude and loss of altitude that may result.

Before demonstrating this stall, the pilot should clear the area for other air traffic while slowly retarding the throttle. Then the landing gear (if retractable gear) should be lowered, the throttle closed, and the altitude maintained until the airspeed approaches the normal glide speed. Because of the possibility of exceeding the airplane's limitations, flaps should not be extended. While the gliding attitude and airspeed are being established, the airplane should be retrimmed. When the glide is stabilized, the airplane should be rolled into a medium-banked turn to simulate a final approach turn which would overshoot the centerline of the runway. During the turn, excessive rudder pressure should be applied in the direction of the turn but the bank held constant by applying opposite aileron pressure. At the same time increased back elevator pressure is required to keep the nose from lowering.

All of these control pressures should be increased until the airplane stalls. When the stall occurs, recovery is made by releasing the control pressures and increasing power as necessary to recover.

In a cross-control stall, the airplane often stalls with little warning. The nose may pitch down, the inside wing may suddenly drop, and the airplane may continue to roll to an inverted position. This is usually the beginning of a spin. It is obvious that close to the ground is no place to allow this to happen.

If recovery can be made *before* the airplane enters an abnormal attitude (vertical spiral or spin), it is a simple matter to return to straight-and-level flight by coordinated use of the controls. If a recovery cannot be completed before the airplane reaches an abnormal or inverted attitude, the control pressures must be released to break the stall and the roll allowed to continue until the airplane reaches straight-and-level flight. Applying power when the nose is pointed toward the ground will result in a greater loss of altitude—with the possibility of impacting the ground if the stall were to occur during an actual landing approach.

The pilot must be able to recognize when this stall is imminent and must take immediate action to prevent a completely stalled condition. It is *imperative* that this type of stall not occur during an actual approach to a landing, since recovery may be impossible before the airplane strikes the ground.

Elevator Trim Stall

This demonstration maneuver shows what can happen when full power is applied for a go-around and positive control of the airplane is not maintained. Such a situation may occur during a go-around procedure from a normal landing approach or a simulated forced landing approach, or immediately after a takeoff (Fig. 11–26). The objective of the demonstration is to show the importance of making smooth power applications, overcoming strong trim forces and maintaining positive control of the airplane to hold safe flight attitudes, and using proper and timely trim techniques.

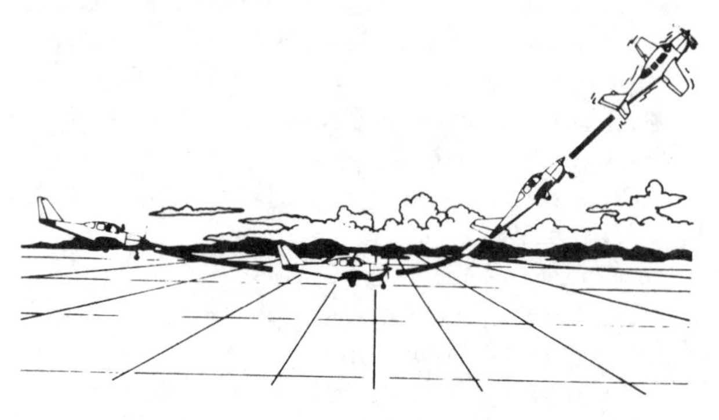

Figure 11–26 Elevator Trim Stall

At a safe altitude after ensuring that the area is clear of other air traffic, the pilot should slowly retard the throttle, and extend the landing gear (if retractable gear). One-half to full flaps should be lowered, the throttle closed, and altitude maintained until the airspeed approaches the normal glide speed. When the normal glide is established, the airplane should be trimmed for the glide just as would be done during a landing approach (nose-up trim).

During this simulated final-approach glide, the throttle is then advanced smoothly to maximum allowable power as would be done in a go-around procedure. The combined forces of thrust, torque, and back-elevator trim will tend to make the nose rise sharply and turn to the left. To demonstrate what could occur if positive control of the airplane were not maintained, no immediate attempt should be made to correct these forces.

When the throttle is fully advanced and the pitch attitude increases above the normal climbing attitude, and it is apparent that a stall is imminent, forward pressure must be applied to return the airplane to the normal climbing attitude. While holding the airplane in this attitude, the trim should then be adjusted to relieve the heavy control pressures and the normal go-around and level-off procedures completed.

The pilot should recognize when the stall is imminent and must take prompt action to prevent a completely stalled condition. It is imperative that a full stall not occur during an actual go-around from a landing approach, since the amount of pitch change necessary for recovery may be such that the airplane would dive into the ground.

Spins

A spin may be described as an aggravated stall that results in what is termed "autorotation" wherein the airplane follows a corkscrew path in a downward direction. The wings are producing some lift and the airplane is forced downward by gravity, wallowing and yawing in a spiral path (Fig. 11–27). It has been estimated that there are actually many factors contributing to spinning, and in fact the spin is not very amenable to theoretical analysis.

Industry and the Federal Aviation Administration have and continue to support studies directed at ways to design safer airplanes. Spin analysis continues to be a major consideration in these studies. Fear of and aversion to spins are deeply rooted in the public's mind, and many pilots have an unconscious aversion to them. If one learns the cause of a spin and the proper techniques to prevent and/or recover from the spin, mental anxiety and many causes of unintentional spins may be removed.

Though instruction in spins is not required of applicants for private or commercial pilot certificates, applicants for a flight instructor certificate with an airplane or glider instructor

Figure 11–27 Two-Turn Spin

rating may be asked to demonstrate that they can recognize and recover from spin situations that might be encountered in poorly executed maneuvers during student training flights. For this reason, a brief discussion of spins is included in this handbook.

The INTENTIONAL SPINNING of an airplane for which the spin maneuver is not specifically approved, is NOT authorized either by this handbook or by Federal Aviation Regulations. Official sources for determining if the spin maneuver is approved for a specific airplane are:

a. In the airplane's Type Certificate and Data Sheets;

b. On a placard located in clear view of the pilot in the airplane, i.e., "NO ACROBATIC MANEUVERS INCLUDING SPINS APPROVED"; and

c. The maneuvers section of the FAA Approved Airplane Flight Manual or Pilot's Operating Handbook.

Increasing occurrences involving airplanes wherein spin restrictions are *intentionally* ignored by pilots, have been brought to the attention of the FAA. Despite the installation of placards prohibiting intentional spins in these airplanes, a number of pilots, including some flight instructors, attempt to justify the maneuver, rationalizing that the spin restriction results merely because of a "technicality" in the airworthiness standards.

Some pilots reason that the airplane was spin tested during its certification process and, therefore, no problem should result from demonstrating or practicing spins. The fact that certification in the Normal category requires only that the airplane recover from a *one* turn spin in not more than one additional turn or 3-seconds, whichever takes longer. This same test of controllability can also be used in certificating an airplane in the Utility category (FAR 23.221(b)).

The point is that 360° of rotation (one turn spin) does not provide a stabilized spin. If the airplane's controllability has not been explored by the engineering test pilot beyond the certification requirements, prolonged spins (inadvertent or intentional) in that airplane places an operating pilot in an unexplored flight situation. Recovery from it may be difficult or even impossible.

In FAR Part 23, "Airworthiness Standards: Normal, Utility, and Acrobatic Category Airplanes," there are no requirements for investigation of *controllability* in a true spinning condition for the Normal category airplanes. The one-turn "margin of safety" is essentially a check of the airplane's controllability in a delayed recovery from a *stall.* Therefore, in airplanes placarded against spins there is absolutely *no assurance whatever* that recovery from a fully developed spin is *possible under any circumstances* (Fig. 11–23).

The occurrence of uncontrollable flat spins in some airplanes, resulting from *intentionally* exceeding the one-turn safety margin, leads the FAA to believe that some pilots are unaware of the inherent risk involved in spinning an airplane which has been placarded to prohibit spins. Based on experience to date, it is possible that a number of pilots will at one time or another, intentionally ignore the spin limitation because of *lack of knowledge, misinformation, or misinterpretation* regarding the intent of applicable airworthiness standards in FAR Part 23.221, relating to spins. Since those pilots may be expected to have little or no knowledge regarding the possible development of an insidious flat spin, they will be unaware of the serious risks incurred. Accordingly, all pilots are cautioned to adhere strictly to the spin restrictions in airplanes placarded against intentional spins.

THE PILOT OF AN AIRPLANE PLACARDED AGAINST INTENTIONAL SPINS SHOULD ASSUME THAT THE AIRPLANE MAY BECOME UNCONTROLLABLE IN A SPIN.

It has been estimated that there are actually several hundred factors that contribute to spinning. From this it is evident that, whether or not spinning is a desirable maneuver or characteristic, it will be a feature of airplanes for some time to come and must be reckoned with in the training of a pilot.

Many modern airplanes have to be forced to spin and require considerable judgment and technique to get the spin started. Paradoxical as it may seem, these same airplanes that have to be forced to spin, may be accidentally put into a spin by mishandling the controls in turns, stalls, and flight at minimum controllable airspeeds. This fact is additional evidence of the necessity for the practice of stalls until the ability to recognize and recover from them is developed.

Often a wing will drop at the beginning of a stall. When this happens the nose will attempt to move (yaw) in the direction of the low wing. This is where use of the rudder is important during a stall. The correct amount of opposite rudder must be applied to keep the nose from yawing toward the low wing. By maintaining directional control and not allowing the nose to yaw toward the low wing, the wing will not drop farther before the stall is broken, and thus a spin will be averted. If the nose is allowed to yaw during the stall, the airplane begins to slip in the direction of the lowered wing, and as it does, the air meeting the side of the fuselage, the vertical fin, and other vertical surfaces, tends to "weathervane" the airplane into the relative wind. This accounts for the continuing yaw which is present in a spin.

At the same time, rolling is also occurring about the longitudinal axis of the airplane. This is caused by the lowered wing having an increasingly greater angle of attack, due to the upward motion of the relative wind against its surfaces. This wing, then, is well beyond the stalling angle of attack, and accordingly suffers an extreme loss of lift. The rising wing, since the relative wind is striking it at a smaller angle, has a smaller angle of attack than the opposite wing. Thus, the rising wing has more lift than the lowering wing, so that the airplane begins to rotate about its longitudinal axis. This rotation, combined with the effects of centrifugal force and the different amount of drag on the two wings, then becomes a spin and the airplane descends vertically, rolling and yawing until recovery is effected.

Continued practice in *stalls* will help the pilot develop a more instinctive and prompt reaction in recognizing an *approaching spin*. It is essential to learn to apply immediate corrective action any time it is apparent that the airplane is nearing spin conditions. If an unintentional spin can be prevented, it should be by all means. This is sound pilot judgment and a positive indication of alertness. If it is impossible to avoid a spin, the pilot should execute an immediate recovery—the controls must not be held with the spin.

The first corrective action taken during any power-on spin is to close the throttle. Power aggravates the spin characteristics and causes an abnormal loss of altitude in the recovery.

Prior to spin demonstrations the FAA approved Airplane Flight Manual or Pilot's Operating Handbook for the airplane being flown should be consulted for the proper recovery technique. In the absence of such recommendations, the following technique is suggested.

To recover from the spin, the pilot should first apply full opposite rudder; then after the rotation slows, apply brisk, positive straightforward movement of the elevator control (forward of the neutral position) (Fig. 11–28). The control should be held firmly in this position. The forceful movement of the elevator will decrease the excessive angle of attack and

CLOSE THROTTLE
FULL OPPOSITE RUDDER
BRISK FORWARD ELEVATOR

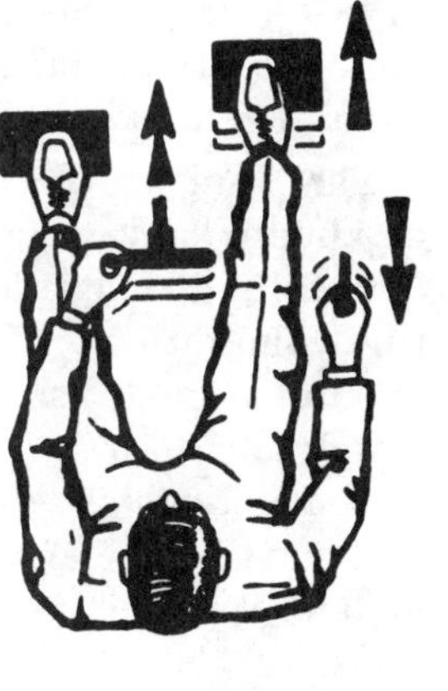

HOLD ELEVATOR FORWARD
NEUTRALIZE RUDDER

EASE ELEVATOR BACK
TOWARD NEUTRAL

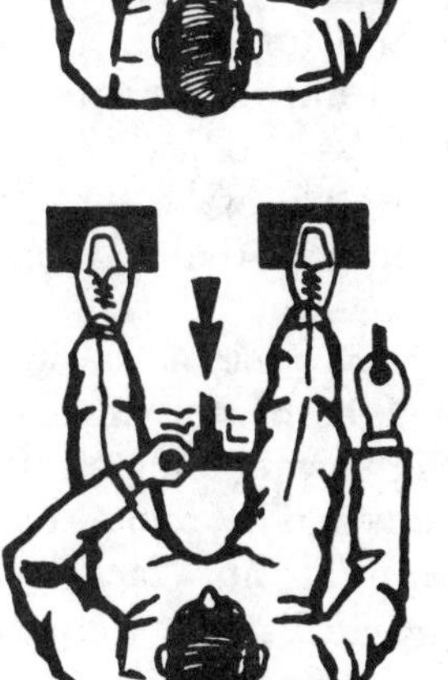

Figure 11–28 Typical Spin Recovery Procedure

thus will break the stall. When the stall is broken the spinning will stop. This straightforward position should be maintained and as the spin rotation stops, the rudder should be neutralized.

If the rudder is not neutralized at the proper time, the ensuing increased airspeed acting upon the fully deflected rudder will cause an excessive and unfavorable yawing effect. This places great strain on the airplane, and may cause a secondary spin in the opposite direction.

Slow and overly-cautious control movements during spin recovery must be avoided. In certain cases it has been found that such movements result in the airplane continuing to spin indefinitely, even with the application of full opposite controls. Brisk and positive operation, on the other hand, results in a more positive recovery.

After the spin rotation stops and the rudder has been neutralized, the pilot should begin applying back elevator pressure to raise the nose to level flight. Caution must be used so as not to apply excessive back pressure after the rotation stops. Sometimes a pilot does this because of being too anxious to stop the descent. To do so will cause a secondary stall and may result in another spin, more violent than the first.

Any time a spin is encountered, regardless of the conditions, the normal spin recovery sequence should be used: (1) retard power; (2) apply opposite rudder to slow rotation; (3) apply positive forward-elevator movement to break stall; (4) neutralize rudder as spinning stops; and (5) return to level flight.

Following training in spins, the pilot should recognize when a spin condition exists and take prompt action to prevent a fully developed spin. Performance is considered unsatisfactory if more than one turn of a spin occurs or if it becomes necessary for the instructor to take control of the airplane to avoid a fully developed spin.

Maximum Performance Flight Maneuvers

Airplane "performance" is a term used to describe the ability of an airplane to accomplish certain things which make it useful for specific purposes. The chief elements of performance include takeoff and landing distance, rate and angle of climb, maneuverability, range, speed, and fuel economy.

Each airplane has its own set of flight characteristics and capabilities. The maximum performance flight maneuvers explained here are not necessarily operational in nature, but they are used effectively in pilot training to develop skills and safe habits in preparation for obtaining the best maneuvering performance from any airplane the pilot flies. A full understanding of the principles involved in the performance of these flight maneuvers will enable the pilot to apply them effectively in the operation of most airplanes. A pilot who is familiar with a given airplane's capabilities and limitations acquires a sixth sense about how it will maneuver, how quickly it turns, accelerates, decelerates, climbs, and descends. Regardless of the airspeeds or flight attitudes, the airplane cannot be operated with complete safety and accuracy unless the pilot is well acquainted with its maneuvering performance. In executing these maximum performance flight maneuvers, either during the training period or later in an operational situation, the value of developing the ability to recognize approaching stalls and to fly at critically slow airspeeds will be immediately realized.

Steep Power Turns

The "steep power turn" maneuver consists of a turn in either direction, using a bank steep enough to cause an "overbanking" tendency during which maximum turning performance is attained and relatively high load factors are imposed. Because of the high load factors imposed, these turns should be performed at an airspeed which does not exceed the airplane's design maneuvering speed (V_A). The principles of an ordinary steep turn apply but as a practice maneuver the steep power turns should be continued until 360° or 720° of turn has been completed, as specified by the instructor or examiner (Fig. 11–29).

The objective of the maneuver is to develop smoothness, coordination, orientation, division of attention, and control techniques while executing high performance turns.

An airplane's maximum turning performance is its fastest *rate* of turn and its shortest

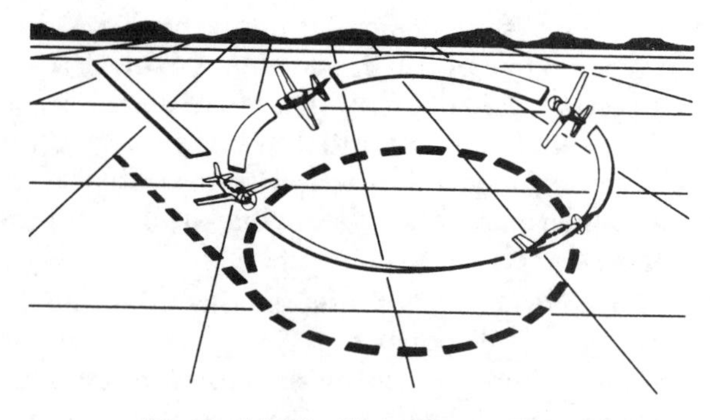

Figure 11–29 Steep Power Turns

radius of turn, which change with both *airspeed* and *angle of bank*. Each airplane's turning performance is limited by the amount of power its engine is developing, its limit load factor (structural strength), and its aerodynamic characteristics.

The limiting load factor determines the maximum bank which can be maintained without stalling or exceeding the airplane's structural limitations. In most light planes, the maximum bank has been found to be approximately 50° to 60°. The steepest bank possible is that in which altitude can be maintained without the airplane displaying any indication of an approaching stall with all available power.

The pilot should realize the tremendous additional load that is imposed on an airplane as the bank is increased beyond 45°. During a coordinated turn with a 70° bank, a load factor of approximately 3 G's is placed on the airplane's structure. Most general aviation type airplanes are stressed for approximately 3.8 G's.

Regardless of the airspeed or the type of airplanes involved, a given angle of bank in a turn during which altitude is maintained, will always produce the same load factor. Pilots must be aware that an additional load factor increases the stalling speed at a significant rate—that is, stalling speed increases with the square root of the load factor. For example, a light plane which stalls at 60 knots in level flight will stall at nearly 85 knots in a 60° bank. The pilot's understanding and observance of this fact is an indispensable safety precaution for the performance of all maneuvers requiring turns, particularly near the ground.

Before starting the steep power turn the pilot should ensure that the area is clear of other air traffic since the rate of turn will be quite rapid. After establishing the manufacturer's recommended entry speed or the design maneuvering speed, the airplane should be smoothly rolled into a coordinated steep turn with at least 50° of bank. As the turn is being established, back pressure on the elevator control should be smoothly increased to increase the angle of attack. This provides the additional wing lift required to compensate for the increasing centrifugal force.

After the bank has reached approximately 50° the pilot will find that considerable force is required on the elevator control to hold the airplane in level flight—that is, to maintain altitude. Because of this increase in the force applied to the elevators, the load factor increases rapidly as the bank is increased. Additional back elevator pressure increases the angle of attack which, of course, results in an increase in drag. Consequently, power must be added to maintain the entry altitude and the airspeed.

Eventually, as the bank approaches the airplane's maximum angle, the maximum performance or structural limit is being reached. If this limit is exceeded the airplane will be subjected to excessive structural loads, and will lose altitude, or stall. The limit load factor must not be exceeded, so as to prevent structural damage.

During the turn the pilot should not stare at any one object. To maintain altitude, as well as orientation, requires an awareness of the relative position of the nose, the horizon, the wings, and the amount of turn. The pilot who turns by watching only the nose, will have trouble holding altitude constant; on the other hand, the pilot who watches the nose, the horizon, and the wings, can usually hold altitude within a few feet. If the altitude begins to increase, the bank should be increased by coordinated use of aileron and rudder. If the altitude begins to decrease, the bank should be decreased by coordinated use of aileron and rudder. Rudder should never be used alone to control the altitude.

The rollout from the turn should be timed so that the wings reach level flight when the airplane is exactly on the heading from which the maneuver was started. While the recovery is being made, back elevator pressure must be gradually released and power reduced as necessary to maintain the altitude and airspeed.

Steep Spirals

A "steep spiral" is nothing more than a continuous gliding turn, during which a constant radius around a point on the ground is maintained similar to the maneuver "turns around a point." The radius should be such that the steepest bank will be approximately 50° to 55°. The objective of the maneuver is to improve pilot techniques for power-off turns, wind drift control, planning, orientation, and division of attention. This spiral is not only a valuable flight training maneuver, but it has practical application in providing a procedure for dissipating altitude while remaining over a selected spot in preparation for landing, especially for emergency forced landings.

Sufficient altitude must be obtained before starting this maneuver so that the spiral may be continued through a series of at least three 360° turns (Fig. 11–30). However, the maneuver should not be continued below a minimum safe altitude.

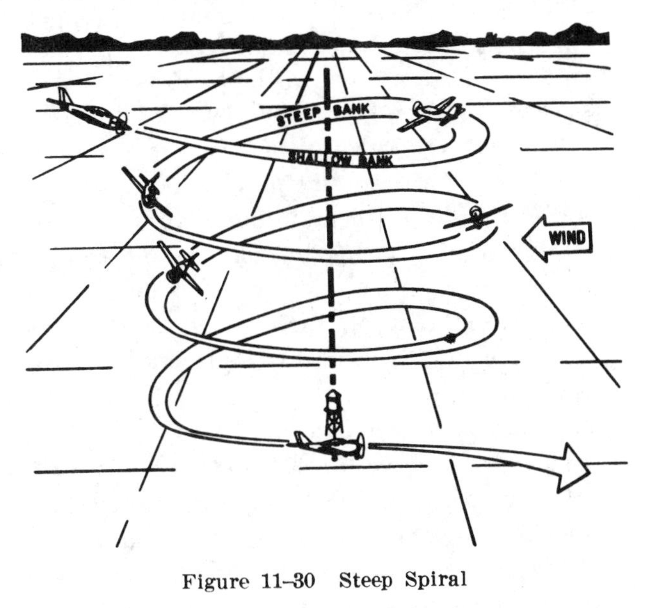

Figure 11–30 Steep Spiral

Operating the engine at idle speed for a prolonged period during the glide may result in excessive engine cooling or spark plug fouling. Therefore, the engine should be cleared periodically by briefly advancing the throttle to normal cruise power, while adjusting the pitch attitude to maintain a constant airspeed. Preferably, this should be done while headed into the wind to minimize any variation in groundspeed and radius of turn.

After the throttle is closed and gliding speed is established, a gliding spiral should be started and a turn of constant radius maintained around the selected spot on the ground. This will require correction for wind drift by steepening the bank on downwind headings and shallowing the bank on upwind headings, just as in the maneuver "turns around a point." During the descending spiral the pilot must judge the direction and speed of the wind at different altitudes and make appropriate changes in the angle of bank to maintain a uniform radius.

A constant airspeed should also be maintained throughout the maneuver. Failure to hold the airspeed constant will cause the radius of turn and necessary angle of bank to vary excessively. On the downwind side of the maneuver, the steeper the bank angle the lower the pitch attitude must be to maintain a given airspeed. Conversely, on the upwind side, as the bank angle becomes shallower, the pitch attitude must be raised to maintain the proper airspeed. This is necessary because the airspeed tends to change as the bank is changed from shallow to steep to shallow.

During practice of the maneuver the pilot should execute a specific number of turns and roll out toward a definite object or on a specific heading. During the rollout, smoothness, is essential, and the use of controls must be so coordinated that no increase or decrease of speed results when the straight glide is resumed.

Descents (Maximum Distance Glides)

The best angle of glide is one that allows the airplane to travel the greatest distance over the ground with the least loss of altitude. This is the airplane's maximum L/D (lift over drag) and is usually expressed as a ratio. For example, an airplane having an L/D or glide ratio of 10:1 will travel 10 feet forward for every foot it descends.

For a particular airplane the manufacturer recommends an airspeed and configuration that will provide the maximum glide distance. This speed (best glide speed) usually found in the Airplane Flight Manual or Pilot's Operating Manual, is of primary importance because if the engine should fail in flight the pilot's chief concern may be whether or not the airplane can glide far enough to reach a suitable landing area.

The objective of this maneuver, then, is to establish a glide that will allow the airplane to travel forward the greatest possible distance from a given altitude.

To establish the glide, the landing gear and flaps should first be retracted to eliminate unwanted drag. The throttle should be reduced to idle, the propeller placed in full high pitch (low RPM) position, and the airplane then eased into a glide until the proper airspeed is established. If the airplane's nose is lowered excessively, the airplane will go into too steep a glide, and naturally will cover very little horizontal distance. On the other hand if the nose is raised too high and too much airspeed is lost, the airplane will settle and descend at a steeper angle than if the nose were somewhat lower.

When practicing the power-off descents, the engine should be cleared periodically, as is done in the steep spiral maneuver, to prevent excessive cooling and fouling, and of course the descent should be terminated at a safe altitude. Care must be exercised when advancing the throttle to avoid overstressing the engine.

Descents (Emergency)

This maneuver is a procedure for establishing the fastest practical rate of descent during emergency conditions which may arise as the result of an uncontrollable fire, a sudden loss of cabin pressurization, or any other situation demanding an immediate and rapid descent. The objective, then, is to descend the airplane as soon and as rapidly as possible, within the limitations of the airplane, to an altitude from which a safe landing can be made, or an altitude where pressurization or supplemental oxygen is not needed.

The simulated emergency descent must be started high enough to permit recovery at a safe altitude. Before entering the maneuver, the area below must be free of other air traffic, since the loss of altitude is quite rapid. *In no case* should the airplane's never-exceed speed (V_{ne}), maximum gear-extended speed (V_{le}), or maximum flap-extended speed (V_{fe}) be exceeded.

Generally, the maneuver should be performed with the airplane configured as recommended by the manufacturer. Except when prohibited by the manufacturer, the power should be reduced to idle, and the propeller control (if so equipped), should be placed in the low pitch (or high RPM) position. This will allow the propeller to act as an aerodynamic brake to help prevent excessive airspeed during the descent. As quickly as practical, the landing gear and full flaps should be extended to provide maximum drag so that a descent as rapidly as possible can be made without excessive airspeed. This, of course, should be done only in accordance with the airplane manufacturer's recommendations.

To maintain *positive* load factors (G forces) and for the purpose of clearing the area below, a 30° to 45° bank should be established for at least a 90° heading change while initiating the descent.

Normally during student training, as soon as all prescribed procedures are completed and the descent is established and stabilized, the maneuver should be terminated. In airplanes with piston engines, a prolonged practice emergency descent should be avoided to prevent excessive cooling of the cylinders.

Chandelles

A "chandelle" is a climbing turn beginning from approximately straight-and-level flight, and ending at the completion of 180° of turn in a wings-level, nose-high attitude at the minimum controllable airspeed (Fig. 11–31). The maneuver demands that the maximum flight performance of the airplane be obtained; that is, the airplane should gain the most altitude possible for a given degree of bank and power setting without stalling. However, since numerous atmospheric variables beyond control of the pilot will affect the specific amount of altitude gained, the altitude gain is not a criterion of the quality of the maneuver.

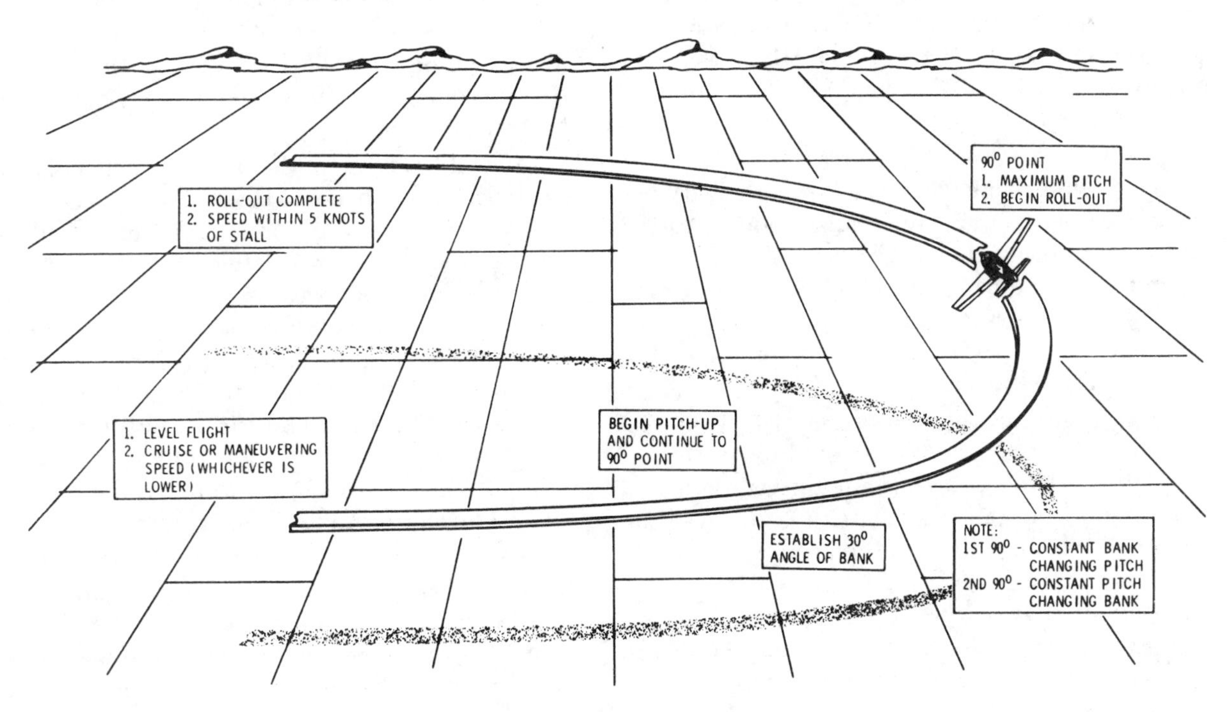

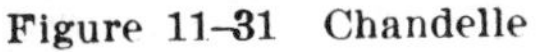
Figure 11–31 Chandelle

The objective of this maneuver is to develop the pilot's coordination, orientation, planning, and feel for maximum-performance flight, and to develop positive control techniques at varying airspeeds and attitudes.

Prior to starting a chandelle, the flaps and gear (if retractable) should be in the UP position, power set to cruise condition, the airspace behind and above clear of other air traffic. A chandelle should be started at any speed no greater than the maximum entry speed recommended by the manufacturer—in most cases not above the airplane's design maneuvering speed.

After the appropriate airspeed and power setting have been established, the chandelle is started by smoothly entering a coordinated turn with an angle of bank appropriate for the airplane being flown. Normally, this angle of bank should not exceed approximately 30°. After the appropriate bank is established, a climbing turn should be started by smoothly applying back elevator pressure to increase the pitch attitude at a constant rate and to attain the highest pitch attitude as 90° of turn is completed. As the climb is initiated in airplanes with fixed-pitch propellers, full throttle may be applied but must be applied *gradually* so that the maximum allowable RPM is not exceeded. In airplanes with constant-speed propellers, power may be left at the normal cruise setting. (Prior to starting the maneuver, RPM may be increased to climb or takeoff setting and then throttle increased as the climb is started.)

Once the bank has been established, the angle of bank should remain constant until 90° of turn is completed. Although the degree of bank is fixed during this climbing turn, it may appear to increase and, in fact, actually will tend to increase if allowed to do so as the maneuver continues.

When the turn has progressed 90° from the original heading, the pilot should begin rolling out of the bank at a constant rate while maintaining a constant-pitch attitude. Since the angle of bank will be decreasing during the rollout, the vertical component of lift will increase slightly. For this reason, it may be necessary to release a slight amount of back elevator pressure in order to keep the nose of the airplane from rising higher.

As the wings are being leveled at the completion of 180° of turn, the pitch attitude should be noted by checking the outside references and the attitude indicator. This pitch attitude should be held momentarily while the airplane is at the minimum controllable airspeed. Then the pitch attitude may be gently reduced to return to straight-and-level cruise flight.

Since the airspeed is constantly decreasing throughout the maneuver, the effects of engine torque become more and more prominent. Therefore, right rudder pressure must be gradually increased to control yaw and maintain a constant rate of turn and to keep the airplane in coordinated flight. The pilot should maintain coordinated flight by the "feel" of pressures being applied on the controls, and by the ball instrument of the turn-and-slip indicator. If coordinated flight is being maintained, the ball will remain in the center of the race.

To roll out of a left chandelle, the left aileron must be lowered to raise the left wing. This creates more drag than the aileron on the ring wing, resulting in a tendency for the airplane to yaw to the left. With the low airspeed at this point, torque effect tries to make the airplane yaw to the left even more. Thus, there are two forces pulling the airplane's nose to the left—aileron drag and torque. To maintain coordinated flight, considerable right rudder pressure must be used during the rollout to overcome the effects of aileron drag and torque.

In a chandelle to the right, when control pressure is applied to begin the rollout, the aileron on the right wing is lowered. This creates more drag on that wing and tends to make the airplane yaw to the right. At the same time, however, the effect of torque at the low airspeed is causing the airplane's nose to yaw to the left. Thus, aileron drag pulling the nose to the right and torque pulling to the left, tend to neutralize each other. If excessive left rudder pressure is applied, the rollout will be uncoordinated.

The rollout to the left can usually be accomplished with very little left rudder, since the effects of aileron drag and torque tend to neutralize each other. *Releasing* some right rudder, which has been applied to correct for torque, will normally give the same effect as applying left rudder pressure. When the wings become level and the ailerons are neutralized, the aileron drag disappears. At this time, however, because of the low airspeed and high power, the effects of torque become the more prominent force and must continue to be controlled with rudder pressure.

A rollout to the left, therefore, is accomplished mainly by applying aileron pressure. During the rollout, right rudder pressure should be gradually released, and left rudder applied only as necessary to maintain coordination. Even when the wings are level and aileron pressure is released, right rudder pressure must be held to counteract torque and hold the nose straight.

Lazy 8

This maneuver derives its name from the manner in which the extended longitudinal axis of the airplane is made to trace a flight pattern in the form of a figure 8 lying on its side (a "Lazy" 8) (Fig. 11–32). Objective of the Lazy 8 is to develop the pilot's feel for varying control forces, and the ability to plan and remain oriented while maneuvering the airplane with positive, accurate control. It requires constantly changing control pressures necessitated by changing combinations of climbing and descending turns at varying airspeeds. This is a maneuver often used to develop and demonstrate the pilot's mastery of the airplane in maximum performance flight situations.

A "Lazy 8" consists of two 180° turns, in opposite directions, while making a climb and a descent in a symmetrical pattern during each of the turns. At no time throughout the Lazy 8 is the airplane flown straight and level; instead, it is rolled directly from one bank to the other with the wings level only at the moment the turn is reversed at the completion of each 180° change in heading.

As an aid to making symmetrical loops of the 8 during each turn, prominent reference points should be selected on the horizon. The reference points selected should be 45°, 90°, and 135° from the direction in which the maneuver is begun.

Prior to performing a Lazy 8, the airspace behind and above should be clear of other air traffic. The maneuver should be entered from

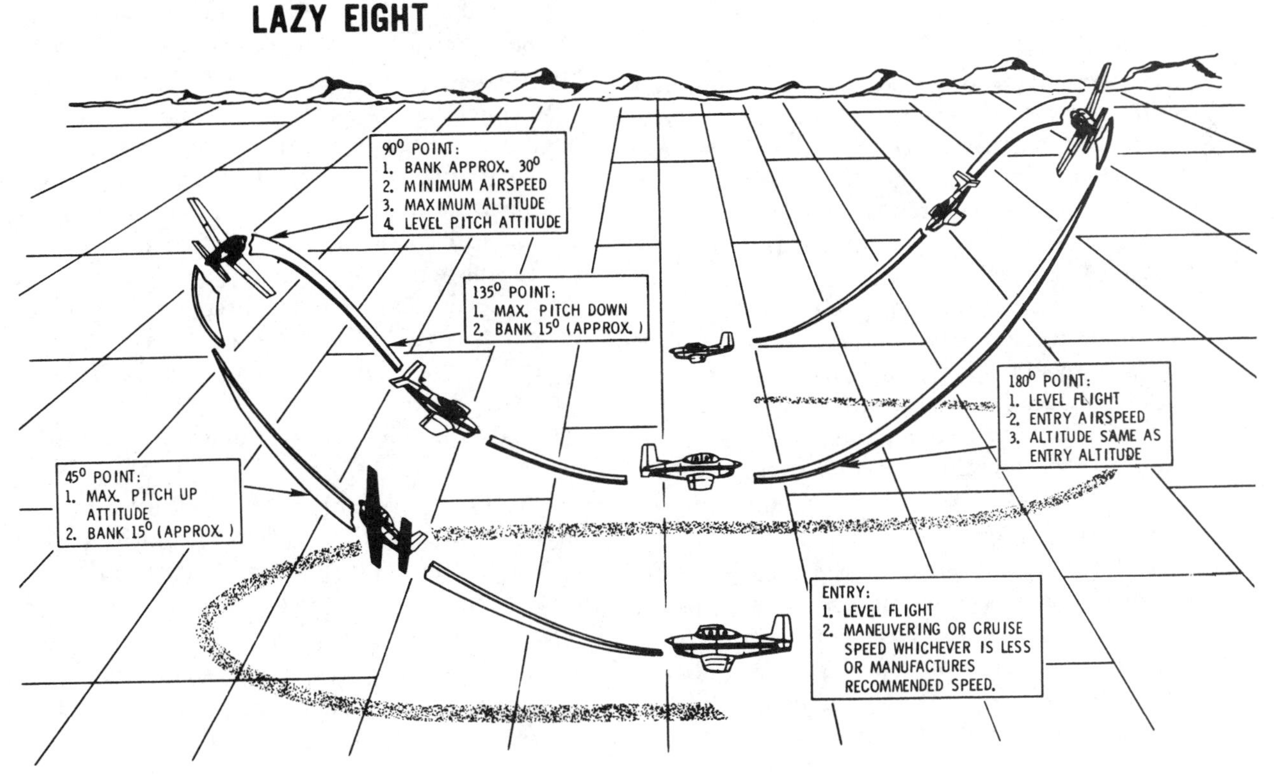

Figure 11–32 Lazy Eight

straight-and-level flight at normal cruise power and at the airspeed recommended by the manufacturer or at the airplane's design maneuvering speed.

The maneuver is started from level flight with a gradual climbing turn in the direction of the 45° reference point. The climbing turn should be planned and controlled so that the maximum pitch-up attitude is reached at the 45° point. The rate of rolling into the bank must be such as to prevent the rate of turn from becoming too rapid. As the pitch attitude is raised the airspeed decreases, causing the rate of turn to increase. Since the bank also is being increased, it too causes the rate of turn to increase. Unless the maneuver is begun with a slow rate of roll, the combination of increasing pitch and increasing bank will cause the rate of turn to be so rapid that the 45° reference point will be reached before the highest pitch attitude is attained.

At the 45° point, the pitch attitude should be at maximum and the angle of bank continuing to increase. Also, at the 45° point, the pitch attitude should start to decrease slowly toward the horizon and the 90° reference point. Since the airspeed is still decreasing, right-rudder pressure will have to be applied to counteract torque.

As the airplane's nose is being lowered toward the 90° reference point, the bank should continue to increase. Due to the decreasing airspeed, a slight amount of opposite aileron pressure may be required to prevent the bank from becoming too steep. When the airplane completes 90° of the turn, the bank should be at the maximum angle (approximately 30°), the airspeed should be at its minimum (5–10 knots above stall speed), and the airplane pitch attitude should be passing through level flight. It is at this time that an imaginary line, extending from the pilot's eye and parallel to the longitudinal axis of the airplane, passes through the 90° reference point.

Lazy 8's normally should be performed with no more than approximately a 30° bank. Steeper banks may be used but control touch and technique must be developed to a much higher degree than when the maneuver is performed with a shallower bank.

The pilot should not hesitate at this point but should continue to fly the airplane into a descending turn so that the airplane's nose describes the same size loop below the horizon as it did above. As the pilot's reference line passes through the 90° point, the bank should be decreased gradually, and the airplane's nose allowed to continue lowering. When the airplane has turned 135°, the nose should be in its lowest pitch attitude. The airspeed will be increasing during this descending turn so it will be necessary to gradually relax rudder and aileron pressure and to simultaneously raise the nose and roll the wings level. As this is being accomplished, the pilot should note the amount of turn remaining and adjust the rate of rollout and pitch change so that the wings become level and the original airspeed is attained in level flight just as the 180° point is reached. Upon reaching that point, a climbing turn should be started immediately in the opposite direction toward the selected reference points to complete the second half of the eight in the same manner as the first half.

Due to the decreasing airspeed considerable right-rudder pressure must be gradually applied to counteract torque at the top of the eight in both the right and left turns. The pressure will be greatest at the point of lowest airspeed.

More right-rudder pressure will be needed during the climbing turn to the right than in the turn to the left because more torque correction is needed to prevent yaw from decreasing the rate of turn. In the left climbing turn the torque will tend to contribute to the turn; consequently, less rudder pressure is needed. It will be noted that the controls are slightly crossed in the right climbing turn because of the need for left aileron pressure to prevent over-banking and right rudder to overcome torque.

The correct power setting for the lazy eight is that which will maintain the altitude for the maximum and minimum airspeeds used during the climbs and descents of the eight. Obviously, if excess power were used, the airplane would have gained altitude when the maneuver is completed, while if insufficient power were used, altitude would have been lost.

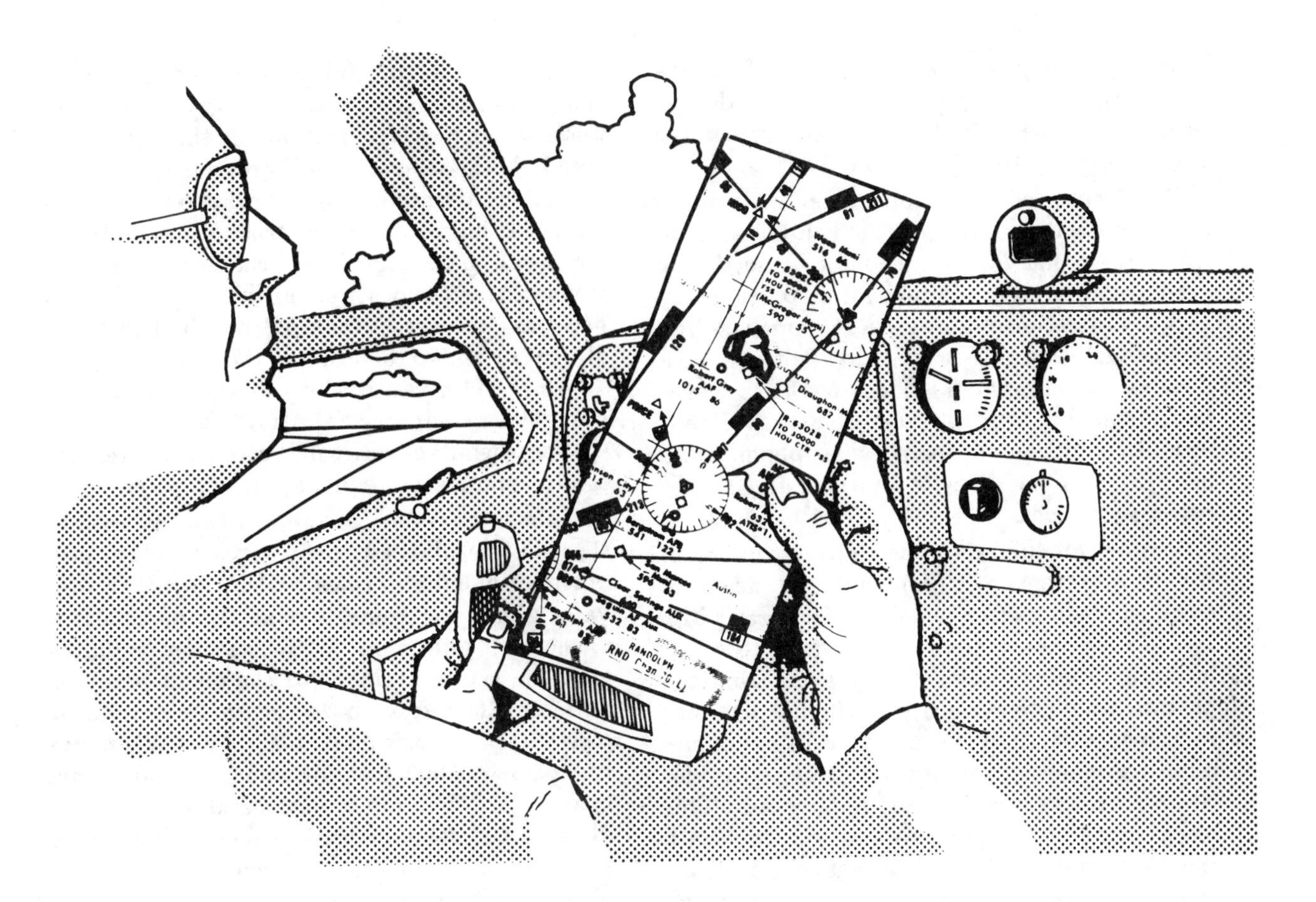

CHAPTER 12

Cross-Country Flying

This chapter discusses the basic elements of cross-country flying, including various methods of air navigation—the art of flying the airplane from one point to another and determining its position along the route. It is not intended to explain in detail the intricacies of air navigation. For that information, the learning pilot is directed to the *Pilot's Handbook of Aeronautical Knowledge*, AC 61–23B, published by the FAA, or to suitable commercially published navigation books.

Preflight Preparation

Air navigation is not limited to the actual guiding of an airplane from one place to another—it begins and ends on the ground. Before starting a cross-country flight, the pilot should plan the flight thoroughly. This includes obtaining pertinent weather information, plotting the course on an aeronautical chart, selecting checkpoints, measuring distances, and computing flight time, headings, and fuel requirements.

The first item for discussion is the aeronautical chart, inasmuch as cross-country flight is possible with this item alone, even without the magnetic compass or wind data. The type of chart recommended for VFR navigation is the Aeronautical Sectional Chart.

A course line should be drawn between the points of intended flight. This line should be dark enough to be seen, but not so heavy as to obscure the symbols on the chart. The line may be scaled off in convenient segments, such as marking every 10 miles or at each prominent checkpoint, with the distance to each recorded. This provides a convenient method of quickly determining the distance from the point of departure to any point along the course line. The angle between the course line and the meridian midway between the departure and destination can then be measured to determine the true course or direction of the flight. Meridians are the vertical lines printed on the chart that converge at the North Pole and the South Pole.

The Department of Commerce publication, Airport/Facility Directory, a valuable aid to the cross-country pilot, should be used for preflight reference. It lists all airports in the U.S. by state, city, and airport name, as well as the facilities and services available at each airport, including maintenance facilities, elevations, runway lengths, fuel grades available, etc. In addition, it includes telephone numbers for Weather Service Stations and Flight Service Stations.

In preflight planning, accurate estimates of the time enroute (ETE) must be made to establish fuel requirements for the flight (Fig. 12–1). Estimated time of arrival (ETA) is based on the computed flying time from departure to destination. If a flight plan has been filed with the FAA, the time of departure must be reported to the local Flight Service Station (FSS) handling the flight plan. This can be done by the pilot via radio contact with the FSS after takeoff. The net enroute time should be the total flight time including any planned enroute stops of less than one hour duration between the departure and destination. The flight plan is then forwarded to the FSS nearest the intended destination.

It is extremely important for the pilot to know the fuel capacity and rate of fuel consumption of the particular airplane being used on a cross-country flight. Fuel capacity and rate of consumption can be found in the Airplane Flight Manual or the Pilot's Operating Handbook, either of which should be in the airplane at all times. The amount of fuel required and any necessary fueling stops enroute can be determined by computing the enroute time versus rate of fuel consumption. The fuel requirements are important enough for safe flight to be included in the regulations. The rules state that before beginning a day VFR flight in an airplane there must be enough fuel available to fly to the first point of intended landing and to fly after that for at least 30 minutes. When conducting a VFR night flight, the minimum fuel reserve is increased to 45 minutes.

The estimate of flying time and fuel consumption in cross-country flying is dependent upon a correct application of the airplane's speed, both airspeed and groundspeed. The indicated airspeed (IAS), as shown on the airspeed indicator, is a true value only at sea level under standard temperature conditions. At altitudes well above sea level, it is necessary to correct the IAS for the less dense air in order to determine the true airspeed (TAS). This can be done by means of an air navigation computer. The TAS, plus or minus the headwind or tailwind component respectively, equals the airplane's groundspeed.

Unless the pilot has an instrument rating, the pilot is allowed to fly cross-country only when visual flight rules (VFR) weather prevails along the route and at the destination. Basically, minimum VFR weather requires that the ceiling (height of the cloud base) be at least 1,000 feet above the surface, with at least 3 miles horizontal visibility. When flying below 10,000 feet MSL in controlled airspace, the VFR pilot must be able to remain at least 500 feet vertically below or 1,000 feet above, and 2,000 feet horizontally from the clouds. However, these are the absolute *minimum* requirements and are not recommended for pilots having limited experience. Much better weather conditions are advisable.

On all cross-country flights, the pilot is required by law to determine that the existing and forecast weather conditions are appropriate for the flight. Therefore, the pilot must learn to obtain, read, and understand aviation weather forecasts and reports (Fig. 12–2).

There are several types of forecasts and reports: (1) area forecasts, (2) terminal forecasts, (3) hourly sequence reports, and (4) winds aloft forecasts and reports. A thorough explanation of these is contained in the Pilot's Handbook of Aeronautical Knowledge, AC 61–23B, and the Aviation Weather Services Publication, AC 00–45A.

FAA Flight Service Stations (FSS) have prime responsibility for preflight pilot briefing, enroute communications with VFR flights, assisting lost VFR aircraft, broadcasting aviation weather information, accepting and closing flight plans, and operating the weather teletypewriter systems. In addition, FSSs

FLIGHT LOG

PREFLIGHT												ENROUTE					
From	To	True Course	Wind Corr.	True Head.	Var.	Mag. Head.	Dist. Leg/Total	Est. GS	Time Leg/Total	Est. Fuel	Act'l. Time	Act'l. GS	Dist. Next Pt.	Est. Arrv'l.	Fuel Used	Fuel Remain

Weather Reports

Winds Aloft

Terminal Forecasts

NOTAMS

FLIGHT PLAN

1. TYPE — VFR / IFR / DVFR
2. AIRCRAFT IDENTIFICATION
3. AIRCRAFT TYPE/SPECIAL EQUIPMENT
4. TRUE AIRSPEED — KTS
5. DEPARTURE POINT
6. DEPARTURE TIME — PROPOSED (Z) / ACTUAL (Z)
7. CRUISING ALTITUDE
8. ROUTE OF FLIGHT
9. DESTINATION (Name of airport and city)
10. EST. TIME ENROUTE — HOURS / MINUTES
11. REMARKS
12. FUEL ON BOARD — HOURS / MINUTES
13. ALTERNATE AIRPORT (S)
14. PILOT'S NAME, ADDRESS & TELEPHONE NUMBER & AIRCRAFT HOME BASE
15. NUMBER ABOARD
16. COLOR OF AIRCRAFT

CLOSE VFR FLIGHT PLAN WITH__________FSS ON ARRIVAL

Figure 12-1 Cross-Country Flight Log

Figure 12–2 Obtain a Thorough Weather Briefing

take local weather observations and issue airport advisories.

The pilot should always consult the local Flight Service Station (FSS), or National Weather Service Office (NWSO) for preflight weather briefing. FSS and NWSO personnel are certificated pilot weather briefers and thoroughly understand the weather needs of pilots.

When telephoning for information, the pilot should use the following procedure:

a. Identify oneself as a pilot. (Many persons calling Weather Service Stations want information for purposes other than flying.)

b. State the intended route, destination, proposed departure time and estimated time enroute.

c. Advise if intending to fly only VFR.

d. State the airplane type and identification.

It is the pilot's responsibility to determine that the airplane is properly certificated and airworthy. There must be displayed in the airplane a registration certificate showing the name of the registered owner, and an airworthiness certificate for the airplane. In addition, the pilot should have available an operations limitations record or airplane flight manual which shows maximum loading conditions of the airplane, allowable speeds, performance capabilities, etc.

Upon selecting the aeronautical charts to be used for the flight, it is important to check their dates of issue to make certain they are current. Prior to takeoff, the charts should be folded in a manner that will allow easy reading in flight without having to unfold or refold the rather large, cumbersome sheets.

On cross-country flights, inexperienced and lackadaisical pilots often allow the cockpit to look like a shambles with charts scattered all over the floor, drop the one-and-only pencil under the seat, and permit the navigation computer to slide to the back of the airplane. Good "cockpit management" is essential to a pleasant, efficient flight just as it is for experienced pilots who conduct their flights in a professional manner.

"A place for everything and everything in its place" is a wise saying which certainly has true meaning in the conduct of a well-executed cross-country flight. Thorough preflight planning, plus efficient cockpit management, contribute substantially to safe and efficient flight.

The charts should be organized in proper sequence according to their order of use. They should be put in a special place away from other charts that definitely won't be used. How to keep charts from falling off the pilot's lap is one thing some pilots have never solved. Laps just aren't big enough, have round edges and lead to everything falling to the floor—and usually out of reach. However, while the top of the instrument panel appears tempting as a storage area, the area should never be used in flight for charts or other equipment.

The pilot should establish a definite, convenient storage place for pencil, navigation computer, flight log, and note pad. After each time they are used, they should be put back where they belong—not loosely on the lap. Rubber bands, clips, clipboard, or a pad that can be strapped to the pilot's leg are useful in holding the needed items in place.

Pilotage

Flying cross-country when using only a chart and flying from one visible landmark to another is known as "pilotage." This method requires that the flight be conducted at comparatively low altitudes so that landmarks ahead may be seen easily. Therefore, it cannot be used effectively in areas which lack prominent landmarks, or under conditions of low visibility. Among the advantages of pilotage are the facts that it is comparatively easy to perform, and it does not require special equipment. The chief disadvantage is that a direct course is usually impractical because it is often necessary to follow a zigzag route to prominent geographical landmarks, often resulting in a longer flight.

Inasmuch as magnetic compasses are installed in all airplanes, pilotage is *not* the more commonly used method for a long cross-country flight. However, pilotage alone may be used on any course which affords plenty of prominent landmarks. It is accomplished by selecting two landmarks on the desired course, and then steering the airplane so that the two objects are kept aligned over the nose. Before the first of the two landmarks is reached, another more distant object should be selected and a second course steered. For the use of landmarks the pilot should take advantage of roads, railroads, and streams, but should beware of those that may vanish in mountains or dense foliage.

There is no set rule for selecting landmarks. Each locality has its own peculiarities. Consequently, a particular type of landmark may be more distinctive in one section of the country than it is in another. The general rule to follow is never to place complete reliance in any single landmark, but to use a combination of at least two or more, if practicable. One cannot depend upon only a silver water tank to identify a particular town, as every adjacent town may also have one. For identification, the pilot should check the time of passing the town against the estimated time of arrival, the number and direction of railroad lines into and out of the town, the adjacent road patterns, any nearby rivers, and the overall layout of ground references. It is well to remember that charts are not infallible. Man-made landmarks are continually being constructed throughout the country, and may not yet appear on the chart being used.

The inexperienced pilot should be impressed with the importance of becoming oriented with the surrounding area, and should be warned against spending too much time trying to find some specific landmark to the exclusion of noting the overall pattern of the area. The general pattern of roads, railroads, mountains, and other large features, give to each locality a certain distinctiveness of its own that constantly should be borne in mind. If this is done, it will be comparatively simple to single out specific landmarks from time to time to obtain accurate checkpoints.

Intersecting lines, such as railways, highways, or rivers which meet near the destination, make excellent reference brackets or boundaries to prevent the pilot from unintentionally passing to one side of the destination. In this regard, to help locate the desired airport, one need only ensure that the airplane does not cross a certain railway, highway, or river that forms the bracket.

When leaving a large city, it is usually poor practice to immediately follow a railroad or highway. It is better to fly a steady course for a few minutes after leaving the airport and not attempt to locate the desired railroad or highway until well clear of the maze of lines on the ground that generally surround a large metropolitan area.

To place dependence upon other than *major* roads is, in general, poor practice unless nothing better is available. Aeronautical charts do not profess to show all roads, as that would increase the clutter on the charts and require constant revisions. It will be found that some roads shown on the charts may be neither accurate, complete, nor up to date. However, the charts do attempt to give the pilot the general road pattern for the vicinity; but it should be remembered that new highways and roads are being constructed continually.

Roads shown on the charts are primarily those which are conspicuous when viewed from the air. It will be noted that surrounding terrain and other factors sometimes have a camouflage effect; the charts take this into consideration and may omit sections of the road which do not serve as good landmarks. In areas where there are few roads, a crossroad properly identified along with other landmarks in the vicinity usually will lead to a fair sized town.

Rivers usually are excellent landmarks. The curves of a winding river offer many good position references, or fixes. However, it must be remembered that rivers also have their peculiarities. In flat, woody country, rivers are sometimes confusing and hard to trace. The water is sometimes hard to detect through the dense foliage unless the sunlight is reflected just right or unless it is directly below. Some rivers have so many tributaries that it is very difficult to trace the main stream. When a river is in flood stage, its appearance and surrounding area may be so changed that it will be unsafe to depend upon it as a visual checkpoint or reference line.

In the southwestern areas, the chances are that a large river shown on the chart will turn out to be nothing more than a dry arroyo during most of the year. Only by careful examination of the topography of the surrounding country is the pilot able to find adequate traces of it.

Except where there are a great many of them in the area, lakes generally offer excellent references or fixes. As with rivers, lakes also may dry up completely and disappear during certain seasons of the year. There are, also, many small artificial lakes not marked on the chart, that show up well when the sunlight reflects from them. In connnection with artificial lakes created as the result of hydroelectric projects, it is well to remember these too are rapidly being developed, and they may not yet appear on the latest chart.

On the sectional charts, lakes are very often shown much larger than they really are, in order to give the general contour. Sometimes what appears to be a small lake on a chart may be just a little farm pond, or it may have dried up entirely.

Dead Reckoning

Dead reckoning, as applied to flying, is the navigation of an airplane solely by means of computations based on airspeed, course, heading, wind direction and speed, groundspeed, and elapsed time. Dead reckoning which, to oversimplify, is a system of "determining where the airplane should be on the basis of where it has been." In other words, it is literally deduced reckoning, which is where the term came from, i.e., ded. or "dead" reckoning. The most common form of VFR navigation is a combination of dead reckoning and pilotage, during which the course flown and the airplane's position are calculated by true dead reckoning and then constantly corrected for error and variables after visually checking nearby landmarks.

The simplest kind of dead reckoning assumes that the air is calm. If the wind were to remain calm, the airplane's track (path) over the ground would be the same as the intended course and the groundspeed would be the same as the airplane's true airspeed.

As shown in Fig. 12–3, an airplane flying eastward at an airspeed of 120 knots in still air, will have a groundspeed exactly the same —120 knots. If the mass of air is moving eastward at 20 knots, the speed of the airplane (airspeed) through the air will not be affected, but the progress of the airplane as measured over the ground will be 120 plus 20, or a groundspeed of 140 knots. On the other hand, if the mass of air is moving westward at 20 knots, the airspeed of the airplane still remains the same but the groundspeed becomes 120 minus 20 or 100 knots.

In preparation for the combination of elementary dead reckoning and pilotage on a cross-country flight, the proposed course line to be followed is drawn on the chart and measured with a protractor to determine the number of degrees it lies from true north. This

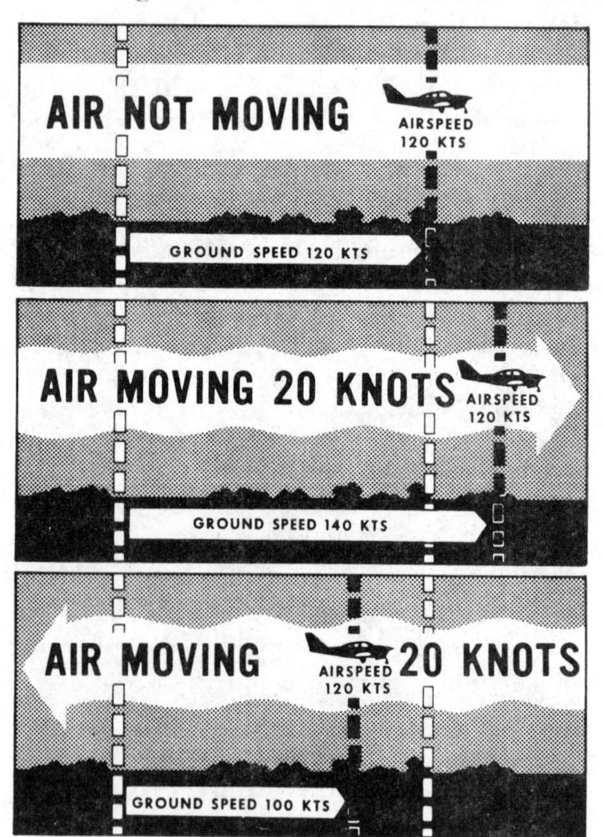

Figure 12–3 Airspeed VS Groundspeed

is called "true course" (TC). To the number of degrees thus measured, the pilot must add or subtract the "magnetic variation" (shown on the chart by a red, dashed, isogonic line). The variation shown is the angular difference between true north and magnetic north in that particular locality. As we know the true North Pole and the magnetic North Pole are not located at the same place. Since the magnetic compass points to the magnetic North Pole, it indicates a magnetic direction. The course line on the chart is measured in relation to true North; therefore, the "True Course" must be converted to the "Magnetic Course." When converting from a "true" to a "magnetic" course, the pilot adds westerly variation, or subtracts easterly variation. The resultant figure should be written down beside the course line and is termed the "Magnetic Course" (MC).

In actual flight, winds may cause considerable deviation from the desired ground track unless some sort of correction is made. An airplane flying within a moving mass of air will move with the air in the same direction and speed that the air is moving over the ground. Consequently, at the end of a given time period, the airplane will be in a position which resulted from a combination of the two motions: the movement of the air mass in reference to the ground, and the forward movement of the airplane though the air mass.

As shown in Fig. 12–4, if the airplane is flying eastward at an airspeed of 120 knots, and the air mass is moving southward at 20 knots, the airplane at the end of 1 hour will be at a point that is approximately 120 miles eastward of its point of departure (due to its progress through the air) and 20 miles southward (due to the motion of the air). Under these circumstances the airspeed remains 120 knots, but the groundspeed is determined by combining the movement of the airplane with the movement of the air mass. Groundspeed can be measured as the distance traveled in 1 hour from the point of departure. The groundspeed can be computed in flight by noting the time required to fly between two points a known distance apart (two checkpoints on the course). It also can be calculated before flight by plotting the airplane's heading and airspeed along with a wind vector. This is done

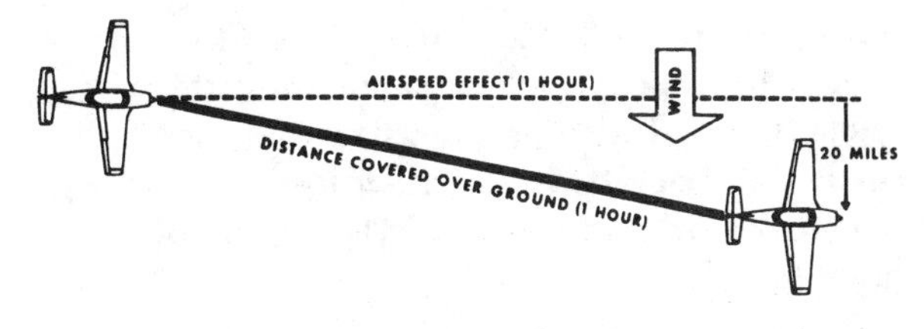

Figure 12–4 Effect of Wind in One Hour

most frequently by means of an air navigation computer—a circular type slide rule.

The direction in which the airplane is pointing as it flies is its *heading* (Fig. 12–5). Its actual path over the ground, a combination of the motion of the airplane and the motion of the air, is the *ground track*. The angle between the heading and the track is termed *drift angle*. As demonstrated in Fig. 12–5, if the airplane is headed along the course line with the wind blowing from the left, the track will not coincide with the desired course. The wind will drift the airplane to the right, so the track will lie to the right of the desired course.

By anticipating the amount of drift, the pilot can counteract the effect of the wind, thereby making the ground track of the airplane coincide with the desired course. If the mass of air is moving across the course from the right, the airplane will drift to the left, and a correction must be made by heading the airplane sufficiently to the right to offset this drift. If the wind is from the left the correction must be made by turning the airplane to the left—into the wind (Fig. 12–5).

Familiarity with a magnetic compass is essential, not only so that it can be read easily and accurately, but in order that the pilot may have complete confidence in the instrument. Despite tales of compass failure, actual

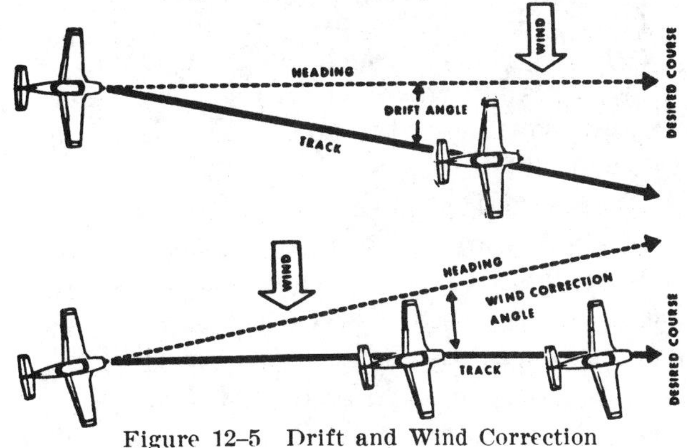

Figure 12–5 Drift and Wind Correction

instances of such are rare. Compasses may spin temporarily, or read inaccurately over areas where there is a natural magnetic disturbance, but its inherent idiosyncrasies are not to be seized upon as the cause of poor navigation.

Trusting the compass implies a knowledge of how it functions. The magnetic compass does not point to true north, but is controlled by the lines of force in the earth's magnetic field. These lines are not parallel to the meridians (lines which meet at the true North and South poles of the earth); they diverge from the meridians at different angles in different locations on the earth's surface. The amount of this difference, or "variation," is indicated on all aeronautical charts for the benefit of pilots.

The compass is affected also by the attraction of metal, such as the engine, wiring, radio, and steel structure of the airplane. Compensation is made for this through a ground procedure referred to as "swinging the compass" and the deviation from various magnetic directions is noted on a compass card installed in the cockpit. Thus, in computing a compass course, both variation and deviation must be allowed for on all headings.

Keeping an airplane flying on a desired heading using only a magnetic compass can be frustrating to a pilot who is not familiar with the compass's characteristics and errors. This lack of understanding can result in the pilot thinking that the compass has malfunctioned and headings are unreliable.

No attempt should be made to read the compass until the airplane has been held straight and level for at least 30 seconds because the compass is deflected during any inclination or banking. This is the result of the Magnetic North Pole (toward which the needle attempts to point) being actually below the horizon, due to the curvature of the earth. This is called magnetic dip.

In level flight it is easier to fly the average compass indication than to try to hold an exact heading. To steer a course after the airplane has been established accurately on the desired course, one should look directly ahead to select some object to head toward, and then note the compass indication to determine whether the proper heading is being maintained. If it is not, a slight change should be made relative to the object.

It will be found that a knowledge of the principles of dead reckoning is very important not only in planning the course and determining the elapsed time required for the flight, but in assisting the pilot in determining the airplane's position after having become disoriented or confused as to the airplane's position. By using information gained on the part of the flight already completed, the pilot who is thoroughly familiar with these principles can determine the airplane's approximate present position. In this way, it is possible to restrict the search for identifiable landmarks to a limited area to verify calculations and to relocate one's self.

Losing Track of Position

Whether a pilot becomes confused depends on whether or not a definite plan is followed. The greatest hazard to a pilot failing to arrive at a given checkpoint at a particular time, is *panic.* The natural reaction is to fly to where it is assumed the checkpoint is located. On arriving at that point and not finding the checkpoint, a second position is usually *assumed*, and the panicked pilot will then fly in another direction for some time. As a result of several of these wanderings, the pilot may have no idea where the airplane is located. Generally, if planning was correct and the pilot uses basic dead reckoning until the estimated time of arrival (ETA) runs out, the airplane is going to be within a reasonable distance of the planned checkpoint.

There are several different actions to be followed when the pilot is unsure of the airplane's position while using pilotage or dead reckoning. The best choice depends upon the circumstances, but usually they should be applied in the following sequence.

When unsure of one's position, the pilot should continue to fly the original heading and watch for recognizable landmarks, while rechecking the calculated position. By plotting the estimated distance and compass direction flown from the last noted checkpoint as though there were no wind, the point so determined will be the center of a circle within which the airplane's position may be located (Fig. 12–6). This is often called a "circle of error." If it is certain that the wind is no more than 30 knots, and it has been 30 minutes since the last known checkpoint was noted, the radius of the circle should be about 15 nautical miles. It is then a matter of continuing straight ahead, and checking the landmarks within this circle. The most likely position will be downwind from the desired course.

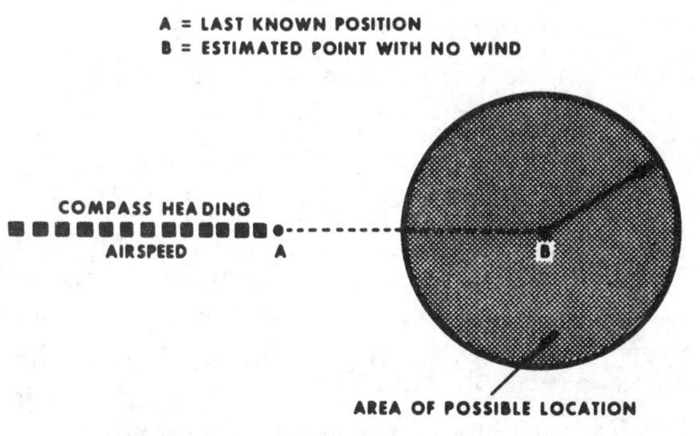

Figure 12–6 Estimate Circle of Error

If this procedure fails to identify the position, the pilot should change course toward the nearest concentration of prominent landmarks shown on the chart. If some town or developed area is sighted, the pilot can circle low to observe identifiable features of markings; but the pilot should comply with the minimum safe altitudes prescribed in FAR Part 91.

In the event these methods are ineffective in locating position, or when fuel exhaustion or darkness is imminent, it is recommended that the pilot make a precautionary landing while adequate fuel and daylight are still available. It is desirable, of course to land at an airport, but if one cannot be found, a suitable field may be used. Prior to landing at other than an established airport, the pilot should first survey the area for obstructions or other hazards.

When a landmark is finally recognized, or a probable fix obtained, the pilot should at first use the information both cautiously and profitably. If there is a pronounced discrepancy, it is well to be dubious of the new fix until it can be positively identified. NO ABRUPT CHANGE in course should be made until a second or third landmark is found to corroborate the first.

It is well to determine the probable cause of getting off the course originally so that the error will not be repeated. Miscalculations in determining the ground track may arise from miscalculating the wind drift, applying the wind forecast at a level at which drift is opposite to that at cruising altitude, applying magnetic variation to the compass incorrectly, or simply from misreading the compass.

When the airplane seems to have made an abnormally high or low groundspeed, the error may be caused by using the wrong mileage scale (nautical versus statute), misreading the clock, skipping mileage marks when scaling the course line on the chart, using improper airspeed indications (knots versus MPH), or from other causes. Whatever the error, once determined it should be borne in mind to avoid repetition on the remainder of the flight. In some cases it may be necessary to reestimate the fuel hours remaining and to change the destination accordingly.

DF (VHF/UHF Direction Finder) equipment is of particular value in locating lost aircraft and in giving guidance to the pilot. DF is a ground-based radio receiver with an azimuth display used by the operator of the ground station where it is located, usually at FSSs and control towers. The pilot needs only two-way VHF radio communications capability to make use of the services available.

The ground equipment consists of a directional antenna system, an azimuth display screen, and a VHF (and UHF) radio receiver. At a radar-equipped tower or center, the cathode-ray tube indications may be superimposed on the radarscope.

The DF display indicates the magnetic direction of the airplane from the station each time the pilot transmits. Where DF equipment is tied into radar, a strobe of light is flashed from the center of the radarscope in the direction of the airplane.

Regardless of the equipment used, all operate on the principle of indicating the bearing from the station to the airplane or from the airplane to the station. It is necessary for the pilot to be transmitting in order for the DF operator to take a bearing. Once this bearing has been established, it is relatively easy for the operator to direct the pilot to the desired point.

When a pilot is in doubt about the airplane's position, or feels apprehensive for the safety of the flight, there should be no hesitation to request assistance from the nearest FSS. Search facilities, including radar, radio, and DF stations, are ready and willing to help, and there is no penalty for using them. Delay in asking for help has often caused accidents and cost lives. Safety is not a luxury that is beyond the pilot's means.

Radio Aids to Navigation

In addition to the navigation methods discussed in the preceding sections, the FAA provides several types of radio aids to air navigation. For example, the Very High Frequency omnirange (VOR) and the Low-Frequency nondirectional radiobeacon (NDB) are particularly useful to VFR pilots for navigation guidance.

For many years, the VOR has been the basic radio aid to navigation. Transmitting frequencies of omnirange stations are in the VHF (very high frequency) band between 108 and 118 MHz. The word "omni" means *all*, and an omnirange is a VHF radio station that projects radials in all directions (360°) from the station, like spokes from the hub of a wheel. Each of these "spokes," or radials, is denoted by the outbound magnetic direction of the "spoke". A radial is defined as "a line of magnetic bearing extending *from* an omnidirectional range (VOR)." However, due to the features of VOR receivers, it is possible to fly either *to* or *from* omniranges in any direction.

A few of the advantages of using omniranges are:

1. A flight may be made *to* a VOR from any direction by flying inbound on the selected radial.
2. A flight may be made *from* the VOR to any destination by flying outbound on the selected radial.
3. When within receiving range of two or more VOR's, a fix or lines of position may be determined quickly and easily by taking bearings on the stations.
4. Static-free reception, and the elimination of complex orientation procedures, provide easy identification and utilization of the VOR facility.

An important fact is that VOR signals, like other VHF transmissions, follow an approximate line-of-sight course. Therefore, reception distance increases with an increase in the airplane's altitude. Conversely, flight at low altitude decreases the reception distance. A means is usually provided on omnireceivers to warn the pilot when the signal is too weak for satisfactory reception.

Though new and improved types of electronic equipment are constantly being developed to make flying safer and easier, VOR and VORTAC (VOR-Tactical Air Navigation) are the basic VHF systems currently in use for radio navigation. In addition to the bearing information obtained from the omnirange, VORTACs supply pilots of airplanes which have distance measuring equipment (DME) with the distance of the airplane from the station. In providing the bearing and distance, the pilots can determine the airplane's exact location, eliminating the need for taking cross bearings on two or more stations. However, pilots of airplanes equipped with only a VOR receiver can still use a VORTAC station for bearing information just as they use a normal VOR station. (Note: For simplicity, the term VOR will be used in this handbook to include both VOR and VORTAC stations.)

VOR stations are assigned three-letter identifications. At most stations these identification letters are broadcast continuously in Morse code. Some stations are also identified by a voice recording (Example: The spoken words, "Kingfisher V-O-R".)

VOR receivers are very simple to operate and consist of three basic components used by the pilot. One component is the omnibearing or course selector (OBS), which enables the pilot to select the course or radial desired. A second component is the TO-FROM indicator, which shows the pilot whether the selected course would take the airplane TO or FROM the station. The third is a course deviation indicator (CDI—often called the "LEFT-RIGHT" indicator), which by means of a vertical needle tells the pilot when the airplane is on the selected course, or left, or right of course. Using these three components, the pilot obtains visual indications which give a variety of information and guidance. The "TO-FROM" indicator, omnibearing or course indicator, and course deviation indicator are usually combined into a single display on the instrument panel (Fig. 12–7)

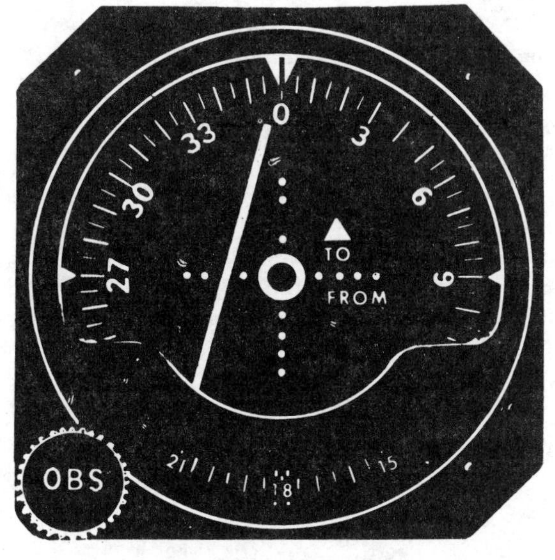

Figure 12–7 VOR Indicator

Because VOR receivers are manufactured with varying degrees of complexity, pilots should familiarize themselves with the receivers installed in the airplanes being used to understand the use and limitations of the equipment.

Since accuracy is an important factor in any navigation equipment, pilots should check their VOR receivers periodically to be sure they are functioning properly. Procedures for checking VOR receivers are published by FAA in the Airman's Information Manual.

VOR Navigation

The location, transmitter frequency, and identification code of VORs and VORTACs are shown on aeronautical charts. Along with a magnetic compass rose at each station to depict the radials, designated airways which emanate from the station are also shown with their outbound magnetic courses (radials) specified.

To fly from one VOR directly to another VOR the pilot should draw a course line (or use the printed centerline of the airway) between the two stations to determine the magnetic course or radial that will be followed (Fig. 12–8). After takeoff the omnireceiver should be tuned to the assigned frequency of the VOR from which the flight will start and the identification of the station verified. Then the omnibearing selector should be set to the planned magnetic course from that station. The TO-FROM indicator will show FROM and the course deviation indicator will be deflected to the right, to the left, or centered, depending on where the airplane is located in relation to the selected course. The airplane should then be turned so that its heading is the same as that set on the omnibearing indicator. The airplane will now be flying parallel to the radial. This will help the pilot determine which way to turn to intercept the radial if not exactly on the radial. If off the radial a considerable distance, a turn should be made to a heading that will intercept the planned course at about a 30° to 45° angle.

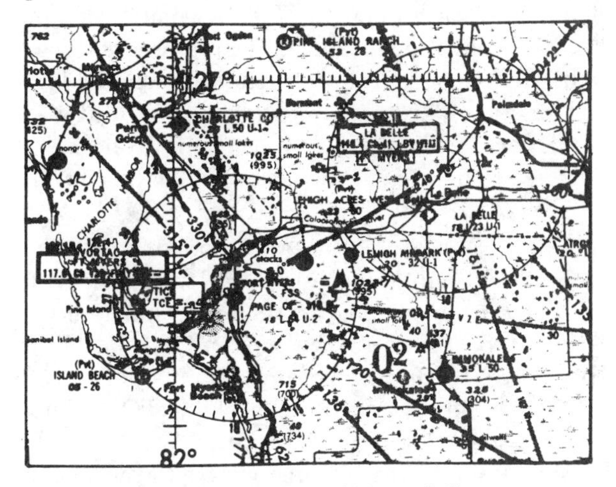

Figure 12–8 VOR Radials and Courses

If the course deviation needle was originally deflected to one side or the other, it will gradually move to the center as the airplane approaches and intercepts the selected radial or course. When the needle is centered, indicating that the airplane is now on course, a turn should be made to the preplanned heading. However, while following the radial the airplane may be affected by crosswinds. Unless a correction is made for the crosswind as the flight continues, it will be noted that the course deviation indicator moves to the upwind side of center. This indicates that the airplane is drifting off course. Instead of making a sharp turn to get back on the desired radial, a turn of 10° to 20° should be made to reintercept the radial. If a turn of more than 20° is made when the needle is off center only slightly, chances are the airplane will overshoot the radial, particularly when close to the station, and the flight will then become a series of time-wasting zigzags. When the deviation indicator needle is again centered, showing that the airplane is back on course, approximately a 5° drift correction should be made, and the flight continued making only slight heading changes as necessary to keep the needle centered.

Arithmetical computations for crosswind corrections when flying cross-country using VOR are actually unnecessary. Flying whatever heading that results in keeping the course deviation indicator needle centered automatically "crabs" the airplane into the wind just the right amount to fly a straight-line course to (or from) the VOR station. As long as the course deviation indicator needle shows the airplane to be on a radial, the airplane is on it. The airplane may be crabbed any number of degrees, and the only concern is to keep the course deviation indicator needle centered.

Upon reaching the approximate halfway point between the departure VOR and the next VOR on the planned course, the omnireceiver should be tuned to the next VOR—but it is essential to positively identify it. The TO-FROM indicator will now show TO. Since the outbound radial of the first VOR usually coincides with the inbound course to the second VOR (Fig. 12–8), it is normally not necessary to reset the bearing selector to a new course. When it is reset for the radial of the approaching VOR, it must be remembered that the radial shown on the chart is the outbound course. To follow it inbound, the *reciprocal* of the shown radial must be set on the omnibearing selector. Otherwise the indicator would provide "reverse sensing"—that is, the course deviation indicator would be on the wrong side and it would be necessary to turn *away* from the needle. That could result in confusion. After setting the OBS to the proper course, the pilot is again set for another period of navigation by simply correcting toward the needle.

Upon nearing the second VOR, the deviation indicator needle will begin to deviate or fluctuate rather rapidly. This means that the airplane is close to the station. Under this condition, rather than make heading corrections for each fluctuation, the compass heading that has been keeping the airplane on the course to the station should be maintained, and the needle fluctuations ignored.

Within a few seconds the airplane will be over the station. The deviation indicator needle will swing sharply back and forth, and the TO-FROM indicator will change from TO to FROM. This means the airplane is now heading away from the station. If there is a slight difference between the previous inbound course and the desired outbound course from this VOR, the bearing selector to the new outbound course should be set and again the heading should be changed to keep the deviation indicator centered.

At certain times a red flag alarm or OFF indication will appear adjacent to the course deviation needle, and the needle itself will swing from side to side as though "hunting." The TO-FROM needle may settle in a neutral or center position and may not move out of the neutral indication, even though the receiver is tuned to the proper frequency. Such visual warnings indicate that the received signal is too weak to give reliable indications.

These warnings appear when obstacles or mountainous terrain interfere with the line-of-sight transmission, or the airplane is too low or too far away from the station to obtain steady, reliable signals. The alarm provides a good safeguard against using erroneous or unreliable bearing readings. In a case like this, the pilot should tune to the next VOR on the course and maintain a constant heading toward the station until the flag alarm disappears and the course deviation needle again comes to rest at or near the center position.

The need for identifying the VOR station cannot be overemphasized. Receivers with hand tunable oscillators are frequently found to be off calibration. Merely tuning to the indicated frequency on the dial is no guarantee of receiving the desired station. Also, frequencies of ground stations are sometimes changed and unless the pilot has current information at hand, the wrong station could be tuned in.

One of the main advantages of VOR is that the pilot can quickly and easily locate the airplane's exact position by taking bearings from two VOR stations. The airplane's position (fix) is where the two bearings cross. For the most accurate fix, the cross bearings should be 90° to each other or as close to a right angle as possible. If several stations are available for a bearing check, it is best to use the one which will give the closest to a right angle crossbearing. Position checks or fixes along a course are most convenient if the pilot has two omnireceivers and keeps one set to maintain the course being flown while using the other to take the crossbearing on the off-course VOR. For uniformity and to avoid confusion, a bearing to establish a fix is best made by using degrees FROM the off-course station rather than degrees TO the station. In this manner the bearings or lines of position can be drawn from the station since the station's location is known and the airplane's location is still unknown.

To determine the airplane's position in relation to intersections of airways while tracking on a radial, the pilot can use the same crossbearing or fix techniques used to pinpoint the airplane's position. Since the bearing of the intersection from the off-course VOR is known, this should be set in advance. As the intersection is approached, the course deviation needle will move towards center from full deflection. When the needle is in the center position, the airplane is at the intersection. Again, this procedure is even simpler if there are two omnireceivers—one for enroute navigation; the other for locating the intersection.

ADF Navigation

Many airplanes are equipped with ADF (Automatic Direction Finder) radios which operate in the low- and medium-frequency bands (Fig. 12–9). By tuning to low-frequency (LF) radio stations such as nondirectional radiobeacons (NDB), or to commercial broadcast (AM) stations, a pilot may use ADF for navigation in cross-country flying. NDB frequency and identification information may be obtained from aeronautical charts and the Airport/Facility Directory. Some major commercial broadcast station locations and frequencies are shown on sectional aeronautical charts. Positive station identification is essential when navigating with the ADF. The Morse codes are used to identify the NDB stations while the commercial broadcast stations are identified at random times by the station's announcer.

Figure 12–9 ADF Receiver

Most ADF radio receivers signals in the frequency spectrum of 190 kHz to 1750 kHz, which includes LF and MF navigation facilities, and the AM commercial broadcast stations. Primarily for air navigation, the LF/MF stations are FAA and privately operated nondirectional radiobeacons. Some broadcast stations operate only during daylight hours, and many of the low-powered stations transmit on identical frequencies and may cause erratic ADF indications.

The ADF has automatic direction seeking qualities which result in the bearing indicator always pointing *to* the station to which it is tuned. That is, when the bearing pointer is on the nose position, the station is directly ahead of the airplane; when the pointer is on the tail position, the station is directly behind the airplane; and when the pointer is 90° to either side (wingtip position), the station is directly off the respective wingtip.

The more commonly used ADF instrument has a stationary azimuth dial graduated from 0° up to 360° (with 0° at the top of the instrument to represent the airplane's nose). In this type, the bearing pointer shows only the station's *relative bearing*, i.e., the angle from the nose of the airplane to the station (Fig. 12–10). A more sophisticated instrument called a Radio Magnetic Indicator (RMI) uses a 360° azimuth dial which, being slaved to a gyro compass, rotates as the airplane turns and continually shows the Magnetic Heading at the top of the instrument (Fig. 12–10). Thus, with this rotating azimuth referenced to a magnetic direction, the bearing pointer superimposed on the azimuth indicates the Magnetic Bearing to the station.

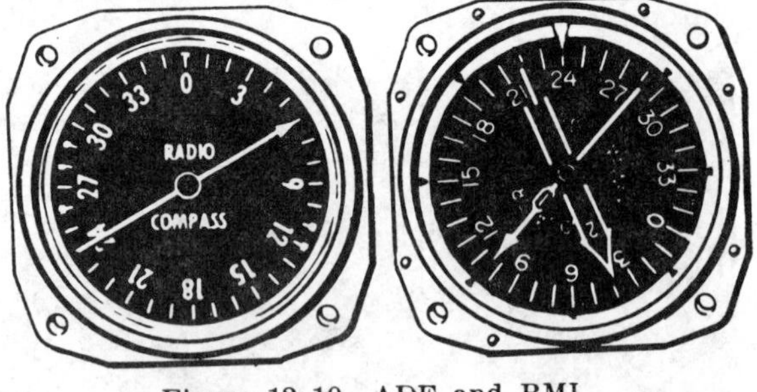

Figure 12–10 ADF and RMI

The easiest, and perhaps the most common method of using ADF, is to "home" to the station. Since the ADF pointer always points to the station, the pilot can simply head the airplane so that the pointer is on the 0° or nose position. The station, then will be directly ahead of the airplane. With a crosswind, however, the airplane would continually drift to the side and, unless a corrective change in heading is made, would no longer be flying straight to the station. This would be indicated by the pointer moving to the upwind side of the nose position on the dial. By periodically turning the airplane into the wind (toward the head of the pointer) so as to continually return the pointer to the 0° position, the airplane can be flown to the station, although in a curving flightpath as a result of wind drift. The lighter the crosswind and the shorter the distance from the station, the less the flightpath curves. Upon arrival at and passing the station, the pointer will swing 180° from a nose position to a tail position.

ADF should be considered as a moving, "fluid" thing. The number to which the bearing indicator points on the *fixed* azimuth dial has no directional meaning to the pilot until it is related to the airplane's heading. To apply this relationship, the magnetic heading must be observed *carefully* when reading the Relative Bearing to the station. Any time the airplane's heading is changed, the Relative Bearing will be changed an equal number of degrees.

To determine the Magnetic Bearing to a station on a fixed ADF azimuth dial, the pilot may imagine the airplane as being in the center of the fixed azimuth, with the nose of the airplane at the 0° position, the tail at the 180° position, and the left and right wingtips at the 270° and 090° positions, respectively. When the pointer is on the nose position, the airplane is heading straight to the station and the Magnetic Bearing can be read directly from the *magnetic compass.* If the pointer is left or right of the nose, the pilot should note the direction and number of degrees of turn that would (if the airplane were to be headed to the station) move the pointer to the nose position, and mentally apply this to the airplane's heading. For example, in Fig. 12–11, when the airplane is headed 090°, the pointer is 60° to the left of the nose position. A turn 60° to the left would place the pointer on the nose position. If the airplane were to be turned 60° to the left, the heading would be 030°. Thus, the magnetic course to the station is 030°. The bearing *from* the station is the reciprocal—or 210°.

One of the several valuable uses of ADF is the determination of the airplane's position along the course being flown. Even though the airplane is following a course along a VOR radial, obtaining an ADF bearing that crosses

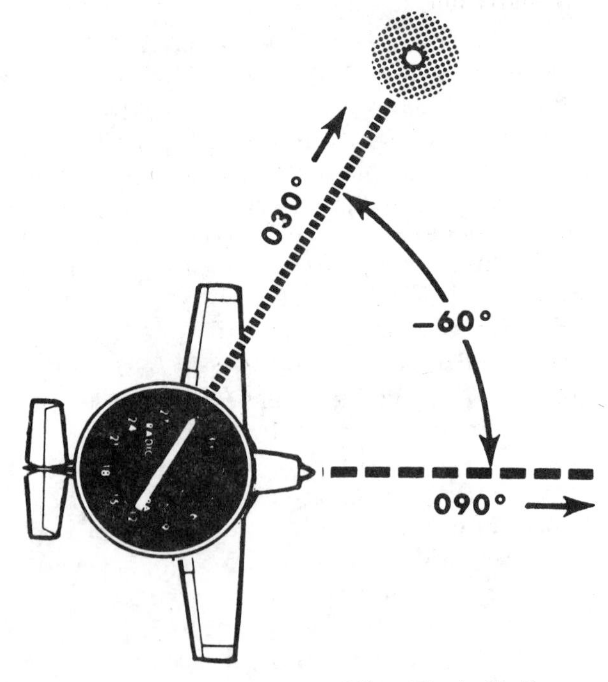

Figure 12–11 Turn Left 60° to Fly to Station

the course will establish a "fix" or position along that course. This is particularly advantageous when an off-course VOR is not available for a cross bearing or when the only VOR receiver must be used as the primary tracking system.

Diversion to an Alternate

Among the aeronautical skills that all pilots must have is the ability to plot courses in flight, to alternate destinations when, for one reason or another, continuation of the flight to the original destination is impracticable. This may be accomplished by means of pilotage, dead reckoning, or radio navigation aids.

Course, time, speed, and distance computations in flight require the same basic procedures as those used in preflight planning. Because of the limitations in cockpit space and available equipment, however, and because the pilot's attention must be divided between solving the problem and operating the airplane, advantage must be taken of all possible shortcuts and rule-of-thumb computations.

It is rarely practical while in flight to actually plot a course line on the chart used, and to mark checkpoints and distance as is usually done on preflight planning. Because the alternate airport selected in an emergency is usually not very far from the original course and known position, such actual plotting is seldom necessary.

Courses to alternates can be measured accurately with a protractor or plotter, but they can also be measured with reasonable accuracy using a straightedge and the compass roses shown at VOR stations on the chart. The VOR radials and airway courses (already oriented to magnetic direction) printed on the chart can be used satisfactorily for approximation of magnetic bearings during VFR flights. This approximation can be made on the basis of the radial of a nearby VOR or airway that most closely parallels the course to the station. The pilot must remember that the VOR radial or printed airway direction is *outbound* from the station. To find the course *to* the station, it may be necessary to determine the reciprocal of the parallel radial or airway. Distances can be determined by using the measurements on a plotter, or by placing a finger at the appropriate place on the straight edge of a piece of paper and then measuring the approximate distance on the mileage scale at the bottom of the chart.

If radio aids are used to divert to an alternate, the pilot should select the appropriate facility, tune to the proper frequency, and determine the course or radial to intercept or follow.

In the event the diversion to an alternate airport results from an emergency, it is important for the pilot to divert to the new course as early as possible. Before changing course, first consider the relative distance to all suitable alternates, select the one most appropriate to the emergency at hand, and then determine the magnetic course to the alternate selected. The heading should be changed to establish this new course immediately, and then later, the wind correction, actual distance, and estimated time and fuel required can be computed accurately while the airplane is proceeding toward the alternate.

To complete all plotting, measuring, and computation involved *before* diverting to the alternate airport may only aggravate an actual emergency. It is important to develop the ability to orient one's self immediately with a chart showing the terrain and landmarks over which the flight is made, and to make rapid and reasonably accurate computations of headings and arrival time estimates.

CHAPTER 13

Emergency Flight By Reference to Instruments

This chapter is intended to provide guidance in developing the ability to maneuver the airplane for limited periods by reference to flight instruments and to follow radar and DF instructions from ATC, when outside visual references are lost due to unexpected adverse weather. In an emergency situation, this ability could save the pilot's life and those of the passengers, but intentional ventures into even marginal weather with no more than this training will eventually end in a serious accident.

During this training, pilots must understand that this emergency use of flight instruments does not prepare them for unrestricted operations in instrument weather conditions. *It is intended as an emergency measure only.* Intentional flight in such conditions should be attempted only by those who have been thoroughly trained and certificated as instrument pilots. Persons interested in pursuing a comprehensive instrument flying program should study the material in AC 61–27B, Instrument Flying Handbook, and other pertinent publications, and complete a suitable instrument flying training course under the guidance of a certificated instrument flight instructor.

Accident investigations reveal that, as a related factor, weather continues to be cited more frequently than any other in general aviation accidents. The data also show that weather-involved accidents are more likely to result in fatal injury than are accidents not involving weather. Low ceilings, rain, and fog continue to head the list in the fatal, weather-involved general aviation accidents. The pilot involvement in this type of accident is usually the result of inadequate preflight preparation and/or planning, continued VFR flight into adverse weather conditions, and attempted operation *beyond the pilot's experience/ability level.* In far too many cases it was determined that the pilot did not obtain a preflight weather briefing. It appears logical, then, to assume that if an adequate preflight briefing had been obtained, unexpected weather conditions would not have been encountered and many of the accidents would not have occurred.

The only way a pilot can control an airplane safely in a low-visibility environment is by using and trusting flight instruments. Man's orientation senses are not designed to cope with flight when external visual references are obscured by clouds, fog, haze, dust, darkness or other phenomena, unless visual reference is transferred to the flight instruments. When the visual sense is provided with reference points such as the earth's horizon or the flight instruments, there is usually no problem with airplane attitude control since the visual sense overrides the other senses.

It is in situations where visual references such as the ground and horizon are obscured that trouble develops, especially for pilots who lack training, experience, and proficiency in instrument flight. The vestibular sense (motion sensing by the inner ear) in particular tends to confuse the pilot. Because of inertia, the sensory areas of the inner ear cannot detect slight changes in the attitude of the airplane, nor can they accurately sense attitude changes which occur at a uniform rate over a period of time. On the other hand, *false* sensations often are generated, leading the pilot to believe the attitude of the airplane has changed when, in fact, it has not. These false sensations result in the pilot experiencing spacial disorientation (vertigo).

When a disoriented pilot actually does make a recovery from a turn, bank, climb, or descent, there is a very strong tendency to feel that the airplane has entered a turn, bank, climb, or descent in the opposite direction. These false sensations may lead to the well-known "graveyard spiral."

All pilots should be aware of these illusions and their consequences. Flight instructors should provide each student with an opportunity to experience these sensations under controlled conditions.

All pilots should be somewhat conservative in judging their own capabilities and should use every means available (weather check, postponed flight, 180° turn-around, precautionary landing at an airport, etc.) to avoid weather situations which overtax one's ability.

If inadvertently caught in poor weather conditions, the VFR pilot should, in addition to maintaining control of the airplane, immediately notify the nearest FAA facility by radio, and follow their instructions. Calmness, patience, and compliance with those instructions represent the best chance for survival.

The use of an airplane equipped with flight instruments and an easy means of simulating instrument flight conditions, such as windshield "slats" or an extended visor cap or hood, are needed for training in flight by reference to instruments.

Instruction in attitude control by reference to instruments should be conducted with the use of all available instruments in the airplane concerned. When an attitude indicator is provided, its use as the primary reference for the control of the attitude of the airplane should be emphasized.

The following discussion of maneuvering the airplane by reference to flight instruments is primarily for student pilots and private pilots who have not received the benefit of integrated flight instruction described briefly in the chapter on Basic Flight Maneuvers.

From the beginning of instruction in maneuvering the airplane by reference to instruments, three important actions should be stressed. First, a person cannot feel control pressure changes with a tight grip on the controls. Relaxing and learning to "control with the eyes and the brain" instead of only the muscles, usually takes considerable conscious effort.

Second, attitude changes should be smooth and small, yet with positive pressure. No attitude changes should be made unless the instruments show a need for change.

Third, with the airplane properly trimmed, all control pressure should be released momentarily when one becomes aware of tenseness. The airplane is inherently stable and, except in turbulent air, will maintain approximate straight-and-level flight if left alone.

It must be reemphasized that the following procedures are intended only as *emergency* means for extracting one's self from a hazardous situation once it has been encountered. The main goal is *not* precision instrument flying; rather, it is to help the VFR pilot keep the airplane under adequate control until suitable visual references are regained.

Straight and Level

To maintain a *straight* flightpath, the pilot must keep the airplane's wings level with the horizon. Any degree of bank (in coordinated flight) will result in a deviation from straight flight, and a change in the airplane's heading. If an attitude indicator is available, straight flight is simplified by merely keeping the wings of the representative airplane level with the representative or artificial horizon (Fig. 13–1). This is accomplished by applying the necessary coordinated aileron and rudder pressures.

Figure 13–1 Straight and Level

The needle of a turn indicator or the representative wings on a turn coordinator will deflect whenever the airplane is turned and will be centered or level when the airplane is in straight flight. Thus, they also can be used to maintain straight flight by applying coordinated aileron and rudder pressures as needed to keep the needle centered or the turn coordinator's airplane wings level.

Regardless of which of these instruments is being used, the heading indicator should be checked frequently to determine whether a straight flightpath is actually being maintained. This is particularly true when flying in turbulent air since every little gust may bank the airplane and make it turn.

Using only the magnetic compass as a reference for maintaining a straight course is difficult but is possible. Therefore, before using the compass, the pilot must learn its characteristics.

When turning from a northerly heading, the compass will lag behind the actual heading of the airplane and in fact, indicate a turn in the opposite direction. The only general compass heading that accurately indicates when and in which direction a turn is started, is southerly. While the magnetic compass may be used to maintain a direction and detect the start of a turn on any heading from approximately 120° to 240°, the pilot will find it easier to detect a turn on a heading of south. On this heading when a turn is inadvertently started, the magnetic compass exaggerates it and correctly shows in which direction the airplane is turning. With practice, even in turbulence, the pilot can hold a reasonably accurate heading in the southern portion of the compass card. Consequently, relatively large movements of the compass should not be regarded with alarm. The compass should be allowed to oscillate if turbulence makes it unavoidable. Smooth and alert control action will result in a surprisingly small actual deviation from desired heading.

At the same time that straight flight is being maintained, the pilot must also control the pitch attitude to keep the airplane level—that is, no gain or loss of altitude. This can be accomplished by reference to several instruments.

The first is the altimeter, which tells when a constant altitude is being maintained or if the flight altitude is changing. A vertical speed indicator, when available, will indicate the rate at which altitude is changing. Either of these instruments, therefore, shows the pilot whether a change in pitch attitude is needed and approximately how much (Fig. 13–1).

Just as when flying by reference to outside visual references, a change in altitude when level flight is desired, requires that the airplane's nose be raised or lowered in relation to the horizon. This can be done most easily with reference to the attitude indicator, by applying elevator pressure to adjust the representative airplane in relation to the horizon bar. The application of elevator pressure should be very slight to prevent overcontrolling. It must be emphasized that turn coordinators provide NO PITCH INFORMATION even though they have an appearance similar to attitude indicators.

In lieu of an attitude indicator, the vertical speed indicator may be used. However, if the instrument shows a climb or descent, the pilot should apply only sufficient elevator pressure to *start* the pointer moving toward the zero indication, since there is a certain amount of lag in the indication. Trying to obtain an immediate zero indication usually results in overcontrolling. When the pointer stabilizes again, additional pressure, if needed, can then be added in increments to get a zero indication and gradually stop the climb or descent. Only after the vertical speed is zero and the altimeter remains constant should an attempt be made to return to the original altitude.

In the case of an airplane having neither attitude indicator nor vertical speed indicator, the airspeed indicator can be used much like the vertical speed indicator to maintain level flight. Remember though that it, too, lags somewhat as a result of the time required for the airplane to accelerate and decelerate after a pitch change is made.

Pilots must be cautioned not to "chase" the pointers on the instruments when flight through turbulent air produces erratic movements.

Straight Descents

When unexpected adverse weather is encountered by the VFR pilot, the most likely situation is that of being trapped in or above a broken or solid layer of clouds or haze, requiring that a descent be made to an altitude where the pilot can reestablish visual reference to the ground. Generally, the descent should be made in straight flight.

A descent can be made at a variety of airspeeds and vertical speeds by reducing power, adding drag (gear and flaps), and lowering the nose to a predetermined attitude. Before beginning the descent, it is recommended that first the descent airspeed and the desired heading be established while holding the wings level. In addition, the landing gear and flaps should be positioned up or down, to help in maintaining either a slow rate of descent, or a fast rate of descent, as desired. Establishing the desired configuration before starting the descent will permit a more stabilized descent and require less division of attention once the descent is started. Rather than attempting to maintain a specific rate of descent, it is recommended that only a *constant airspeed* be maintained.

The following method for entering a descent is effective either with or without an attitude indicator. First the airspeed is reduced to the desired airspeed by reducing power while maintaining straight-and-level flight. When the descent speed is established, a further reduction in power is made, and simultaneously the nose is lowered to maintain a constant airspeed (Fig. 13–2). The power should remain at a fixed (constant) setting and deviations in airspeed corrected by making pitch changes. Jockeying the throttle to control airspeed only adds to the pilot's workload.

Figure 13–2 Straight Descent

If an attitude indicator is available, the pitch attitude can be adjusted by reference to the representative airplane and the artificial horizon, and then checking the airspeed indicator to determine if the attitude is correct. Deviations from the desired airspeed are corrected by again adjusting the pitch attitude. If no attitude indicator is available and the airspeed is too high or too low, the pilot should apply only sufficient elevator pressure to *start* the airspeed pointer moving toward the desired airspeed, since it takes a little time for the airspeed to stabilize. Trying to "nail down" the airspeed immediately will only result in overcontrolling. Additional pressure can then be added as necessary, to attain the desired airspeed.

In any case, the pilot need not be concerned with slight deviations in airspeed. The main objective is to descend at a safe airspeed—well above the stall but not more than the airplane's design maneuvering speed.

While descending, directional control should be maintained by reference to the directional instruments just as described for straight-and-level flight. Pilots are cautioned against "chasing" the instrument pointers.

If any thought was given to the matter before starting the flight, the pilot will have at least a rough idea of the height of obstructions and terrain in the vicinity of the descent. Before starting the descent, then, a decision must be made regarding the minimum altitude to which the descent will be made.

Straight Climbs

Generally, when adverse weather is encountered, a climb by reference to flight instruments is required primarily to assure clearance of obstructions or terrain. However, it may sometimes be advisable to climb to a clear area above a layer of fog, haze, or low clouds.

As in straight descents, the climb should be made at a constant airspeed—one that is well above the stall and yet results in a positive climb. The power setting and the pitch attitude determine the airspeed.

Again, the attitude indicator provides the greatest help in visualizing the pitch attitude. To enter a constant airspeed climb from cruising airspeed (the most likely entry speed), the nose of the representative airplane is raised in relation to the artificial horizon to the approximate climbing attitude (Fig. 13-3). Only a small amount of elevator back pressure should be added to initiate and maintain the climb attitude. The power setting may be advanced

Figure 13-3 Straight Climb

to climb power simultaneously with the pitch change, or, after the pitch change is established and the airspeed approaches the desired climb speed.

If no attitude indicator is available, the pilot need apply only sufficient elevator pressure to *start* the airspeed pointer moving toward the desired climb airspeed and to cause the altimeter to show an upward trend. Because of inertia, speed will not be reduced immediately to the climb speed. The pilot must give the airspeed time to stabilize, then should apply whatever additional elevator pressure is needed to attain and maintain the desired airspeed. During the climb, the power should remain at a fixed (constant) setting and deviation in airspeed should be corrected by making pitch changes. Making throttle adjustments to correct for airspeed is unnecessary and burdensome.

As in straight descents during the emergency use of flight instruments, maintaining a precise airspeed is not important. The primary objective is keep the airplane climbing—and *not allow a stall to occur.*

While climbing, the directional instruments should be scanned to detect any lapse of directional control just as in straight-and-level flight and straight descents. Unless a specific heading is required, slight deviations in heading, particularly in gusty air, should be of little concern—just keep the wings as level as possible.

Turns to Headings

Sometimes upon encountering adverse weather conditions, it is advisable for the pilot to use radio navigation aids, or to obtain directional guidance from ATC facilities. This usually requires that turns be made and/or specific headings be maintained.

When making turns in adverse weather conditions, there is nothing to be gained by maneuvering the airplane faster than the pilot's ability to keep up with the changes that occur in the flight instrument indications. It is advisable then, to limit all turns to *no more* than a standard rate. A standard rate turn is one during which the heading changes three degrees per second. On most turn indicators this is shown when the needle is deflected one needle width; on turn coordinators this is shown when the wing tip of the representative airplane is opposite the standard rate marker.

The rate at which a turn should be made is dictated generally by the amount of turn desired—a slow turn for small changes (less than 30°) in heading, a faster turn (up to a standard rate) for larger changes (more than 30°) in heading. The actual rate at which the airplane is being turned can be determined directly by the deflection of the turn indicator needle (or the turn coordinator) and indirectly by the bank angle shown on the attitude indicator (Fig. 13–4).

Figure 13–4 Level Turn

Before starting the turn to any new heading the pilot should hold the airplane straight and level and determine in which direction the turn is to be made. Then, based upon the amount of turn needed to reach the new heading, the rate or angle of bank should be decided upon. When using the turn indicator, the needle should be deflected either one-third or one needle width; when using a turn coordinator the representative airplane's wings should be banked no more than the standard marker.

When an attitude indicator is available, the pilot should roll into the turn by using coordinated aileron and rudder pressure in the direction of the desired turn to establish the desired bank angle. The amount and direction of the bank will be shown by the angle formed between the wings of the representative airplane and the line representing the horizon. If only a turn indicator is available, control pressures should be applied until the needle is deflected the desired amount; then the bank angle or turn needle deflection should be maintained until just before the desired heading is reached. Throughout the turn, the pitch attitude and altitude must be controlled as previously described.

While making turns for large heading changes, there may be a tendency to gain or lose altitude. If the bank is controlled adequately, the altitude deviation usually will be only slight. The pilot should not be concerned about small deviations—they can be corrected after the rollout. If the bank becomes too steep, however, altitude may be lost rapidly. In this case, the bank should be shallowed rather than adding more back elevator pressure.

As long as the airplane is in a coordinated bank, it continues to turn. Thus, the rollout to a desired heading must be started before the heading is reached. Therefore it is important to refer to the heading indicator to determine the progress being made toward the desired heading, and when the roll-out should be started.

At approximately 10° before reaching the desired heading (less lead for small heading changes), coordinated aileron and rudder pressures should be applied to roll the wings level and stop the turn. This is accomplished best by reference to the attitude indicator. If only a turn indicator or turn coordinator is available, the needle should be centered or the representative wings leveled as appropriate. Failure to roll out exactly on the desired heading should cause no great alarm—final corrections can be made after the airplane is in straight-and-level flight and the pilot is assured of having positive control. Remember, the airplane's nose will tend to rise as the wings are being returned to the level attitude. Sufficient forward elevator pressure must be applied to maintain a constant altitude.

Once again, the pilot is cautioned against "chasing" the pointers on the instruments. The pointers should be allowed to settle down and then adjustments made as needed.

Critical Flight Attitudes

When outside visual references are inadequate or lost, the VFR pilot is apt to unintentionally let the airplane enter a critical attitude (sometimes called an "unusual attitude"). In general, this involves an excessively nose-high attitude in which the airplane may be approaching a stall, or an extremely steep bank which may result in a steep downward spiral.

Since such attitudes are not intentional, they are often unexpected, and the reaction of an inexperienced or inadequately trained pilot is usually instinctive rather than intelligent and deliberate. However, with practice, the techniques for rapid and safe recovery from these critical attitudes can be learned.

During dual instruction flights, the pilot should be instructed to take the hands and feet off the controls and to close the eyes. The airplane then should be put into a critical attitude by the instructor. The attitude may be an approach to a stall, or a well developed spiral dive. At this point, the pilot should be told to open the eyes, take the controls and effect a recovery by reference to the flight instruments. IN ALL CASES, recoveries should be made to straight-and-level flight.

When a critical attitude is noted on the flight instruments, the immediate problem is to recognize what the airplane is doing and decide how to return it to straight-and-level flight as quickly as possible (Fig. 13–5).

Figure 13–5 Unusual Attitude (Critical)

Nose-high attitudes are shown by the rate and direction of movement of the altimeter, vertical speed, and airspeed indicator, as well as the immediately recognizable indication on the attitude indicator. Nose-low attitudes are shown by the same instruments, but pointer movement is in the opposite direction.

Since many critical attitudes involve a rather steep bank, it is important to determine the direction of the turn. This can be accomplished best by reference to the attitude indicator. In the absence of an attitude indicator, it will be necessary in the recovery to refer to the turn needle or turn coordinator to determine the direction of turn. Coordinated aileron and rudder pressure should be applied to level the wings of the representative airplane and center the turn needle.

Unlike the control applications in normal maneuvers, larger control movements in recoveries from critical attitudes may be necessary to bring the airplane under control. Nevertheless, such control applications must be smooth, positive, and prompt. To avoid aggravating the critical attitude with a control application in the wrong direction, the initial interpretation of the instruments *must* be accurate. Once the airplane is returned to approximately straight-and-level flight, control movements should be limited to small adjustments.

If the airspeed is decreasing rapidly and the altimeter indication is increasing faster than desired, the airplane's nose is too high. To prevent a stall from occurring, it is important to lower the nose as quickly as possible while simultaneously increasing power to prevent a further loss of airspeed. If an attitude indicator is available, the representative airplane should be lowered in relation to the artificial horizon by applying positive forward elevator pressure. If no attitude indicator is available, sufficient forward pressure should be applied to stop the movement of the pointers on the altimeter and airspeed indicators.

After the airplane has been returned to straight-and-level flight and the airspeed returns to normal, the power can be reduced to the normal setting.

If the airspeed is increasing rapidly and the altimeter indication is decreasing faster than desired, the airplane's nose is too low. To prevent losing too much altitude or exceeding the speed limitations of the airplane, power must be reduced and the nose must be raised. With the higher-than-normal airspeed, it is vital to raise the nose very smoothly to avoid overstressing the airplane. Back elevator pressure must not be applied too suddenly. If the airplane is in a steep bank while descending, the wings should be leveled *before* attempting to raise the nose. Increasing elevator back pressure before the wings are leveled will tend to increase the bank and only further aggravate the situation, leading to what is called a "graveyard spiral." Furthermore, excessive G-loads may be imposed, resulting in structural failure.

During initial training, students should be required to make the recovery from a nose-low spiral attitude by taking actions in the following sequence: (1) reduce power; (2) level the wings; (3) raise the nose. After proficiency is attained, all recovery actions may be taken simultaneously.

To level the wings, coordinated aileron and rudder pressure should be applied until the wings of the representative airplane are approximately parallel to the horizon bar on the attitude indicator. If only a turn indicator or turn coordinator is available, pressures should be applied to center the needle or to level the wings of the representative airplane as appropriate. Then smooth back-elevator pressure is necessary to bring the representative airplane on the attitude indicator up to the horizon bar. Remember, the turn coordinator provides NO PITCH INFORMATION. If no attitude indicator is available, sufficient back pressure must be applied to *start* the airspeed pointer moving toward a lower airspeed and to stop the movement of the altimeter pointer. After the airplane is in level flight and the airspeed returns to normal, the power can be adjusted to the normal setting. If considerable altitude has been lost, a gradual climb to the original altitude may be necessary to ensure safe terrain clearance.

Use of Radio Navigation Aids

Ordinarily, VFR flights will at least begin in good weather conditions. Most often it is only after the flight progresses from good weather into deteriorating weather and the pilot "continues" in the hope that conditions will improve, that the need for navigational help arises. Since the area from which the pilot came is where the good weather was, naturally it is advisable to turn around and head back to that area when deteriorating weather first is encountered. In most cases there will be some type of radio navigation aid available to help the pilot return to the good weather area. Unless hopelessly disoriented, the pilot should determine the location and transmitting frequency of a VOR or NDB (ADF) that can be used for guiding the airplane back to the better weather area.

When a VOR is chosen, the omnireceiver should be tuned to the assigned frequency of the selected VOR station. After the station is identified the omnibearing selector should be turned until the TO-FROM indicator shows TO and the course deviation needle is centered.

The omnibearing selector then will be indicating the magnetic course to fly directly to the station. The airplane then should be turned to the corresponding magnetic heading, and heading adjustments made as required to follow the course to the station. Procedures for maintaining the course are explained in the chapter on Cross-Country Flying.

If an NDB is to be used, the ADF receiver should be tuned to the frequency of the selected NDB and the station positively identified. Then the airplane should be turned until the ADF pointer is on the nose position of the instrument. Keeping the pointer on this position will result in the airplane flying to the station although the ground track may be slightly curved due to a crosswind. This "homing" procedure is also explained in the chapter on Cross-Country Flying.

It should be remembered that the use of radio navigation aids when in unfavorable weather conditions requires additional division of attention while attempting to maintain control of the airplane. The pilot's main concern, of course, must be airplane control.

Use of DF or Radar Services

While radio navigation aids can be very useful in following a specific course, they do add to the pilot's workload, both mentally and manually, particularly under dire circumstances. The pilot must divide attention between control of the airplane and the operation of the navigation equipment. Under the emergency conditions of flying the airplane by reference to flight instruments, this is often a difficult task for experienced pilots.

However, there is another option available to pilots of airplanes with or without radio navigation equipment if they have VHF radio communications capability. All pilots should be aware of the VHF Direction Finding (DF) and Radar service available all over the U.S., either through civil or military radio facilities. This service permits a control tower, FSS, or radar facility to give the pilot a heading to fly to reach a nearby airport or to an area of good weather. In this way, the pilot need only communicate and follow instructions while giving almost full attention to flying the airplane.

To obtain assistance by radio and apply it effectively, some preparation and training is necessary. All pilots should become familiar with the appropriate emergency procedures and practice them when the situation permits. The *actual use* of designated emergency radio frequencies *for training exercises is not permitted*, but FAA facilities are often able to provide practice orientation and radar guidance procedures using their regular communications frequencies. In any case, the pilot must inform the ground station when the request is only for practice purposes.

When a pilot is in doubt about the airplane's position, or feels apprehensive about the safety of the flight, there should be no hesitation to ask for help. That is the first means of declaring an emergency—use the radio transmitter and ask for help. If in actual distress, and help is needed immediately, the pilot should transmit the word MAYDAY several times before transmitting the emergency message. This will get immediate attention from all who hear.

Emergency messages may be transmitted on any radio frequency; however, there is a frequency especially designated for such messages. The designated emergency VHF frequency is 121.5 MHz and is available on most radios installed in general aviation type airplanes. This is usually the best frequency on which to transmit and receive because almost all direction finding (DF) stations, radar facilities, and Flight Service Stations monitor this frequency. Regardless of which type of facility is contacted, that facility can help, even if only by alerting other facilities to the emergency.

Since frequent communications may be necessary when using this service, it is recommended that the microphone be continually held in the hand. This will eliminate the need to take the eyes away from the flight instruments every time the microphone is removed from or replaced in its receptacle. After some practice, loosely holding the "mike" in one hand should not create any difficulty in using the flight controls. It is important though, that the "mike" button not be depressed accidentally, which would block the frequency and prevent the reception of further assistance. The initial request for assistance can be made on the regular communications frequency of the facility, or on the emergency frequency 121.5 MHz. Regardless of which frequency is used, it is essential that the pilot not change frequency unless instructed to do so by the operator or unless absolutely necessary.

A direction-finding (DF) station is a ground-based receiver capable of indicating the bearing from its antenna to the transmitting airplane. There are HF, VHF, and UHF direction-finding stations. However, only VHF stations are discussed in this handbook, since this is the type of radio equipment most likely to be in the airplanes flown by the average general aviation pilot.

If a pilot is unable to establish communication with a VHF/DF facility, or if there is doubt about whether this service is available at a particular station, the service may be obtained by calling any Flight Service Station or control tower. The request then will be relayed immediately to the appropriate DF facility. The pilot must remember, though, that VHF transmissions follow line-of-sight; therefore, the higher the altitude, the better the chance of obtaining this service. Depending on terrain conditions and altitude, DF service is effective up to a radius of 100 miles.

The VHF/DF operator on the ground can note the airplane's bearing from the facility by looking at a scope, quite similar to a radar scope. Each time the pilot transmits, it shows up on the scope as a line radiating out from the center (Fig. 13–6). From this the operator reads the course the pilot should fly (with zero wind conditions) to reach the facility. The pilot need not be concerned about wind drift since subsequent heading instructions will consider the position to which the airplane has drifted and a new course plotted.

When DF services are requested by the pilot, the operator will ask if the airplane is in VFR or IFR weather conditions, the amount of fuel remaining, the altitude, and the heading. Also, the operator should be informed whether the pilot is instrument rated. If the airplane is in IFR weather conditions, the pilot will be informed of the minimum safe altitude and the current local altimeter setting will be provided. The pilot will then be instructed to transmit (by keying or voice) for 10 seconds.

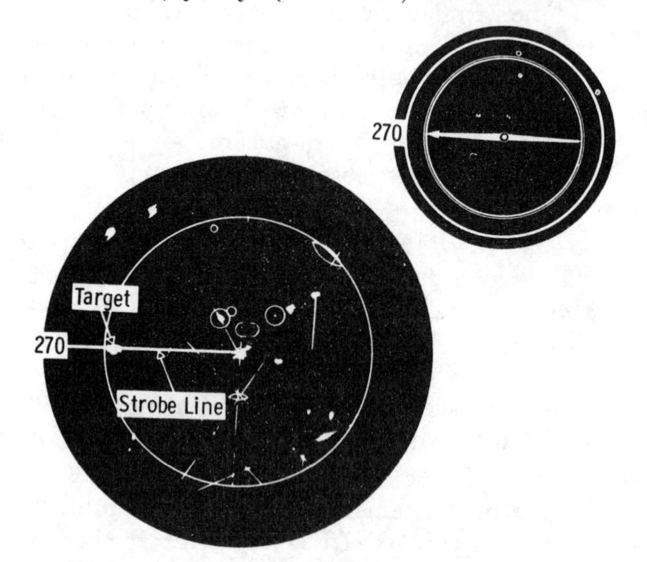

Figure 13–6 VHF Direction Finder Scope

To get a good indication on the scope, two continuous transmissions of approximately 10 seconds each from the airplane are required. The volume of the transmission should remain as nearly constant as possible. This can be done by depressing the "mike" key for 10 seconds or by making a voice transmission, "Aaaah," followed by the airplane's call sign and the word "Over."

When the airplane's bearing has been determined, the DF operator will specify the direction to turn and the magnetic heading to be flown. (i.e., "TURN LEFT, HEADING ZERO ONE ZERO, FOR DF GUIDANCE TO ARDMORE AIRPORT, REPORT AIRPORT IN SIGHT.") The number of times the transmissions must be made will vary, depending on frequency congestion, distance, and wind.

Radar equipped ATC facilities also provide assistance and navigation service, provided the airplane has appropriate communications equipment, is within radar coverage, and can be radar identified.

Primary radar relies on a signal being transmitted from the radar antenna site and for this signal to be reflected or "bounced back" from an airplane. This reflected signal is then automatically displayed as a "target" or blip on the controller's radarscope. For identification purposes, the pilot may be asked to turn the airplane in a certain direction to a specific heading. If the airplane is transponder equipped, this identifying turn is not necessary (Fig. 13–7).

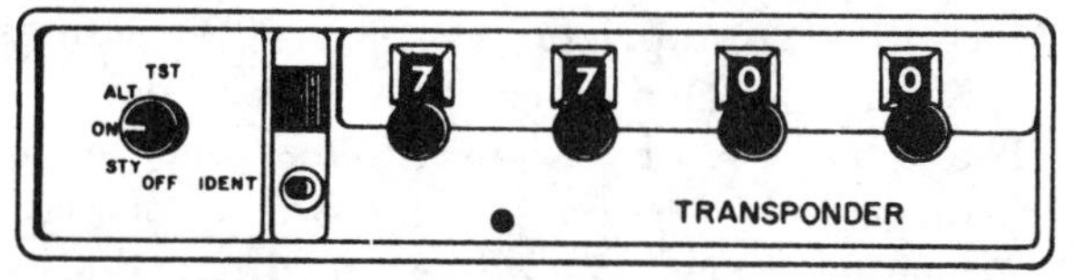

Figure 13–7 Typical Transponder Squawking "Mayday"

A transponder is an airborne radar beacon transmitter-receiver which automatically receives the signals from the ground-based radar and selectively replies with a specific code. These replies are independent of and much stronger than the primary radar return and are displayed uniquely on the radarscope. Thus, by having the pilot set the transponder to a certain code, the radar controller can identify the airplane immediately. After the controller positively identifies the airplane on the radarscope, the pilot will be notified of the airplane's position and will then be given a heading to an airport or other specific point in the radius of coverage of the radar station.

The procedures used in providing radar navigational guidance to pilots are very similar to those used for DF service. There are, however, certain terms in the use of transponders which the pilot must understand. When instructed to "Squawk" a certain code (number), the pilot should first ensure that the transponder is set to the specified 4-digit number and turned ON (not standby). If told to "Ident," the pilot should press the button marked as such on the transponder. When told to "Squawk MAYDAY" (the emergency position), the transponder should be set to 7700. The term "Vector" simply means the heading to fly to reach a certain location.

Pilots should understand clearly that authorization to proceed in accordance with such radar navigational assistance *does not constitute authorization* for the pilot to violate Federal Aviation Regulations. The controller must be informed whether the pilot is instrument rated and whether the airplane is equipped for instrument flight.

To avoid possible hazards resulting from being vectored into IFR conditions, a VFR pilot in difficulty should keep the controller advised of the weather conditions in which the airplane is operating and along the course ahead.

If the airplane has already encountered IFR conditions, the controller will inform the pilot of the minimum safe altitude. If the airplane is below the minimum safe altitude and sufficiently accurate position information has been received or radar identification is established, a heading or VOR radial on which to climb to reach the minimum safe altitude will be furnished.

Summary of Airplane Control

In summarizing the control of the airplane during flight in unfavorable weather conditions, the following elements are essential in attaining the main objectives of the desired flight condition:

1. Straight and level;—bank control to keep the wings level and the heading constant; pitch control to maintain altitude; CHECK HEADING AND ALTITUDE.
2. Straight climbs;—bank control to keep the wings level and the heading constant; pitch control to maintain airspeed; CHECK HEADING AND AIRSPEED.
3. Straight descents;—bank control to keep the wings level and the heading constant; pitch and power control to maintain airspeed; CHECK HEADING AND AIRSPEED.
4. Turns to headings;—bank control to keep bank constant; pitch control to maintain altitude; CHECK HEADING AND AIRSPEED.
5. Flying to radio station;—bank control to make turns; pitch control to maintain altitude; CHECK HEADING, NAVIGATION INDICATOR, AND ALTITUDE.
6. DF and radar headings;—bank control to make turns; pitch control to maintain altitude; CHECK HEADING AND ALTITUDE.

The *practice* of these procedures should be accomplished only on dual instruction flights or when someone is along to watch for other aircraft.

The most important point to be stressed is that the *pilot must not panic.* When the task at hand seems to be overwhelming by so many things to do or items to note, the best procedure for any pilot is to make a conscious effort to relax and take one thing at a time. It may be best to let some things go and concentrate mainly on the more urgent matters. In most cases the pilot's primary concern is to keep the wings level. An uncontrolled turn or bank usually leads to difficulty in achieving the objectives of any desired flight condition. It will be found that good bank control makes pitch control so much easier.

Next in importance is to *believe* what the instruments show about the airplane's attitude regardless of what the natural senses tell the pilot. Instrument flying has no place for "seat of the pants flying." This is often a difficult fact for many pilots to learn, but it must be stressed. Remember, these are emergency procedures, and without outside visual references, the pilot has little choice but to rely on the flight instruments.

When using DF or radar services, it is very important that the pilot *listen carefully,* and as accurately as possible, *follow the instructions* of the controllers; otherwise the task becomes most difficult for both the pilot and the controller. If instructions are not understood or cannot be complied with, a clarification or different instruction should be requested. While concentrating on flying the airplane, the pilot must also make a conscious effort to remember the latest heading and altitude instruction given by ATC.

CHAPTER 14

Night Flying

The intent of this chapter is to introduce only the major aspects of night flying to the pilot who has flown only in daylight. These elements include a brief description of night visual perception, suggested pilot equipment, airplane lighting and equipment, airport lighting, and the various night-flight operations, as well as basic pilot techniques.

Night flying is considered to be an important phase in the complete training of a pilot. Proficiency in night flying not only increases utilization of the airplane but it provides important experience in case an intended day flight inadvertently extends into darkness.

Night flying is really not difficult but it differs from daylight operation, in that vision is restricted at night. This instills a certain amount of anxiety in pilots who lack night-flight experience. These apprehensions can be overcome only by acquiring the necessary knowledge and experience in night operation. As confidence is gained through experience, many pilots prefer night flying over day flying because the air is usually smoother and generally there is less air traffic to contend with.

Before attempting night operations it is recommended that a complete and thorough checkout by a competent flight instructor be accomplished. This checkout should include a night cross-country flight with landings at airports other than the home field.

Night Vision

Generally, many persons are completely uninformed about night vision, while others feel there is nothing to be learned about the subject. Human eyes never function as effectively at night as the eyes of animals with nocturnal habits, but if a human learns to use the eyes correctly, night vision can be improved greatly. There are several reasons for the training and practice necessary to use the eyes correctly.

One reason is that the mind and eyes act as a team for a person to see well; both team members must be used effectively. Also, the construction of the eyes is such that to see at night they must be used differently than during the daytime. Therefore, it is important for the pilot to understand the eye's construction and how the eye is affected by darkness.

Innumerable light-sensitive nerves, called "cones" and "rods," are located at the back of the eye or retina, a layer upon which all images are focused (Fig. 14–1). These nerves connect to the cells of the optic nerve which transmit messages directly to the brain. The cones are located in the center of the retina, and the rods are concentrated in a ring around the cones.

The function of the cones is to detect color, details, and far-away objects. The rods function when something is seen out of the corner of the eye—peripheral vision. They detect objects, particularly those which are moving, but do not give detail or color—only shades of gray. Both the cones and the rods are used for vision during daylight.

Although there is no clear-cut division of function, generally speaking, the rods make night vision possible. The rods and cones function in daylight and in moonlight, but in

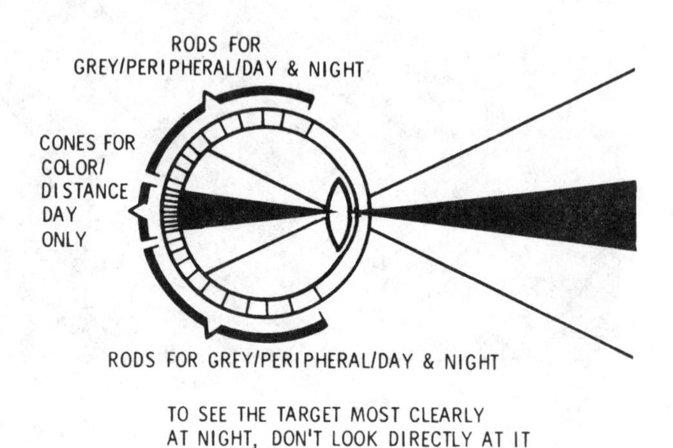

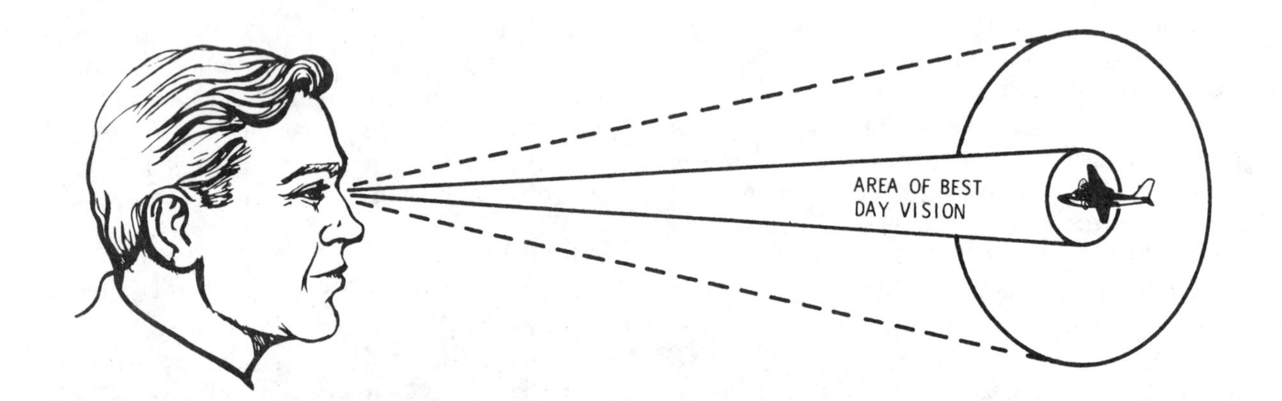

Figure 14–1 Cones and Rods Affect Vision

the absence of these the process of vision is placed almost entirely on the rods.

The fact that the rods are distributed in a band around the cones and do not lie directly behind the pupils, makes "off center" viewing (looking to one side of an object) important during night flight (Fig. 14–1). During daylight an object can be seen best by looking directly at it, but at night a scanning procedure to permit "off center" viewing of the object is more effective. Therefore, the pilot should consciously practice this scanning procedure to improve night vision.

The eye's adaptation to darkness is another important aspect of night vision. When a dark room is entered, it is difficult to see anything until the eyes become adjusted to the darkness. Most everyone has experienced this after entering a darkened movie theater. In this process, the pupils of the eyes first enlarge to receive as much of the available light as possible. After approximately 5 to 10 minutes, the *cones* become adjusted to the dim light and the eyes become 100 times more sensitive to the light than they were before the dark room was entered. Much more time, about 30 minutes, is needed for the *rods* to become adjusted to darkness; but when they do adjust they are about 100,000 times more sensitive to light than they were in the lighted area. After the adaptation process is complete much more can be seen, especially if the eyes are used correctly.

This entire process is reversed when a bright area is entered from a dark room. The eyes are first dazzled by the brightness, but become completely adjusted in a very few seconds, thereby losing their adaptation to the dark. Now, if the dark room is reentered, the eyes must again go through the long process of adapting to the darkness.

The adaptation process of the eyes must be considered by the pilot before and during night flight. First the eyes must be allowed to adapt to the low level of light and then they must be kept adapted. After the eyes have become adapted to darkness, the pilot must avoid exposing them to any bright white light which will cause temporary blindness and could result in serious consequences.

Temporary blindness caused by an unusually bright light, may result in illusions or "after images" during the time the eyes are recovering from the brightness. The brain creates these illusions reported by the eyes. This results in misjudging or incorrectly identifying objects, such as mistaking slanted clouds for the horizon or populated areas for a landing field. Vertigo is experienced by the pilot as a feeling of dizziness and imbalance which in itself can create or increase illusions. The illusions seem very real and pilots of any level of experience and skill can be affected. Recognizing that the brain and eyes can play tricks in this manner is the best protection for the pilot flying at night.

Good eyesight depends upon physical condition. Fatigue, colds, vitamin deficiency, alcohol, stimulants, smoking, or medication can seriously impair the pilot's vision. Keeping these facts in mind and taking adequate precautions should safeguard the pilot's night vision.

In review and in addition to the principles previously discussed, the following will aid the pilot in increasing night vision effectiveness:

1. Adapt the eyes to darkness prior to flight and keep them adapted. About 30 minutes is needed to adjust the eyes to maximum efficiency after exposure to a bright light.
2. Close one eye when exposed to bright light to help avoid the blinding effect.
3. Do not wear sunglasses after sunset.
4. Move the eyes more slowly than in daylight.
5. Blink the eyes if they become blurred.
6. Concentrate on seeing objects.
7. Force the eyes to view off center.
8. Maintain good physical condition.
9. Avoid smoking, drinking, and using drugs which may be harmful.

Pilot Equipment

The pilot, before beginning a night flight, should carefully consider certain personal equipment that should be readily available during the flight. This equipment may not differ greatly from that needed for a day flight, but the importance of its availability when needed at night cannot be over-emphasized.

At least one reliable flashlight is recommended as standard equipment on all night flights. A "D" cell size flashlight with a bulb switching mechanism that can be used to select white or red light, is preferable. The white light is for use while performing the preflight visual inspection of the airplane, and the red light for use in performing cockpit operations. Since the red light is nonglaring, it will not impair night vision. Some pilots prefer two flashlights, one with a white light for preflight, and the other a penlight type with a red light. The latter can be suspended by a string from around the neck to ensure that the light is always readily available during flight. One word of caution; if a red light is used for reading an aeronautical chart the red features of the chart will not show up.

Just as for daylight flights aeronautical charts are essential for night cross-country flight and, if the intended course is near the edge of the chart, the adjacent chart also should be available. The lights of cities and towns can be seen at surprising distances at night, and if this adjacent chart is not available to identify those landmarks which lie off the primary chart confusion could result, particularly if the pilot strays off course.

To prevent losing essential items in the dark cockpit, the pilot should have a clipboard or mapboard on which charts, navigation logs, and other essentials can be fastened. Map cases in which to store needed materials should also be considered.

A reliable clock is essential for both day and night flights.

Regardless of what is used, organization of the equipment and material in the cockpit into a simple well arranged manner, eases the burden on the pilot and certainly enhances safety.

Airplane Lighting and Equipment

Federal Aviation Regulations, Part 91, specify certain minimum airplane equipment required for night flight. This equipment includes instruments, lights, electrical energy source, and spare fuses.

Though not required by FAR Part 91, it is recommended, because of the limited outside visual references during night flight, that other flight instruments supplement those required. An operable attitude indicator, heading indicator, and sensitive altimeter are very valuable in controlling the airplane at night. Individual instrument lights along with adequate cockpit illumination, aid significantly in making night flying a safe and pleasurable operation.

An anticollision light system, including a flashing or rotating beacon and position lights, is required airplane equipment. Airplane position lights are arranged similar to those of boats and ships. A red light is positioned on the left wingtip, a green light on the right wingtip, and a white light on the tail (Fig. 14–2). This arrangement provides a means by which pilots can determine the general direction of movement of other airplanes in flight just as sailors determine the direction of boats and ships. If both a red and green light of another airplane are observed, the airplane would be flying in a general direction toward the pilot seeing the lights, and could be on a collision course. If only a red light is seen, the airplane is traveling from right to left in relation to the observing pilot. On the other hand, if only a green light is observed, the airplane is traveling from left to right.

Landing lights are not only useful for taxi, takeoffs, and landings, but they also provide a means by which your airplane can be seen by other pilots during flight. FAA has initiated a voluntary pilot safety program, "Operation Lights On" to enhance the "see-and-be-seen"

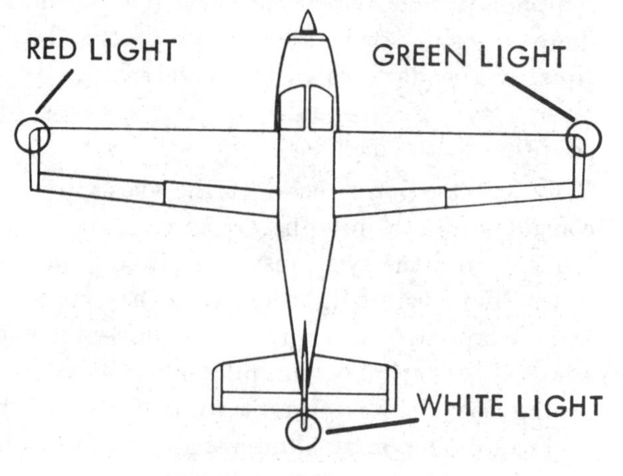

Fig. 14–2 Position Lights

concept of averting collisions both in the air and on the ground, and to reduce bird strikes. All pilots are encouraged to turn on their landing lights when operating within 10 miles of any airport (day and night), in conditions of reduced visibility and in areas where flocks of birds may be expected.

Although turning on aircraft lights does enhance the "see-and-be-seen" concept, pilots should not become complacent about keeping a sharp lookout for other aircraft. At times aircraft lights blend in with stars in the sky or the lights of the cities and go unnoticed unless a conscious effort is made to distinguish them from other lights.

With the increase in the number of radio navigation aids and communication facilities, it is also recommended that the airplane be equipped with at least one radio navigation receiver and indicator, and two-way radio communication capability.

Most personal general aviation airplanes manufactured today are equipped with all items recommended in this chapter and perhaps more, and add much to the safety of night flight.

Airport and Navigation Lighting Aids

The light systems used for airports, runways, obstructions, and other visual aids at night are other important aspects of night flying.

Lighted airports located away from congested areas can be identified readily at night by the lights outlining the runways; however, airports located near or within large cities are often difficult to identify in the maze of lights. It is important not only to know the exact location of these airports relative to the city, but also to be able to identify these airports by the characteristics of their lighting pattern.

Aeronautical lights are designed and installed in a variety of colors and configurations, each having its own purpose. Although some are used during low ceiling and visibility (IFR) conditions, this discussion includes only the lights that are fundamental to VFR night operation (Fig. 14–3).

It is recommended that prior to night flight, particularly cross-country, the pilot check the availability and status of lighting systems at airports along the planned route and the destination airport. This information can be found

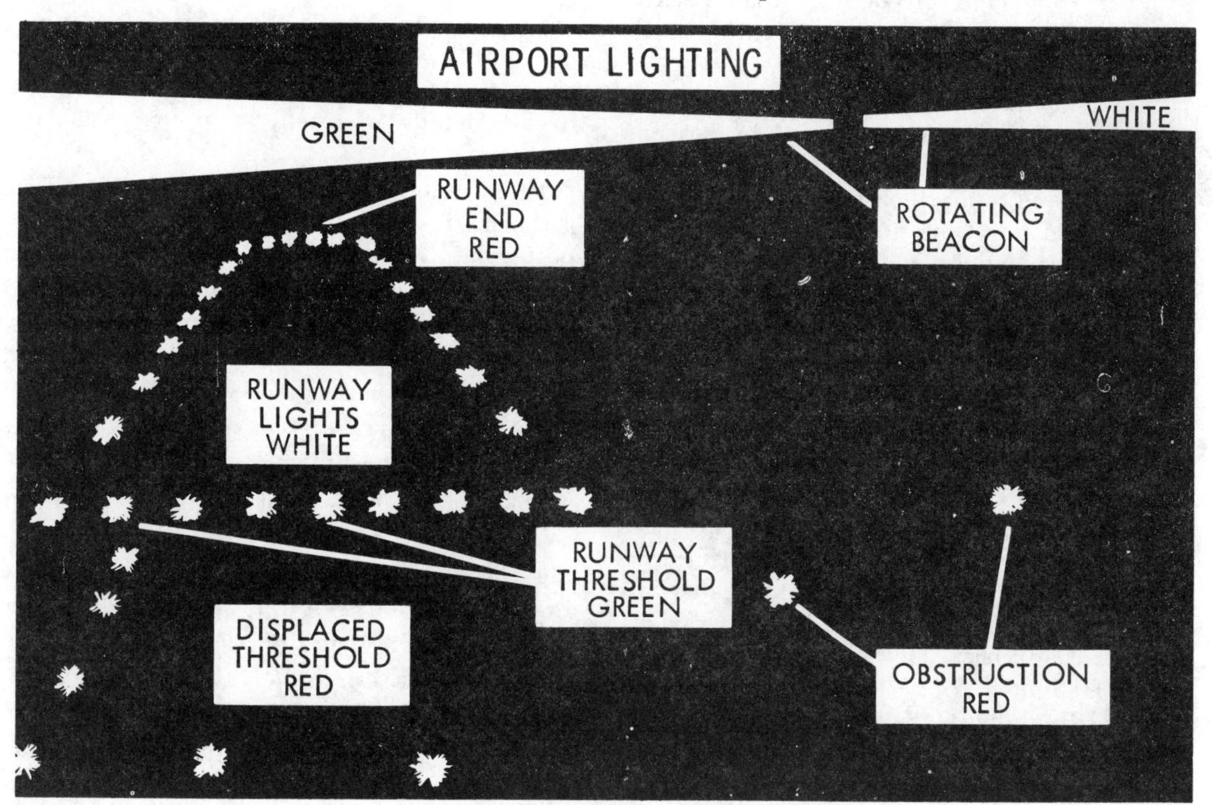

Figure 14–3 Airport Lighting

on aeronautical charts and in the Airport/Facility Directory, and the status of each facility can be determined by reviewing pertinent Notices to Airmen.

A rotating beacon is used to indicate the location of an airport. Its vertical light distribution is such that it can be seen most effectively from one to three degrees above the horizontal; however, the light can also be seen well above and below this range. The beacon rotates at a constant speed, thus producing what appears to be a series of light flashes at regular intervals. These flashes may be one or two different colors which are used to identify various types of landing areas. For example:

Alternating white and green—lighted land airports.

White alone—unlighted land airports.

Green alone—sometimes installed as an aid to find a not-too-distant white and green beacon.

Military airports also have beacons flashing alternately white and green, but are differentiated from civil airports by dual peaked (two quick) white flashes between the intermittent green flashes.

Obstructions or areas considered hazardous to aerial navigation are indicated by beacons producing red flashes. Steady burning red lights are used to mark obstructions on or near airports and sometimes to supplement flashing lights on enroute obstructions. High intensity flashing white lights are now being used to mark some supporting structures of overhead transmission lines which stretch across rivers, chasms, and gorges. These high intensity lights are used also to identify tall structures, such as chimneys and towers, that are considered to be obstructions to air navigation.

As a result of the technological advancements in aviation, many runway lighting systems have become quite sophisticated to accommodate takeoffs and landings in various weather conditions. However, the pilot whose flying is limited to VFR conditions need be concerned only with the following basic lighting of runways and taxiways.

The basic runway lighting system consists of two straight parallel lines of runway-edge lights defining the lateral limits of the runway. These lights are aviation white, although aviation yellow may be substituted for a distance of 2,000 feet from the far end of the runway to indicate a caution zone. At some airports the intensity of the runway-edge lights can be adjusted to satisfy the individual needs of the pilot. The longitudinal limits of the runway are defined by straight lines of lights across the runway ends. At some airports the runway threshold lights are aviation green; and the runway end lights are aviation red.

At many airports the taxiways also are lighted. A taxiway-edge lighting system consists of omnidirectional blue lights which outline the usable limits of taxi paths. Caution must be observed when taxiing, particularly at airports that are not equipped with taxiway lights. The use of the airplane's landing or taxiing lights is essential under these conditions.

Preparation and Preflight

Night flying requires that pilots have a complete realization of their abilities and limitations, and observe more caution than during day operations. Although careful planning of any flight is essential for maximum safety and efficiency, night flying demands more attention to all details of preflight preparation and planning.

Preparation for a night flight should include a thorough study of the available weather reports and forecasts, with particular attention given to temperature/dewpoint spread because of the possibility of formation of ground fog during the night flight. Also, emphasis should be placed on awareness of wind direction and speed, since drifting cannot be detected as readily at night as during the day.

On night cross-country flights pertinent aeronautical charts should be selected, including the appropriate adjacent charts. Course lines should be drawn in black so as to be more distinguishable, and direction, distances, and time estimates accurately recorded. Aeronautical charts should be folded and systematically arranged prior to flight in a manner that they will be convenient to use in the cockpit.

Prominently lighted checkpoints along the prepared course should be noted. Rotating beacons at airports, lighted obstructions, lights of cities or towns, and lights from major highway traffic all provide excellent visual checkpoints. The use of radio navigation aids and communication facilities add significantly to the safety and efficiency of the night flight and should be considered in preflight planning.

All personal equipment should be checked prior to flight to ensure proper functioning. It is very disconcerting to find, at the time of need, that a flashlight for example, doesn't work.

A thorough preflight check of the airplane, and a review of its systems and emergency procedures, is of particular importance for night operations. Since each airplane has its own checklist, it is not intended in this chapter to cover all the specific points. However, there are some general areas, in addition to those involved on all flights, that should be included on the night preflight check.

All airplane lights should be turned "on" momentarily and checked for operation. Position lights can be checked for loose connections by tapping the light fixture while the light is "on." If the lights blink while being "tapped" further investigation to determine the cause should be initiated.

The parking ramp should be examined prior to entering the airplane. During the day it is quite easy to see stepladders, chuckholes, stray wheel chocks, and other obstructions, but at night it is more difficult, and a check of the area can prevent taxiing mishaps.

Starting, Taxiing, and Runup

After the pilot is seated in the cockpit and prior to starting the engine, all items and materials to be used on the flight should be arranged in such a manner that they will be readily available and convenient to use.

The airplane manufacturer's recommended engine starting procedures should always be followed; however, *extra caution* should be taken at night to assure that the propeller area is "clear." Turning the rotating beacon "on," or flashing other airplane lights will serve to alert any person nearby to remain clear of the propeller. To avoid excessive drain of electrical current from the battery, it is recommended that unnecessary electrical equipment should be kept "off" until after the engine has been started.

Before starting the engine the aircraft's position lights should be turned "on," and checked for operation. The check can be made by observing the lights' glow on the ground under the wingtips and the tail. After starting and before taxiing, the taxi or landing light should be turned "on." However, continuous use of the landing light with the low power settings normally used for taxiing may place an excessive burden on the airplane's electrical system. Also, overheating of the landing light bulbs could become a problem because of inadequate airflow to carry away the excessive heat generated. Therefore, it is recommended generally that landing lights be used only intermittently while taxiing, but sufficiently to assure that the taxi area is clear. While using the lights during taxiing, consideration should be given to other airplanes so as to not blind the pilots with the landing lights. Taxi slowly, particularly in congested areas, and if taxi lines are painted on the ramp or taxiway, these lines should be followed to ensure a proper path along the route.

The pretakeoff runup should be performed using the airplane's checklist, and each item should be checked carefully—the proper functioning of any of the airplane components must never be taken for granted. During the day unintended forward movement of the airplane can be detected quite easily, but at night the airplane could creep forward without being noticed unless the pilot is alert, especially for this possibility. Therefore, it is important to lock the brakes during the runup and attention be given to any unintentional forward movement.

Takeoff and Departure Climb

Although night flying is very little different from day flying, it does demand more attention of the pilot. The most impressive difference is the limited availability of outside visual references. Therefore, the flight instruments should be used to a greater degree in controlling the airplane. This is particularly true on night takeoffs and departure climbs. The cockpit lights should be adjusted to a minimum brightness that will allow the pilot to read the instruments and switches and yet not hinder the pilot's outside vision. This also will eliminate light reflections on the windshield and windows.

Before taxiing onto an active runway for takeoff, the pilot should exercise extreme caution to prevent conflict with other aircraft. Even at controlled airports where the control tower issues the clearance for takeoff, it is recommended that the pilot check the final approach course for approaching aircraft. At uncontrolled airports, it is recommended that a slow 360° turn be made in the same direction as the flow of air traffic while closely searching for other aircraft in the vicinity.

After ensuring that the final approach and runway are clear of other air traffic, the airplane should be lined up with the centerline of the runway. If the runway has no painted centerline, the pilot can use the runway lighting and align the airplane midway between and parallel to the two rows of runway edge lights. After the airplane is aligned, the heading indicator should be noted or set to correspond to the known runway direction. To begin the takeoff, the brakes should be released and the throttle smoothly advanced to takeoff power. As the airplane accelerates, it should be kept moving between and parallel to the runway edge lights. This can best be done by looking at the more distant runway lights rather than those close in and to the side.

The technique for night takeoffs is the same as for normal daytime takeoffs, but the flight instruments should be monitored more closely. As the airspeed reaches the normal lift-off speed, the pitch attitude should be adjusted to that which will establish a normal climb by referring to both outside visual references such as lights, *and* to the attitude indicator. The airplane should not be forcibly pulled off the ground; it is best to let it fly off in the lift-off attitude while cross-checking the attitude indicator against any outside visual references that may be available.

After becoming airborne, the darkness of night often makes it difficult to note whether the airplane is getting closer to or farther from the surface. It is extremely important, then, to ensure that the airplane continues in a positive climb and does not settle back to the runway. This can be accomplished by ensuring that there is a climb rate on the vertical speed indicator and a gradual but continual increase in the altimeter indication (Fig. 14–4). It is also important to note that the airspeed is well above the stall speed and that it continues to accelerate.

Figure 14–4 Establish Positive Climb

Necessary pitch adjustments to establish a stabilized climb should be made with reference to the attitude indicator. At the same time the wings should be checked for a level attitude using the attitude indicator and the heading indicator. It is recommended that no turn be made until reaching a safe maneuvering altitude.

Although the use of the landing lights provides help during the takeoff roll, they become ineffective after the airplane has climbed to an altitude where the light beam no longer extends to the surface. The light can also be deceptive when it is reflected by haze, smoke, or fog that might exist in the takeoff climb. Therefore, if the landing light is used for the takeoff roll, it may be turned off after the climb is well established.

Orientation and Navigation

On night flights, pilots should be aware of the importance of being alert and looking for other aircraft. The pilot should be able to recognize the airplane's position relative to other aircraft by the color combination of the other aircraft's position lights.

Generally, at night it is difficult to see clouds and restrictions to visibility, particularly on dark nights or under an overcast. The pilot flying under visual flight rules (VFR) must exercise caution to avoid flying into clouds or a layer of fog. Usually, the first indication of flying into restricted visibility conditions is the gradual disappearance of lights on the ground. If the lights begin to take on an appearance of being surrounded by a "cotton ball" or glow, the pilot should use caution in attempting further flight in that same direction. Remember that if a descent must be made through any fog, smoke, or haze in order to land, the visibility is considerably less when looking horizontally through the restriction than it is when looking straight down through it from above. Under no circumstances should a VFR night flight be made during poor or marginal weather conditions.

The practice of maneuvers at night should be conducted within a designated practice area, or at least in an area that is known to be comparatively free from other air traffic. The learning pilot should practice and acquire competency in straight-and-level flight, straight climbs and descents, level turns, and climbing and descending turns. Recovery from unusual attitudes also should be practiced, but only on dual flights with a competent flight instructor.

In spite of fewer landmarks or checkpoints, night cross-country flights present no particular problem, if preplanning is adequate and the pilot continues to monitor position, time estimates, and fuel consumed.

Crossing large bodies of water on night flights could be potentially hazardous, not only from the standpoint of landing (ditching) in the water should it become necessary, but also because the horizon may blend in with the water, in which case, control of the airplane becomes difficult. During haze conditions over open water the horizon will become obscure, and may result in loss of spacial orientation. Even on clear nights the stars may be reflected on the water surface, which could appear as a continuous array of lights, thus making the horizon difficult to identify.

Lighted runways, buildings, or other objects may cause illusions to the pilot when seen from different altitudes. At an altitude of 2,000 feet a group of lights on an object may be seen individually, while at 5,000 feet or higher the same lights could appear to be one solid light mass. These illusions may become quite acute with altitude changes and if not overcome could present problems in respect to approaches to lighted runways.

Landing Approaches and Landings

When arriving at the airport to enter the traffic pattern and land, it is important that the runway lights and other airport lighting be identified as early as possible. If the airport layout is unfamiliar to the pilot, sighting of the runway may be difficult until very close-in due to the maze of lights observed in the area (Fig. 14–5). The pilot should, therefore, fly towards the airport beacon light until the lights outlining the runway are distin-

Figure 14–5 Use Light Pattern for Orientation

guishable. To fly a traffic pattern of the proper size and direction when there is little to see but a group of lights, the runway threshold and runway edge lights must be positively identified. Once seen, the approach threshold lights should be kept in sight throughout the airport traffic pattern and approach.

Distance may be deceptive at night due to limited lighting conditions, lack of intervening references on the ground, and the inability of the pilot to compare the size and location of different ground objects. This also applies to the estimation of altitude and speed. Consequently, more dependence must be placed on flight instruments, particularly the altimeter and the airspeed indicator.

Inexperienced pilots often have a tendency to make approaches and landings at night with excessive airspeed. Every effort should be made to execute the approach and landing in the same manner as during the day. A low, shallow, approach is definitely inappropriate during a night operation. The altimeter and vertical speed indicator should be constantly cross-checked against the airplane's position along the base leg and final approach.

After turning onto the final approach and aligning the airplane midway between the two rows of runway edge lights, the pilot should note and correct for any wind drift. Throughout the final approach, power should be used with coordinated pitch changes to provide positive control of the airplane, enabling the pilot to accurately adjust airspeed and descent angle. Where a Visual Approach Slope Indicator is installed, it is helpful in maintaining the proper approach angle (Fig. 14–6). Usually, when approximately halfway along the final approach, the landing light should be turned on. Earlier use of the landing light may be ineffective since the light's beam will usually not reach the ground from higher altitudes, and may be reflected back into the pilot's eyes by any existing haze, smoke, or fog. However, this disadvantage may be overshadowed by the safety advantage provided by using the "Operation Lights On" procedure in the terminal area.

The roundout and touchdown should be made in the same manner as in day landings. Judgment of height, speed, and sink rate is impaired, however, by the scarcity of observable objects in the landing area. The inexperienced pilot may have a tendency to roundout too high until attaining familiarity with the apparent height for the correct roundout position. To aid in determining the proper roundout point, it may be well to continue a constant approach descent until the landing light reflects on the runway, and tire marks on the runway, or runway expansion joints, can be seen clearly (Fig. 14–7). At that point

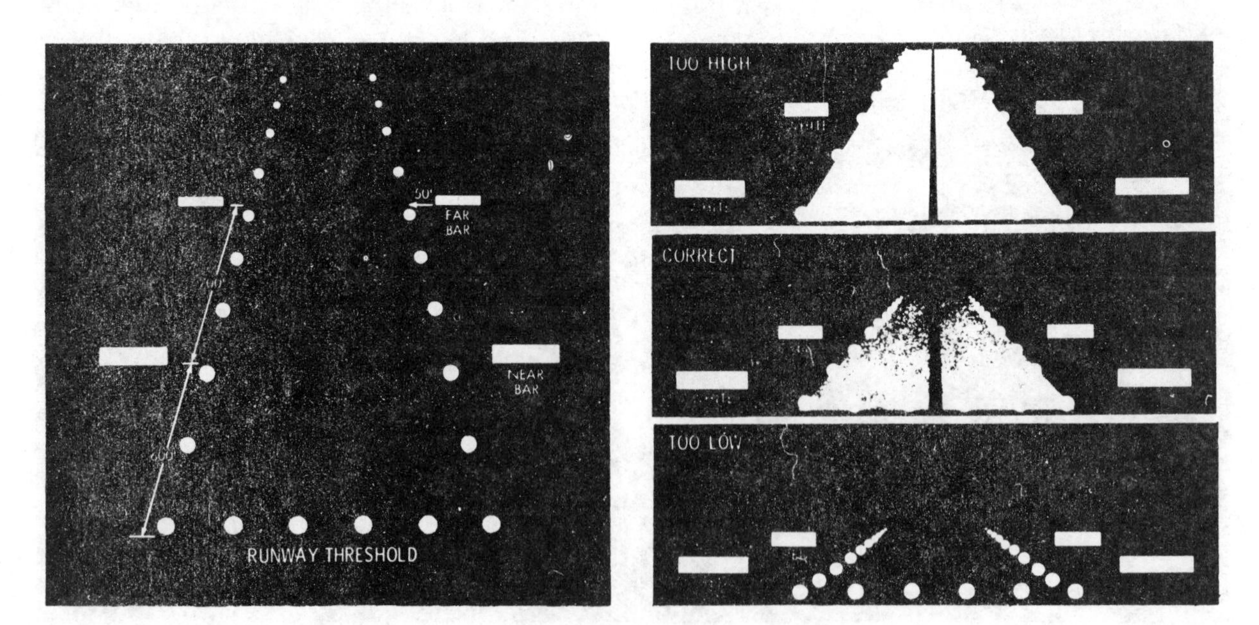

Figure 14–6 Visual Approach Slope Indicator

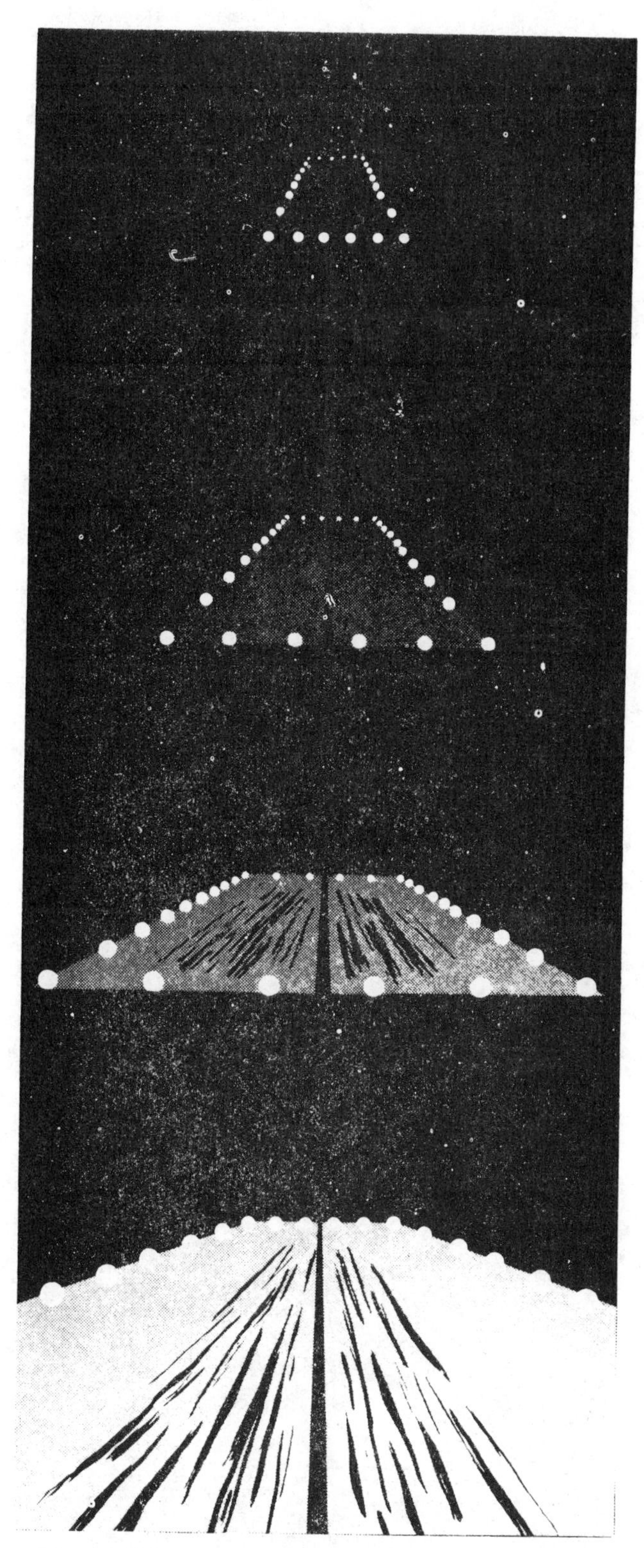

Figure 14-7 Roundout when Tire Marks are Visible

the roundout for touchdown should be started smoothly and the throttle gradually reduced to idle as the airplane is touching down. During landings without the use of landing lights or where marks on the runway are indiscernible, the roundout may be started when the runway lights at the far end of the runway first *appear* to be rising higher than the airplane. This, of course, demands a smooth and very timely roundout, and requires, in effect, that the pilot "feel" for the runway surface, using power and pitch changes as necessary for the airplane to settle softly on the runway.

Night Emergencies

Perhaps the pilot's greatest concern about flying a single-engine airplane at night is complete engine failure, even though adverse weather and poor pilot judgment account for most serious accidents.

If the engine fails at night, the first step is to maintain *positive control* of the airplane—do not panic. A normal glide should be established and maintained and the airplane turned toward an airport or away from congested areas. A check should be made to determine the cause of the engine failure such as the position of magneto switches, fuel selectors, or primer. If possible, the cause of the malfunction should be corrected immediately and the engine restarted. Orientation with the wind direction also should be maintained to avoid a downwind landing. The landing lights should be checked at altitude and turned on in sufficient time to illuminate the terrain or obstacles along the flightpath. If the landing lights are unusable and outside visual references are not available, the airplane should be held in level-landing attitude until the ground is contacted. Most important of all, positive control of the airplane must be maintained at all times—do not allow a stall to occur.

CHAPTER 15

Seaplane Operations

This chapter is intended to introduce seaplane flying, as well as to provide a general review for experienced seaplane pilots. It contains general explanations of commonly accepted techniques and procedures for operating seaplanes on the water, with special emphasis on those which are different from landplane flying.

The explanations herein apply to light single-engine and multiengine seaplanes typical of those used in general aviation operations. For information regarding specific types and models of airplanes approved as seaplanes, reference should be made to that airplane's operating manual and the manufacturer's recommendations.

In addition to material contained herein, there are numerous commercially produced publications relating to water operations that contain additional valuable information. All this information used collectively with good training and practice will result in a safe and pleasurable experience during water based operations.

The operation of an airplane on water is somewhat different than operating one on land, but should be no more difficult if the pilot acquires the essential knowledge and skill in the techniques involved. This is particularly important because of the widely varying and constantly changing conditions of the water surface.

Water Characteristics

The competent seaplane pilot must be knowledgeable in the characteristics of water to understand its effects on the seaplane. Water is a fluid, and although it is much heavier than air it behaves in a manner similar to air.

Since it is a fluid, water seeks its own level and, if not disturbed, lies flat and glassy. It yields, however, if disturbed by such forces as winds, undercurrents, and objects traveling on its surface, creating waves or movements.

Because of its weight, water can exert a tremendous force. This force, a result of resistance, produces drag as the water flows around or under an object being propelled through it or on its surface. The force of drag imposed by the water increases as the square of the speed. This means that as the speed of the object traveling on the water is doubled, the force exerted is four times as great.

Forces created when operating an airplane on water are more complex than those created on land. When a landplane's wheels contact the ground, the force of friction or drag acts at a fixed point on the airplane; however, the water forces act along the entire length of a seaplane's hull or floats with the center of pressure constantly changing depending upon the pitch attitude, dynamic hull or float motion, and action of the waves.

Since the surface condition of water varies constantly, it becomes important that the seaplane pilot be able to recognize and understand the effects of these various conditions of the water surface.

Under calm wind conditions, the waveless water surface is perhaps the most dangerous to the seaplane pilot and requires precise piloting techniques. Glassy water presents a uniform mirror-like appearance from above, and with no other visual references from which to judge height, it can be extremely deceptive. Also, if waves are decaying and setting up certain patterns, or if clouds are reflected from the water surface, distortions result that are even more confusing for inexperienced as well as experienced pilots.

Wave conditions on the surface of the water are a very important factor in seaplane operation. Wind provides the force that generates waves, and the velocity of the wind governs the size of the waves or the roughness of the water surface (Fig. 15–1).

Calm water resists wave motion until a wind velocity of about 2 knots is attained; then patches of ripples are formed. If the wind velocity increases to 4 knots, the ripples change to small waves that continue to persist for some time even after the wind stops blowing. If this gentle breeze diminishes, the water viscosity dampens the ripples and the surface promptly returns to a flat and glassy condition.

As the wind velocity increases above 4 knots, the water surface becomes covered with a complicated pattern of waves, the characteristics of which vary continuously between wide limits. This is referred to as the generating area. This generating area remains disarranged so long as the wind velocity is increasing. With a further increase in wind velocity, the waves become larger and travel faster. When the wind reaches a constant velocity and remains constant, waves develop into a series of equidistant parallel crests of the same height.

An object floating on the water surface where simple waves are present, will show that the water itself does not actually move along with waves. The floating object will describe a circle in a vertical plane, moving upward as

SURFACE WIND FORCE TABLE

Terms Used by U.S. Weather Service	Velocity mph	Estimating Velocities on Land	Estimating Velocities on Sea	
Calm	Less than 1	Smoke rises vertically	Sea like a mirror	Check your glassy water technique before water flying under these conditions
Light air	1 - 3	Smoke drifts; wind vanes unmoved.	Ripples with the appearance of scales are formed but without foam crests.	
Light Breeze	4 - 7	Wind felt on face; leaves rustle; ordinary vane moves by wind.	Small wavelets, still short but more pronounced; crests have a glassy appearance and do not break.	
Gentle Breeze	8 - 12	Leaves and small twigs in constant motion; wind extends light flag	Large wavelets; crests begin to break. Foam of glassy appearance. (Perhaps scattered white-caps.	Ideal water flying characteristics in protected water.
Moderate Breeze	13 - 18	Dust and loose paper raised; small branches are moved.	Small waves, becoming longer; fairly frequent whitecaps.	
Fresh Breeze	19 - 24	Small trees in leaf begin to sway; crested wavelets form in inland water.	Moderate waves; taking a more pronounced long form; many whitecaps are formed. (Chance of some spray.)	This is considered rough water for seaplanes and small amphibians,especially in open water.
Strong Breeze	25 - 31	Large branches in motion; whistling hear in telegraph wires; umbrellas used with difficulty.	Large waves begin to form; white foam crests are more extensive everywhere. Probably some spray.)	
Moderate Gale	32 - 38	Whole trees in motion; inconvenience felt in walking against the wind.	Sea heaps up and white foam from breaking waves begins to be blown in streaks along the direction of the wind.	This type of water condition is for emergency only in small aircraft in inland waters and for the expert pilot.

Figure 15–1 Wind Force Table

the crest approaches, forward and downward as the crest passes, and backward as the trough between the waves passes. After the passage of each wave the object stays at almost the same point at which it started. Consequently, the actual movement of the object is a vertical circle whose diameter is equal to the height of the wave. This theory must be slightly modified however, because the friction of the wind will cause a slow downwind flow of water resulting in drift. Therefore, a nearly submerged object, such as a hull or float, will slowly drift with the waves.

When the wind increases to a velocity of 12 knots, waves will no longer maintain smooth curves. The waves will break at their crest and create foam—whitecaps. When the wind decreases, the whitecaps disappear. However, lines or streaks form which can be used as an accurate indication of the path of the wind. Generally, it will be found that waves generated by wind velocities up to 10 knots do not reach a height of more than one foot.

A great amount of wind energy is needed to produce large waves. When the wind ceases, the energy in the wave persists and is reduced only by a very slight internal friction in the water. As a result, the wave patterns continue for long distances from their source and diminish at a barely perceptible rate. These waves are known as swells, and gradually lengthen, become less high, but increase in speed.

If the wind changes direction during the diminishing process, an entirely separate wave pattern will form which is superimposed on the swell. These patterns are easily detected by the pilot from above, but are difficult to see from the surface.

Islands, shoals, and tidal currents also affect the size of waves. An island with steep shores and sharply pointed extremities allows the water at some distance from the shore to pass with little disturbance or wave motion. This creates a relatively calm surface on the lee side. If the island has rounded extremities and a shallow slope and outlying shoals where the water shallows and then becomes deep again, the waves will break and slow down. This breaking will cause a considerable loss of wave height on the lee side of the shoal. However, if the water is too deep above the shoal, the waves will not break.

When waves are generated in nonflowing water and travel into moving water such as a current, they undergo important changes. If the current is moving in the same direction as the waves, they increase in speed and length but lose their height. If the current is moving opposite to the waves, they will decrease in speed and length, but will increase in height and steepness. This explains "tidal rips" which are formed where strong streams run against the waves. A current traveling at 6 miles per hour will break almost all waves traveling against it. When waves break, a considerable loss in wave height occurs to the leeward side of the breaking.

Another characteristic of water that should be mentioned is the ability of water to provide buoyancy and cause some objects to float on the surface. Some of these floating objects can be seen from the air, while others is partially submerged and are very difficult to see. Consequently, seaplane pilots must be constantly aware of the possibility of floating debris to avoid striking these objects during operation on the water.

Characteristics of Seaplanes

A seaplane is defined as "an airplane designed to take off from and land on water." Seaplanes can be generally classified as either flying boats, or floatplanes. Those that can be operated on both land and water are called amphibians.

The floatplane is ordinarily understood to be a conventional landplane equipped with separate floats instead of wheels, as opposed to a flying boat in which the hull serves the dual purpose of providing buoyancy in the water and space for the pilot, crew, and passengers. The float type is the more common seaplane, particularly those with relatively low horsepower. It may be equipped with either single float or twin floats; however, most seaplanes are the twin-float variety. Though there is considerable difference between handling a floatplane and handling a flying boat, the theory on which the techniques are based is similar. Therefore, with few exceptions, the explanations given here for one type may be considered to apply to the other.

In the air the seaplane is operated and controlled in much the same manner as the landplane, since the only major difference between the floatplane and the landplane is the installation of floats instead of wheels. Generally, because of the float's greater weight, replacing wheels with floats increases the airplane's empty weight and thus decreases its useful load, and rate of climb.

On many floatplanes, the directional stability will be affected to some extent by the installation of the floats. This is caused by the length of the floats and the location of their mass in relation to the airplane's CG. To help restore directional stability, an auxiliary fin is often added to the tail. The pilot will also find that less aileron pressure is needed to hold the floatplane in a slip and holding some rudder pressure during in-flight turns is usually required. This is due to the water rudder being connected to the air rudder or rudder pedals by cables and springs which tend to prevent the air rudder from streamlining in a turn.

Research and experience have improved float and hull designs throughout the years. Figs. 15-2 and 15-3 illustrate the basic construction of a float and a flying boat. The primary consideration in float construction is the use of sturdy, lightweight material, designed hydrodynamically and aerodynamically for optimum performance.

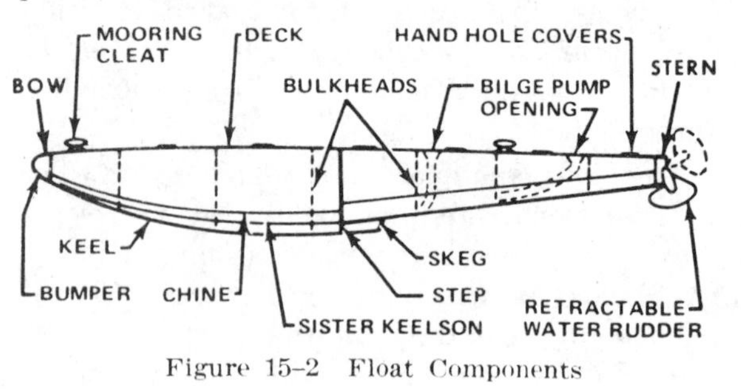

Figure 15-2 Float Components

All floats and hulls now being used have multiple watertight compartments which make the seaplane virtually unsinkable, and prevent the entire float or hull from becoming filled with water in the event it is ruptured at any point.

Both the lateral and longitudinal lines of a float or hull are designed to achieve a maximum lifting force by diverting the water and the air downward. The forward bottom por-

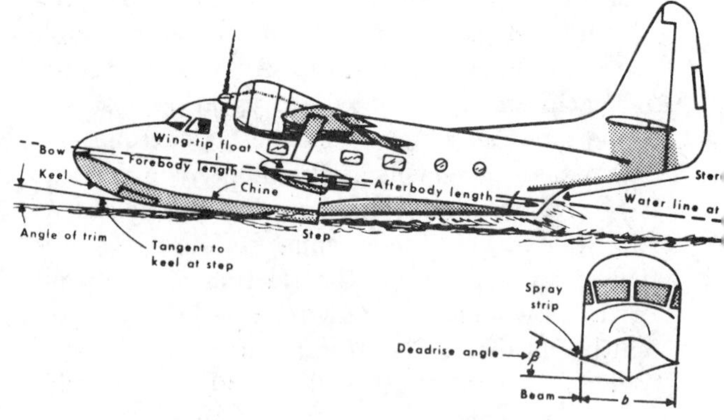

Figure 15-3 Hull Components

tion of the float (and a hull) is designed very much like the bottom surface of a speedboat. The rearward portion, however, differs significantly from a speedboat.

A speedboat is designed for travel at an almost constant pitch angle and, therefore, the contour of the entire bottom is constructed in approximately a continuous straight line. However, a seaplane float or hull must be designed to permit the seaplane to be rotated or pitched up to increase the wing's angle of attack and gain the most lift for takeoffs and landings. Thus, the underside of the float or hull has a sudden break in its longitudinal lines at the approximate point around which the seaplane rotates into the lift off attitude. This break, called a "step," also provides a means of interrupting the capillary or adhesive properties of the water. The water can then flow freely behind the step, resulting in minimum surface friction so the seaplane can lift out of the water.

The steps are located slightly behind the airplane's center of gravity, approximately at the point where the main wheels of a landplane are located. If the steps were located too far aft or forward of this point, it would be difficult, if not impossible, to rotate the airplane into a pitch-up attitude prior to planing (rising partly out of the water while moving at high speed) or lift off.

Although steps are necessary, the sharp break along the float's or hull's underside causes structural stress concentration, and in flight produces considerable drag because of the eddying turbulence it creates in the airflow.

Seaplane Bases

With few exceptions seaplane operations are authorized on U.S. Army Corps of Engineer lakes. Some states and cities are very liberal in the laws regarding the operation of seaplanes on their lakes and waterways, while other states and cities may impose stringent restrictions. It is recommended that before operating a seaplane on public waters, the Parks and Wildlife Department of the state, the State Aeronautics Department, or the FAA General Aviation District Office nearest the site of planned operation be contacted concerning the local requirements. In any case, seaplane pilots should always avoid creating a nuisance in any area, particularly in congested marine areas or near swimming or boating facilities.

The location of established seaplane bases is symbolized on aeronautical charts by depicting an anchor inside a circle. They are also listed in Airport/Facility Directories. The facilities provided at seaplane bases vary greatly, but most include a hard surface ramp for launching, servicing facilities, and an area for mooring or hangaring seaplanes. Many marinas designed for boats also provide seaplane facilities.

In many cases seaplane operations are conducted in "bush country" where regular or emergency facilities are either poor or nonexistent. The terrain is often hazardous, the waterways treacherous, and servicing must be the individual pilot's responsibility.

Too many times pilots receive their water training in the "lower 48" states where facilities are mostly excellent, and shortly after receiving a seaplane rating head north to Alaska or the north woods of Canada. The results are frequently tragic. Prior to operating in the "bush," it is recommended that pilots obtain the advice of FAA appointed Accident Prevention Counselors who are familiar with the area.

Rules of the Road

In addition to Federal Aviation Regulations, the "Rules of the Road" applicable to the operation of boats, apply also to seaplane operations. "Inland Rules of the Road" apply to all vessels navigating upon certain waters inshore of the boundary line which divides the inland waterways from the high seas; "International Rules of the Road" apply to all public or private vessels navigating on the high seas outside the boundary line. The U.S. Coast Guard, of course, has jurisdiction over operations on the high seas. It is strongly recommended that seaplane pilots acquire copies of the pertinent rules, become thoroughly familiar with their contents, and comply with the requirements during all operations.

In the interest of safety, it is particularly important that seaplane pilots become familiar with such navigation aids as buoys, day and night beacons, light and sound signals, and also steering and sailing rules.

Preflight Inspection

Generally, with a few exceptions, the preflight inspection of a seaplane is similar to that of a landplane. The major difference is the checking of floats or hull. The manufacturer's manual or handbook should be used in conducting the inspection.

The pilot should first note how the seaplane is setting in the water prior to each flight. If the sterns of the floats are very low in the water, consideration should be given to how the seaplane is loaded. Also, if lower than normal for a given load, a rear compartment may have a leak.

Floats and hulls should be inspected for obvious or apparent defects and damage, such as loose rivets, corrosion, separation of seams, punctures, and general condition of the metal skin. Because of the rigidity of the float installation, fittings and adjacent structure should be checked for cracks, defective welds, proper attachment, alignment, and safetying. All hinged points should be examined for wear and *corrosion*, particularly if the seaplane is operated in salt water. If water rudders are installed, they should be inspected for free and proper movement.

It is important to check each compartment of the floats or hull for any accumulation of water before flight. Even a small amount of water, such as a cup full, is not unusual and can occur from condensation or normal leakage. All water should be removed before flight, because the water may critically affect the location of the seaplane's center of gravity.

If an excessive amount of water is found, a thorough search for the leak should be made. If drain plugs and inspection plates are installed, a systematic method of removing and reinstalling these plugs and plates securely should be used. Naturally, it is extremely important to ensure that all drain plugs and inspection plates are securely in place before launching the seaplane onto the water. It is recommended that each plug and plate be counted and placed in a receptical upon removal and counted again when reinstalled.

Float compartments, water rudders, etc., should be inspected for ice if near freezing temperatures are encountered. Airframe icing, resulting from water spray during a takeoff or landing, must also be considered. Part of the preflight inspection should include a cabin inspection. All items must be secured, such as anchors and paddles prior to takeoff. Flotation gear should be available for each occupant.

During the preflight and boarding of passengers, a thorough passenger briefing is very important. Evacuation of a seaplane causes a few problems not encountered with the landplane. Location and operation of regular and emergency exits should be known by all persons on board. The pilot should assure that all passengers are familiar with operation of seatbelts and shoulder harnesses, most especially that all persons can *UNFASTEN* their own seatbelts and shoulder harnesses in the event an accident occurs on the water.

Before beginning any seaplane operation, it is especially advisable to consider the existing and expected water condition, and the windspeed and direction to determine their combined effects on the operation.

Taxiing

One of the major differences between the operation of a seaplane and that of a landplane is the method of maneuvering the aircraft on the surface. The landplane will usually remain motionless with the engine idling, particularly with the brakes applied, but a seaplane, since it is free-floating, will invariably move in some direction, depending upon the forces exerted by wind, water currents, propeller thrust, and inertia. Because a seaplane has no brakes, it is important that the pilot be familiar with the existing wind and water conditions, effectively plan the course of action, and mentally stay "ahead" of the aircraft.

There are three positions or attitudes in which a seaplane can be moved about on the water: (1) the "idling" position, (2) the "plowing" position, and (3) the "planing" or "on the step" position (Fig. 15-4).

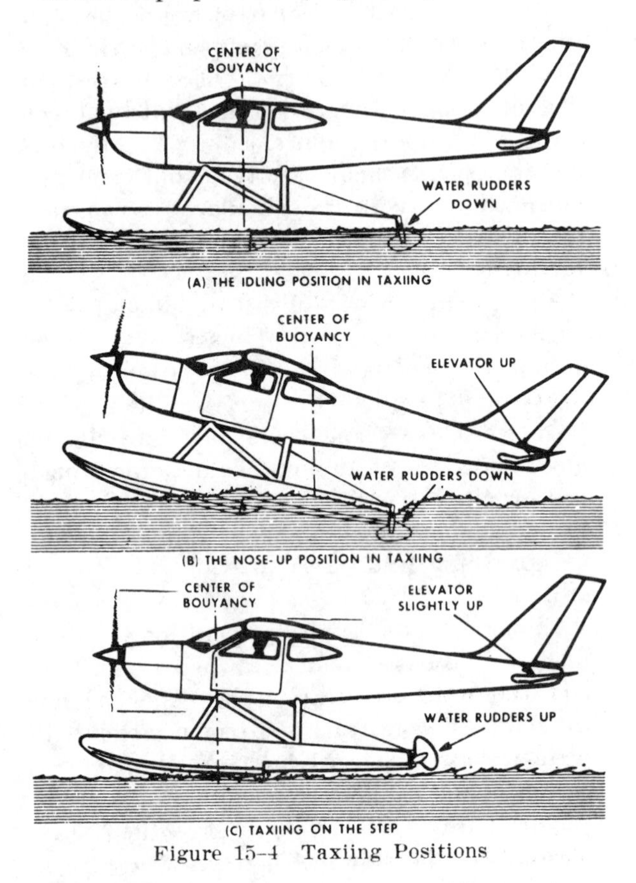

Figure 15-4 Taxiing Positions

Idling Position

When taxiing with the engine idling or at a low RPM, the seaplane will remain in what is considered a displacement condition similar to being at rest on the water (Fig. 15-4). This is the "idling" position. The recommended taxi speed is usually below 6 or 7 knots so that the propeller will not pick up water spray which causes serious erosion of the propeller blades. In calm or light wind conditions, the elevator control should be held full back to raise the seaplane's nose and further reduce the possibility of water spray on the propeller, and to improve overall maneuverability of the seaplane. This is particularly true if it is equipped with water rudders because more rudder area is kept in the water. Since seaplanes have no brakes, it is especially important to taxi at this slow speed in congested or confined areas because inertia forces build up rapidly, making the seaplane vulnerable to serious damage even in minor collisions.

Plowing Position

When the power is increased significantly above idling, the seaplane will usually assume a nose-up or "plowing" position (Fig. 15-4). Most seaplane experts do not recommend the plowing position for taxiing, except in rough water when it would be desirable to raise the propeller clear of the spray, or when turning the seaplane downwind during strong wind conditions. To attain this position, full power should be applied and the elevator control held in the full aft position. Sea planes that have a high thrust line will tend to nose down upon application of power, in which case it is imperative that the elevator control be held in the full aft position. The "plowing" position is brought about by the combination of the propeller slipstream striking the elevator and the hydrodynamic force of water exerted on the underside of the float's or hull's bow. After the planing position is attained, the power should be reduced to maintain the proper speed.

If the water conditions are favorable and there is a long distance to travel, the seaplane may be taxied at high speed "on the step." This position (Fig. 15-4) is reached by accelerating the seaplane to the degree that it passes through the plowing phase until the floats or hull are literally riding on the water in a level position.

Planing Position

Basically, the planing or step position is best attained by holding the elevator control full aft and advancing the throttle to full power. As the seaplane accelerates it will then gradually assume a nose-high pitch attitude, raising the bow of the float or hull and causing the weight of the seaplane to be transferred toward the aft portion of the float or hull. At the time the seaplane attains its highest pitch attitude, back pressure should be gradually relaxed, causing the weight to be transferred from the aft portion of the float or hull onto the step area. This can be compared to a speedboat's occupants moving forward in the boat to aid in attaining a planing attitude. In the seaplane we do essentially the same thing by use of aerodynamics (elevators). As a result of aerodynamic and hydrodynamic lifting, the seaplane is raised higher in the water, allowing the floats or hull to ride on top of rather than in the water.

The entire process of planing a seaplane is similar to that of water skiing. The skier cannot make the transition from a submerged condition to that of being supported on the surface of the water unless a sufficiently high speed is attained and maintained.

As further acceleration takes place the flight controls become more responsive just as in the landplane and elevator deflection must be reduced in order to hold the required planing/pitch attitude. This of course is accomplished by further relaxing back pressure, increasing forward pressure, or using forward elevator trim depending on the aircraft flight characteristics.

Throughout the acceleration, the transfer of weight and the hydrodynamic lifting of the float or hull may be seen from the cockpit. When the seaplane is taxiing slowly, the water line is quite high on the floats or hull as compared to "on the step" (Fig. 15-3). At slow taxi speeds a small wake is created close to the bow of the float or hull and moves outward at a very shallow angle. As acceleration commences, the wake starts to move from the bow aft toward the step area and the wake now turns into an outward spray pattern. As speed and lifting action increase, the spray pattern continues to move aft toward the step position and increases in intensity, i.e., slow speed spray may be approximately one foot outboard compared to about a 20-foot outboard spray at higher speed on the step position. Some seaplane pilots use the spray pattern as an additional visual reference in aiding them in determining when the seaplane has accelerated sufficiently to start easing it over onto the step.

After the planing position has been attained, proper control pressures must be used to control the proper pitch attitude/trim angle. Usually this will be maintained with slight back pressure. As for the amount of pressure to be held, the beginner will find a very "thin line" between easing off back pressure too much or too little. It can perhaps best be described as finding the "slippery spot" on the float or hull. Too much back pressure, acceleration rate decreases. Not enough back pressure or too much forward pressure also decreases acceleration rate. So that fine line or "slippery spot" is that position between not enough or too much back pressure.

If one does not want to take off and just wants to continue to taxi on the step, a reduction in power is initiated at approximately the time the seaplane is eased over onto the step. Power requirements to maintain the proper speed with wind, load and current action will vary. More power will be required taxiing into the wind or an upcurrent or with a heavy load. However, 65 to 70 percent of maximum power can be used as a starting point.

From either the plowing or on-the-step position, if power is reduced to idle, the seaplane will decelerate quite rapidly and eventually assume the displacement or idle position. Care must be taken to use proper flight control pressures during the deceleration phase because weight is now being transferred toward the bow and drag is increasing; hence, some aircraft have a nose-over tendency. This is of course controllable by proper use of the elevator controls.

Turns

If water rudders have the proper amount of movement, a seaplane can be turned within a radius less than the span of the wing during calm conditions or a light breeze. It will be found that water rudders are usually more effective at slow speeds because they are acting in comparatively undisturbed water. At high speeds, however, the stern of the float churns the adjacent water, thereby causing the water rudder to become less efficient. Furthermore, because of the high speed, the water's impact on the rudders may tend to force them to swing up or retract.

Particular attention should always be given to the risks involved in making turns in a strong wind or at high speeds. Any seaplane will tend to weathervane into a strong wind if the controls are not positioned to prevent it. In a single-engine seaplane rudder should be applied as necessary to control the turn while aileron is held into the wind. On twin-engine seaplanes this tendency can be overcome through the use of differential power—using a higher setting on the upwind side. The rate at which the seaplane turns when it weathervanes is directly proportional to the speed of the crosswind. When taxiing downwind or crosswind, the seaplane will swing into the wind as soon as the flight controls are neutralized or power is reduced.

During a high speed taxiing turn, centrifugal force tends to tip the seaplane toward the outside of the turn (Fig. 15–5). Simultaneously, when turning from a downwind heading to an upwind heading, the wind force striking the fuselage and the under side of the wing increases the tendency for the seaplane to lean to the outside of the turn. If an abrupt turn is made, the combination of these two forces may be sufficient to tip the seaplane to the extent that the outside wing drags in the water, and may even tip the seaplane onto its back (Fig. 15–6). Obviously,

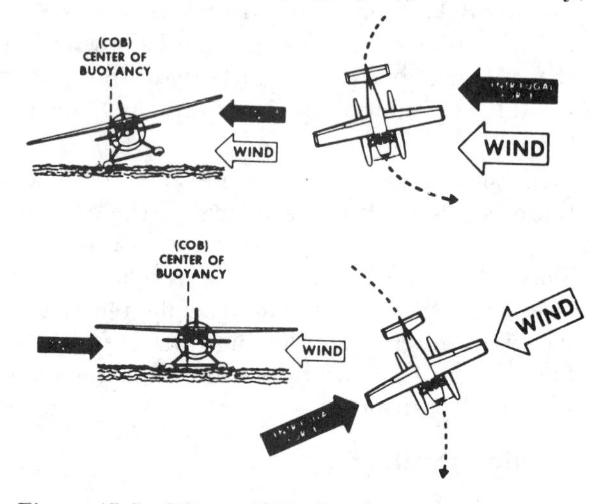

Figure 15–5 Effect of Wind and Centrifugal Force

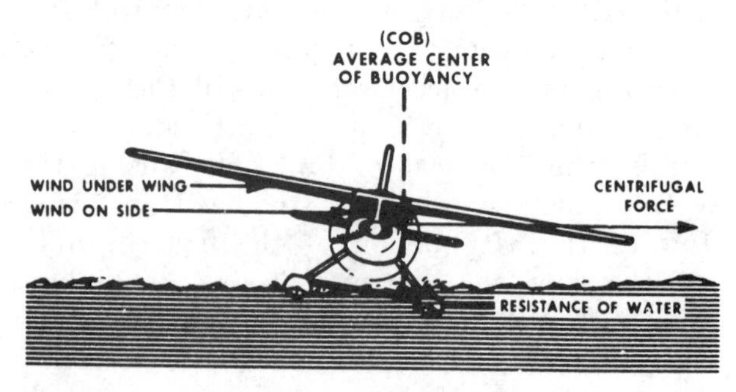

Figure 15–6 Effect of Wind

the further the seaplane tips, the greater will be the effect of wind, since more wing area on the windward side is exposed to the wind force.

When making a turn into the wind from a crosswind condition, the air-rudder may be neutralized and the seaplane allowed to weathervane into the wind. If taxiing directly downwind, a turn into the wind may be started by deflecting the air rudder in the same direction that the turn is desired. As soon as the seaplane begins to turn, the rudder should be neutralized; if the wind is strong, some opposite rudder may be needed during the turn. The amount of opposite rudder depends upon the rate at which the seaplane turns. The greater the amount of opposite rudder, the slower the rate of turn. Normally, the power should be reduced to idle when the turn begins because with the power on the left turning tendency of the seaplane may become excessive. Short bursts of power are best for turning in a small radius, but sustained excessive power causes a buildup of speed and a larger turning radius.

The seaplane tends to use its center of buoyancy (COB) as a pivot point wherever it may be. Center of buoyancy moves laterally, as well as forward and aft. Each object in water has a point of center of buoyancy. In a twin-float installation the effects of wind, power and flight controls are shared by the two floats and the average COB is free to move significantly.

The center of buoyancy (COB), or average point of support, moves aft when the seaplane is placed in a nose-up or plowing position (Fig. 15–4). This position exposes to the wind a considerable amount of float and fuselage side area forward of the center of buoyancy. Therefore, when taxiing crosswind in this position, many seaplanes will show a tendency to turn downwind because of the wind force on the exposed area of the float and the fuselage. For this reason it is sometimes helpful to place the seaplane in a nose-up position when turning downwind, particularly if the wind velocity is high. Under high wind conditions, the throttle may be used as a turning device by increasing power to cause a nose-up position when turning downwind, and decreasing power to allow the seaplane to weathervane into the wind.

Sailing

Occasions often arise when it is advisable to move the seaplane backward or to one side because wind or water conditions, or limited space make it impractical to attempt a turn (Fig. 15–7). In this situation, particularly if there is a significant wind, the seaplane can be "sailed" into a space which to an inexperienced pilot might seem extremely cramped. Even if the wind is calm and the space is inadequate for making a normal turn, a paddle (which should be part of every seaplane's equipment) may be used to propel the seaplane or to turn the nose in the desired direction.

In light wind conditions with the engine idling or shut down, the seaplane will naturally weathervane into the wind and then sail in the direction the tail is pointed (Fig. 15–8). With a stronger wind and a slight amount of power, the seaplane will usually sail downwind toward the side in which the nose is pointed. Rudder and aileron can be deflected to create drag on the appropriate side to control the direction of movement. Positioning the controls for the desired direction of motion in light or strong winds is illustrated in Fig. 15–8. Lowering the wing flaps and opening the cabin doors will increase the air resistance and thus add to the effect of the wind; however, the effect of the air rudder may be reduced in this configuration. Since water rudders have little or no effect in controlling direction while sailing, they should be lifted.

With the engine shut down, most flying boats will sail backward and toward the side to which the nose is pointed, much as a sailboat tacks, regardless of wind velocity because the hull does not provide as much keel (side area) as do floats in proportion to the size of the seaplane. To sail directly backward in a seaplane having a hull, the controls should be released and the wind allowed to steer the seaplane.

Sailing is an essential part of seaplane operation. Since each type of seaplane has its own minor peculiarities, depending on the design of the floats or hull, it should be prac-

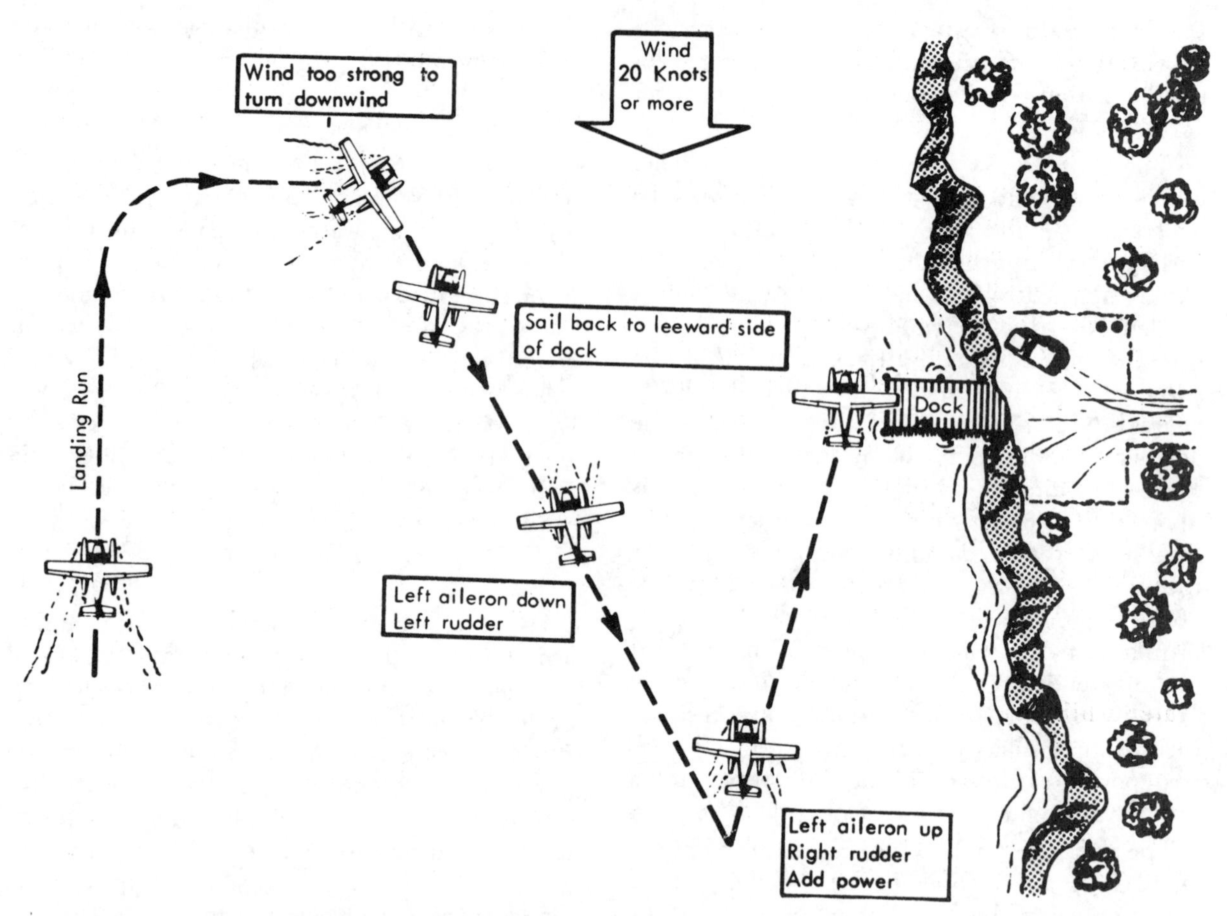

Figure 15–7 Taxiing/Sailing in Strong Wind

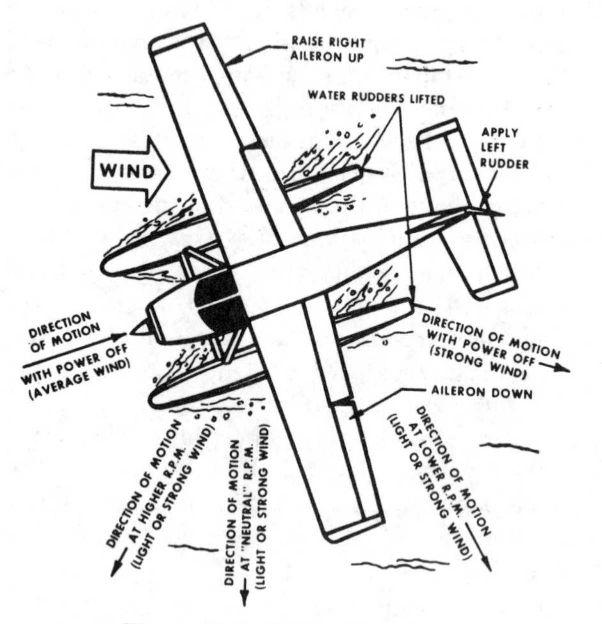

Figure 15–8 Sailing Procedures

ticed until thorough familiarization with that particular type is gained.

During initial seaplane training, sailing should be practiced in large bodies of water such as lakes or bays, but sufficiently close to a prominent object in order to evaluate performance. Where there are strong tides or a rapidly flowing current, such as in rivers, care must be taken in observing the relative effect of both the wind and the water current. Often the force of the current will be equal to or greater than the force of the wind.

Before taxiing into a confined area, the effect of wind and the current should be considered carefully. Otherwise, the seaplane may be carried into obstructions with resulting damage to the wings, tail surfaces, floats, hull, or other parts of the seaplane. Generally, with a seaplane of average size and power at idle, a water current of 5 knots will more than offset a wind velocity of 25 knots. This means that the seaplane will move against the wind.

When operating multiengine seaplanes, differential power can be used to aid in steering the seaplane along a desired course.

Porpoising

Porpoising in a seaplane is much like the antics of a dolphin—a rhythmic pitching and heaving while in the water. Porpoising is a dynamic instability of the seaplane and may occur when the seaplane is moving acoss the water while on the step during takeoff or landing. It occurs when the angle between the float or hull, and the water surface exceeds the upper or lower limit of the seaplane's pitch angle. Improper use of the elevator, resulting in attaining too high or too low a pitch (trim angle) sets off a cyclic oscillation which steadily increases in amplitude unless the proper trim angle or pitch attitude is reestablished.

A seaplane will travel smoothly across the water while on the step, so long as the floats or hull remain within a moderately tolerant range of trim angles. If the trim angle is held too low during planing, water pressure in the form of a small crest or wall is built up under the bow or forward part of the floats or hull. As the seaplane's forward speed is increased to a certain point, the bow of the floats or hull will no longer remain behind this crest, and is abruptly forced upward as the seaplane rides over the crest. As the crest passes the step and on to the stern or aft portion of the floats or hull, the bow abruptly drops into a low position. This again builds a crest or wall of water in front of the bow resulting in another oscillation. Each oscillation becomes increasingly severe, and if not corrected will cause the seaplane to nose into the water, resulting in extensive damage or possible capsizing. Porpoising can also cause a premature lift off with an extremely high angle of attack, resulting in a stall or being in the area of reverse command and unable to climb over obstructions.

Porpoising will occur during the takeoff run if the trim angle is not properly controlled with proper elevator pressure just after passing through the "hump" speed, or when the highest trim angle before the planing attitude is attained; that is, if up-elevator is held too long and the angle reaches the upper limits. On the other hand, if the seaplane is nosed down too sharply, the lower trim range can be entered and will also result in porpoising. Usually, porpoising does not start until a degree or two after the seaplane has passed into the critical trim angle range, and does not cease until a degree or two after the seaplane has passed out of the critical range.

If porpoising does occur, it can be stopped by applying timely back pressure on the elevator control to prevent the bow of the floats or hull from digging into the water. The back pressure must be applied and maintained until porpoising is damped. If porpoising is not damped by the time the second oscillation occurs, it is recommended that the power be reduced to idle and elevator control held firmly back so the seaplane will settle into the water with no further instability.

The correct trim angle for takeoff, planing and landing applicable to each type of seaplane must be learned by the pilot and practiced until there is no doubt as to the proper angles for the various maneuvers.

The upper and lower trim angles are established by the design of the aircraft; however, changing the seaplane's gross weight, wing flap position, and center of gravity location will also change these limits. Increased weight increases the displacement of the floats or hull and raises the lower limit considerably. Extending the wing flaps frequently trims the seaplane to the lower limit at lower speeds, and may lower the upper limit at high speeds. A forward center of gravity location raises the lower trim limit at high speeds, and an aft location increases the possibility of high angle porpoising especially during landing.

Skipping

Skipping is a form of instability which may occur when landing with excessive speed at a nose-up trim angle. This nose-up attitude places the seaplane at the upper trim limit of stability, and causes the seaplane to enter a cyclic oscillation when touching the water, resulting in the seaplane skipping across the surface. This action may be compared to "skipping" flat stones across the water.

Skipping can also occur by crossing a boat wake while fast taxiing on the step or during a takeoff. Sometimes the new seaplane pilot will confuse a skip with a porpoise. Pilot's body feelings can quickly determine whether a skip or a porpoise has been encountered. A skip will give the body vertical "G" forces similar to bouncing a landplane. The porpoise is a rocking chair type forward and aft motion feeling.

Correction for skipping is made by first increasing back pressure on the elevator control and adding sufficient power to prevent the floats from contacting the water. Then pressure on the elevator must be adjusted to attain the proper trim angle and the power gradually reduced to allow the seaplane to settle gently onto the water.

Skipping will not continue increasing its oscillations, as in porpoising, because of the lack of forward thrust with reduced power.

Takeoffs

Unlike landplane operations at airports, seaplane operations are often conducted on water areas at which other activities are permitted. Therefore, the seaplane pilot is constantly confronted with floating objects, some of which are almost submerged and difficult to see—swimmers, skiers, and a variety of watercraft. Before beginning the takeoff, it is advisable to taxi along the intended takeoff path to check for the presence of any hazardous objects or obstructions. Thorough scrutiny should be given to the area to assure not only that it is clear, but that it will remain clear throughout the takeoff. Operators of motorboats and sailboats often do not realize the hazard resulting from moving their vessels into the takeoff path of a seaplane.

To accelerate during takeoff in a landplane, propeller thrust must overcome only the surface friction of the wheels and the increasing aerodynamic drag. During a seaplane takeoff, however, hydrodynamic or water drag becomes the major part of the forces resisting acceleration. This resistance reaches its peak at a speed of about 27 knots, and just before the floats or hull are placed into a planing attitude.

The hydrodynamic forces at work during a seaplane takeoff are shown in Fig. 15–9. The point of greatest resistance is referred to as the "hump" because the increasing and decreasing effect of water drag causes a hump in the resisting curve. After the hump is passed and the seaplane is traveling on the step, water resistance decreases.

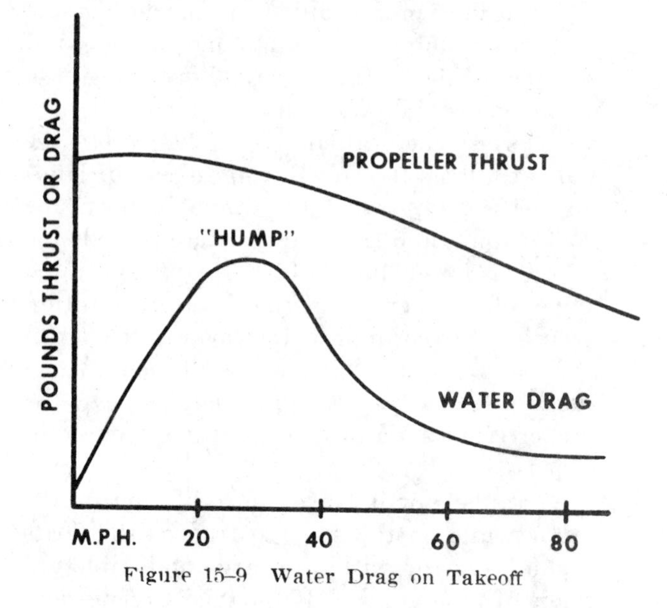

Figure 15–9 Water Drag on Takeoff

Several factors greatly increase the water drag or resistance; heavy loading of the aircraft, or glassy water conditions in which no air bubbles slide under the floats or hull, as they do during a choppy water condition. In extreme cases, the drag may exceed the available thrust and prevent the seaplane from becoming airborne. This is particularly true when operating in areas with high density altitudes (high elevations/high temperatures) where the engine cannot develop full rated power. For this reason the pilot should also practice takeoffs using only partial power to simulate the long takeoff run usually needed when operating at water areas where the density altitude is high and/or the seaplane is heavily loaded.

The seaplane takeoff may be divided into four distinct phases: (1) The "displacement" phase, (2) the "hump" or "plowing" phase, (3) the "planing" or "on the step" phase, and (4) the "lift off." The first three phases were previously described in the section on taxiing. The "lift off" is merely transferring support of the seaplane from the floats or hull to the wings by applying back elevator pressure. This results in the seaplane lifting off the water and becoming airborne.

To avoid porpoising during the takeoff run, it is important to maintain the proper pitch angles. Too much back elevator pressure during the planing or lift off phases will force the stern of the floats or hull deeper into the water, creating a strong resistance and appreciably retarding the takeoff. Conversely, insufficient back elevator pressure will cause the bows to remain in the water, which also results in excessive water drag. Experience will determine the best angle to maintain during takeoff for each seaplane, and if held at this angle, the seaplane will take off smoothly.

Because the seaplane is not supported on a solid surface and the float or one side of the hull can be forced deeper into the water, right aileron control is usually required to offset the effect of torque when full power is applied during takeoff.

The spray pattern for each particular seaplane should also be considered during takeoff. During acceleration the water is increasingly sprayed upward, outward, and rearward from the bow portion of the floats or hull, and on some seaplanes will be directed into the propeller, eventually causing erosion of the blades. This water spray is greater during the hump phase. The spray can be reduced during takeoff, however, by first increasing the planing speed about 10 knots, then opening the throttle as rapidly as practical. This method shortens the time that propellers are exposed to the spray. Again, the best technique must be learned through experience with each particular seaplane. Bear in mind that a rough water condition creates more spray than does smooth water.

Glassy water takeoffs in a low-powered seaplane loaded to its maximum authorized weight presents a difficult, but not necessarily a dangerous, problem. Under these conditions the seaplane may assume a "plowing" or nose-up position, but may not "unstick" or get "on the step" because of the adhesive action of smooth water; consequently, always plan ahead and consider the possibility of aborting the takeoff. Nonetheless, if these conditions are not too excessive, the takeoff can be accomplished using the following procedure.

After the bow has risen to the highest point in the plowing position with full back elevator pressure, it should be lowered by decreasing back elevator pressure. The bow will drop if the seaplane has attained enough speed to be on the verge of attaining the step position. After a few seconds, the bow will rise again. At the instant it starts to rise, the rebound should be caught by again applying firm back elevator pressure, and as soon as the bow reaches its maximum height, the entire routine should be repeated. After several repetitions, it will be noted that the bow attains greater height and that the speed is increasing. If the elevator control is then pushed well forward and held there, the seaplane will slowly flatten out "on the step" and the controls may then be eased back to the neutral position.

Whenever the water is glassy smooth, a takeoff can be made with less difficulty by making the takeoff run across the wakes created by motorboats. If boats are not operating in the area, it is possible to create wakes by taxiing the seaplane in a circle and then taking off across these self-made wakes.

On seaplanes with twin floats water drag can be reduced by applying sufficient aileron pressure to raise the wing and lift one float out of the water after the seaplane is on the step. By allowing the seaplane to turn slightly in the direction the aileron is being held rather than holding opposite rudder to maintain a straight course, considerable aerodynamic drag can be eliminated, aiding acceleration and lift off. When using this technique, great care must be exercised so as not to lift the wing to the extent that the opposite wing strikes the water. Naturally, this would result in serious consequences.

In most cases an experienced seaplane pilot can safely take off in rough water, but a beginner should not attempt to takeoff if the waves are high. Using the proper procedure during rough water operation lessens the abuse of the floats, as well as the entire seaplane.

During rough water takeoffs, the throttle should be opened to takeoff power just as the bow is rising on a wave. This prevents the bow from digging into the water and helps keep the spray from the propeller. Slightly more back elevator pressure should be applied to the elevator than on a smooth water takeoff. This raises the bow to a higher angle.

After planing has begun, the seaplane will bounce from one wave crest to the next, raising the nose higher with each bounce, and each successive wave will be struck with increasing severity. To correct this situation and to prevent a stall, smooth elevator pressures should be used to set up a fairly constant pitch attitude that will allow the aircraft to "skim" across each successive wave as speed increases. Remember, in waves, the length of the float is very important. It is important that control pressure be maintained to prevent the bow from being pushed under the water surface or "stubbing its toe," which could result in capsizing the seaplane. Fortunately, a takeoff in rough water is accomplished within a short time because if there is sufficient wind to make the water rough, the wind would also be strong enough to produce aerodynamic lift earlier and enable the seaplane to become airborne quickly.

With respect to water roughness, one condition that seaplane pilots should be aware of is the effect of a strong water current flowing against the wind. For example, if the velocity of the current is moving at 10 knots, and the wind is blowing at 15 knots, the relative velocity between the water and the wind is 25 knots. In other words, the waves will be as high as those produced in still water by a wind of 25 knots.

The advisability of canceling a proposed flight because of rough water depends upon the size of the seaplane, wing loading, power loading, and, most important, the pilot's ability. As a general rule, if the height of the waves from trough to crest is more than 20 percent of the length of the floats, takeoffs should not be attempted except by the most experienced and expert seaplane pilots.

Downwind takeoffs are possible, and at times preferable, if the wind velocity is light and normal takeoffs would involve clearing hazardous obstructions, or flying over congested areas before adequate altitude can be attained. The technique used for downwind takeoffs is almost identical to that used for upwind takeoffs. The only difference is that the elevator control should be held further aft, if possible. When downwind takeoffs are made, it should be kept in mind that more space is needed for the takeoff. If operating from a small body of water, an acceptable technique may be to begin the takeoff run while headed downwind, and then turning so as to complete the takeoff into the wind. This may be done by planing the seaplane while on a downwind heading then making a step turn into the wind to complete the takeoff. Caution must be exercised when using this technique since wind and centrifugal force will be acting in the same direction and could result in the seaplane tipping over.

Crosswind takeoff techniques will be discussed later in the chapter.

Landings

In comparison, the land surfaces of all airports are of firm, static matter, whereas the surface of water is changing continually as a fluid. Floating obstacles and various activities frequently present on the water surface may present serious hazards during seaplane landings, especially to the careless pilot. For these reasons, it is advisable to circle the area of intended landing and examine it thoroughly for obstructions such as buoys or floating debris, and to note the direction of movement of any boats which may be operating at the intended landing site.

Most established seaplane bases are equipped with a wind sock to indicate wind direction, but if one is not available the wind can still be determined prior to landing. The following are but a few of the methods by which to determine the wind direction.

If there are no strong tides or water currents, boats lying at anchor will weathervane and automatically point into the wind. It is also true that sea gulls and other water fowl usually land facing the wind. Smoke, flags, and the set of sails on sailboats also provide the pilot with a fair approximation of wind direction. If there is an appreciable wind velocity, streaks parallel to the wind are formed on the water. During strong winds, these streaks form distinct white lines. However, wind direction cannot always be determined from these streaks alone. If there are whitecaps or foam on top of the waves, the foam appears to move into the wind. This illusion is caused by the waves moving under the foam.

In seaplanes equipped with retractable landing gear (amphibians), it is extremely important to make certain that the wheels are in the *retracted* position when landing on water. Wherever possible, a visual check of the wheels themselves is recommended, in addition to checking the landing gear position indicating devices. A wheels-down landing on water is almost certain to capsize the seaplane, and is far more serious than landing the seaplane wheels-up on land. The water rudder should also be in the retracted position during landings.

The landing approach procedure in a seaplane is very similar to that of a landplane and is governed to a large extent by pilot preference, wind, and water conditions.

Under normal conditions a seaplane can be landed either power-off or power-on; however, power-on landings are recommended in most cases, because this technique gives the pilot more positive control of the seaplane and provides a means for correcting errors in judgment during the approach and landing. So that the slowest possible airspeed can be maintained, the power-on landing should be accomplished with maximum flaps extended. The seaplane should be trimmed to the manufacturer's recommended approach speed, and the approach made similar to that of a landplane.

Touchdown on the water should be made in a pitch attitude that is correct for taxiing "on the step," or perhaps a slightly higher attitude (Fig. 15–10). This attitude will result in the floats or hull first contacting the water at a point aft of the step. Once water contact is made, the throttle should be closed and back elevator pressure gradually applied. The application of back pressure reduces the tendency for the seaplane to nose down and the bows to dig in due to increased drag of the floats as they contact the water. The faster the speed at which a seaplane is landed, the more water drag is encountered, resulting in a greater nose-down attitude after touchdown. If the seaplane has a tendency to nose down excessively with full flaps extended, it is recommended that subsequent approaches and landings be made with less flaps. Remember, the objective is to land the seaplane at the slowest possible speed in a slightly nose-up attitude.

Figure 15–10 Touchdown Attitude

After contacting the water, gradually increase back elevator pressure. It may be desirable at times to remain on the step after touchdown. To do so, merely add sufficient power and maintain the planing attitude immediately after touchdown.

Flat, calm, glassy water is perhaps the most deceptive condition that a seaplane pilot will experience. The calmness of the water has a psychological effect in that it tends to overly relax the pilot when there should be special alertness. Consequently, this surface condition is frequently the most dangerous for seaplane operation.

From above, the mirror-like appearance of smooth water looks most inviting and easy to land on but as many pilots have suddenly learned, adequate depth perception may be lacking. Even experienced pilots misjudge height above the water, making timely roundouts difficult. This results in either flying bow first into the water or stalling the seaplane at too great a height above the water. When the water is crystal clear and glassy, pilots often attempt to judge height by using the bottom of the lake as a reference, rather than the water surface.

An accurately-set altimeter may be used as an aid in determining height above the glassy water. However, a more effective means is to make the approach and landing near the shoreline so it can be used as a reference for judging height above the water. Another method is to cross the shoreline on final approach at the lowest possible safe altitude so that a height reference is maintained to within a few feet of the water surface.

Glassy water landings should always be made power-on, and the need for this type of landing should be recognized in ample time to set up the proper final approach.

During the final approach the seaplane should be flown at the best nose-high attitude, using flaps as required or as recommended by the manufacturer. A power setting and pitch attitude should be established that will result in a rate of descent not to exceed 150 feet per minute and at an airspeed approximately 10 knots above stall speed. With a constant-power setting and a constant-pitch attitude, the airspeed will stabilize, and remain so if no changes are made. The power or pitch should be changed only if the airspeed or rate of descent deviates from that which is desired. Throughout the approach the seaplane performance should be closely monitored by cross checking the instruments until contact is made with the water.

Upon touchdown, back elevator control pressure should be applied as necessary to maintain the same pitch attitude. Throttle should be reduced or closed only after the pilot is sure that the aircraft is firmly on the water. Several indications should be used.

1. A slight deceleration force will be felt.
2. A slight downward pitching moment will be seen.
3. The sound of water spray striking the floats, hull, or other parts of the aircraft will be heard.

All three cues should be used because accidents have resulted from cutting the power rapidly after initially touching the water. To the pilot's surprise a skip had taken place and it was found that when the power was cut, the aircraft was 10 to 15 feet in the air and not on the water, resulting in a stall and substantial damage.

Maintaining a nose-up, wings-level attitude, at the correct speed and a small rate of descent, are imperative for a successful glassy water landing. All aspects of this approach and landing should be considered prior to its execution. Bear in mind that this type of approach and landing will usually consume considerable landing distance. Landing near unfamiliar shorelines increases the possibility of encountering submerged objects and debris.

It is impractical to describe an ideal rough water procedure because of the varying conditions of the surface. In most instances, though, the approach is made the same as for any other water landing. It may be better however, to level off just above the water surface and increase the power sufficiently to maintain a rather flat attitude until conditions appear to be more acceptable, and then reduce the power to touchdown. If severe bounces occur, power should be increased and a search made for more ideal landing spot.

Generally, it is recommended that night water landings in seaplanes be avoided, since they can be extremely dangerous due to the difficulty or almost impossibility of seeing objects in the water. If it becomes necessary to land at night in a seaplane, serious consideration should be given to landing at a lighted airport. An emergency landing can be made on land in seaplanes with little or no damage to the floats or hull. Touchdown should be made with the keel of the floats or hull as nearly parallel to the surface as possible. After touchdown, full back elevator must be applied and additional power applied to lessen the rapid deceleration and nose-over tendency. Don't worry about getting stopped with additional power applied after touchdown. It will stop! The power is applied only for increasing elevator effectiveness.

Crosswind Techniques

Because of restricted or limited areas of operation, it is not always possible to take off or land the seaplane directly into the wind. Such restricted areas may be canals or narrow rivers. Therefore, skill must be acquired in crosswind techniques to enhance the safety of seaplane operation.

The forces developed by crosswinds during takeoffs or landings on water are almost the same as those developed during similar operations on land. Directional control is more difficult because of the more yielding properties of water, less surface friction, and lack of nosewheel, tailwheel, or brakes. Though water surface is more yielding than solid land, a seaplane has no shock absorbing capability, so all the shock is absorbed by the hull or floats and transmitted to the aircraft structure.

As shown in Fig. 15–11, a crosswind tends to push the seaplane sideways. The drifting force, acting through the seaplane's center of

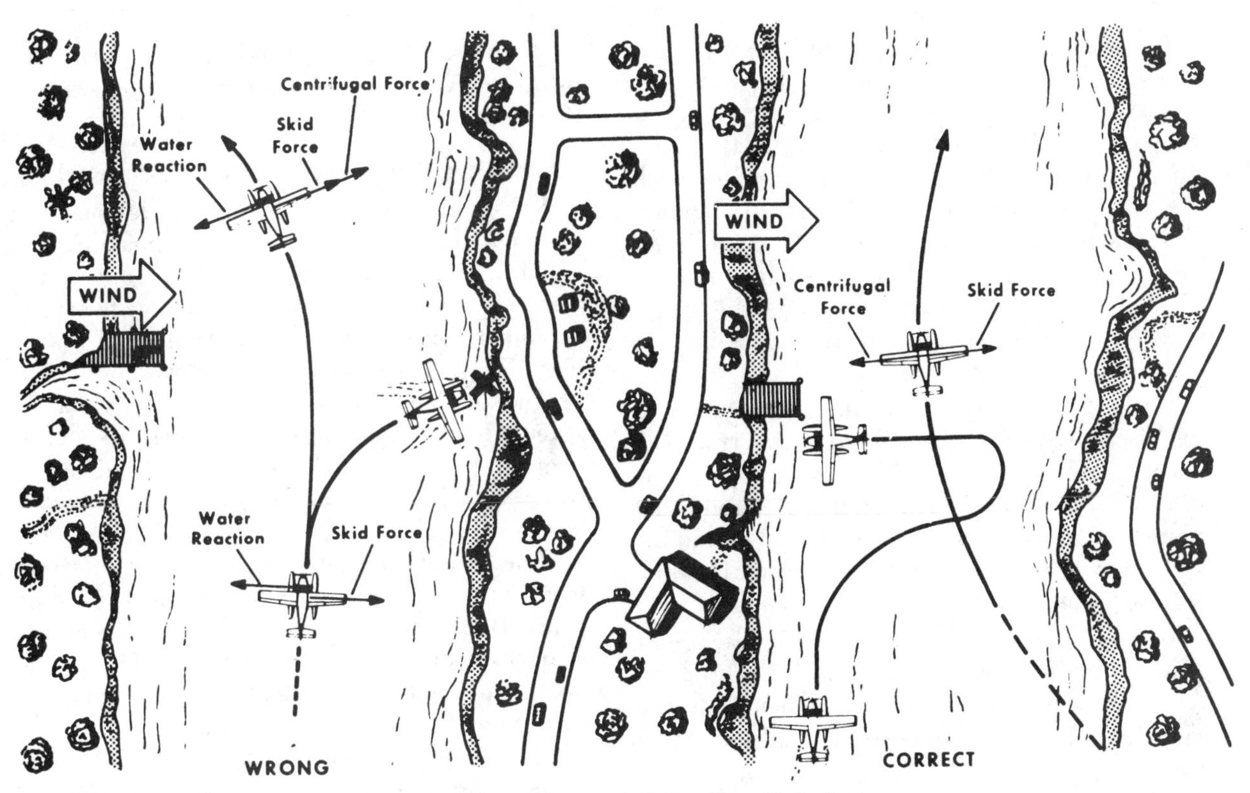

Figure 15–11 Crosswind Landing Technique

gravity, is opposed by the water reacting on the area of the floats or hull in contact with the water. This results in a tendency to weathervane into the wind. Once this weathervaning has started, the turn continues and is further aggravated by the addition of centrifugal force acting outward from the turn, which again is opposed by the water reaction on the floats or hull. If strong enough, the combination of the wind and centrifugal force may tip the seaplane to the point where the downwind float will submerge and subsequently the wingtip may strike the water and capsize the seaplane. This is known as a "waterloop" similar to a "groundloop" on land.

Because of the lack of clear reference lines for directional guidance, such as are found on airport runways, it is difficult to quickly detect sidedrift on water. Fortunately, early detection of sidedrift is not really essential, because the seaplane takeoff and landing can be made without maintaining a straight line while in contact with the water. A turn should be made toward the downwind side after landing. This will allow the seaplane to dissipate its forward speed prior to its weathervaning into the wind. By doing this, centrifugal force while weathervaning will be kept to a minimum and better aircraft control will result with less turnover tendency.

One technique sometimes used to compensate for crosswinds during water operations is the same as that used on land; that is, by lowering the upwind wing while holding a straight course with rudder. This creates a slip into the wind to offset the drifting tendency. The upwind wing is held in the lowered position throughout the touchdown and until completion of the landing.

Another technique used to compensate for crosswinds (preferred by many seaplane pilots) is the downwind arc method. Using this method, the pilot creates a sideward force (centrifugal force) that will offset the crosswind force. This is accomplished by steering the seaplane in a downwind arc as shown in Fig. 15–11. The pilot merely plans an arced path and follows this arc to produce sufficient centrifugal force so that the seaplane will tend to lean outward against the wind force. During the run, the pilot can adjust the rate of turn by varying rudder pressure, thereby increasing or decreasing the centrifugal force to compensate for a changing wind force.

In practice, it is quite simple to plan sufficient curvature of the takeoff path to cancel out strong crosswinds, even on very narrow rivers. As illustrated in Fig. 15–12, the takeoff is started at the lee side of the river with the seaplane heading slightly into the wind. The takeoff path is then gradually made in an arc away from the wind and the liftoff accomplished on the downwind edge of the river. This pattern also allows for more climbout space into the wind.

It should be noted that the greatest degree of the downwind arc is during the time the seaplane is traveling at the slower speeds of takeoff or landing. At the faster speeds, the crosswind effect lessens considerably, and at very slow speeds the seaplane can weathervane into the wind with no ill effect.

Unless the current is extremely swift, crosswind or calm wind takeoffs and landings in rivers or tidal flows should be made in the same direction as the current. This reduces the water forces on the floats or hull of the seaplane.

Again, experience will play an important part in successful seaplane operation during crosswinds. It is essential that all seaplane pilots have thorough knowledge and skill in these maneuvers.

Anchoring, Mooring, Docking, and Beaching

Anchoring the seaplane is the easiest method of securing it on the water surface after a flight. The area selected should be out of the way of moving vessels, and in water deep enough to ensure that the seaplane will not be left high and dry during low tide. The length of the anchor line should be approximately seven times the depth of the water. After dropping anchor with the seaplane headed into the wind, allow the seaplane to drift backward so the anchor is set. To determine that the anchor is holding the seaplane at the desired location, select two fixed objects nearby or on shore that are lined up, and check to assure that these objects remain aligned. If they do not, it means that the seaplane is drifting and dragging the anchor on the bottom.

The effects of a wind shift must also be considered and sufficient room should be allowed in which the seaplane can swing around without striking other anchored vessels or nearby obstacles.

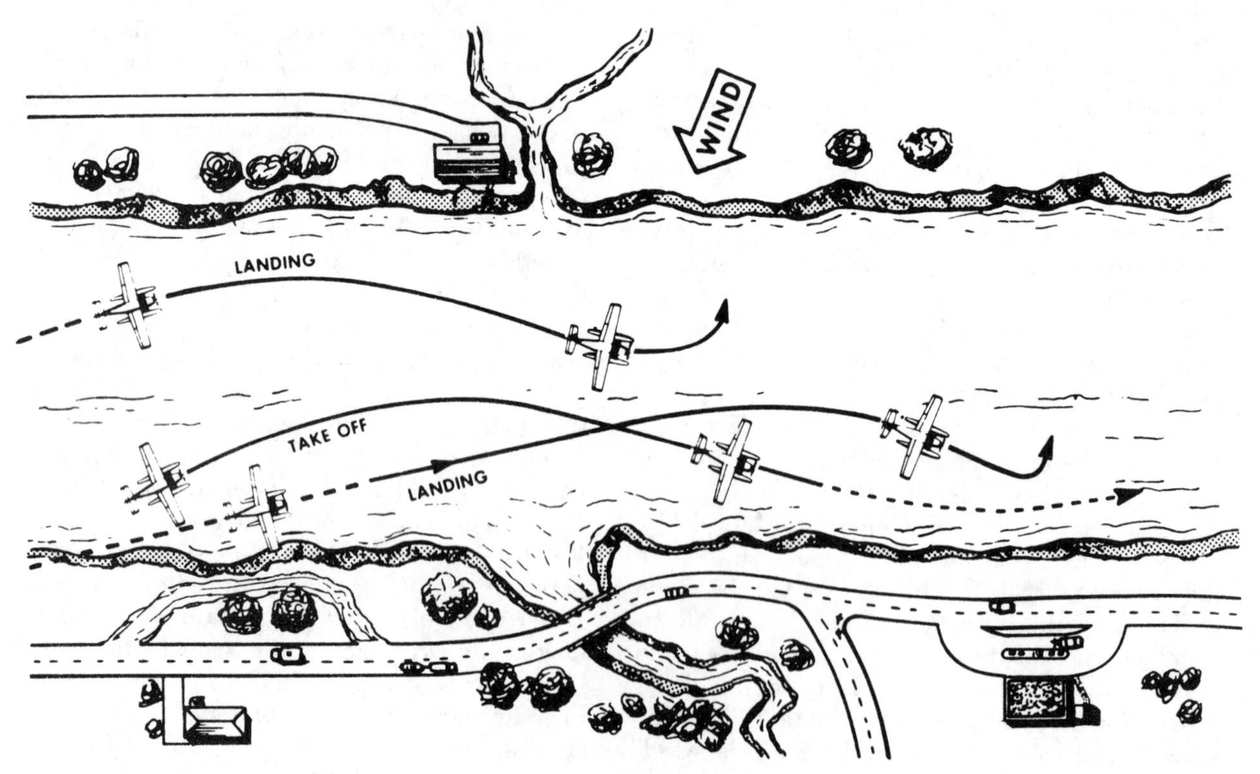

Figure 15–12 Crosswind Takeoff and Landing Technique

If anchoring the seaplane overnight or for longer periods of time, an additional, heavier anchor should be used. This anchor should be dropped about twice as far ahead as the first anchor and about thirty degrees to one side of the seaplane.

Mooring a seaplane eliminates the problem of anchor dragging. A permanent mooring installation consists of a firmly implanted anchor or heavy weight connected by a wire or chain to a floating buoy.

A mooring should be approached at a very low speed and straight into the wind. To avoid the possibility of overrunning the mooring, the engine should be shut down early and the seaplane allowed to coast to the mooring. The engine can always be started again if needed for better positioning. *Never straddle the buoy* with a twin float installation. Always approach so as to have the buoy on the outside of the float to avoid damage to the propeller and underside of the fuselage. It is recommended that initial contact with the mooring be made with a boathook or a person standing on the deck of one float.

If a person is on the float, the seaplane should be taxied right or left of the mooring so that the float on which the person is standing is brought directly alongside the buoy. A short line, which has one end already secured to a strut, can then be secured to the mooring.

It is very important to exercise extreme caution whenever a person is assisting in securing the seaplane. Numerous accidents have been caused by the helper being struck by the propeller.

The procedure for docking is essentially the same as that used for mooring. Properly planning the approach to the dock under existing conditions, and skill in handling the seaplane in congested areas are essential to successful docking. Bear in mind that a seaplane is fragile and striking an obstruction could result in extensive damage to the seaplane.

Beaching the seaplane is easy. Success in beaching depends primarily upon the type and firmness of the shoreline. Inspect the beach before using it. If this is impossible, the approach to the beach should be made at an oblique angle so that the seaplane can be turned out into deeper water in the event the beach is not satisfactory. The hardest packed sand is usually found near the water's edge and becomes softer further from the water's edge where it is dry. Mud bottoms are usually not desirable for beaching.

CHAPTER 16

Transition to Other Airplanes

This chapter is devoted to the factors associated with, and the basic operating practices applicable to, a pilot checkout on airplanes which have significantly different flight characteristics, performance capabilities, and operating procedures from those airplanes which the pilot has previously flown. Accident records indicate that some pilots take unnecessary risks when they attempt to fly a different type of airplane without familiarizing themselves with its peculiarities, limitations, and systems. A knowledge and observance of the basic practices discussed in this chapter may save many lives and unnecessary accidents.

Checkout in Different Makes or Models of Airplanes

The increasing complexity of modern airplanes emphasizes the importance of a thorough checkout for pilots who change from one make or model airplane to another with which they are not familiar. The similarity of the operating controls in most airplanes leads many persons to believe that full pilot competency can be carried from one type of airplane to another, regardless of its weight, speed, performance characteristics and limitations, and operating procedures.

The importance of acquiring a thorough knowledge of an unfamiliar airplane and the inefficiency of trial-and-error methods of learning to fly that airplane have been well established. It is just as important for pilots who regularly fly transport-type airplanes to obtain a checkout in smaller airplanes they propose to fly, as it is for them to have a checkout when advancing to even larger transport airplanes.

Size alone, of course, is not the important consideration. Different airplanes are as different as people, and the only safe and sure way to know them is to be properly introduced. In the case of airplanes, the following are considered the important points of introduction:

1. *Before Flight.* The transitioning pilot should study and understand the airplane's flight and operation manual. A thorough understanding of the fuel system, electrical

and/or hydraulic system, empty and maximum allowable weights, loading schedule, normal and emergency landing gear and flap operations, and preflight inspection procedures, is essential.

2. *Learn the Cockpit Arrangement.* The transitioning pilot should study the engine and flight controls, engine and flight instruments, fuel management controls, wing flaps and landing gear controls and indicators, and radio equipment until proficient enough to pass a blindfold cockpit check in the airplane in which qualification is sought.

3. *Engage a Checkout Pilot.* The transitioning pilot should obtain the services of a checkout pilot who is fully qualified in the airplane concerned. THIS IS VERY IMPORTANT. The checkout pilot not only should be well qualified in the airplane to be used, but also should be capable of communicating effectively to the pilot the techniques essential for the safe operation of the airplane.

4. *Learn the Flight and Operating Characteristics.* The transitioning pilot should not limit familiarization flights to the practice of normal takeoffs and landings: It is extremely important to learn the "V-speeds," and become thoroughly familiar with the stall and minimum controllability characteristics, maximum performance techniques, and all pertinent emergency procedures, as well as all normal operating procedures.

The transition from training type airplanes to larger and faster airplanes may be the pilot's first experience in airplanes equipped with a constant-speed propeller, a retractable landing gear, and wing flaps (Fig. 16–1). All airplanes having a constant-speed propeller require that the pilot have a thorough understanding of the need for *proper* combinations of manifold pressure (MP) and engine or propeller RPM which are prescribed in the airplane manufacturer's manuals.

When transitioning to high performance or complex airplanes, the pilot must be cautioned not to exceed the specified combinations of power settings since the engine can be damaged by using excessively high MP with a low engine RPM. If this situation were to occur, the BMEP (Brake Mean Effective Pressure) might be exceeded. BMEP refers to the average internal pressure exerted upon the cylinder walls and pistons in the combustion chamber during the power stroke. To preclude excessive stress on the engine when increasing power, the pilot should first move the propeller

Figure 16–1 Transitioning to a More Complex Airplane

control forward (increased engine RPM), and then advance the throttle. When reducing power, the throttle should be retarded first, then the engine RPM reduced.

Before applying full power during takeoff, the propeller control should be placed full forward (high RPM, low pitch) to protect the engine from excessive internal pressures (BMEP). After takeoff the MP should be reduced first and then the RPM reduced to normal climb setting. This procedure should never be performed in the reverse order.

On the approach to a landing when the airplane is committed to a landing, the propeller control should be placed to a high RPM position so that should it be necessary to advance the power for a go-around, the propeller will have the correct pitch for maximum thrust.

5. *Learn the Gross Weight and CG Limitations.* The transitioning pilot should include in the checkout at least a demonstration of takeoffs, landings, and flight maneuvers with the airplane *fully loaded.* Most four-place and larger airplanes handle quite differently when loaded to near maximum gross weight, as compared with operation with only two occupants in the pilot seats. Weight and balance should be made for various loading conditions.

6. *Rely on the Checkout Pilot.* The transitioning pilot should accept the checkout pilot's evaluation of performance during the checkout process. It is inadvisable to consider oneself qualified to accept responsibility for the airplane before the checkout is completed—half a checkout may prove more dangerous than none at all.

Checkout in a Multiengine Airplane

Modern design, engineering, and manufacturing technology have produced outstanding multiengine airplanes. Their utility and acceptance have more than fulfilled the expectations of their builders. As a result of this rapid development and increasing use, many pilots have found it necessary to make the transition from single-engine airplanes to those with two or more engines and complex equipment. Good basic flying habits formed during earlier training, and carried forward to these new sophisticated airplanes, will make this transition relatively easy, but only *if* the transition is properly directed.

The following paragraphs discuss several important operational differences which must be considered in progressing from the simpler single-engine airplanes to the more complex multiengine airplanes.

1. *Preflight Preparation.* The increased complexity of multiengine airplanes demands the conduct of a more systematic inspection of the airplane before entering the cockpit, and the use of a more complete and appropriate checklist for each ground and flight operation.

Preflight visual inspections of the exterior of the airplane should be conducted in accordance with the manufacturer's operating manual. The procedures set up in these manuals usually provide for a comprehensive inspection, item by item in an orderly sequence, to be covered on a complete check of the airplane. The transitioning pilot should have a thorough briefing in this inspection procedure, and should understand the reason for checking each item.

2. *Checklists.* Essentially, all modern multiengine airplanes are provided with checklists, which may be very brief or extremely comprehensive. A pilot who desires to operate a modern multiengine airplane safely has no alternative but to use the checklist pertinent to that particular airplane. Such a checklist normally is divided under separate headings for common operations, such as before starting, takeoff, cruise, in-range, landing, system malfunctions, and engine-out operation.

The transitioning pilot must realize that multiengine airplanes characteristically have many more controls, switches, instruments, and indicators. Failure to position or check any of these items may have much more serious results than would a similar error in a single-engine airplane. Only definite procedures, systematically planned and executed can ensure safe and efficient operation. The cockpit checklist provided by the manufacturer in the operations manual must be used, with only those modifications made necessary by subsequent alterations or additions to the airplane and its equipment.

In airplanes which require a copilot, or in which a second pilot is available, it is good practice for the second pilot to read the checklist, and the pilot in command to check each item by actually touching the control or device and repeating the instrument reading or prescribed control position in question, under the careful observation of the pilot calling out the items on the checklist (Fig. 16–2).

Even when no copilot is present, the pilot should form the habit of touching, pointing to, or operating each item as it is read from the checklist.

In the event of an in-flight emergency, the pilot should be sufficiently familiar with emergency procedures to take immediate action instinctively to prevent more serious situations. However, as soon as circumstances permit, the emergency checklist should be reviewed to ensure that all required items have been checked.

3. *Taxiing.* The basic principles of taxiing which apply to single-engine airplanes are generally applicable to multiengine airplanes. Although ground operation of multiengine airplanes may differ in some respects from the operation of single-engine airplanes, the taxiing procedures also vary somewhat between those airplanes with a nosewheel and those with a tailwheel-type landing gear. With either of these landing gear arrangements, the difference in taxiing multiengine airplanes that is most obvious to a transitioning pilot is the capability of using power differential between individual engines to assist in directional control.

Tailwheel-type multiengine airplanes are usually equipped with tailwheel locks which can be used to advantage for taxiing in a straight line especially in a crosswind. The tendency to weathervane can also be neutralized to a great extent in these airplanes by using more power on the upwind engine, with the tailwheel lock engaged and the brakes used as necessary.

On nosewheel-type multiengine airplanes, the brakes and throttles are used mainly to control the momentum, and steering is done principally with the steerable nosewheel. The steerable nosewheel is usually actuated by the rudder pedals, or in some airplanes by a separate hand-operated steering mechanism.

No airplane should be pivoted on one wheel when making sharp turns, as this can damage the landing gear, tires, and even the airport pavement. All turns should be made with the inside wheel rolling, even if only slightly.

Figure 16–2 Teamwork in Multiengine Airplane

Brakes may be used, as with any airplane, to start and stop turns while taxiing. When initiating a turn though, they should be used cautiously to prevent overcontrolling of the turn. Brakes should be used as lightly as practicable while taxiing to prevent undue wear and heating of the brakes and wheels, and possible loss of ground control. When brakes are used repeatedly or constantly they tend to heat to the point that they may either lock or fail completely. Also, tires may be weakened or blown out by extremely hot brakes. Abrupt use of brakes in multiengine as well as single-engine airplanes, is evidence of poor pilot technique; it not only abuses the airplane, but may even result in loss of control.

Due to the greater weight of multiengine airplanes, effective braking is particularly essential. Therefore, as the airplane begins to move forward when taxiing is started, the brakes should be tested immediately by depressing each brake pedal. If the brakes are weak, taxiing should be discontinued and the engines shut down.

Looking outside the cockpit while taxiing becomes even more important in multiengine airplanes. Since these airplanes are usually somewhat heavier, larger, and more powerful than single-engine airplanes they often require more time and distance to accelerate or stop, and provide a different perspective for the pilot. While it usually is not necessary to make S-turns to observe the taxiing path, additional vigilance is necessary to avoid obstacles, other aircraft, or bystanders.

4. *Use of Trim Tabs.* The trim tabs in a multiengine airplane serve the same purpose as in a single-engine airplane, but their function is usually more important to safe and efficient flight. This is because of the greater control forces, weight, power, asymmetrical thrust with one engine inoperative, range of operating speeds, and range of center-of-gravity location. In some multiengine airplanes it taxes the pilot's strength to overpower an improperly set elevator trim tab on takeoff or go-around. Many fatal accidents have occurred when pilots took off or attempted a go-around with the airplane trimmed "full nose up" for the landing configuration. Therefore, prompt retrimming of the elevator trim tab in the event of an emergency go-around from a landing approach is essential to the success of the flight.

Multiengine airplanes should be retrimmed in flight for each change of attitude, airspeed, power setting, and loading. Without such changes, constant application of firm forces on the flight controls is necessary to maintain any desired flight attitude.

5. *Normal Takeoffs.* There is virtually little difference between a takeoff in a multiengine airplane and one in a single-engine airplane. The controls of each class of airplane are operated the same; the multiple throttles of the multiengine airplane normally are treated as one compact power control and can be operated simultaneously with one hand.

In the interest of safety it is important that the flight crew have a plan of action to cope with engine failure during takeoff. It is recommended that just prior to takeoff the pilot in command review, or brief the copilot on takeoff procedures. This briefing should consist of at least the engine out minimum control speed, best all-engine rate of climb speed, best single-engine rate of climb speed, and what procedures will be followed if an engine fails prior to reaching minimum control speed. This latter speed is the minimum airspeed at which safe directional control can be maintained with one engine inoperative and one engine operating at full power.

The multiengine (light twin) pilot's primary concern on all takeoffs is the attainment of the engine-out minimum control speed prior to liftoff. Until this speed is achieved, directional control of the airplane in flight will be impossible after the failure of an engine, unless power is reduced immediately on the operating engine. If an engine fails before the engine-out minimum control speed is attained, THE PILOT HAS NO CHOICE BUT TO CLOSE BOTH THROTTLES, ABANDON THE TAKEOFF, AND DIRECT COMPLETE ATTENTION TO BRINGING THE AIRPLANE TO A SAFE STOP ON THE GROUND.

The multiengine (light-twin) pilot's second concern on takeoff is the attainment of the single-engine best rate-of-climb speed in the least amount of time. This is the airspeed which will provide the greatest rate of climb when operating with one engine out and feathered (if possible), or the slowest rate of descent. In the event of an engine failure, the single-engine best rate-of-climb speed must be held until a safe maneuvering altitude is reached, or until a landing approach is initiated. When takeoff is made over obstructions the best angle-of-climb speed should be maintained until the obstacles are passed, then the best rate of climb maintained.

The engine-out minimum control speed and the single-engine best rate-of-climb speed are published in the airplane's FAA approved flight manual, or the Pilot's Operating Handbook. These speeds should be considered by the pilot on every takeoff, and are discussed in later sections of this chapter.

6. *Crosswind Takeoffs.* Crosswind takeoffs are performed in multiengine airplanes in basically the same manner as those in single-engine airplanes. Less power may be used on the downwind engine to overcome the tendency of the airplane to weathervane at the beginning of the takeoff, and then full power applied to both engines as the airplane accelerates to a speed where better rudder control is attained.

7. *Stalls and Flight Maneuvers at Critically Slow Speeds.* As with single-engine airplanes, the pilot should be familiar with the stall and minimum controllability characteristics of the multiengine airplane being flown. The larger and heavier airplanes have slower responses in stall recoveries and in maneuvering at critically slow speeds due to their greater weight. The practice of stalls in multiengine airplanes, therefore, should be performed at altitudes sufficiently high to allow recoveries to be completed at least 3,000 feet above the ground.

It usually is inadvisable to execute full stalls in multiengine airplanes because of their relatively high wing loading; therefore, practice should be limited to approaches to stalls (imminent), with recoveries initiated at the first physical indication of the stall. As a general rule, however, full stalls in multiengine airplanes are not necessarily violent or hazardous.

The pilot should become familiar with imminent stalls entered with various flap settings, power settings, and landing gear positions. It should be noted that the extension of the landing gear will cause little difference in the stalling speed, but it will cause a more rapid loss of speed in a stall approach.

Power-on stalls should be entered with both engines set at approximately 65 percent power. Takeoff power may be used provided the entry speed is not greater than the normal lift-off speed. Stalls in airplanes with relative low power loading using maximum climb power usually result in an excessive nose-high attitude and make the recovery more difficult.

Because of possible loss of control, stalls with one engine inoperative or at idle power and the other developing effective power are *not* to be performed during multiengine flight tests nor should they be practiced by applicants for multiengine class ratings.

The same techniques used in recognition and avoidance of stalls of single-engine airplanes apply to stalls in multiengine airplanes. The transitioning pilot must become familiar with the characteristics which announce an approaching or imminent stall, the indicated airspeed at which it occurs, and the proper technique for recovery.

The increase in pitch attitude for stall entries should be gradual to prevent momentum from carrying the airplane into an abnormally high nose-up attitude with a resulting deceptively low indicated airspeed at the time the stall occurs. It is recommended that the rate of pitch change result in a 1 knot-per-second decrease in airspeed. In all stall recoveries the controls should be used very smoothly, avoiding abrupt pitch changes. Because of high gyroscopic stresses, this is particularly true in airplanes with extensions between the engines and propellers.

Smooth control manipulation is particularly a requisite of flight at minimum or critically slow airspeeds. As with all piloting operations, a smooth technique permits the development of a more sensitive feel of the controls with a keener sense of stall anticipation. Flight at minimum or critically slow airspeeds gives the pilot an understanding of the relationship between the attitude of an airplane, the feel of its control reactions and the approach to an actual stall.

Generally, the technique of flight at minimum airspeeds is the same in a multiengine airplane as it is in a single-engine airplane. Because of the additional equipment in the multiengine airplane, the transitioning pilot has more to do and observe, and the usually slower control reaction requires better anticipation. Care must be taken to observe engine temperature indications for possible overheating, and to make necessary power adjustments smoothly on both engines at the same time.

8. *Approaches and Landings.* Multiengine airplanes characteristically have steeper gliding angles because of their relatively high wing loading, and greater drag of wing flaps and landing gear when extended. For this reason, power is normally used throughout the approach to shallow the approach angle and prevent a high rate of sink.

The accepted technique for making stabilized landing approaches is to reduce the power to a predetermined setting during the arrival descent so the appropriate landing gear extension speed will be attained in level flight as the downwind leg of the approach pattern is entered (Fig. 16–3). With this power setting, the extension of the landing gear (when the airplane is on the downwind leg opposite the intended point of touchdown) will further reduce the airspeed to the desired traffic pattern airspeed. The manufacturer's recommended speed should be used throughout the pattern. When practicable, however, the speed should be compatible with other air traffic in the traffic pattern. When within the maximum speed for flap extension, the flaps may be partially lowered if desired, to aid in reducing the airspeed to traffic pattern speed. The angle of bank normally should not exceed 30° while turning onto the legs of the traffic pattern.

The prelanding checklist should be completed by the time the airplane is on base leg so that the pilot may direct full attention to the approach and landing. In a power approach, the airplane should descend at a stabilized rate, allowing the pilot to plan and control the approach path to the point of touchdown. Further extension of the flaps and slight adjustment of power and pitch should be accomplished as necessary to establish and maintain a stabilized approach path. Power and pitch changes during approaches should in all cases be smooth and gradual.

The airspeed of the final approach should be as recommended by the manufacturer; if a

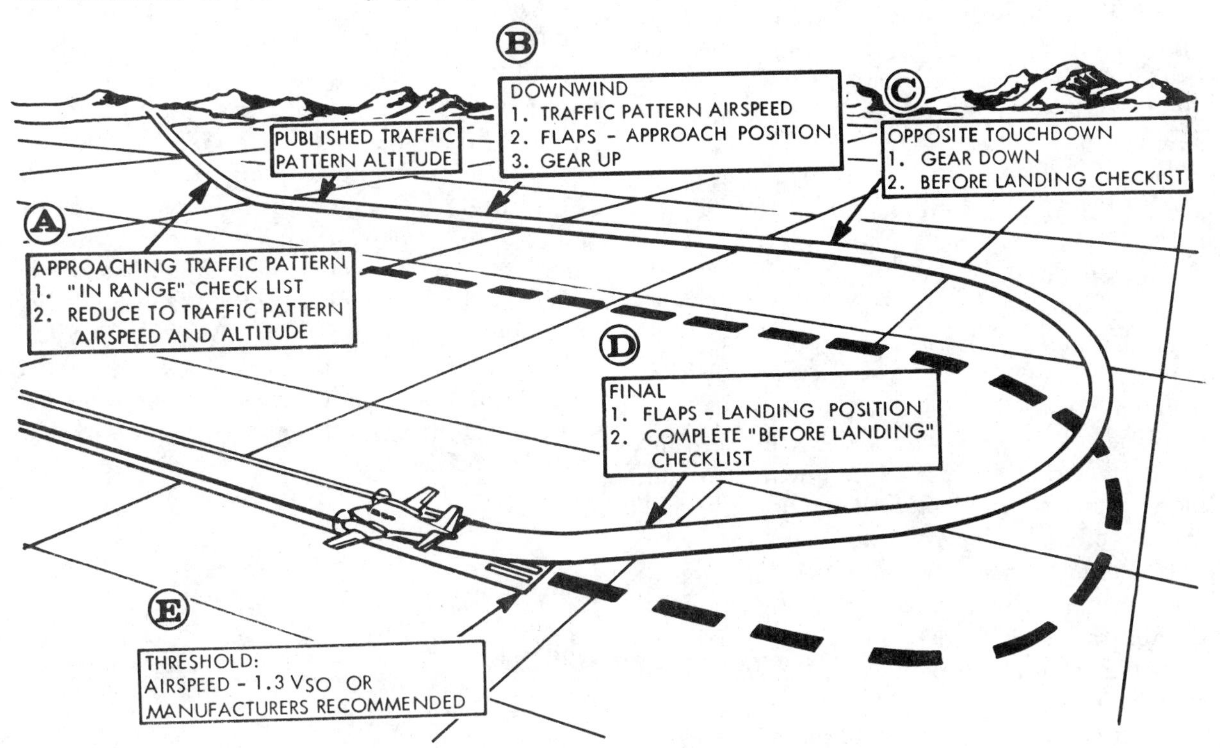

Figure 16–3 Normal Two-Engine Approach and Landing

recommended speed is not furnished, the airspeed should be not less than the engine-out best rate-of-climb speed (V_{yse}) until the landing is assured, because that is the minimum speed at which a single-engine go-around can be made if necessary. IN NO CASE SHOULD THE APPROACH SPEED BE LESS THAN THE CRITICAL ENGINE-OUT MINIMUM CONTROL SPEED. If an engine should fail suddenly and it is necessary to make a go-around from a final approach at less than that speed, a catastrophic loss of control could occur. As a rule of thumb, after the wing flaps are extended the final approach speed should be gradually reduced to 1.3 times the power-off stalling speed (1.3 V_{so}).

The roundout or flare should be started at sufficient altitude to allow a smooth transition from the approach to the landing attitude. The touchdown should be smooth, with the airplane touching down on the main wheels and the airplane in a tail-low attitude, with or without power as desired. The actual attitude at touchdown is very little different in nosewheel- and tailwheel-type airplanes. Although airplanes with nosewheels should touch down in a tail-low attitude, it should not be so low as to drag the tail on the runway. On the other hand, since the nosewheel is not designed to absorb the impact of the full weight of the airplane, level or nose-low attitudes must be avoided.

Directional control on the rollout should be accomplished primarily with the rudder and the steerable nosewheel, with discrete use of the brakes applied only as necessary for crosswinds or other factors.

9. *Crosswind Landings.* Crosswind landing technique in multiengine airplanes is very little different from that required in single-engine airplanes. The only significant difference lies in the fact that because of the greater weight, more positive drift correction must be maintained before the touchdown.

It should be remembered that FAA requires that most airplanes have satisfactory control capabilities when landing in a direct crosswind of not more than 20 percent of the stall speed (0.2 V_{so}). Thus, an airplane with a power-off stalling speed of 60 knots has been designed for a maximum direct crosswind of 12 knots (.2×60) on landings. Though skillful pilots may successfully land in much stronger winds, poor pilot technique is apt to cause serious damage in even more gentle winds. Some light and medium multiengine airplanes have demonstrated satisfactory control with crosswind components greater than .2 V_{so}. If this has been done it will be noted in the Pilot's Operating Handbook under operations limitations.

The two basic methods of making crosswind landings, *the slipping approach* (wing-low) and *the crabbing approach* may be combined. These are discussed in the chapter on Approaches and Landings.

The essential factor in all crosswind landing procedures is touching down without drift, with the heading of the airplane parallel to its direction of motion. This will result in minimum side loads on the landing gear.

10. *Go-Around Procedure.* The complexity of modern multiengine airplanes makes a knowledge of and proficiency in emergency go-around procedures particularly essential for safe piloting. The emergency go-around during a landing approach is inherently critical because it is usually initiated at a very low altitude and airspeed with the airplane's configuration and trim adjustments set for landing.

Unless absolutely necessary, the decision to go around should not be delayed to the point where the airplane is ready to touch down (Fig. 16–4). The more altitude and time available to apply power, establish a climb, retrim, and set up a go-around configuration, the easier and safer the maneuver becomes. When the pilot has decided to go around, immediate action should be taken without hesitation, while maintaining positive control and accurately following the manufacturer's recommended procedures.

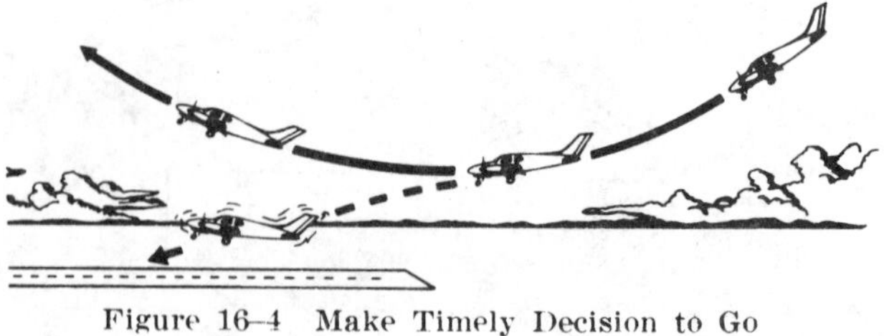

Figure 16–4 Make Timely Decision to Go Around or Land

Go-around procedures vary with different airplanes, depending on their weight, flight characteristics, flap and retractable gear systems, and flight performance. Specific procedures must be learned by the transitioning pilot from the Pilot's Operating Handbook, which should always be available in the cockpit.

There are several general go-around procedures which apply to most airplanes, and are worth pointing out:

a. When the decision to go around is reached, takeoff power should be applied immediately and the descent stopped by adjusting the pitch attitude to avoid further loss of altitude.

b. The flaps should be retracted only in accordance with the procedure prescribed in the airplane's operating manual. Usually this will require the flaps to be positioned as for takeoff.

c. After a positive rate of climb is established the landing gear should be retracted, best rate-of-climb airspeed obtained and maintained, and the airplane trimmed for this climb. The procedure for a normal takeoff climb should then be followed.

The basic requirements of a successful go-around, then, are the prompt arrest of the descent, and the attainment and maintenance of the best rate-of-climb airspeed.

At any time the airspeed is faster than the flaps-up stalling speed, the flaps may be retracted completely without losing altitude if simultaneously the angle of attack is increased sufficiently. At critically slow airspeeds, however, retracting the flaps prematurely or suddenly can cause a stall or an unanticipated loss of altitude. Rapid or premature retraction of the flaps should be avoided on go-arounds, especially when close to the ground, because of the careful attention and exercise of precise pilot technique necessary to prevent a sudden loss of altitude. It generally will be found that retracting the flaps only halfway or to the specified approach setting decreases the drag a relatively greater amount than it decreases the lift.

The FAA approved Airplane Flight Manual or Pilot's Operating Handbook should be consulted regarding landing gear and flap retraction procedures because in some installations simultaneous retraction of the gear and flaps may increase the flap retraction time, and full flaps create more drag than the extended landing gear.

Light-Twin Performance Characteristics

From the transitioning pilot's point of view, the basic difference between a light-twin and single-engine airplane is the potential problem involving engine failure. The information that follows is confined to that one basic difference.

The term "light-twin" as used here pertains to the propeller driven airplane having a maximum certificated gross weight of less than 12,500 pounds, and which has two reciprocating engines mounted on the wings.

Before the subject of operating technique in light twin-engine airplanes can be thoroughly discussed, there are several terms that need to be reviewed. "V" speeds such as V_x, V_{xse}, V_y, V_{yse}, and V_{mc} are the main performance speeds the light-twin pilot needs to know in addition to the other performance speeds common to both twin-engine and single-engine airplanes. The airspeed indicator in twin-engine airplanes is marked (in addition to other normally marked speeds) with a red radial line at the minimum controllable airspeed with the critical engine inoperative, and a blue radial line at the best rate-of-climb airspeed with one engine inoperative (Fig. 16–5).

Figure 16–5 Airspeed Markings for Twin-Engine Airplane

V_x —The speed for best angle of climb. At this speed the airplane will gain the greatest height for a given distance of forward travel. This speed is used for obstacle clearance with all engines operating. However, this speed is different when one engine is inoperative, and in this handbook is referred to as V_{xse} (single-engine).

V_y —The speed for the best rate of climb. This speed will provide the maximum altitude gain for a given period of time with all engines operating. However, this speed too will be different when one engine is inoperative and in this handbook is referred to as V_{yse} (single-engine).

V_{mc}—The minimum control speed with the critical engine inoperative. The term V_{mc} can be defined as the minimum airspeed at which the airplane is controllable when the critical engine is suddenly made inoperative, and the remaining engine is producing takeoff power. The Federal Aviation Regulations under which the airplane was certificated, stipulate that at V_{mc} the certificating test pilot must be able to: (1) stop the turn which results when the critical engine is suddenly made inoperative within 20° of the original heading, using maximum rudder deflection and a maximum of 5° bank into the operative engine, and (2) after recovery, maintain the airplane in straight flight with not more than a 5° bank (wing lowered toward the operating engine). This does not mean that the airplane must be able to climb or even hold altitude. It only means that a heading can be maintained. The principle of V_{mc} is not at all mysterious. It is simply that at any airspeed less than V_{mc}, air flowing along the rudder is such that application of rudder forces cannot overcome the asymmetrical yawing forces caused by takeoff power on one engine and a powerless windmilling propeller on the other. The demonstration of V_{mc} is discussed in a later section of this handbook.

Many pilots erroneously believe that because a light-twin has two engines, it will continue to perform at least half as well with only one of those engines operating. There is nothing in FAR, Part 23, governing the certification of light-twins, which requires an airplane to maintain altitude while in the takeoff configuration and with one engine inoperative. In fact, many of the current light-twins are not required to do this with one engine inoperative in any configuration, even at sea level. This is of major significance in the operations of light-twins certificated under Part 23. With regard to performance (but not controllability) in the takeoff or landing configuration, the light twin-engine airplane is, in concept, merely a single-engine airplane with its power divided into two individual units. The following discussion should help the pilot to eliminate any misconceptions of single-engine operation of light-twin airplanes.

When one engine fails on a light-twin, performance is not really halved, but is actually reduced by 80 percent or more. The performance loss is greater than 50 percent because an airplane's climb performance is a function of the thrust horsepower which is in excess of that required for level flight. When power is increased in both engines in level flight and the airspeed is held constant, the airplane will start climbing—the rate of climb depending on the power added (which is power in excess of that required for straight-and-level flight). When one engine fails, however, it not only loses power but the drag increases considerably because of asymmetric thrust and the operating engine must then carry the full burden alone. To do this, it must produce 75 percent or more of its rated power. This leaves very little excess power for climb performance.

As an example, an airplane which has an all-engine rate of climb of 1,860 FPM and a single engine rate of climb of 190 FPM would lose almost 90 percent of its climb performance when one engine fails.

Nonetheless, the light-twin does offer obvious safety advantages over the single-engine airplane (especially in the enroute phase) but *only if* the pilot fully understands the real options offered by that second engine in the takeoff and approach phase of flight.

It is essential then that the light-twin pilot take proficiency training periodically from a competent flight instructor.

Engine-Out Emergencies

In general, the operating and flight characteristics of modern light-twins with one engine inoperative are excellent. These airplanes can be controlled and maneuvered safely as long as sufficient airspeed is maintained. However, to utilize the safety and performance characteristics effectively, the pilot must have a sound understanding of the single-engine performance and the limitations resulting from an unbalanced of power.

A pilot checking out for the first time in any multiengine airplane should practice and become thoroughly familiar with the control and performance problems which result from the failure of one engine during any flight condition. Practice in all the control operations and precautions is necessary and demonstration of these is required on multiengine rating flight tests. Practice should be continued as long as the pilot engages in flying a twin-engine airplane, so that corrective action will be instinctive and the ability to control airspeed, heading, and altitude will be retained.

The feathering of a propeller should be demonstrated and practiced in all airplanes equipped with propellers which can be feathered and unfeathered safely in flight. If the airplane used is not equipped with feathering propellers, or is equipped with propellers which cannot be feathered and unfeathered safely in flight, one engine should be secured (shut down) in accordance with the procedures in the FAA approved Airplane Flight Manual or the Pilot's Operating Handbook. The recommended propeller setting should be used, and the emergency setting of all ignition, electrical, hydraulic, and fire extinguisher systems should be demonstrated.

Propeller Feathering

When an engine fails in flight the movement of the airplane through the air tends to keep the propeller rotating, much like a windmill. Since the failed engine is no longer delivering power to the propeller to produce thrust but instead, may be absorbing energy to overcome friction and compression of the engine, the drag of the windmilling propeller is significant and causes the airplane to yaw toward the failed engine (Fig. 16–6). Most multiengine airplanes are equipped with "full-feathering propellers" to minimize that yawing tendency.

Figure 16–6 Windmilling Propeller Creates Drag

The blades of a feathering propeller may be positioned by the pilot to such a high angle that they are streamlined in the direction of flight. In this feathered position, the blades act as powerful brakes to assist engine friction and compression in stopping the windmilling rotation of the propeller. This is of particular advantage in case of a damaged engine, since further damage, caused by a windmilling propeller creates the least possible drag on the airplane and reduces the yawing tendency. As a result, multiengine airplanes are easier to control in flight when the propeller of an inoperative engine is feathered.

Feathering of propellers for training and checkout purposes should be performed only under such conditions and at such altitudes and locations that a safe landing on an established airport could be accomplished readily in the event of difficulty in unfeathering the propeller.

Engine-Out Procedures

The following procedures are recommended to develop in the transitioning pilot the habit of using proper procedures and proficiency in coping with an inoperative engine.

At a safe altitude (minimum 3,000 feet above terrain) and within landing distance of a suitable airport, an engine may be shut down with the mixture control or fuel selector. At lower altitudes, however, shut down should be *simulated by reducing power* by means of the throttle to the zero thrust setting. The following procedures should then be followed:

(1) Set mixture and propeller controls as required; both power controls should be positioned for maximum power to maintain at least V_{mc}.

(2) Retract wing flaps and landing gear.

(3) Determine which engine failed, and *verify* it by closing the throttle on the dead engine.

(4) Bank at least 5° into the operative engine.

(5) Determine the cause of failure, or feather the inoperative engine.

(6) Turn toward the nearest airport.

(7) Secure (shut down) the inoperative engine in accordance with the manufacturer's approved procedures and check for engine fire.

(8) Monitor the engine instruments on the operating engine; and adjust power, cowl flaps, and airspeed as necessary.

(9) Maintain altitude and an airspeed of at least V_{yse} if possible.

The pilot must be proficient in the control of heading, airspeed, and altitude, in the prompt identification of a power failure, and in the accuracy of shutdown and restart procedures as prescribed in the FAA approved Airplane Flight Manual or Pilot's Operating Handbook.

There is no better way to develop skill in single-engine emergencies than by *continued* practice. The fact that the techniques and procedures of single-engine operation are mastered thoroughly at one time during a pilot's career is no assurance of being able to cope successfully with an engine-out emergency *unless review and practice are continued.* Some engine-out emergencies may be so critical that there may be no safety margin for lack of skill or knowledge. Unfortunately, many light-twin pilots never practice single-engine operation after receiving their multiengine rating.

The pilot should practice and demonstrate the effects (on engine-out performance) of various configurations of gear, flaps, and both; the use of carburetor heat; and the failure to feather the propeller on an inoperative engine. Each configuration should be maintained, at best engine-out rate-of-climb speed long enough to determine its effect on the climb (or sink) achieved. Prolonged use of carburetor heat, if so equipped, at high power settings should be avoided.

The Critical Engine

"P-factor" is present in multiengine airplanes just as it is in single-engine airplanes. Remember, P-factor is caused by the dissimilar thrust of the rotating propeller blades when in certain flight conditions. It is the result of the downward moving blade having a greater angle of attack than the upward moving blade when the relative wind striking the blades is not aligned with the thrust line (as in a nose-high attitude).

In most U.S. designed light-twins, both engines rotate to the right (clockwise) when viewed from the rear, and both engines develop an equal amount of thrust. At low airspeed and high power conditions, the downward moving propeller blade of each engine develops more thrust than the upward moving blade. This asymmetric propeller thrust or "P-factor," results in a center of thrust at the right side of each engine as indicated by lines D1 and D2 in Fig. 16-7. The turning (or yawing) force of the right engine is greater than the left engine because the center of thrust (D2) is much farther away from the center line (CL) of the fuselage—it has a longer level arm. Thus, when the right engine is operative and the left engine is inoperative, the turning (or yawing) force is greater than in the opposite situation of a "good" left engine and a "bad" right engine. In other words, directional control may be difficult when the left engine (the *critical* engine) is suddenly made inoperative.

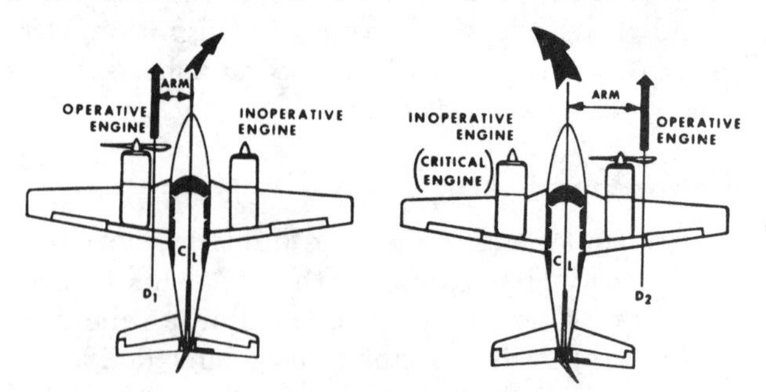

Figure 16-7 Forces Created During Single-Engine Operation

It should be noted that some light-twin engine airplanes are equipped with engines turning in opposite directions; that is, the left engine and propeller turn clockwise and the right engine and propeller turn counterclockwise. With this arrangement, the thrust line of either engine is the same distance from the center line of the fuselage, so there will be no difference in yaw effect between loss of left or right engine.

V_{mc} Demonstrations

Every light-twin engine airplane checkout should include a demonstration of the airplane's engine-out minimum control speed. The engine-out minimum control speed given in the FAA approved Airplane Flight Manual, Pilot's Operating Handbook, or other manufacturer's published limitations is determined during original airplane certification under conditions specified in the Federal Aviation Regulations. These conditions normally are not duplicated during pilot training or testing because they consist of the most adverse situations for airplane type certification purposes. Prior to a pilot checkout, a thorough discussion of the factors affecting engine-out minimum control speed is essential.

Basically, when one engine fails the pilot must overcome the asymmetrical thrust (except on airplanes with center line thrust) created by the operating engine by setting up a counteracting moment with the rudder. When the rudder is fully deflected, its yawing power will depend on the velocity of airflow across the rudder—which in turn is dependent on the airspeed. As the airplane decelerates it will reach a speed below which the rudder moment will no longer balance the thrust moment and directional control will be lost.

During engine-out flight the large rudder deflection required to counteract the asymmetric thrust also results in a "lateral lift" force on the vertical fin. This lateral "lift" represents an unbalanced side force on the airplane which must be counteracted either by allowing the airplane to accelerate sideways until the lateral drag caused by the sideslip equals the rudder "lift" force or by banking into the operative engine and using a component of the airplane weight to counteract the rudder-induced side force.

In the first case, the wings will be level, the ball in the turn-and-slip indicator will be centered and the airplane will be in a moderate sideslip toward the inoperative engine. In the second case, the wings will be banked 3–5° into the good engine, the ball will be deflected one diameter toward the operative engine, and the airplane will be at zero sideslip.

The sideslipping method has several major disadvantages: (1) the relative wind blowing on the inoperative engine side of the vertical fin tends to increase the asymmetric moment caused by the failure of one engine; (2) the resulting sideslip severely degrades stall characteristics; and (3) the greater rudder deflection required to balance the extra moment and the sideslip drag cause a significant reduction in climb and/or acceleration capability.

Flight tests have shown that holding the ball of the turn-and-slip indicator in the center while maintaining heading with wings level drastically increases V_{mc} as much as 20 knots in some airplanes. (Remember, the value of V_{mc} given in the FAA approved flight manual for the airplane is based on a maximum 5° bank into the operative engine.) Banking into the operative engine *reduces* V_{mc}, whereas decreasing the bank angle away from the operative engine *increases* V_{mc} at the rate of approximately 3 knots per degree of bank angle.

Flight tests have also shown that the high drag caused by the wings level, ball centered configuration can reduce single-engine climb performance by as much as 300 FPM, which is just about all that is available at sea level in a nonturbocharged light twin.

Banking at least 5° into the good engine ensures that the airplane will be controllable at any speed above the certificated V_{mc}, that the airplane will be in a minimum drag configuration for best climb performance, and that the stall characteristics will not be degraded. Engine-out flight with the ball centered is *never* correct.

The magnitude of these effects will vary from airplane to airplane, but the principles are applicable in all cases.

> NOTE.—A bank limitation of up to 5° during V_{mc} demonstration is applicable only to certification tests of the airplane and is not intended as a limit in training or testing a pilot's ability to extract maximum performance from the airplane.

For an airplane with nonsupercharged engines, V_{mc} decreases as altitude is increased. Consequently, directional control can be maintained at a *lower* airspeed than at sea level. The reason for this is that since power decreases with altitude the thrust moment of the operating engine becomes less, thereby lessening the need for the rudder's yawing force. Since V_{mc} is a function of power (which decreases with altitude), it is possible for the airplane to reach a stall speed prior to loss of directional control.

It must be understood, therefore, that there is a certain density altitude above which the stalling speed is higher than the engine-out minimum control speed. When this density altitude exists close to the ground because of high elevations or temperatures, an effective flight demonstration is impossible and should *not* be attempted. When a flight demonstration is impossible, the check pilot should emphasize orally the significance of the engine-out minimum control speed, including the results of attempting flight below this speed with one engine inoperative, the recognition of the imminent loss of control, and the recovery techniques involved.

V_{mc} is greater when the center of gravity is at the rearmost allowable position. Since the airplane rotates around its center of gravity, the moments are measured using that point as a reference. A rearward CG would not affect the thrust moment, but would shorten the arm to the center of the rudder's horizontal "lift" which would mean that a higher force (airspeed) would be required to counteract the engine-out yaw. Figure 16–8 shows an exaggerated view of the effects of a rearward CG.

Generally, the center of gravity range of most light twins is short enough so that the effect on the V_{mc} is relatively small, but it is a factor that should be considered. Many pilots

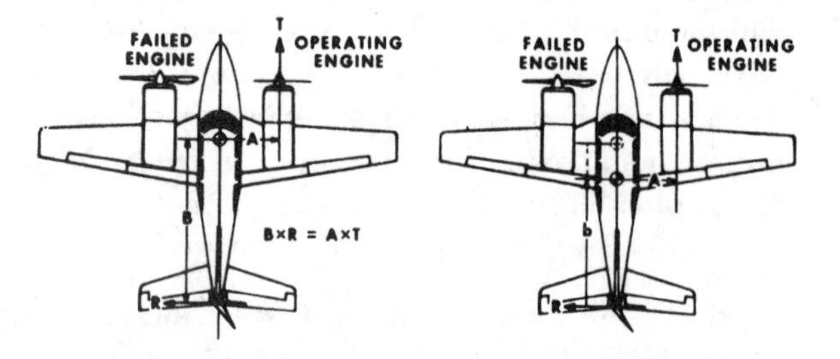

Figure 16–8 Effect of CG Location on Yaw

would only consider the rear CG of their light-twin as a factor for pitch stability, not realizing that it could affect the controllability with one engine out.

There are many light-twin pilots who think that the only control problem experienced in flight below V_{mc} is a yaw toward the inoperative engine. Unfortunately, this is not the whole story.

With full power applied to the operative engine, as the airspeed drops below V_{mc}, the airplane tends to roll as well as yaw into the inoperative engine. This tendency becomes greater as the airspeed is further reduced. since this tendency must be counteracted by aileron control, the yaw condition is aggravated by aileron yaw (the "down" aileron creates more drag than the "up" aileron). If a stall should occur in this condition, a violent roll into the dead engine may be experienced. Such an event occurring close to the ground could be disastrous. This may be avoided by maintaining airspeed above V_{mc} at all times during single-engine operation. If the airspeed should fall below V_{mc}—for whatever reason—*then power must be reduced on the operative engine and the airplane must be banked at least 5° toward the operative engine if the airplane is to be safely controlled.*

The V_{mc} demonstrations should be performed at an altitude from which recovery from loss of control could be made safely. One demonstration should be made while holding the wings level and the ball centered, and another demonstration should be made while banking the airplane at least 5° toward the operating engine to establish "zero sideslip." These maneuvers will demonstrate the engine-out minimum control speed for the existing conditions and will emphasize the necessity of banking into the operative engine. *No attempt should be made to duplicate V_{mc} as determined for airplane certification.*

After the propellers are set to high RPM, the landing gear is retracted, and the flaps are in the takeoff position, the airplane should be placed in a climb attitude and airspeed representative of that following a normal takeoff. With both engines developing as near rated takeoff power as possible, power on the critical engine (usually the left) should then be reduced to idle (windmilling, not shut down). After this is accomplished, the airspeed should be reduced slowly with the elevators until directional control no longer can be maintained. At this point, recovery should be initiated by simultaneously reducing power on the operating engine and reducing the angle of attack by lowering the nose. Should indications of a stall occur prior to reaching this point, recovery should be initiated immediately by reducing the angle of attack. In this case, a minimum engine-out control speed demonstration is not possible under existing conditions.

If it is found that the minimum engine-out control speed is reached before indications of a stall are encountered, the pilot should demonstrate the ability to control the airplane and initiate a safe climb in the event of a power failure at the published engine-out minimum control speed.

Accelerate/Stop Distance

The most critical time for an engine-out condition in a twin-engine airplane is during the two or three-second period immediately following the takeoff roll while the airplane is accelerating to a safe engine-failure speed.

Although most twin-engine airplanes are controllable at a speed close to the engine-out minimum control speed, the performance is often so far below optimum that continued flight following takeoff may be marginal or impossible. A more suitable recommended speed, termed by some aircraft manufacturers as minimum safe single-engine speed, is that at which altitude can be maintained while the landing gear is being retracted and the propeller is being feathered.

Upon engine failure after reaching the safe single-engine speed on takeoff, the twin-engine pilot (having lost one-half of the normal power) usually has a significant advantage over the pilot of a single-engine airplane, because, if the particular airplane has single-engine climb capability at the existing gross weight and density altitude, there may be the choice of stopping or continuing the takeoff. This compares with the only choice facing a single-engine airplane pilot who suddenly has lost half of the normal takeoff power—that is stop!

If one engine fails prior to reaching V_{mc}, there is no choice but to close both throttles and bring the airplane to a stop. If engine failure occurs after becoming airborne, the pilot must decide immediately to land or to continue the takeoff.

If the decision is made to continue the takeoff, the airplane must be able to gain altitude with one engine inoperative. This requires acceleration to V_{yse} if no obstacles are involved, or to V_{xse} if obstacles are a factor.

To make a correct decision in an emergency of this type, the pilot must consider the runway length, field elevation, density altitude, obstruction height, headwind component, and the airplane's gross weight. (For simplification purposes, additional factors such as runway contaminants [rubber, soot, water, ice, snow] and runway slope will not be discussed here.) The flightpaths illustrated in Fig. 16–9 indicate that the "area of decision" is bounded by: (1) the point at which V_y is reached and (2) the point where the obstruction altitude is reached. An engine failure in this area demands an immediate decision. Beyond this decision area, the airplane, within the limitations of engine-out climb performance, can usually be maneuvered to a landing at the departure airport.

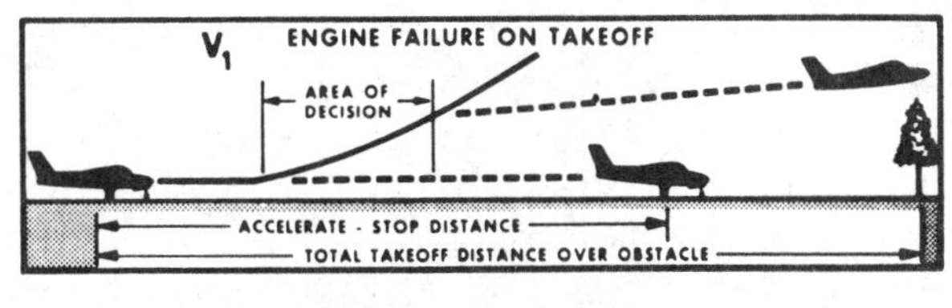

Figure 16–9 Area of Decision

The "accelerate-stop distance" is the total distance required to accelerate the twin-engine airplane to a specified speed and, assuming failure of an engine at the instant that speed is attained, to bring the airplane to a stop on the remaining runway. The "accelerate-go distance" is the total distance required to accelerate the airplane to a specified speed and, assuming failure of an engine at the instant that speed is attained, continue takeoff on the remaining engine to a height of 50 feet.

For example, use the chart in Fig. 16–10 and assume that with a temperature of 80° F., a calm wind at a pressure altitude of 2,000 feet, a gross weight of 4,800 pounds, and all engines operating, the airplane being flown requires 3,525 feet to accelerate to 105 MPH and then be brought to a stop. Assume also that the airplane under the same conditions requires a distance of 3,830 feet to take off and climb over a 50-foot obstacle (Fig. 16–11) when one engine fails at 105 MPH.

With such a slight margin of safety (305 feet) it would be better to discontinue the takeoff and stop if the runway is of adequate length, since any slight mismanagement of the engine-out procedure would more than outweigh the small advantage offered by continuing the takeoff. At higher field elevations the advantage becomes less and less until at very high density altitudes a successful continuation of the takeoff is extremely improbable.

Factors in Takeoff Planning

Competent pilots of light-twins will plan the takeoff in sufficient detail to be able to take immediate action if and when one engine fails during the takeoff process. They will be thoroughly familiar with the airplane's performance capabilities and limitations, including accelerate-stop distance, as well as the distance available for takeoff, and will include such factors in their plan of action. For example, if it has been determined that the airplane cannot maintain altitude with one engine inoperative (considering the gross weight and density altitude), the seasoned pilot will be well aware that should an engine fail right after lift-off, an immediate landing may have to be made in the most suitable area available. *The competent pilot will make no attempt to maintain altitude at the expense of a safe airspeed.*

Consideration will also be given to surrounding terrain, obstructions, and nearby landing areas so that a definite direction of flight can be established immediately if an engine fails at a critical point during the climb after takeoff. It is imperative then, that the takeoff and

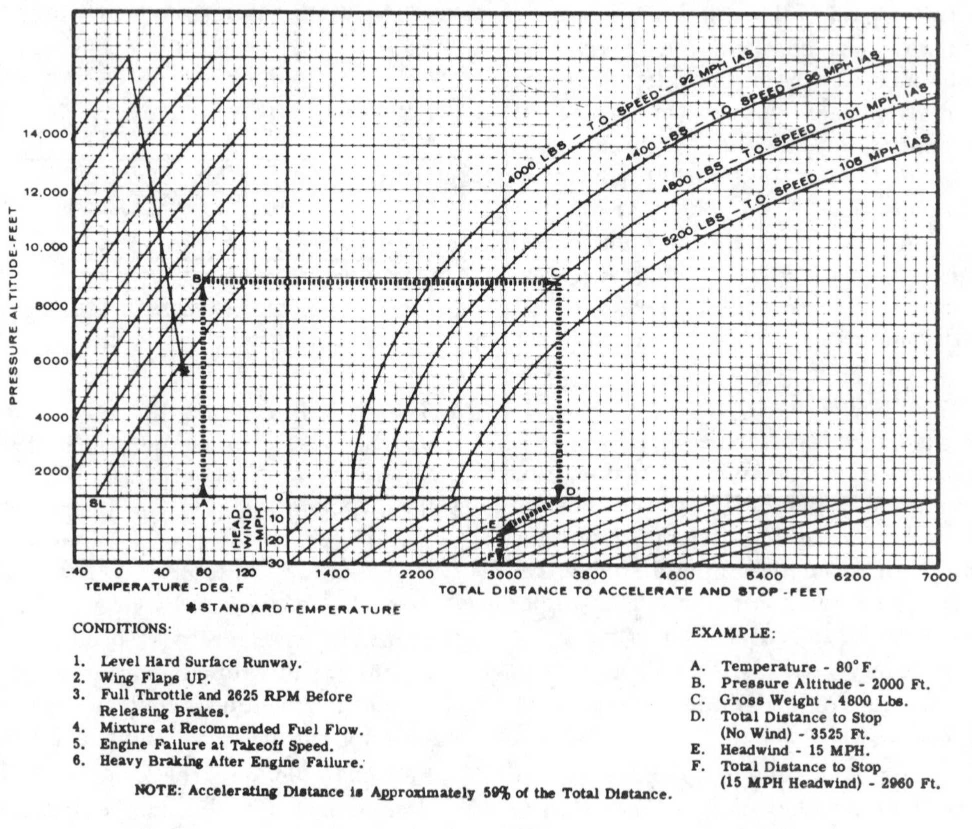

Figure 16–10 Accelerate/Stop Distance

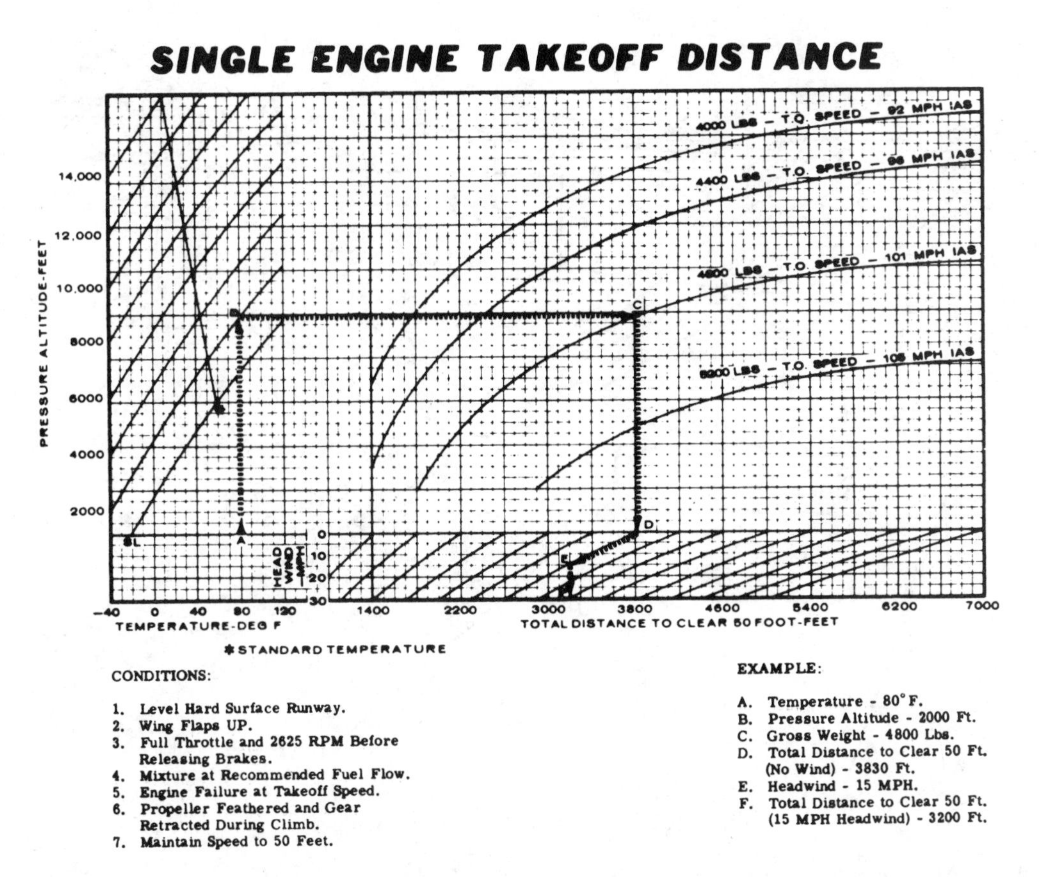

Figure 16–11 SE Takeoff Distance

climb path be planned so that all obstacles between the point of takeoff and the available areas of landing can be cleared if one engine suddenly becomes inoperative.

In addition, a competent light-twin pilot knows that the twin-engine airplane must be flown with precision if maximum takeoff performance and safety are to be obtained. For example, the airplane must lift off at a specific airspeed, accelerate to a definite climbing airspeed, and climb with maximum permissible power on both engines to a safe single-engine maneuvering altitude. In the meantime, if an engine fails, a different airspeed must be attained *immediately*. This airspeed must be held precisely because only at this airspeed will the pilot be able to obtain maximum performance from the airplane. To understand the factors involved in proper takeoff planning, a further explanation of this critical speed follows, beginning with the lift-off.

The light-twin can be controlled satisfactorily while firmly on the ground when one engine fails prior to reaching V_{mc} during the takeoff roll. This is possible by closing both throttles, by proper use of rudder and brakes, and with many airplanes, by use of nosewheel steering. *If the airplane is airborne at less than* V_{mc}, *however, and suddenly loses all power on one engine, it cannot be controlled satisfactorily.* Thus, on normal takeoffs, lift-off should never take place until the airspeed reaches and exceeds V_{mc}. The FAA recommends a minimum speed of V_{mc} plus 5 knots before lift-off. From this point, an efficient climb procedure should be followed (Fig. 16–12).

An efficient climb procedure is one in which the airplane leaves the ground slightly above V_{mc}, accelerates quickly to V_y (best rate-of-climb speed) and climbs at V_y. The climb at V_y should be made with both engines set to maximum takeoff power until reaching a safe single-engine maneuvering altitude (minimum of approximately 500′ above field elevation or as dictated by airplane performance capability and/or local obstacles). At this point, power may be reduced to the allowable maximum continuous power setting (METO—maximum

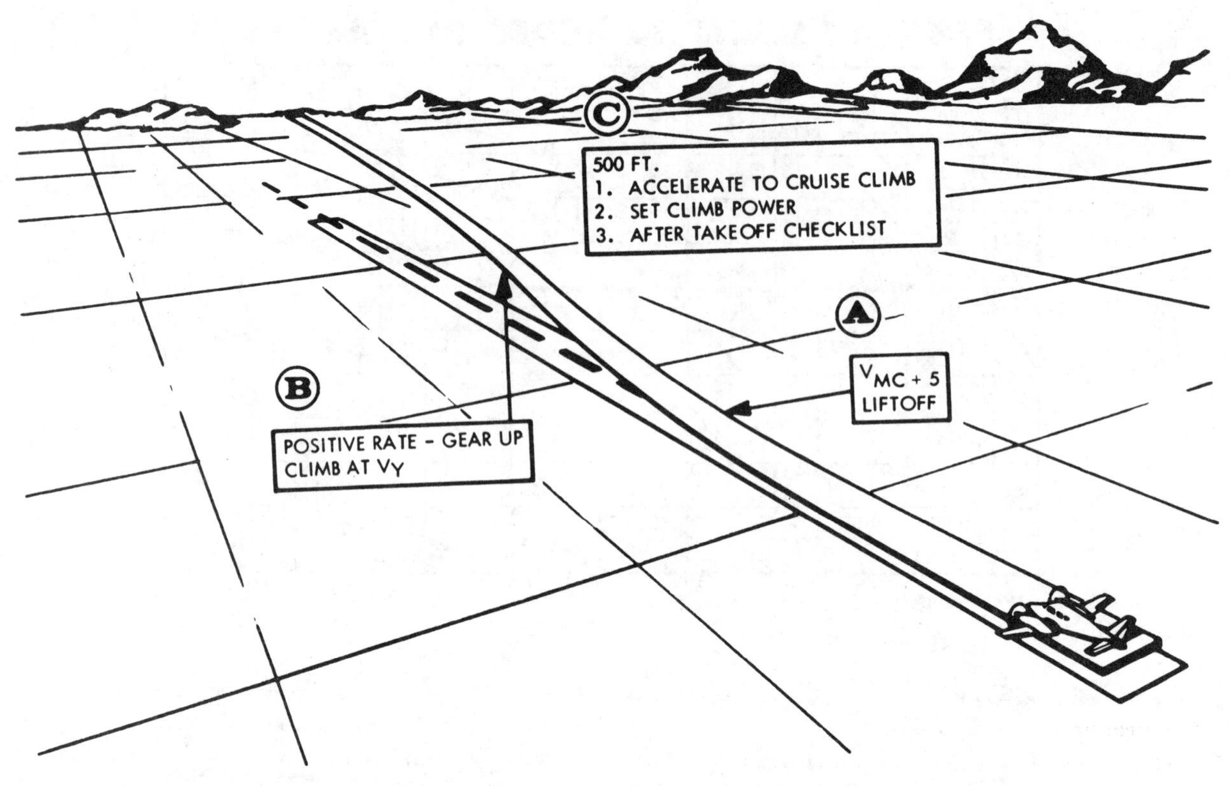

Figure 16–12 Normal Takeoff Procedure

except takeoff) or less, and any desired enroute climb speed then may be established. The following discussion explains why V_y is recommended for the initial climb.

To improperly trained pilots, the extremes in takeoff technique may suggest "hold it down" to accelerate the airplane to near cruise speed before climbing, or "pull it off" below V_{mc} and climb as steeply as possible. If one considers the possibility of an engine failure somewhere during the takeoff, neither of these procedures makes much sense for the following reasons: Remember, drag increases as the square of the speed; so for any increase in speed over and above the best rate-of-climb speed, V_y, the greater the drag and the less climb performance the airplane will have. At 123 knots the drag is approximately one and one-half times greater than it is at 100 knots. At 141 knots the drag is doubled, and at 200 knots the drag is approximately four times as great as at 100 knots. While the drag is increasing as the square of the velocity (V^2), the power required to maintain a velocity increases as the cube of that velocity (V^3).

In the event of engine failure, a pilot who uses excessive speed on takeoff will discover suddenly that all the energy produced by the engines has been converted into speed. Improperly trained pilots often believe that the excess speed can always be converted to altitude, but this theory is not valid. Available power is only wasted in accelerating the airplane to an unnecessary speed. Also, experience has shown that an unexpected engine failure so surprises the unseasoned pilot that proper reactions are extremely lagging. By the time the initial shock wears off and the pilot is ready to take control of the situation, the excess speed has dissipated and the airplane is still barely off the ground. From this low altitude, the pilot would still have to climb, with an engine inoperative, to whatever height is needed to clear all obstacles and get back to the approach end of the runway. Excess speed cannot be converted readily to the altitude or distance necessary to reach a landing area safely.

In contrast, however, an airplane will fly in level flight much easier than it will climb. Therefore, if the total energy of both engines is initially converted to enough height above the ground to permit clearance of all obstacles while in level flight (safe maneuvering altitude), the problem is much simpler in the event an engine fails. If some extra height is available, it usually can be traded for velocity or gliding distance when needed.

Simply stated then, altitude is more essential to safety after takeoff than is excess airspeed. On the other hand, trying to gain height too fast in the takeoff also can be very dangerous because of control problems. If the airplane has just become airborne and the airspeed is at or below V_{mc} when an engine fails, the pilot could avoid a serious accident by retarding both throttles immediately. If this action is not taken immediately, the pilot will be unable to control the airplane.

Consequently, the pilot always should keep one hand on the control wheel (when not operating hand-controlled nose steering) and the other hand on the throttles throughout the takeoff roll. The airplane should remain on the ground until adequate speed is reached so that a smooth transition to the proper climb speed can be made. THE AIRPLANE SHOULD NEVER LEAVE THE GROUND BEFORE V_{mc} IS REACHED. Preferably, $V_{mc}+5$ knots should be attained.

If an engine fails before leaving the ground it is advisable to discontinue the takeoff and STOP. If an engine fails after lift-off, the pilot will have to decide immediately whether to continue flight, or to close both throttles and land. However, waiting until the engine failure occurs is not the time for the pilot to plan the correct action. The action must be planned before the airplane is taxied onto the runway. The plan of action must consider the density altitude, length of the runway, weight of the airplane, and the airplane's accelerate-stop distance, and accelerate-go distance under these conditions. Only on the basis of these factors can the pilot decide intelligently what course to follow if an engine should fail. When the flight crew consists of two pilots, it is recommended that the pilot in command brief the second pilot on what course of action will be taken should the need arise.

To reach a safe single-engine maneuvering altitude as safely and quickly as possible, the climb with all engines operating must be made at the proper airspeed. That speed should provide for:

1. Good control of the airplane in case an engine fails.
2. Quick and easy transition to the single-engine best rate-of-climb speed if one engine fails.
3. A fast rate of climb to attain an altitude which permits adequate time for analyzing the situation and making decisions.

To make a quick and easy transition to the single-engine best rate-of-climb speed, in case an engine fails, the pilot should climb at some speed greater than V_{yse}. If an engine fails at less than V_{yse}, it would be necessary for the pilot to lower the nose to increase the speed to V_{yse} in order to obtain the best climb performance. If the airspeed is considerably less than this speed, it might be necessary to lose valuable altitude to increase the speed to V_{yse}. Another factor to consider is the loss of airspeed that may occur because of erratic pilot technique after a sudden, unexpected power loss. Consequently, the normal initial two-engine climb speed should not be less than V_y.

In summary then, the initial climb speed for a normal takeoff with both engines operating should permit the attainment of a safe single-engine maneuvering altitude as quickly as possible; it should provide for good control capabilities in the event of a sudden power loss on one engine; and it should be a speed sufficiently above V_{yse} to permit attainment of that speed quickly and easily in the event power is suddenly lost on one engine. The only speed that meets all of these requirements for a normal takeoff is the best rate-of-climb speed with both engines operating (V_y).

Normal Takeoff—Both Engines Operating

After runup and pretakeoff checks have been completed, the airplane should be taxied into takeoff position and aligned with the runway. If it is a tailwheel-type, the tailwheel lock (if installed) should be engaged only after the airplane has been allowed to roll straight a few feet along the intended takeoff path to center the tailwheel.

If the crew consists of two pilots, it is recommended that the pilot in command brief the other pilot on takeoff procedures prior to receiving clearance for takeoff. This briefing consists of at least the following: minimum control speed (V_{mc}), rotation speed (V_r), lift-off speed (V_{lof}), single-engine best rate-of-climb speed (V_{yse}), all engine best rate-of-climb speed (V_y), and what procedures will be followed if an engine failure occurs prior to V_{mc} (Fig. 16–12).

Both throttles then should be advanced simultaneously to takeoff power, and directional control maintained by the use of the steerable nosewheel and the rudder. Brakes should be used for directional control only during the initial portion of the takeoff roll when the rudder and steerable nosewheel are ineffective. During the initial takeoff roll it is advisable to monitor the engine instruments.

As the takeoff progresses, flight controls are used as necessary to compensate for wind conditions. Lift-off should be made at no less than $V_{mc}+5$. After lift-off, the airplane should be allowed to accelerate to the all-engine best rate-of-climb speed V_y, and then the climb maintained at this speed with takeoff power until a safe maneuvering altitude is attained.

The landing gear may be raised as soon as practicable but not before reaching the point from which a safe landing can no longer be made on the remaining portion of the runway. The flaps (if used) should be retracted as directed in the airplane's operating manual.

Upon reaching safe maneuvering altitude, the airplane should be allowed to accelerate to cruise climb speed before power is reduced to normal climb power.

Short Field or Obstacle Clearance Takeoff

If it is necessary to take off over an obstacle or from a critically short field, the procedures should be altered slightly. For example, the initial climb speed that should provide the best angle of climb for obstacle clearance is V_x rather than V_y. However, V_x in some light twins is below V_{mc}. In this case, if the climb were made at V_x and a sudden power failure occurred on one engine, the pilot would not be able to control the airplane unless power were reduced on the operating engine. This would create an impossible situation because it would not be likely that the airplane could clear an obstacle with one engine inoperative and the other at some reduced power setting. In any case, if an engine fails and the climb is to be continued over an obstacle, V_{xse} must be established if maximum performance is to be obtained.

Generally, the short field or obstacle clearance takeoff will be much the same as a normal takeoff using the manufacturer's recommended flap settings, power settings, and speeds. However, if the published best angle-of-climb speed (V_x) is less than $V_{mc}+5$, then it is recommended that no less than $V_{mc}+5$ be used.

During the takeoff roll as the airspeed reaches the best angle-of-climb speed, or $V_{mc}+5$, whichever is higher, the airplane should be rotated to establish an angle of attack that will cause the airplane to lift off and climb at that specified speed. At an altitude of approximately 50 feet or after clearing the obstacle, the pitch attitude can be lowered gradually to allow the airspeed to increase to the all engine best rate-of-climb speed. Upon reaching safe maneuvering altitude, the airplane should be allowed to accelerate to normal or enroute climb speed and the power controls reduced to the normal climb power settings.

Engine Failure on Takeoff

If an engine should fail during the takeoff roll before becoming airborne, it is advisable to close both throttles immediately and bring the airplane to a stop. The same procedure is recommended if after becoming airborne an engine should fail prior to having reached the single-engine best rate-of-climb speed (V_{yse}). An immediate landing is usually inevitable because of the altitude loss required to increase the speed to V_{yse}.

The pilot must have determined before takeoff what altitude, airspeed, and airplane configuration must exist to permit the flight to continue in event of an engine failure—the pilot also should be ready to accept the fact that if engine failure occurs before these required factors are established, both throttles must be closed and the situation treated the same as engine failure on a single-engine airplane. If it has been predetermined that the engine-out rate of climb under existing circumstances will be at least 50 feet per minute at

1,000 feet above the airport, and that at least the engine-out best angle-of-climb speed has been attained, the pilot may decide to continue the takeoff.

If the airspeed is below the engine-out best angle-of-climb speed (V_{xse}) and the landing gear has *not* been retracted, the takeoff should be abandoned immediately.

If the engine-out best angle-of-climb speed (V_{xse}) has been obtained and the landing gear is in the retract cycle, the pilot should climb at the engine-out best angle-of-climb speed (V_{xse}) to clear any obstructions, and thereafter stabilize the airspeed at the engine-out best rate-of-climb speed (V_{yse}) while retracting the landing gear and flaps and resetting all appropriate systems.

When the decision is made to continue flight, the single-engine best rate-of-climb speed should be attained and maintained (Fig. 16–13). Even if altitude cannot be maintained, it is best to continue to hold that speed because it would result in the slowest rate of descent and provide the most time for executing the emergency landing. After the decision is made to continue flight and a positive rate of climb is attained, the landing gear should be retracted as soon as practical.

If the airplane is just barely able to maintain altitude and airspeed, a turn requiring a bank greater than approximately 15° should not be attempted. When such a turn is made under these conditions, *both* lift and airspeed will decrease. Consequently, it is advisable to continue straight ahead whenever possible, until reaching a safe maneuvering altitude. At that time a steeper bank may be made safely—and in either direction. There is nothing wrong with banking toward a "dead" engine if a safe speed and zero sideslip are maintained.

When an engine fails after becoming airborne, the pilot should hold heading with rudder and simultaneously roll into a bank of at least 5° toward the operating engine. In this attitude the airplane will tend to turn toward the operating engine, but at the same time, the asymmetrical power resulting from the engine failure will tend to turn the airplane toward the "dead" engine. The result is a partial balance of those tendencies and provides for an increase in airplane performance as well as easier directional control.

NOTE.—In this situation the ball in the turn-and-bank indicator will be approximately one ball width off center toward the good engine.

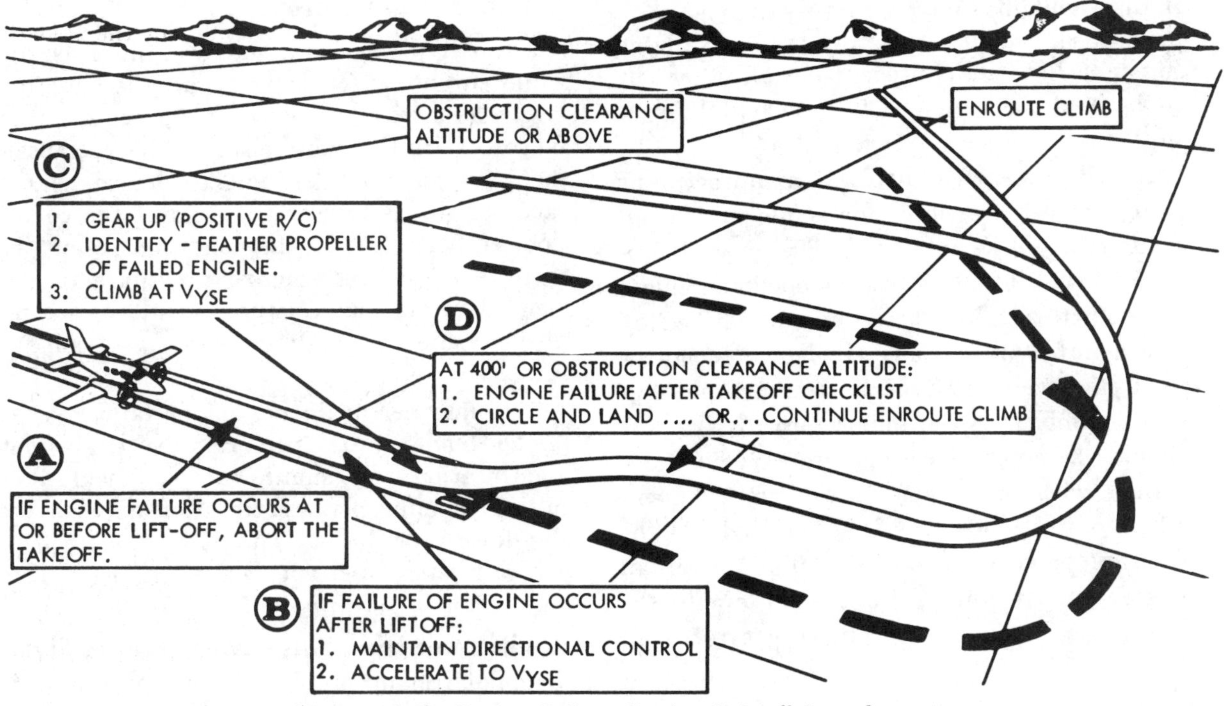

Figure 16–13 Engine Failure During Takeoff Procedures

The best way to identify the inoperative engine is to note the direction of yaw and the rudder pressure required to maintain heading. To counteract the asymmetrical thrust, extra rudder pressure will have to be exerted on the operating engine side. To aid in identifying the failed engine, some pilots use the expressions "Best Foot Forward," or "Dead Foot Dead Engine." Never rely on tachometer or manifold pressure readings to determine which engine has failed. After power has been lost on an engine, the tachometer will often indicate the correct r.p.m. and the manifold pressure gauge will indicate the approximate atmospheric pressure or above.

Experience has shown that the biggest problem is not in identifying the inoperative engine, but rather in the pilot's actions *after* the inoperative engine has been identified. In other words, a pilot may identify the "dead" engine and then attempt to shut down the wrong one—resulting in no power at all. To avoid this mistake, the pilot should *verify* that the dead engine has been identified by retarding the throttle of the suspected engine before shutting it down.

When demonstrating or practicing procedures for engine failure on takeoff, the feathering of the propeller and securing of the engine should be *simulated* rather than actually performed, so that the engine may be available for immediate use if needed; but all other settings should be made just as in an actual power failure.

In all cases, the airplane manufacturer's recommended procedure for single-engine operation should be followed. The general procedure listed below is not intended to replace or conflict with any procedure established by the manufacturer of any airplane. It can be used effectively for general training purposes and to emphasize the importance of V_{yse}. It should be noted that this procedure is concerned with an engine failure on a takeoff where obstacle clearance is not critical. If the decision is made to continue flight after an engine failure during the takeoff climb, the pilot should maintain directional control at all times and:

1. Maintain V_{yse}.
2. Check that all mixture controls, prop controls, and throttles (in that order) are at maximum permissible power settings.
3. Maintain V_{yse}.
4. Check that the flaps and landing gear have been retracted.
5. Maintain V_{yse}.
6. Decide which engine is inoperative (dead).
7. Maintain V_{yse}.
8. Raise the wing on the suspected "dead" engine side *at least* 5°.
9. Maintain V_{yse}.
10. Verify the "dead" engine by retarding the throttle of the suspected engine. (If there is no change in rudder forces, then that is the inoperative engine.)
11. Maintain V_{yse}.
12. Feather the prop on the "dead" engine (verified by the retarded throttle).
13. Maintain V_{yse}.
14. Declare an emergency if operating from a tower controlled airport. Advise the tower of your intentions.
15. Maintain V_{yse}.

Engine Failure Enroute

Normally, when an engine failure occurs while enroute in cruising flight, the situation is not as critical as when an engine fails on takeoff. With the more leisurely circumstances, the pilot should take time to determine the cause of the failure and to correct the condition, if possible. If the condition cannot be corrected, the single-engine procedure recommended by the manufacturer should be accomplished and a landing made as soon as practical.

A primary error during engine failure is the pilot's tendency to perform the engine-out identification and shutdown too quickly, resulting in improper identification or incorrect shutdown procedures. The element of surprise generally associated with actual engine failure may result in confused and hasty reactions.

When an engine fails during cruising flight, the pilot's main problem is to maintain sufficient altitude to be able to continue flight to the point of intended landing. This is dependent on the density altitude, gross weight of the airplane, and elevation of the terrain and obstructions. When the airplane is above its single-engine service ceiling, altitude will be lost. The single-engine service ceiling is the maximum density altitude at which the single-engine best rate-of-climb speed will produce 50 FPM rate of climb. This ceiling is determined by the manufacturer on the basis of the airplane's maximum gross weight, flaps and landing gear retracted, the critical engine inoperative, and the propeller feathered.

Although engine failure while enroute in normal cruise conditions may not be critical, it is a recommended practice to add maximum permissible power to the operating engine before securing or shutting down the failed engine. If it is determined later that maximum permissible power on the operating engine is not needed to maintain altitude, it is a simple matter to reduce the power. Conversely, if maximum permissible power is not applied, the airspeed may decrease much farther and more rapidly than expected. This condition could present a serious performance problem, especially if the airspeed should drop below V_{yse}.

The altitude should be maintained if it is within the capability of the airplane. In an airplane not capable of maintaining altitude with an engine inoperative under existing circumstances, the airspeed should be maintained within ±5 knots of the engine-out best rate-of-climb speed (V_{yse}) so as to conserve altitude as long as possible to reach a suitable landing area.

After the landing gear and flaps are retracted and the failed engine is shut down and everything is under control (including heading and altitude), it is recommended that the pilot communicate with the nearest ground facility to let them know the flight is being conducted with one engine inoperative. FAA facilities are able to give valuable assistance if needed, particularly when the flight is conducted under IFR or a landing is to be made at a tower-controlled airport. Good judgment would dictate, of course, that a landing be made at the nearest suitable airport as soon as practical rather than continuing flight.

During engine-out practice using zero thrust power settings, the engine may cool to temperatures considerably below the normal operating range. This factor requires caution when advancing the power at the termination of single-engine practice. If the power is advanced rapidly, the engine may not respond and an actual engine failure may be encountered. This is particularly important when practicing engine-out approaches and landings. A good procedure is to slowly advance the throttle to approximately one-half power, then allow it to respond and stabilize before advancing to higher power settings. This procedure also results in less wear on the engines of the training aircraft.

Restarts after feathering require the same amount of care, primarily to avoid engine damage. Following the restart, the engine power should be maintained at the idle setting or slightly above until the engine is sufficiently warm and is receiving adequate lubrication.

Although each make and model of airplane must be operated in accordance with the manufacturer's instructions, the following typical checklist is presented to familiarize the transitioning pilot with the actions that may be required when an engine fails.

ENGINE FAILURE DURING FLIGHT

1. Mixtures—AS REQUIRED for flight altitude.
2. Propellers—FULL FORWARD.
3. Throttles—FULL FORWARD.
4. Landing Gear—RETRACTED.
5. Wing Flaps—RETRACTED.
6. Inoperative Engine — DETERMINE. Idle engine same side as idle foot.
7. Establish at least 5° Bank -- TOWARD OPERATIVE ENGINE.
8. *Inoperative Engine—SECURE.*
   a. Throttle—CLOSE.
   b. Mixture—IDLE CUT-OFF.
   c. Propeller—FEATHER.
   d. Fuel Selector—OFF.
   e. Auxiliary Fuel Pump—OFF.
   f. Magneto Switches—OFF.
   g. Alternator Switch—OFF.
   h. Cowl Flap—CLOSE.

9. *Operative Engine—ADJUST.*
   a. Power—AS REQUIRED.
   b. Mixture—AS REQUIRED for flight altitude.
   c. Fuel Selector—AS REQUIRED.
   d. Auxiliary Fuel Pump—ON.
   e. Cowl Flap—AS REQUIRED.
10. Trim Tabs—ADJUST bank toward operative engine.
11. Electrical Load—DECREASE to minimum required.
12. As Soon As Practical—LAND.

AIRSTART (After Shutdown)

Airplanes Without Propeller Unfeathering System:

1. Magneto Switches—ON.
2. Fuel Selector—MAIN TANK (Feel For Detent).
3. Throttle — FORWARD approximately one inch.
4. Mixture—AS REQUIRED for flight altitude.
5. Propeller—FORWARD of detent.
6. Starter Button—PRESS.
7. Primer Switch—ACTIVATE.
8. Starter and Primer Switch—RELEASE when engine fires.
9. Mixture—AS REQUIRED.
10. Power—INCREASE after cylinder head temperature reaches 200° F.
11. Cowl Flap—AS REQUIRED.
12. Alternator—ON.

Airplanes With Propeller Unfeathering System:

1. Magneto Switches—ON.
2. Fuel Selector—MAIN TANK (Feel For Detent).
3. Throttle — FORWARD approximately one inch.
4. Mixture—AS REQUIRED for flight altitude.
5. Propeller—FULL FORWARD.
6. Propeller—RETARD to detent when propeller reaches 1000 RPM.
7. Mixture—AS REQUIRED.
8. Power—INCREASE after cylinder head temperature reaches 200° F.
9. Cowl Flap—AS REQUIRED.
10. Alternator—ON.

Engine-Out Approach and Landing

Essentially, an engine-out approach and landing is the same as a normal approach and landing. Long, flat approaches with high power output on the operating engine and/or excessive threshold speed that results in floating and unnecessary runway use should be avoided. Due to variations in the performance, limitations, etc., of many light twins, no specific flightpath or procedure can be proposed that would be adequate in all engine-out approaches. In most light twins, however, a single-engine approach can be accomplished with the flightpath and procedures almost identical to a normal approach and landing (Fig. 16–14). The light-twin manufacturers include a recommended single-engine landing procedure in the airplane's operating manual.

During the checkout, the transitioning pilot should perform approaches and landings with the power of one engine set to simulate the drag of a feathered propeller (zero thrust), or if feathering propellers are not installed, the throttle of the simulated failed engine set to idling. With the "dead" engine feathered or set to "zero thrust," normal drag is considerably reduced, resulting in a longer landing roll. Allowances should be made accordingly for the final approach and landing.

The final approach speed should not be less than V_{yse} until the landing is assured; thereafter, it should be at the speed commensurate with the flap position until beginning the roundout for landing. Under normal conditions the approach should be made with full flaps; however, neither full flaps nor the landing gear should be extended until the landing is assured. With full flaps the approach speed should be 1.3 V_{so} or as recommended by the manufacturer.

The pilot should be particularly judicious in lowering the flaps. Once they have been extended it may not be possible to retract them in time to initiate a go-around. Most of the light twins are not capable of making a single-engine go-around with full flaps.

Figure 16–14 Feathered Propeller Reduces Drag

Pressurized Airplanes

When an airplane is flown at a high altitude, it consumes less fuel for a given airspeed than it does for the same speed at a lower altitude. In other words, the airplane is more efficient at a high altitude. In addition, bad weather and turbulence can be avoided by flying in the relatively smooth air above the storms. Airplanes which do not have pressurization and air conditioning systems are usually limited to the lower altitudes. Because of the advantages of flying at high altitudes, many modern general aviation type airplanes are being designed to operate in that environment. It is important then, that pilots transitioning to such sophisticated equipment be familiar with at least the basic operating principles.

A cabin pressurization system accomplishes several functions in providing adequate passenger comfort and safety. It maintains a cabin pressure altitude of approximately 8,000 feet at the maximum designed cruising altitude of the airplane, and prevents rapid changes of cabin altitude which may be uncomfortable or injurious to passengers and crew. In addition, the pressurization system permits a reasonably fast exchange of air from inside to outside the cabin. This is necessary to eliminate odors and to remove stale air.

Pressurization of the airplane cabin is now the accepted method of protecting persons against the effects of hypoxia. Within a pressurized cabin, people can be transported comfortably and safely for long periods of time, particularly if the cabin altitude is maintained at 8,000 feet or below, where the use of oxygen equipment is not required. However, the flight crew in this type of airplane must be aware of the danger of accidental loss of cabin pressure and must be prepared to meet such an emergency whenever it occurs.

In this typical pressurization system, the cabin, flight compartment, and baggage compartments are incorporated into a sealed unit which is capable of containing air under a pressure higher than outside atmospheric pressure. Pressurized air is pumped into this sealed fuselage by cabin superchargers which deliver a relatively constant volume of air at all altitudes up to a designed maximum. Air is released from the fuselage by a device called an outflow valve. Since the superchargers provide a constant inflow of air to the pressurized area, the outflow valve, by regulating the air exit, is the major controlling element in the pressurization system.

It is necessary to become familiar with some terms and definitions to understand the operating principles of pressurization and air conditioning systems. These are:

1. *Aircraft altitude.* The actual height above sea level at which the airplane is flying.
2. *Ambient temperature.* The temperature in the area immediately surrounding the airplane.
3. *Ambient pressure.* The pressure in the area immediately surrounding the airplane.
4. *Cabin altitude.* Used to express cabin pressure in terms of equivalent altitude above sea level.
5. *Differential pressure.* The difference in pressure between the pressure acting on one side of a wall and the pressure acting on the other side of the wall. In aircraft air conditioning and pressurizing systems, it is the difference between cabin pressure and atmospheric pressure.

The cabin pressure control system provides cabin pressure regulation, pressure relief, vacuum relief, and the means for selecting the desired cabin altitude in the isobaric and differential range. In addition, dumping of the cabin pressure is a function of the pressure control system. A cabin pressure regulator, an outflow valve, and a safety valve are used to accomplish these functions.

The cabin pressure regulator controls cabin pressure to a selected value in the isobaric range and limits cabin pressure to a preset differential value in the differential range. When the airplane reaches the altitude at which the difference between the pressure inside and outside the cabin is equal to the highest differential pressure for which the fuselage structure is designed and further increase in airplane altitude will result in a corresponding increase in cabin altitude. Differential control is used to prevent the maximum differential pressure, for which the fuselage was designed, from being exceeded. This differential pressure is determined by the structural strength of the cabin and often by the relationship of the cabin size to the probable areas of rupture, such as window areas and doors.

The cabin air pressure safety valve is a combination pressure relief, vacuum relief, and dump valve. The pressure relief valve prevents cabin pressure from exceeding a predetermined differential pressure above ambient pressure. The vacuum relief prevents ambient pressure from exceeding cabin pressure by allowing external air to enter the cabin when ambient pressure exceeds cabin pressure. The dump valve is actuated by the cockpit control switch. When this switch is positioned to "ram," a solenoid valve opens, causing the valve to dump cabin air to atmosphere.

The degree of pressurization and, therefore, the operating altitude of the aircraft are limited by several critical design factors. Primarily the fuselage is designed to withstand a particular maximum cabin differential pressure.

Several instruments are used in conjunction with the pressurization controller. The cabin differential pressure gauge indicates the difference between inside and outside pressure. This gauge should be monitored to assure that the cabin does not exceed the maximum allowable differential pressure. A cabin altimeter is also provided as a check on the performance of the system. In some cases, these two instruments are combined into one. A third instrument indicates the cabin rate of climb or descent. A cabin rate of climb instrument and a cabin altimeter are illustrated in Fig. 16–15.

Figure 16–15 Cabin Pressurization Instruments

Decompression is defined as the inability of the airplane's pressurization system to maintain its designed pressure differential. This can be caused by a malfunction in the pressurization system or structural damage to the airplane. Physiologically, decompressions fall into two categories:

1. *Explosive Decompression.* Explosive decompression is defined as a change in cabin pressure faster than the lungs can

decompress. Therefore, it is possible that lung damage may occur. Normally, the time required to release air from the lungs where no restrictions exist, such as masks, etc., is 0.2 seconds. Most authorities consider any decompression which occurs in less than 0.5 seconds as explosive and potentially dangerous.

2. *Rapid Decompression.* Rapid decompression is defined as a change in cabin pressure where the lungs can decompress faster than the cabin. Therefore there is no likelihood of lung damage.

During a decompression there may be noise, and for a split second one may feel dazed. The cabin air will fill with fog, dust or flying debris. Fog occurs due to the rapid drop in temperature and the change of relative humidity. Normally, the ears clear automatically. Belching or passage of intestinal gas may occur. Air will rush from the mouth and nose due to the escape of air from the lungs, and may be noticed by some individuals.

The primary danger of decompression is hypoxia. Unless proper utilization of oxygen equipment is accomplished quickly, unconsciousness may occur in a very short time. The period of useful consciousness is considerably shortened when a person is subjected to a rapid decompression. This is due to the rapid reduction of pressure on the body. Thus, oxygen in the lungs is exhaled rapidly. This in effect reduces the partial pressure of oxygen in the blood and thus reduces the pilot's effective performance time by ⅓ to ¼ its normal time. It is for this reason the oxygen mask should be worn on the face when flying at very high altitudes (35,000 feet or higher). It is recommended that the crewmembers select the 100% oxygen setting on the oxygen regulator at high altitude if the airplane is equipped with a demand or pressure demand oxygen system.

Another hazard is that of being tossed or blown out of the airplane if near an opening. For this reason, individuals near such openings should wear safety harnesses or seatbelts at all times when the airplane is pressurized and they are seated.

Still another potential hazard on high altitude decompressions is the possibility of evolved gas decompression sicknesses. Exposure to windblast and extremely cold temperatures are other hazards one might have to face.

Rapid descent from altitude is indicated if these problems are to be minimized. This descent is briefly discussed in Chapter 11. Automatic visual and aural warning systems are included in the equipment of all pressurized airplanes so that slow decompressions will not occur and overwhelm the crew before they are detected.

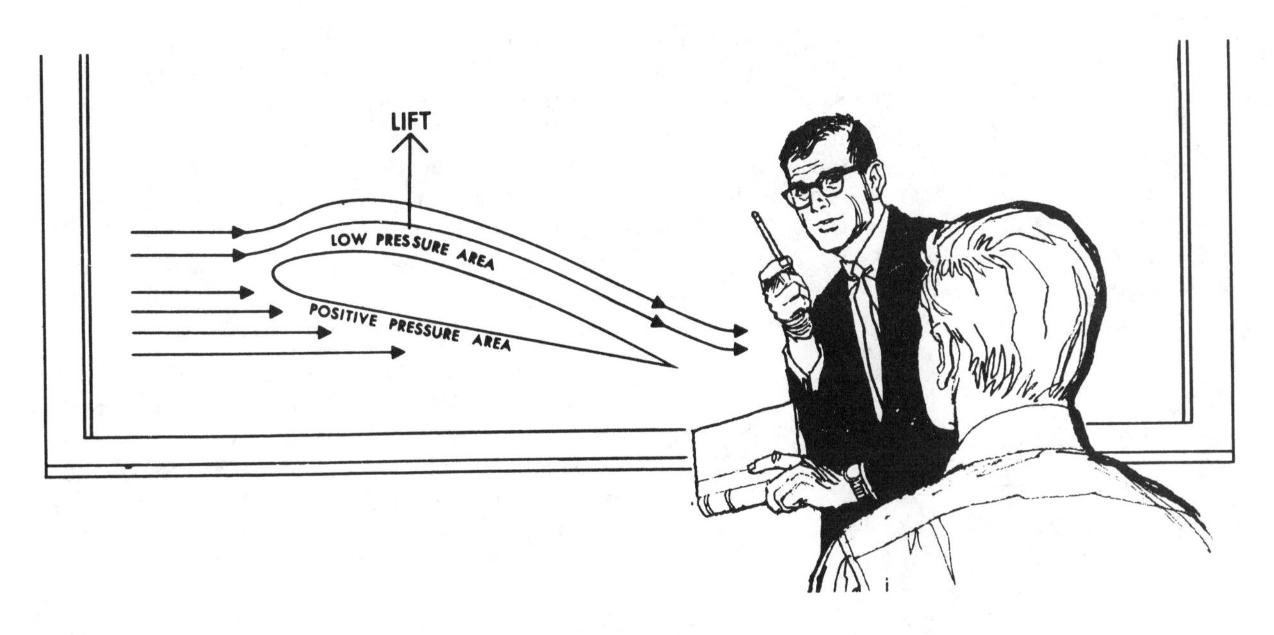

CHAPTER 17

Principles of Flight and Performance Characteristics

This chapter discusses the fundamental physical laws governing the forces acting on an airplane in flight, and what effect these natural laws and forces have on the performance characteristics of airplanes. To competently control the airplane, the pilot must understand the principles involved and learn to utilize or counteract these natural forces.

Modern general aviation airplanes have what may be considered high performance characteristics. Therefore, it is increasingly necessary that pilots appreciate and understand the principles upon which the art of flying is based. The first men to fly learned *why* they were able to fly, before they were successful in their attempts to do so. In other words, they started on the ground and worked up. It is equally essential that the newcomer to today's realm of powered flight do likewise.

The principles of flight spelled out in this handbook represent the essence of the topics selected, and about which pilots should be well informed. Such information, when understood, can make one a safer and more effective pilot. It must be understood though that the material presented is not intended to cover *all* the parameters of powered flight.

Structure of the Atmosphere

The atmosphere in which we fly is an envelope of air which surrounds the earth and rests upon its surface. It is as much a part of the earth as the seas or the land. However, air differs from land and water inasmuch as it is a mixture of gases. It has mass weight and indefinite shape.

Air, like any other fluid, is able to flow, and change its shape when subjected to even minute pressures because of the lack of strong molecular cohesion. For example, gas will completely fill any container into which it is placed, expanding or contracting to adjust its shape to the limits of the container.

The atmosphere is composed of 78 percent nitrogen, 21 percent oxygen, and 1 percent other gases, such as argon, helium, etc. As some of these elements are heavier than others, there is a natural tendency of these heavier elements, such as oxygen, to settle to the surface of the earth, while the lighter elements are lifted up to the region of higher altitude. This explains why most of the oxygen is contained below 35,000 feet altitude.

Because air has mass and weight, it is a *body*, and as a body, it reacts to the scientific laws of bodies in the same manner as other gaseous bodies. This body of air resting upon the surface of the earth has weight (Fig. 17–1) and at sea level develops an average pressure of 14.7 pounds on each square inch of surface, or 29.92 inches of mercury, — but as its thickness is limited, the higher we go the less air there is above us. For this reason, the weight of the atmosphere at 18,000 feet is only one-half what it is at sea level.

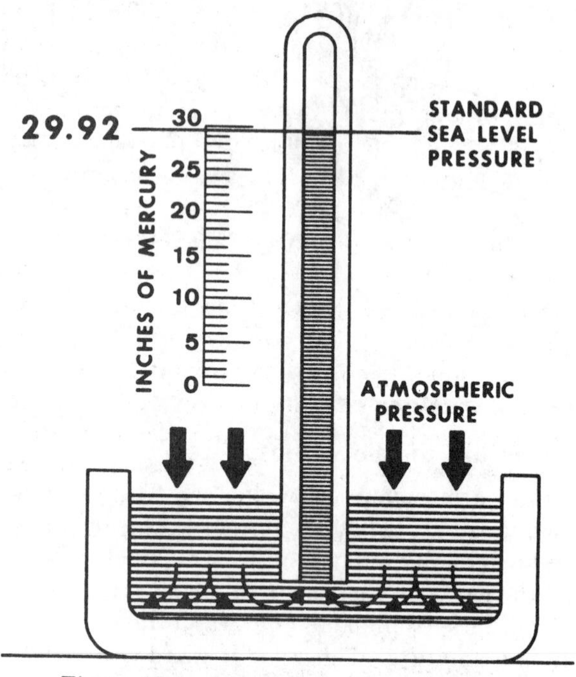

Figure 17–1 Standard Sea Level Pressure

Atmospheric Pressure

Though there are various kinds of pressure, we are mainly concerned with *atmospheric pressure.* It is one of the basic factors in weather changes, helps to lift the airplane, and actuates some of the important flight instruments in the airplane. These instruments are the altimeter, the airspeed indicator, the rate-of-climb indicator, and the manifold pressure gauge.

Though air is very light, it has *mass* and is affected by the attraction of gravity. Therefore, like any other substance, it has weight, and because of its weight, it has force. Since it is a fluid substance, this force is exerted equally in all directions, and its effect on bodies within the air is called *pressure.* Under standard conditions at sea level, the average pressure exerted on the human body by the weight of the atmosphere around it is approximately 14.7 lb./in. The density of air in which we fly has significant effects on the airplane's capability. As air becomes less dense it reduces (1) power because the engine takes in less air, (2) thrust because the propeller is less efficient in thin air, and (3) lift because the thin air exerts less force on the airfoils.

Effects of Pressure on Density

Since air is a gas it can be compressed or expanded. When air is compressed a greater amount of air can occupy a given volume. Conversely, when pressure on a given volume of air is decreased, the air expands and occupies a greater space. That is, the original column of air at a lower pressure, contains a smaller mass of air. In other words, the density is decreased. In fact, density is directly proportoinal to pressure. If the pressure is doubled the density is doubled, and if the pressure is lowered, so is the density. *This statement is true, only at a constant temperature.*

Effect of Temperature on Density

The effect of increasing the temperature of a substance is to decrease its density. Conversely, deceasing the temperature has the effect of increasing the density. Thus, the density of air varies inversely as the absolute temperature varies. *This statement is true only at a constant pressure.*

In the atmosphere, both temperature and pressure decrease with altitude, and have conflicting effects upon density. However, the fairly rapid drop in pressure as altitude is increased usually has the dominating effect. Hence, we can expect the density to decrease with altitude.

Effect of Humidity on Density

The preceding paragraphs have assumed that the air was perfectly dry. In reality, it is never completely dry. The small amount of water vapor suspended in the atmosphere may be almost negligible under certain conditions, but in other conditions humidity may become an important factor in the performance of an aircraft. Water vapor is lighter than air; consequently *moist air is lighter than dry air.* It is lightest or least dense when, in a given set of conditions, it contains the maximum amount of water vapor. The higher the temperature, the greater amount of water vapor that the air can hold. When comparing two separate air masses, the first warm and moist (both qualities tending to lighten the air) and the second cold and dry (both qualities making it heavier), the first necessarily must be less dense than the second. Pressure, temperature, and humidity have a great influence on airplane performance, because of their effect upon density.

Newton's Laws of Motion and Force

In the 17th century, a philosopher and mathematician, Sir Isaac Newton, propounded three basic laws of motion. It is certain that he did not have the airplane in mind when he did so, but almost everything we know about motion goes back to his three simple laws. These laws, named after Newton, are as follows:

Newton's first law states, in part, that:

> A body at rest tends to remain at rest, and a body in motion tends to remain moving at the same speed and in the same direction.

This simply means that, in nature, nothing starts or stops moving until some outside force causes it to do so. An airplane at rest on the ramp will remain at rest unless a force strong enough to overcome its inertia is applied. Once it is moving, however, its inertia keeps it moving, subject to the various other forces acting on it. These forces may add to its motion, slow it down, or change its direction.

Newton's second law implies that:

> When a body is acted upon by a constant force, its resulting acceleration is inversely proportional to the mass of the body and is directly proportional to the applied force.

What we are dealing with here are the factors involved in overcoming Newton's First Law of Inertia. It covers both changes in direction and speed, including starting up from rest (positive acceleration) and coming to a stop (negative acceleration, or deceleration).

Newton's third law states that:

> Whenever one body exerts a force on another, the second body always exerts on the first, a force which is equal in magnitude but opposite in direction.

The firing of a shotgun or large caliber rifle provides a "bang" in more ways than one. It may not be the most pleasant way to get your "kicks," but the recoil of a gun as it is fired is a graphic example of Newton's Third Law (Fig. 17-2). The champion swimmer who pushes against the side of the pool during the turn-around, or the infant learning to walk—both would fail but for the phenomena expressed in this law. The charge leaving the gun, or any push against some surface which "pushes back," illustrates this mutual action-reaction. In an airplane, the propeller moves and pushes back the air; consequently, the air pushes the propeller (and thus the airplane) in the opposite direction—forward. In a jet airplane, the engine pushes a blast of hot gases backward; the force of equal and opposite reaction pushes against the engine and forces the airplane forward. The movement of all vehicles is a graphic illustration of Newton's Third Law.

Figure 17-2 Opposite and Equal Reaction

Bernoulli's Principle of Pressure

A half century after Sir Newton presented his laws, Mr. Daniel Bernoulli, a Swiss mathematician, explained how the pressure of a moving fluid (liquid or gas) varies with its speed of motion. Specifically, he stated that an increase in the speed of movement or flow would cause a decrease in the fluid's pressure. This is exactly what happens to air passing over the curved top of the airplane wing.

An appropriate analogy can be made with water flowing through a garden hose. Water moving through a hose of constant diameter exerts a uniform pressure on the hose; but if the diameter of a section of the hose is increased or decreased, it is certain to change the pressure of the water at that point. Suppose we were to pinch the hose, thereby constricting the area through which the water flows. Assuming that the same volume of water flows through the constricted portion of the hose in the *same period of time* as before the hose was pinched, it follows that the speed of flow must increase at that point.

Therefore, if we constrict a portion of the hose, we not only *increase* the speed of the flow, but we also *decrease* the pressure at that point. We could achieve like results if we were to introduce streamlined solids (airfoils) at the same point in the hose. This same principle is the basis for the measurement of airspeed (fluid flow) and for analyzing the airfoil's ability to produce lift.

A practical application of Bernoulli's theorem is the venturi tube (Fig. 17–3). The venturi tube has an air inlet which narrows to a throat (constricted point) and an outlet section which increases in diameter toward the rear. The diameter of the outlet is the same as that of the inlet. At the throat the airflow speeds up and the pressure decreases; at the outlet the airflow slows and the pressure increases.

If we recognize air as a body and therefore admit that it must follow the above laws, we can begin to see how and why an airplane wing develops lift as it moves through the air.

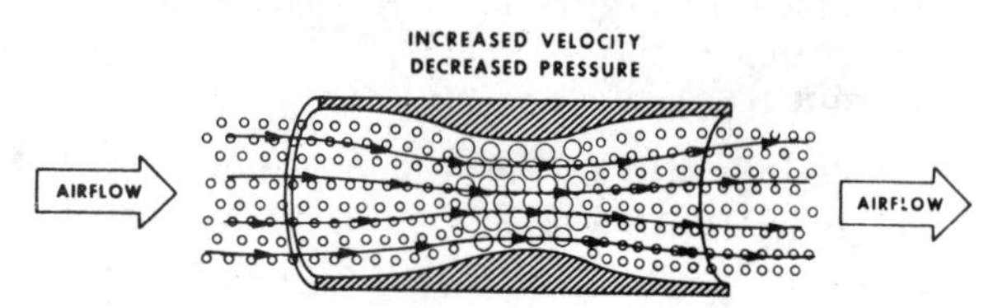

Figure 17–3 Air Pressure Decreases in a Venturi

Airfoil Design

It is far from accidental that there is a basic similarity between the wings of birds and the wings of airplanes. A bird's wing is nothing more than an airfoil, and man has merely copied its shape, modified its design and structure, and developed mechanical power sources as substitutes for his own inadequacies in this area—and so he too flies.

In the sections devoted to Newton's and Bernoulli's discoveries, we have already discussed in general terms the question of how a bird's wings or the wings of man's flying machine sustain flight when both the bird and the machine are heavier than air. Perhaps the explanation to the whole riddle can best be reduced to its most elementary concept by stating that lift (flight) is simply the result of fluid flow (air) about an airfoil—or in everyday language, the result of moving an air foil (wing), by whatever means, through the air.

Since it is the airfoil which harnesses the force developed by its movement through the air, we will further discuss and explain this "magic" structure, as well as some of the material presented in previous discussions on Messrs. Newton's and Bernoulli's laws. We can, in this way, emphasize the principles that are basic to an understanding of airfoils and airflow.

An airfoil is a structure designed to obtain reaction upon its surface from the air through which it moves or that moves past such a structure. Air acts in various ways when submitted to different pressures and velocities; but we will confine this discussion to the parts of an airplane that we are most concerned with in flight; namely, the airfoils designed to produce lift. By looking at a typical airfoil profile such as the cross section of a wing you can see several obvious characteristics of design (Fig. 17–4). Notice that there is a difference in the curvatures of the upper and lower surfaces of the airfoil (the curvature is called camber). The camber of the upper surface is more pronounced than that of the lower surface, which is somewhat flat in most instances.

In Fig. 17–4, note that the two extremities of the airfoil profile differ in appearance. The

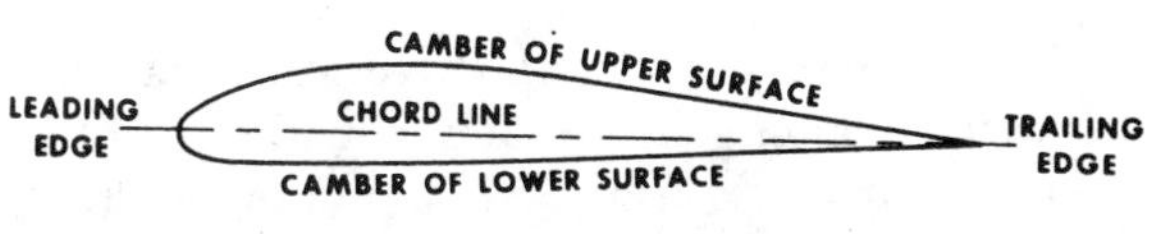

Figure 17–4 Typical Airfoil Section

end which faces forward in flight is called the *leading edge*, and is rounded, while the other end, the *trailing edge*, is quite narrow and tapered.

A reference line often used in discussing the airfoil is the *chord line*, a straight line drawn through the profile connecting the extremities of the leading and trailing edges. The distance from this chord line to the upper and lower surfaces of the wing denotes the magnitude of the upper and lower camber at any point. Another reference line, drawn from the leading edge to the trailing edge, is the "mean camber line." This mean line is equidistant at all points from the upper and lower contours.

The construction of the wing so as to provide actions greater than its weight, is done by shaping the wing (Fig. 17–4) so that advantage can be taken of the air's response to certain physical laws, and thus develop two actions from the air mass; a positive-pressure lifting action from the air mass below the wing, and a negative-pressure lifting action from lowered pressure above the wing.

As the airstream strikes the relatively flat lower surface of the wing when inclined at a small angle to its direction of motion, the air is forced to rebound downward and therefore causes an upward reaction in positive lift, while at the same time airstream striking the upper curved section of the "leading edge" of the wing is deflected upward. In other words, a wing shaped to cause an action on the air, and forcing it downward, will provide an equal reaction from the air, forcing the wing upward. If a wing is constructed in such form that it will cause a lift force greater than the weight of the airplane, the airplane will fly.

Probably you have held your flattened hand out of the window of a moving automobile. As you inclined your hand to the flow of air, the force of air against it pushed it up. Without realizing it you were demonstrating how an airplane gets a portion of its lift. The "wing," in this case your hand, pushed against the air, the air pushed back—remember Newton's third law? As we have already learned, the upward component of this force is called *lift*.

However, if all the lift required were obtained merely from the deflection of air by the lower surface of the wing, an airplane would need only a flat wing like a kite. This, of course, is not the case at all; under certain conditions disturbed air currents circulating at the trailing edge of the wing could be so excessive as to make the airplane lose speed and lift. The balance of the lift needed to support the airplane comes from the flow of air above the wing. Herein lies the key to flight. The fact that most lift is the result of the airflow's downwash from above the wing, must be thoroughly understood in order to continue further in the study of flight. It is neither accurate nor does it serve a useful purpose, however, to assign specific values to the percentage of lift generated by the upper surface of an airfoil versus that generated by the lower surface. These are not constant values, and will vary, not only with flight conditions, but with different wing designs.

It should be understood that different airfoils have different flight characteristics. Many thousands of airfoils have been tested in wind tunnels and in actual flight, but no one airfoil has been found that satisfies every flight requirement. The weight, speed, and purpose of each airplane dictate the shape of its airfoil. It was learned many years ago that the most efficient airfoil for producing the greatest lift was one that had a concave, or "scooped out" lower surface. Later it was also learned that as a fixed design, this type of airfoil sacrificed too much speed while producing lift and, therefore, was not suitable for high speed flight. It is interesting to note, however, that through advanced progress in engineering, today's high speed jets can again take advantage of the concave airfoil's high lift characteristics. Leading edge (Kreuger) flaps and trailing edge (Fowler) flaps, when extended from the basic wing structure, literally change the airfoil shape into the classic concave form, thereby generating much greater lift during slow flight conditions.

On the other hand, an airfoil that is perfectly streamlined and offers little wind resistance sometimes does not have enough lifting power to take the airplane off the ground. Thus, modern airplanes have airfoils which strike a medium between extremes in design, the shape varying according to the needs of the airplane for which it is designed. Figure 17–5 shows some of the more common airfoil sections.

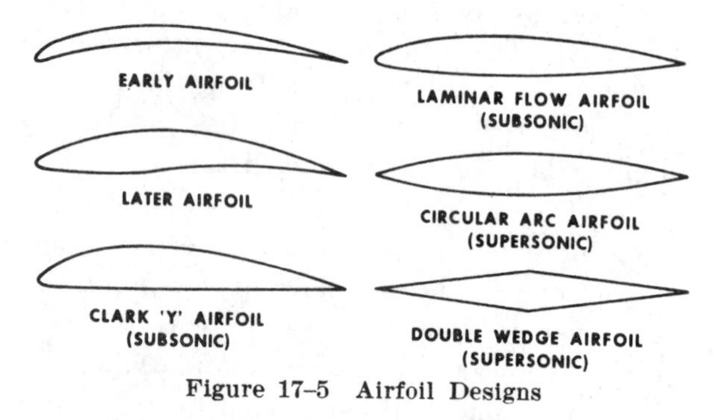

Figure 17–5 Airfoil Designs

Low Pressure Above

In a wind tunnel or in flight, an airfoil is simply a streamlined object inserted into a moving stream of air. If the airfoil profile were in the shape of a teardrop, the speed and the pressure changes of the air passing over the top and bottom would be the same on both sides. But if the teardrop-shaped airfoil were cut in half lengthwise, a form resembling the basic airfoil (wing) section would result. If the airfoil were then inclined so the airflow strikes it at an angle (angle of attack), the air molecules moving over the upper surface would be forced to move faster than would the molecules moving along the bottom of the air foil, since the upper molecules must travel a greater distance due to the curvature of the upper surface. This increased velocity reduces the pressure above the airfoil.

Bernoulli's principle of pressure by itself does not explain the distribution of pressure over the upper surface of the airfoil. To understand more fully, let us investigate the influence of momentum of the air as it flows in various curved paths near the airfoil (Fig. 17–6). Momentum is the resistance a moving body offers to having its direction or amount of motion changed. When a body is forced to move in a circular path it offers resistance in the direction away from the center of the curved path. This is "centrifugal force." While the particles of air move in the curved path AB, centrifugal force tends to throw them in the direction of the arrows between A and B and hence, causes the air to exert more than normal pressure on the leading edge of the airfoil. But after the air particles pass B (the point of reversal of the curvature of the path) the centrifugal force tends to throw them in the direction of the arrows between B and C (causing reduced pressure on the airfoil). This effect is held until the particles reach C, the second point of reversal of curvature of the airflow. Again the centrifugal force is reversed and the particles may even tend to give slightly more than normal pressure on the trailing edge of the airfoil, as indicated by the short arrows between C and D.

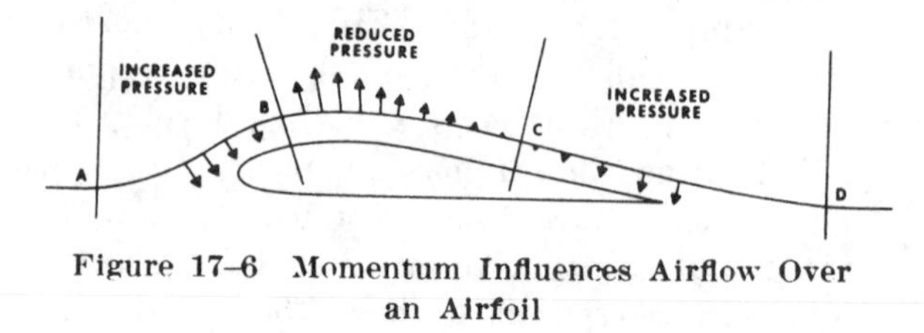

Figure 17–6 Momentum Influences Airflow Over an Airfoil

Therefore, the air pressure on the upper surface of the airfoil is distributed so that the pressure is much greater on the leading edge than the surrounding atmospheric pressure, causing strong resistance to foward motion; but the air pressure is *less* than surrounding atmospheric pressure over a large portion of the top surface (B to C).

As we have seen in the application of Bernoulli's theorem to a venturi, the speedup of air on the top of an airfoil produces a drop in pressure. This lowered pressure is a component of total lift. It is a mistake, however, to assume that the pressure difference between the upper and lower surface of a wing alone accounts for the total lift force produced.

One must also bear in mind that associated with the lowered pressure is downwash; a downward-backward flow from the top surface of the wing. As we have already seen from our previous discussion relative to the dynamic action of the air as it strikes the lower surface of the wing, the reaction of this downward-backward flow results in an upward-forward force on the wing. This same reaction applies to the flow of air over the top of the airfoil as well as to the bottom, and Newton's third law is again in the picture.

High Pressure Below

In the section dealing with Newton's laws as they apply to lift we have already discussed how a certain amount of lift is generated by pressure conditions underneath the wing. Because of the manner in which air flows underneath the wing, a positive pressure results, particularly at higher angles of attack. But there is another aspect to this airflow which must be considered. At a point close to the leading edge, the airflow is virtually stopped (stagnation point) and then gradually increases speed. At some point near the trailing edge it has again reached a velocity equal to that on the upper surface. In conformance with Bernoulli's principles, where the airflow was slowed beneath the wing, a positive upward pressure was created against the wing; i.e., as the fluid speed decreases, the pressure must increase. In essence, this simply "accentuates the positive," since it increases the pressure differential between the upper and lower surface of the airfoil, and therefore increases total lift over that which would have resulted had there been no increase of pressure at the lower surface. Both Bernoulli's principle and Newton's laws are in operation whenever lift is being generated by an airfoil.

Fluid flow or airflow then, is the basis for flight in airplanes, and is a product of the velocity of the airplane. The velocity of the airplane is very important to the pilot since it affects the lift and drag forces of the airplane, as we shall see later in the section on "Forces Acting on an Airplane." The pilot uses the velocity (airspeed) to fly at a minimum glide angle, at maximum endurance, and for a number of other flight maneuvers. Airspeed is the velocity of the airplane relative to the air mass through which it is flying.

Pressure Distribution

From experiments conducted on wind-tunnel models and on full-size airplanes, it has been determined that as air flows along the surface of a wing at different angles of attack there are regions along the surface where the pressure is negative, or less than atmospheric, and regions where the pressure is positive, or greater than atmospheric. This negative pressure on the upper surface creates a relatively larger force on the wing than is caused by the positive pressure resulting from the air striking the lower wing surface. Figure 17–7 shows the pressure distribution along an airfoil at three different angles of attack. In general, at high angles of attack the center of pressure moves forward, while at low angles of attack the center of pressure moves aft. In the design of wing structures this center of pressure travel is very important, since it affects the position of the airloads imposed on the wing structure in low angle-of-attack conditions and high angle-of-attack conditions. The airplane's aerodynamic balance and controllability are governed by changes in the center of pressure.

The center of pressure is determined through calculation and wind tunnel tests by varying the airfoil's angle of attack through normal

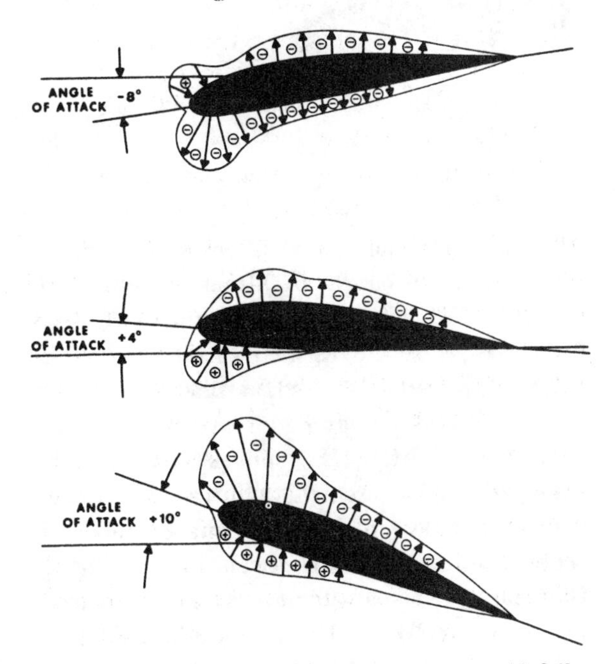

Figure 17–7 Pressure Distribution on an Airfoil

operating extremes. As the angle of attack is changed, so are the various pressure distribution characteristics (Fig. 17–7). Positive (+) and negative (−) pressure forces are totaled for each angle of attack and the resultant force is obtained. The total resultant pressure is represented by the resultant force vector shown in Fig. 17–8. The point of application of this force vector is termed the "center of pressure" (CP). For any given angle of attack, the center of pressure is the point where the resultant force crosses the chord line. This point is expressed as a percentage of the chord of the airfoil. A center of pressure at 30 percent of a 60-inch chord would be 18 inches aft of the wing's leading edge. It would appear then that if the designer would place the wing so that its center of pressure was at the airplane's center of gravity, the airplane would always balance. The difficulty arises, however, that the location of the center of pressure changes with change in the airfoil's angle of attack (Fig. 17–9).

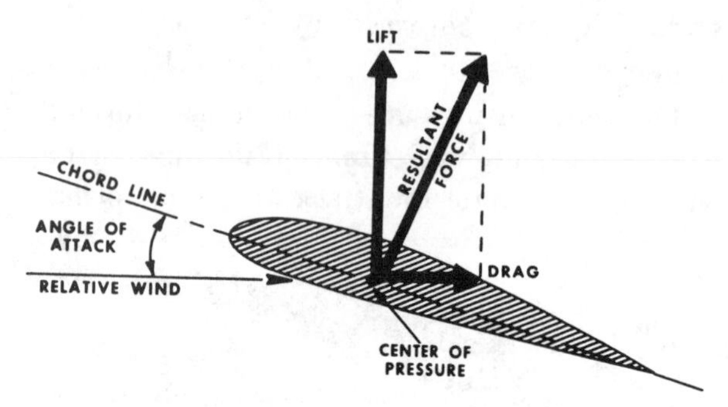

Figure 17–8 Force Vectors on an Airfoil

In the airplane's normal range of flight attitudes, if the angle of attack is increased, the center of pressure moves forward; and if decreased, it moves rearward. Since the center of gravity is fixed at one point, it is evident that as the angle of attack increases, the center of lift (CP) moves ahead of the center of gravity, creating a force which tends to raise the nose of the airplane or tends to increase the angle of attack still more. On the other hand, if the angle of attack is decreased, the center of lift (CP) moves aft and tends to decrease the angle a greater amount. It

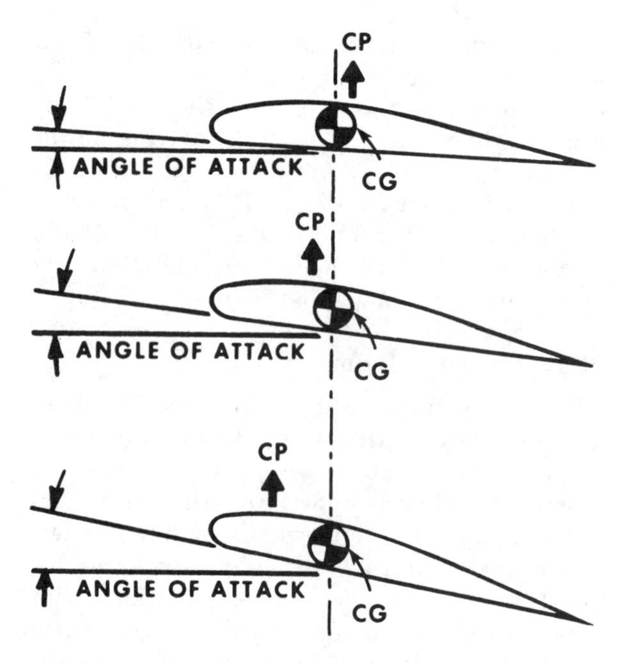

Figure 17–9 CP Changes with Angle of Attack

is seen then, that the ordinary airfoil is inherently unstable, and that an auxiliary device, such as the horizontal tail surface, must be added to make the airplane balance longitudinally. The balancing forces imparted by the tail surfaces will be discussed later in the chapter.

The balance of an airplane in flight depends, therefore, on the relative position of the center of gravity (CG) and the center of pressure (CP) of the airfoil. Experience has shown that an airplane with the center of gravity in the vicinity of 20 percent of the wing chord can be made to balance and fly satisfactorily.

The tapered wing presents a variety of wing chords throughout the span of the wing. It becomes necessary then, to specify some chord about which the point of balance can be expressed. This chord, known as the mean aerodynamic chord (MAC), usually is defined as the chord of an imaginary untapered wing which would have the same center of pressure characteristics as the wing in question.

Airplane loading and weight distribution also affect center of gravity and cause additional forces which in turn affect airplane balance. This will be discussed in a later section.

Forces on an Airfoil

Air through which the wing moves creates a force the components of which are referred to as "lift" and "drag" (Fig. 17-8). The resultant force, or "force vector," is resolved trigonometrically into these components of lift and drag, perpendicular and parallel, respectively, to the direction of the undisturbed "relative wind." The lift is the upward force which sustains the airplane in flight. The drag which retards the airplane's forward motion determines the thrust necessary to propel the wing through the air. It is evident, therefore, that the airplane designer is interested in obtaining an airfoil that will produce high lift and low drag over the airplane's flying range of "angles of attack," (the angle of attack being the acute angle between the chord of an airfoil and the direction of the relative wind). The airplane designer's choice of an airfoil is influenced primarily by structural considerations, and consequently it may not be possible to use an airfoil section that offers the greatest aerodynamic efficiency.

From aerodynamic theory and from wind-tunnel tests using smoke so that the actual airflow can be seen, it has been found that the direction of the airstream, after passing over a wing, is changed slightly. The air, as it leaves the trailing edge of the wing, has imparted to it a downward velocity; and this mass of air, therefore, has downward momentum (downwash). As we have learned from Newton's third law of motion that "for every action there is an equal and opposite reaction," an upward reaction of lift is imparted to the wing. Further, the air passing over the top of an airfoil is accelerated. This results in reduction of pressure on top of the wing. Bernoulli's principle, ". . . as velocity increases, the pressure is reduced . . . ," explains this additional factor in the development of a lifting force.

The drag of a wing is a rearward force which acts opposite to the direction of the airplane's forward motion and is made up of two components, the profile drag and the induced drag.

The profile drag is the resistance, or skin friction, due to the viscosity (stickiness) of the air as it passes along the surface of the wing, in combination with a form drag which is due to the eddying and turbulent wake of air left behind. Profile drag may be thought of as a pure resistance, such as is always encountered when an object is pulled through a viscous medium.

Parasite drag, which applies to the entire airplane, is composed of forces caused by protuberances extending into the airstream which do not contribute to lift. Items such as radio antennas, struts, fittings, landing gear, etc., produce parasite drag.

Induced drag is the direct result of the aerodynamic force resulting from the downward velocity imparted to the air as discussed previously. It may be said induced drag is the result of creating lift.

These forms of drag are more fully explained in later sections of this chapter. The lift and drag of a wing are forces that will vary with the density of the air, the area of the wing, the square of the air velocity, and the angle of attack of the wing as well as other parameters.

The manner in which the lift and drag will vary with the air density, wing area, and velocity factors is the same for any airfoil, but the variation of lift and drag with different angles of attack is a definite characteristic of each individual airfoil section. Graphs which show the variation of lift and drag with angle of attack are drawn for each airfoil design. (Fig. 17-10).

The fact that these graphs give the lift and drag in terms of coefficients should not be confusing. Coefficients are numbers indicating the amount of some change under certain specified conditions, often expressed as a ratio. These coefficients include the mathematical calculations that make it possible to multiply the coefficient by the density, area, and velocity factors corresponding to any particular condition to determine the lift and drag forces in pounds.

Relative Wind

Some pilots seem to find the term "relative wind" difficult to understand. Perhaps Figure 17-11 will help clarify it. The smoke from the factory moves because of natural wind. The

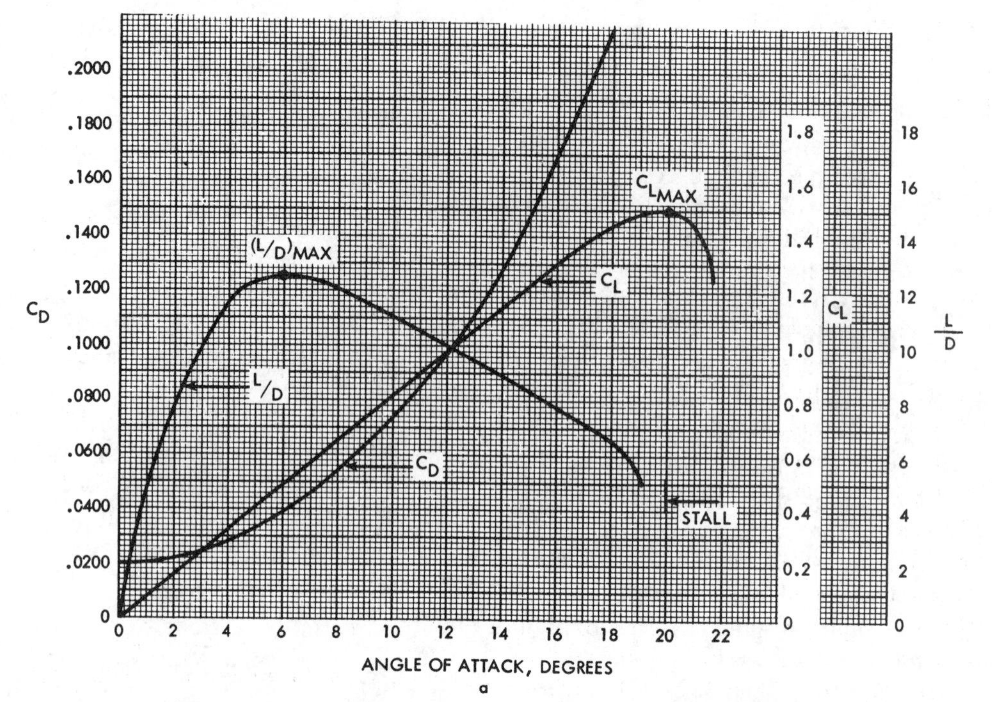

Figure 17–10 Lift Coefficients at Various Angles of Attack

pennant on the car traveling along the highway near the smoke stack flutters in a "relative" wind—that is, a flow of air which is created, not by the natural wind but by the auto's forward speed. As can be seen, the pennant on the car shows a "wind" whose direction is opposite to that of the smoke, though both are in the same natural wind. Relative wind is really not a wind at all in terms of our everyday understanding of the word. When dealing with principles of flight, it would no doubt be more accurate and descriptive to simply call it "the wind of flight," since it is the direction of airflow with respect to the wing as it moves through the air. If a wing is moving forward but settling downward, the relative wind is moving backward and upward.

Figure 17–11 Natural Wind VS Relative Wind

Relative wind can be created by the motion of a body through the air, the motion of air past a stationary body, or the combined motions of the air and the body.

For example, an airplane in flight creates relative wind by virtue of its motion. Likewise, an airplane parked on the ramp with a mass of air flowing over its surfaces is subject to relative wind. Also, on a takeoff roll, an airplane is subject to a relative wind which is the resultant of two motions—that of the aircraft along the ground and that of the moving mass of air. It is for this reason that birds and airplanes, when given a choice, take off directly into the wind—so that the relative wind flowing along the wings will be greatest and provide as much lift as possible.

When airborne, the actual flightpath of the airplane determines the direction of the relative wind. An understanding of the relative wind concept is fundamental to understanding "angle of attack."

Angle of Attack and Lift

Angle of attack must not be confused with an airplane's attitude in relation to the earth's surface, or with "angle of incidence" (the angle at which the wing is attached relative to the longitudinal axis of the airplane). Angle of attack is most frequently defined as the angle between the chord line of the wing, and the *relative wind.* Generally, it is sufficient to say that angle of attack is simply the angular difference between where the wing is headed and where it is actually going. As can be seen from Figure 17–12, this angle may be precisely the same for climbs, descents, and level flight, or can be quite different even when maintaining the same altitude.

The feature that complicates this problem is that with certain exceptions, we have no way of actually seeing the angle at which the wing meets the relative wind. Angle of attack indicators usually are found only in the turbojet powered airplanes.

In a very real sense the angle of attack is what flight in airplanes is all about. By changing the angle of attack the pilot can control lift, airspeed, and drag. Even the total load supported in flight by the wing, may be modified by variations in angle of attack, and when coordinated with power changes, and auxiliary devices such as flaps, slots, slats, etc., is the essence of airplane control.

The angle of attack of an airfoil directly controls the distribution of pressure below and above it. When a wing is at a *low* but positive angle of attack, most of the lift is due to the wing's negative pressure (upper surface) and downwash. (Negative pressure is any pressure less than atmospheric, and positive pressure is pressure greater than atmospheric.)

From Fig. 17–7 it can be seen that the positive pressure below the wing at a low angle of attack is very slight, and it can be noted also that the negative pressure above the wing is quite strong by comparison.

At any angle of attack, other than the angle at zero lift, all the forces acting on the wing as a result of the pressure distribution surrounding it may be summed up and represented as one force—the center of pressure.

When the angle of attack increases to approximately 18° to 20° (on most wings), the air can no longer flow smoothly over the top wing surface. Because the airflow cannot make such a great change in direction so quickly, it becomes impossible for the air to follow the contour of the wing. This is the *stalling or critical angle of attack*, and is often called the *burble point.* The burbling or turbulent flow of air which begins near the trailing edge of the wing, suddenly spreads forward over the entire upper wing surface. The negative pressure above the wing suddenly becomes almost equal to atmospheric pressure in value with a resulting loss of lift and a sudden increase in resistance or drag. These events show that Bernoulli's principle is true only in a streamline or smooth airflow—not in a turbulent airflow. The center of pressure at the point of stall is at its maximum forward

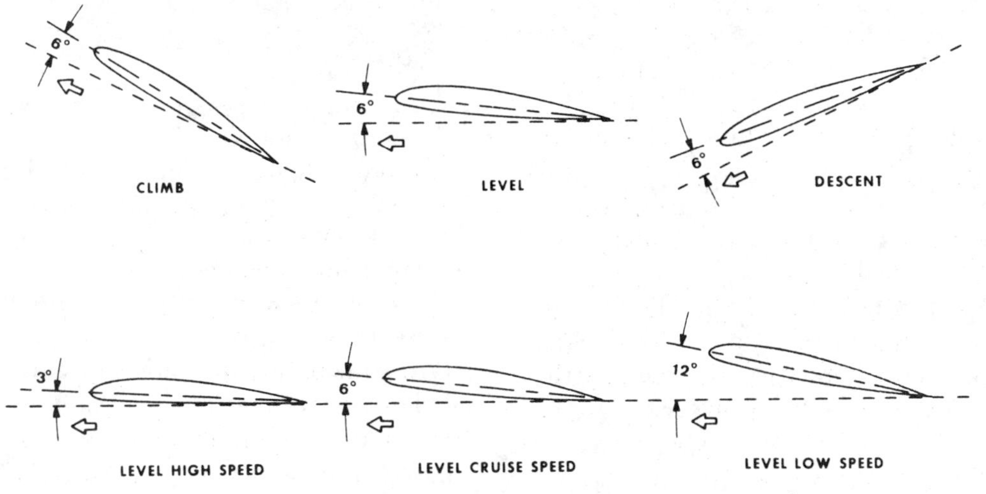

Figure 17–12 Angles of Attack VS Attitude and Speed

position, and the resultant force tilts sharply backward.

One of the most important things a pilot should understand about angle of attack is that for any given airplane the stalling or critical angle of attack remains constant regardless of weight, dynamic pressure, bank angle, or pitch attitude. These factors certainly will affect the *speed* at which the stall occurs, but not the *angle*. The aerodynamicist may say that the stalling angle of attack is not always an absolute constant, but for our purposes here it is a valid, useful, and safe concept.

Wing Planform

The previous discussions on wings have dealt only with airfoil *section* properties and two-dimensional airflow. Wing *planform*—the shape of the wing as viewed from directly above—deals with airflow in three dimensions, and is very important to understanding wing performance and airplane flight characteristics. *Aspect ratio*, *taper ratio*, and *sweepback* are factors in planform design that are very important to the overall aerodynamic characteristic of a wing (Fig. 17–13).

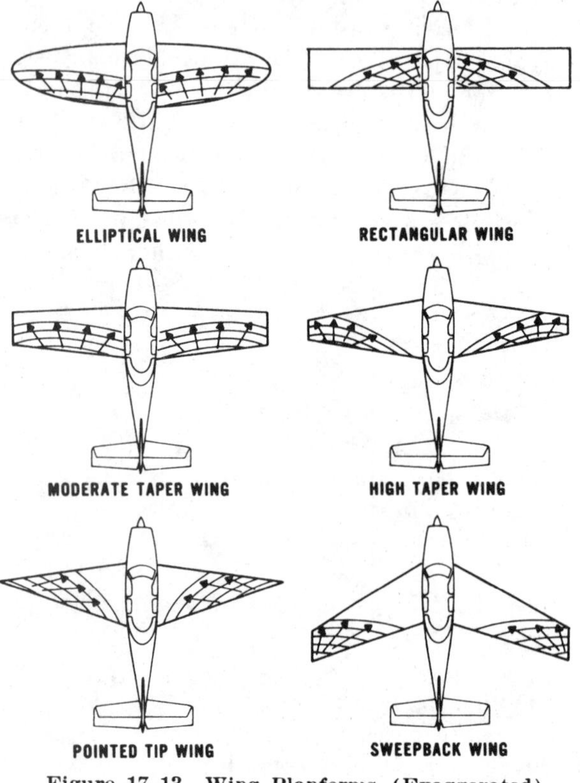

Figure 17–13 Wing Planforms (Exaggerated) and Stall Patterns

Aspect ratio is the ratio of wing span to wing chord.

Taper ratio can be either in planform or thickness, or both. In its simplest terms, it is a decrease from wing root to wingtip in wing chord or wing thickness.

Sweepback is the rearward slant of a wing, horizontal tail, or other airfoil surface.

There are two general means by which the designer can change the planform of a wing, either of which will affect aerodynamic characteristics of the wing. The first is to effect a change in the aspect ratio. Aspect ratio is the primary factor in determining the three-dimensional characteristics of the ordinary wing and its lift/drag ratio. An increase in aspect ratio with constant velocity will decrease the drag, especially at high angles of attack, improving the performance of the wing when in a climbing attitude. A decrease in aspect ratio will give a corresponding increase in drag. It should be noted, however, that with an increase in aspect ratio there is an increase in the length of span, with a corresponding increase in the weight of the wing structure, which means the wing must be heavier to carry the same load. For this reason, part of the gain (due to a decrease in drag) is lost because of the increased weight, and a compromise in design is necessary to obtain the best results from these two conflicting conditions. The second means of changing the planform is by "tapering" (decreasing the length of chord from the root to the tip of the wing). In general, tapering will cause a decrease in drag (most effective at high speeds) and an increase in lift. There is also a structural benefit due to a saving in weight of the wing.

Most training and general aviation type airplanes are operated at high lift coefficients, and therefore require comparatively high aspect ratios. Airplanes which are developed to operate at very high speeds demand greater aerodynamic cleanness, and greater strength—therefore low aspect ratios. Very low aspect ratios result in high wing loadings and high stall speeds. When sweepback is combined with low aspect ratio, it results in flying qualities very different from a more "conventional" high aspect ratio airplane configuration. Such airplanes require very precise and professional flying techniques, especially at slow speeds, while airplanes with a high aspect ratio are usually more forgiving of improper pilot techniques.

The elliptical wing is the ideal subsonic planform since it provides for a minimum of induced drag for a given aspect ratio, though as we shall see, its stall characteristics in some respects are inferior to the rectangular wing. It is also comparatively difficult to construct. The tapered airfoil is desirable from the standpoint of weight and stiffness, but again is not as efficient aerodynamically as the elliptical wing. In order to preserve the aerodynamic efficiency of the elliptical wing, rectangular and tapered wings are sometimes "tailored" through use of wing twist and variation in airfoil sections until they provide as nearly as possible the elliptical wing's lift distribution.

While it is true that the elliptical wing provides the best lift coefficients before reaching an incipient stall, it gives little advance warning of a complete stall, and lateral control may be difficult because of poor aileron effectiveness.

In comparison, the rectangular wing has a tendency to stall first at the wing root and provides adequate stall warning, adequate aileron effectiveness, and is usually quite stable. It is, therefore, favored in the design of low-cost, low-speed airplanes.

Stall progression patterns for various wing planforms are graphically depicted in Figure 17–13. Note that it is possible for the trailing edge of the inboard portion of the rectangular wing to be stalled while the rest of the wing is developing lift. This is a very desirable characteristic, and along with simplicity of construction is the reason why this type of wing is so popular in light airplanes, despite certain structural and aerodynamic inefficiencies.

Forces Acting on the Airplane

In some respects at least, how well a pilot performs in flight depends upon the ability to plan and coordinate the use of the power and flight controls for changing the forces of thrust, drag, lift, and weight. It is the balance between these forces which the pilot must always control. The better the understanding of the forces and means of controlling them, the greater will be the pilot's skill at doing so.

First, we should define these forces in relation to straight-and-level, unaccelerated flight.

Thrust is the forward force produced by the powerplant/propeller. It opposes or overcomes the force of drag. As a general rule it is said to act parallel to the longitudinal axis. However, this is not always the case as will be explained later.

Drag is a rearward, retarding force, and is caused by disruption of airflow by the wing, fuselage, and other protruding objects. Drag opposes thrust, and acts rearward parallel to the relative wind.

Weight is the combined load of the airplane itself, the crew, the fuel, and the cargo or baggage. Weight pulls the airplane downward because of the force of gravity. It opposes lift, and acts vertically downward through the airplane's center of gravity.

Lift opposes the downward force of weight, is produced by the dynamic effect of the air acting on the wing, and acts perpendicular to the flightpath through the wing's center of lift.

In steady flight the sum of these opposing forces is equal to zero. There can be no unbalanced forces in steady, straight flight (Newton's Third Law). This is true whether flying level or when climbing or descending. This is not the same thing as saying that the four forces are *all* equal. It simply means that the *opposing* forces are equal to, and thereby cancel the effects of, each other. Often the relationship between the four forces has been erroneously explained or illustrated in such a way that this point is obscured. Consider Figure 17–14, for example. In the upper illustration the force vectors of thrust, drag, lift, and weight appear to be equal in value. The usual explanation states (without stipulating that thrust and drag *do not* equal weight and lift) that thrust equals drag and lift equals weight as shown in the lower illustration. This basically true statement must be understood or it can be misleading. It should be understood that in straight, level, unaccelerated flight, it is true that the opposing lift/weight forces are equal but they are also *greater than* the opposing forces of thrust/drag that are equal only to each other; not to lift/weight. If we are to be quite correct about this matter, we must say that in steady flight

a. The sum of *all upward* forces (not just lift) equals the sum of *all downward* forces (not just weight).

b. The sum of *all forward* forces (not just thrust) equals the sum of *all backward* forces (not just drag).

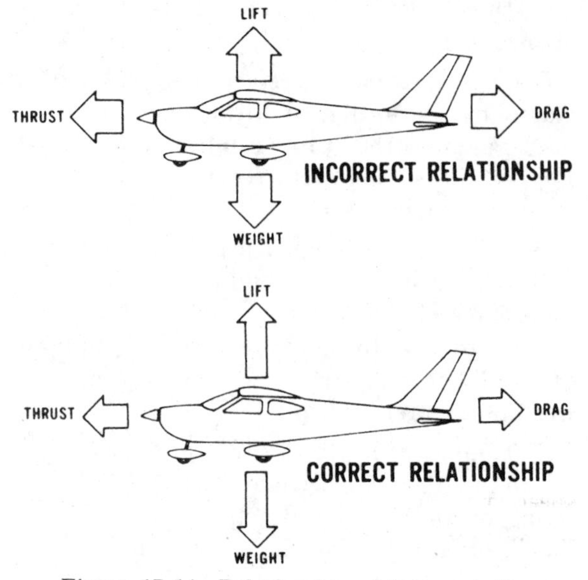

Figure 17–14 Relationship of Forces Acting on an Airplane

This refinement of the old "thrust equals drag; lift equals weight" formula takes into account the fact that in climbs a portion of thrust, since it is directed upward, acts as if it were lift, and a portion of weight since it is directed backward acts as if it were drag. In glides, a portion of the weight vector is directed forward, and therefore acts as thrust. In other words, any time the *flightpath* of the airplane is not horizontal, lift, weight, thrust, and drag vectors must each be broken down into two components (Fig. 17–15).

Discussions of the preceding concepts are frequently omitted in aeronautical texts/handbooks/manuals. The reason is not that they are of no consequence, but because by omitting such discussions, the main ideas with respect to the aerodynamic forces acting upon an airplane in flight can be presented in their most essential elements without being involved in the technicalities of the aerodynamicist. In point of fact, as long as we consider only level flight, and normal climbs and glides in a steady state, it is still true that wing lift is the really *important upward* force, and weight is the really *important downward* force.

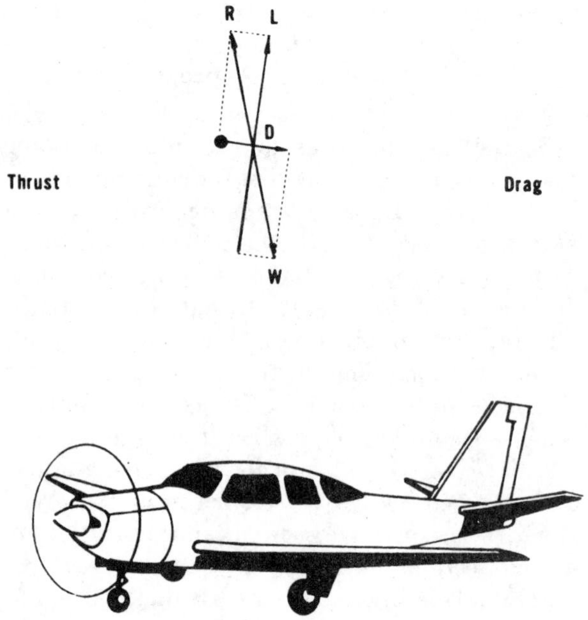

Figure 17–15 Force Vectors During Climb

Frequently much of the difficulty encountered in explaining the forces that act upon an airplane is largely a matter of language and its meaning. For example, pilots have long believed that an airplane climbs because of *excess* lift. This is not true if one is thinking in terms of wing lift alone. It is true, however, if by lift we mean the sum total of all "upward forces." But when we begin to talk about the "lift of thrust" or the "thrust of weight," the definitions we have previously established for these forces are no longer valid and we have simply further complicated matters. It is this impreciseness in language that affords the excuse to engage in arguments, largely academic, over refinements to basic principles.

Though we have already defined the forces acting on an airplane, we should discuss them in more detail to establish how the pilot uses them to produce controlled flight.

Thrust

Before the airplane begins to move, thrust must be exerted. It continues to move and gain speed until thrust and drag are equal. In order to maintain a *constant* airspeed, thrust and drag must remain equal, just as lift and weight must be equal to maintain a *constant* altitude. If in level flight, the engine power is reduced, the thrust is lessened and the airplane slows down. As long as the thrust is less than the drag, the airplane continues to decelerate until its airspeed is insufficient to support it in the air.

Likewise, if the engine power is increased, thrust becomes greater than drag and the airspeed increases. As long as the thrust continues to be greater than the drag, the airplane continues to accelerate. When drag equals thrust, the airplane flies at a constant airspeed.

Straight-and-level flight may be sustained at speeds from very slow to very fast. The pilot must coordinate angle of attack and thrust in all speed regimes if the airplane is to be held in level flight. Roughly, these regimes can be grouped in three categories; low-speed flight, cruising flight, and high-speed flight.

When the airspeed is low, the angle of attack must be relatively high to increase lift if the balance between lift and weight is to be maintained (Fig. 17–16). If thrust decreases and airspeed decreases, lift becomes less than weight and the airplane will start to descend. To maintain level flight, the pilot can increase the angle of attack an amount which will generate a lift force again equal to the weight of the airplane and while the airplane will be flying more slowly, it will still maintain level flight if the pilot has properly coordinated thrust and angle of attack.

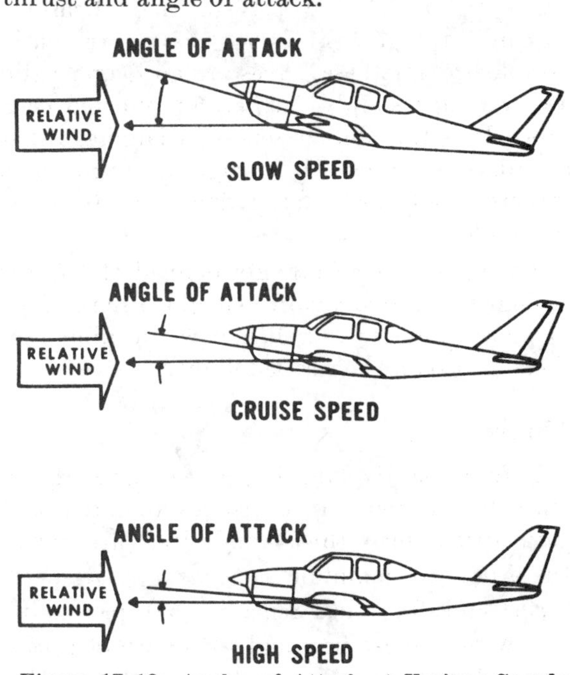

Figure 17–16 Angles of Attack at Various Speeds

Straight-and-level flight in the slow speed regime provides some interesting conditions relative to the equilibrium of forces, because with the airplane in a nose-high attitude there is a vertical component of thrust which helps support the airplane. For one thing, wing loading tends to be less than would be expected. Most pilots are aware that an airplane will stall, other conditions being equal, at a slower speed with the power on than with the power off. (Induced airflow over the wings from the propeller contributes to this, too.) However, if we restrict our analysis to the four forces as they are usually defined, one can say that in *straight-and-level slow speed flight* the thrust is equal to drag, and lift is equal to weight.

During straight-and-level flight when thrust is increased and the airspeed increases, the angle of attack must be decreased. That is, if changes have been coordinated, the airplane will still remain in level flight but at a higher speed when the proper relationship between thrust and angle of attack is established.

If the angle of attack were not coordinated (decreased) with this increase of thrust the airplane would climb. But decreasing the angle of attack modifies the lift, keeping it equal to the weight, and if properly done, the airplane still remains in level flight. Level flight at even slightly negative angles of attack is possible at very high speed. It is evident then, that level flight can be performed with any angle of attack between stalling angle and the relatively small negative angles found at high speed.

Drag

Drag in flight is of two basic types: parasite drag and induced drag. The first is called parasite because it in no way functions to aid flight while the second is induced or created as a result of the wing developing lift.

Parasite drag is composed of two basic elements: form drag, resulting from the disruption of the streamline flow; and the resistance of skin friction.

Of the two components of parasite drag, form drag is the easier to reduce when designing an airplane. In general, a more streamlined object produces the best form to reduce parasite drag.

Skin friction is the type of parasite drag that is most difficult to reduce. No surface is perfectly smooth. Even machined surfaces, when inspected through magnification, have a ragged, uneven appearance. This rough surface will deflect the streamlines of air on the surface, causing resistance to smooth airflow. Skin friction can be minimized by employing a glossy, flat finish to surfaces, and by eliminating protruding rivet heads, roughness, and other irregularities.

Another element must be added to the consideration of parasite drag when designing an airplane. This drag combines the effects of form drag and skin friction and is called interference drag. If we place two objects adjacent to one another the resulting turbulence produced may be 50 to 200 percent greater than the parts tested separately.

The three elements, form drag, skin friction, and interference drag, are all computed to determine parasite drag on an airplane.

Shape of an object is a big factor in parasite drag. However, indicated airspeed is an equally important factor when speaking of parasite drag. The profile drag of a streamlined object held in a fixed position relative to the airflow increases approximately as the square of the velocity; thus, doubling the airspeed increases the drag four times, and tripling the airspeed increases the drag nine times. This relationship, however, holds good only at comparatively low subsonic speeds. At some higher airspeeds the rate at which profile drag has been increased with speed suddenly begins to increase more rapidly.

The second basic type of drag is *induced drag*. It is an established physical fact that no system which does work in the mechanical sense can be 100% efficient. This means that whatever the nature of the system, the required work is obtained at the expense of certain additional work that is dissipated or lost in the system. The more efficient the system the smaller this loss.

In level flight the aerodynamic properties of the wing produce a required lift, but this can be obtained only at the expense of a certain penalty. The name given to this penalty is *induced drag*. Induced drag is inherent whenever a wing is producing lift and, in fact, this type of drag is inseparable from the production of lift. Consequently, it is always present if lift is produced.

The wing produces the lift force by making use of the energy of the free airstream. Whenever the wing is producing lift we have seen that the pressure on the lower surface of the wing is greater than that on the upper surface. As a result, the air tends to flow from the high pressure area below the wingtip upward to the low pressure area above the wing. In the vicinity of the wingtips there is a tendency for these pressures to equalize, resulting in a lateral flow outward from the underside to the upper surface of the wing. This lateral flow imparts a rotational velocity to the air at the wingtips and trails behind the wing. Therefore, flow about the wingtips will be in the form of two vortices trailing behind as the wings moves on.

When the airplane is viewed from the tail, these vortices will circulate counterclockwise about the right wingtip and clockwise about the left wingtip (Fig. 17–17). Bearing in mind the direction of rotation of these vortices, it can be seen that they induce an upward flow of air beyond the wingtip, and a downwash

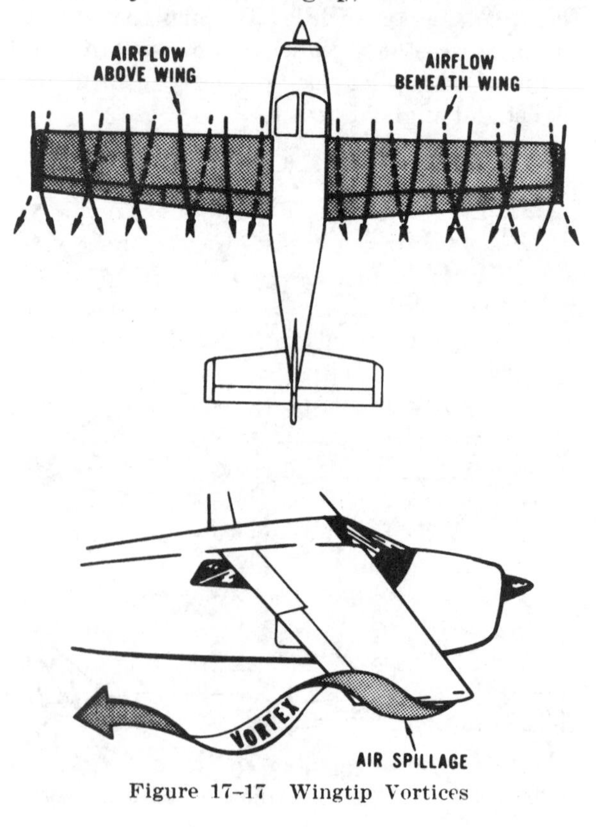

Figure 17–17 Wingtip Vortices

flow behind the wing's trailing edge. This induced downwash has nothing in common with the downwash that is necessary to produce lift. It is, in fact, the source of induced drag. The greater the size and strength of the vortices and consequent downwash component on the net airflow over the wing, the greater the induced drag effect becomes.

This downwash over the top of the wing at the tip has the same effect as bending the lift vector rearward; therefore, the lift is slightly aft of perpendicular to the relative wind, creating a rearward lift component. This is induced drag.

It should be remembered also that in order to create a greater negative pressure on the top of the wing, the wing can be inclined to a higher angle of attack; also, that if the angle of attack of an asymmetrical wing were zero, there would be no pressure differential and consequently no downwash component; therefore, no induced drag. In any case, as angle of attack increases, induced drag increases proportionally.

To state this another way—the lower the airspeed the greater the angle of attack required to produce lift equal to the airplane's weight and consequently, the greater will be the induced drag. The amount of induced drag varies inversely as the square of the airspeed.

From the foregoing discussion we have seen that parasite drag increases as the square of the airspeed, and induced drag varies inversely as the square of the airspeed. It can be seen that as airspeed decreases to near the stalling speed the total drag becomes greater, due mainly to the sharp rise in induced drag. Similarly, as the airspeed reaches the terminal velocity of the airplane, the total drag again increases rapidly, due to the sharp increase of parasite drag. As we can see in Figure 17–18 at some given airspeed, total drag is at its maximum amount. This is very important in figuring the maximum endurance and range of airplanes. For when drag is at a minimum, power required to overcome drag is also at a minimum. The drag factors will play an important part in a later section on airplane performance.

To understand the effect of lift and drag on an airplane in flight we must combine them

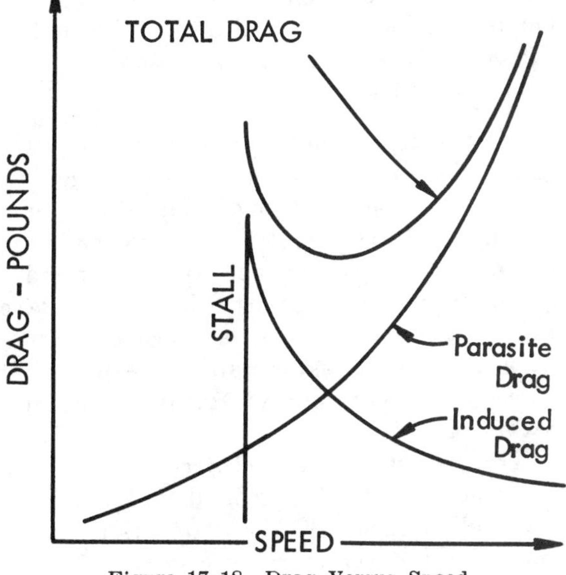

Figure 17–18 Drag Versus Speed

and consider the lift-drag ratio. With the lift and drag data available for various airspeeds of the airplane in steady, unaccelerated flight, the proportions of C_L (Coefficient of Lift) and C_D (Coefficient of Drag) can be calculated for each specific angle of attack. The resulting plot for lift-drag ratio with angle of attack shows that L/D increases to some maximum then decreases at the higher lift coefficients and angles of attack, as shown in Fig. 17–10. Note that the maximum lift-drag ratio, (L/D max) occurs at one specific angle of attack and lift coefficient. If the airplane is operated in steady flight at L/D max, the total drag is at a minimum. Any angle of attack lower or higher than that for L/D max reduces the lift-drag ratio and consequently increases the total drag for a given airplane's lift.

The location of the center of gravity (CG) is determined by the general design of each particular airplane. The designers determine how far the center of pressure (CP) will travel. They then fix the center of gravity forward of the center of pressure for the corresponding flight speed in order to provide an adequate restoring moment to retain flight equilibrium.

The configuration of an airplane has a great effect on the lift-drag ratio. The high performance sailplane may have extremely high lift-drag ratios. The supersonic fighter may have seemingly low lift-drag ratios in subsonic flight, but the airplane configurations required for supersonic flight (and high L/D's at high Mach numbers) cause this situation.

Weight

Gravity is the pulling force that tends to draw all bodies to the center of the earth. The center of gravity may be considered as a point at which all the weight of the airplane is concentrated. If the airplane were supported at its exact center of gravity, it would balance in any attitude. It will be noted that center of gravity is of major importance in an airplane, for its position has a great bearing upon stability.

The location of the center gravity (CG) is determined by the general design of each particular airplane. The designers determine how far the center of pressure (CP) will travel. They then fix the center of gravity forward of the center of pressure for the corresponding flight speed in order to provide an adequate restoring moment to retain flight equilibrium.

Weight has a definite relationship with lift, and thrust with drag. This relationship is simple, but important in understanding the aerodynamics of flying. As stated previously, lift is the upward force on the wing acting perpendicular to the relative wind. Lift is required to counteract the airplane's weight (which is caused by the force of gravity acting on the mass of the airplane). This weight (gravity) force acts downward through the airplane's center of gravity. In stabilized level flight, when the lift force is equal to the weight force, the airplane is in a state of equilibrium and neither gains nor loses altitude. If lift becomes less than weight, the airplane loses altitude. When the lift is greater than weight, the airplane gains altitude.

Lift

The pilot can control the lift. Any time the control wheel is more fore or aft, the angle of attack is changed. As angle of attack increases, lift increases (all other factors being equal). When the airplane reaches the maximum angle of attack, lift begins to diminish rapidly. This is the stalling angle of attack, or burble point.

Before proceeding further with lift and how it can be controlled, we must now interject velocity. The shape of the wing cannot be effective unless it continually keeps "attacking" new air. If an airplane is to keep flying, *it must keep moving.* Lift is proportional to the square of the airplane's velocity. For example, an airplane traveling at 200 knots has four times the lift as the same airplane traveling at 100 knots, *if* the angle of attack and other factors remain constant.

Actually, the airplane could not continue to travel in level flight at a constant altitude and maintain the same angle of attack if the velocity is increased. The lift would increase and the airplane would climb as a result of the increased lift force. Therefore, to maintain the lift and weight forces in balance, and to keep the airplane "straight and level" (not accelerating upward) in a state of equilibrium, as we increase velocity we must decrease lift. This is normally accomplished by reducing the angle of attack; i.e., lowering the nose. Conversely, as we slow the airplane, the decreasing velocity requires increasing the angle of attack to maintain lift sufficient to maintain flight. There is, of course, a limit to how far we can go in this direction, if a stall is to be avoided.

Therefore, it may be concluded that for every angle of attack there is a corresponding indicated airspeed required to maintain altitude in steady, unaccelerated flight—all other factors being constant. (Bear in mind this is only true if we are maintaining "level flight.") Since an airfoil will always stall at the same angle of attack, if we increase weight we must also increase lift, and our only method for doing so is by increased velocity if our angle of attack is held constant just short of the "critical" or stalling angle of attack.

Lift and drag also vary directly with the density of the air. As previously discussed, density is affected by several factors: pressure, temperature, and humidity. Remember, at an altitude of 18,000 feet the density of the air has one-half the density of air at sea level. Therefore, in order to maintain its lift at a higher altitude an airplane must fly at a greater true airspeed for any given angle of attack.

Furthermore, warm air is less dense than cool air, and moist air is less dense than dry air. Thus, on a hot humid day, an airplane must be flown at a greater true airspeed for any given angle of attack than on a cool, dry day.

If the density factor is decreased and the total lift must equal the total weight to remain in flight, it follows that one of the other factors must be increased. The factors usually increased are the airspeed or the angle of attack, because these factors can be controlled directly by the pilot.

It should also be pointed out that lift varies directly with the wing *area*, provided there is no change in the wing's *planform*. If the wings have the same proportion and airfoil sections, a wing with a planform area of 200 square feet lifts twice as much at the same angle of attack as a wing with an area of 100 square feet.

As can be seen, two major factors from the pilot's viewpoint are lift and velocity because these are the two that can be controlled most readily and accurately. Of course, the pilot can also control density by adjusting the altitude and can control wing area if the airplane happens to have flaps of the type that enlarge wing area. However, for most situations, the pilot is controlling lift and velocity to maneuver the airplane. For instance, in straight-and-level flight, cruising along at a constant altitude, altitude is maintained by adjusting lift to match the airplane's velocity or cruise airspeed, while maintaining a state of equilibrium where lift equals weight. In an approach to landing, when the pilot wishes to land as slowly as practical, it is necessary to increase lift to near maximum to maintain lift equal to the weight of the airplane.

Wingtip Vortices

As we have already learned the action of the airfoil that gives an airplane lift also causes induced drag. The explanation of this was discussed in preceding sections of this chapter. It was determined that when a wing is flown at a positive angle of attack, a pressure differential exists between the upper and lower surfaces of the wing; that is—the pressure above the wing is less than atmospheric pressure and the pressure below the wing is equal to or greater than atmospheric pressure. Since air always moves from high pressure toward low pressure, and the path of least resistance is toward the airplane's wingtips, there is a spanwise movement of air from the bottom of the wing outward from the fuselage and upward around the wingtips. This flow of air results in "spillage" over the wingtips, thereby setting up a whirlpool of air called a "vortex" (Fig. 17–17). At the same time the air on the upper surface of the wing has a tendency to flow in toward the fuselage and off the trailing edge. This air current forms a similar vortex at the inboard portion of the trailing edge of the wing, but because the fuselage limits the inward flow, the vortex is insignificant. Consequently, the deviation in flow direction is greatest at the wingtips where the unrestricted lateral flow is the strongest. As the air curls upward around the wingtip it combines with the wing's downwash to form a fast-spinning trailing vortex. These vortices increase drag because of energy spent in producing the turbulence. It can be seen, then, that whenever the wing is producing lift, induced drag occurs, and wingtip vortices are created.

Just as lift increases with an increase in angle of attack, induced drag also increases. This occurs because as the angle of attack is increased, there is a greater pressure difference between the top and bottom of the wing, and a greater lateral flow of air; consequently, this causes more violent vortices to be set up, resulting in more turbulence and more induced drag.

The intensity or strength of the wingtip vortices is directly proportional to the weight of the airplane and inversely proportional to the wingspan and speed of the airplane. The heavier and slower the airplane, the greater the angle of attack and the stronger the wingtip vortices. Thus, an airplane will create wingtip vortices with maximum strength occurring during the takeoff, climb, and landing phases of flight.

Ground Effect

It is possible to fly an airplane just clear of the ground (or water) at a slightly slower airspeed than that required to sustain level flight at higher altitudes. This is the result of a phenomenon which is better known than understood even by some experienced pilots.

When an airplane in flight gets within several feet from the ground surface, a change occurs in the three dimensional flow pattern around the airplane because the vertical component of the airflow around the wing is restricted by the ground surface. This alters the wing's upwash, downwash, and wingtip vortices (Fig. 17-19). These general effects due to the presence of the ground are referred to as "ground effect." Ground effect, then, is due to the interference of the ground (or water) surface with the airflow patterns about the airplane in flight.

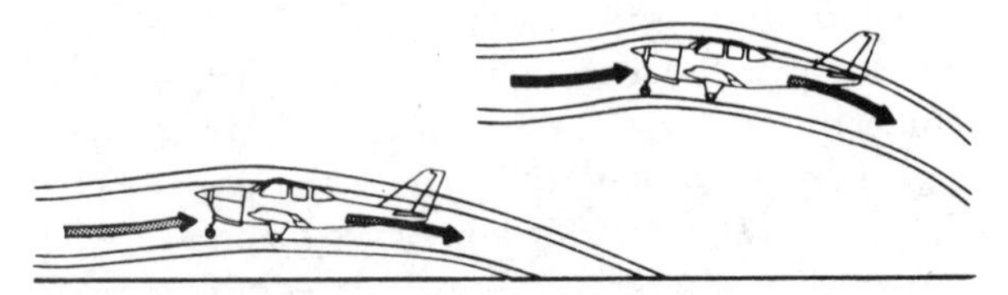

Figure 17-19 Ground Effect Changes Airflow

While the aerodynamic characteristics of the tail surfaces and the fuselage are altered by ground effects, the principal effects due to proximity of the ground are the changes in the aerodynamic characteristics of the wing. As the wing encounters ground effect and is maintained at a constant lift coefficient, there is consequent reduction in the upwash, downwash, and the wingtip vortices.

We have learned earlier that induced drag is a result of the wing's work of sustaining the airplane and that the wing lifts the airplane simply by accelerating a mass of air downward. It is true that reduced pressure on top of an airfoil is essential to lift, but that is but one of the things that contributes to the overall effect of pushing an air mass downward. The more downwash there is, the harder the wing is pushing the mass of air down. At high angles of attack, the amount of induced drag is high and since this corresponds to lower airspeeds in actual flight, it can be said that induced drag predominates at low speed.

However, the reduction of the wingtip vortices due to ground effect alters the spanwise lift distribution and reduces the induced angle of attack and induced drag. Therefore, the wing will require a lower angle of attack in ground effect to produce the same lift coefficient or, if a constant angle of attack is maintained, an increase in lift coefficient will result (Fig. 17-20).

Ground effect also will alter the thrust required versus velocity. Since induced drag predominates at low speeds, the reduction of induced drag due to ground effect will cause the most significant reduction of thrust required (parasite plus induced drag) at low speeds.

The reduction in induced flow due to ground effect causes a significant reduction in *induced drag* but causes no direct effect on parasite drag. As a result of the reduction in induced drag, the thrust required at low speeds will be reduced.

Due to the change in upwash, downwash, and wingtip vortices, there may be a change in position (installation) error of the airspeed system, associated with ground effect. In the majority of cases, ground effect will cause an increase in the local pressure at the static source and produce a lower indication of airspeed and altitude. Thus, the airplane may be airborne at an indicated airspeed less than that normally required.

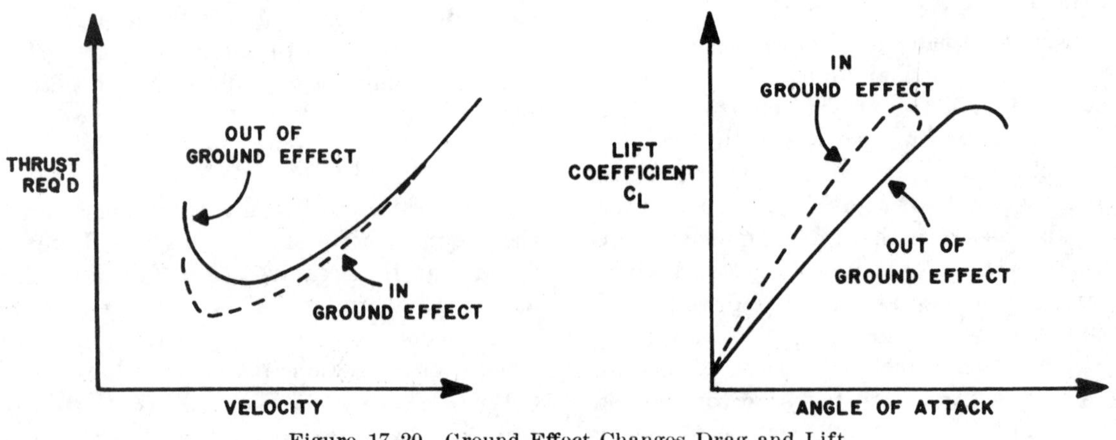

Figure 17-20 Ground Effect Changes Drag and Lift

In order for ground effect to be of significant magnitude, the wing must be quite close to the ground. As we have found, one of the direct results of ground effect is the variation of induced drag with wing height above the ground at an constant lift coefficient. When the wing is at a height equal to its span, the reduction in induced drag is only 1.4 percent. However, when the wing is at a height equal to one-fourth its span, the reduction in induced drag is 23.5 percent and, when the wing is at a height equal to one-tenth its span, the reduction in induced drag is 47.6 percent. Thus, a large reduction in induced drag will take place only when the wing is very close to the ground. Because of this variation, ground effect is most usually recognized during the liftoff for takeoff or just prior to touchdown when landing.

During the takeoff phase of flight, ground effect produces some important relationships. The airplane leaving ground effect after takeoff encounters just the reverse of the airplane entering ground effect during landing; i.e., the airplane *leaving* ground effect will (1) require an increase in angle of attack to maintain the same lift coefficient, (2) experience an increase in induced drag and thrust required, (3) experienced a decrease in stability and a nose-up change in moment, (4) produce a reduction in static source pressure and increase in indicated airspeed. These general effects should point out the possible danger in attempting takeoff prior to achieving the recommended takeoff speed. Due to the reduced drag in ground effect the airplane may seem capable of takeoff well below the recommended speed. However, as the airplane rises out of ground effect with a deficiency of speed, the greater induced drag may result very marginal initial climb performance. In the extreme conditions such as high gross weight, high density altitude, and high temperature, a deficiency of airspeed during takeoff may permit the airplane to become airborne but be incapable of flying out of ground effect. In this case, the airplane may become airborne initially with a deficiency of speed, and then settle back to the runway. It is important that no attempt be made to force the airplane to become airborne with a deficiency of speed; the recommended takeoff speed is necessary to provide adequate initial climb performance. For this reason, it is imperative that a definite climb be established *before* retracting the landing gear or flaps.

During the landing phase of flight, the effect of proximity to the ground also must be understood and appreciated. If the airplane is brought into ground effect with a constant angle of attack, the airplane will experience an increase in lift coefficient and a reduction in the thrust required. Hence, a "floating" effect may occur. Because of the reduced drag and power-off deceleration in ground effect, any excess speed at the point of flare may incur a considerable "float" distance. As the airplane nears the point of touchdown, ground effect will be most realized at altitudes less than the wingspan. During the final phases of the approach as the airplane nears the ground, a reduced power setting is necessary or the reduced thrust required would allow the airplane to climb above the desired glidepath.

Axes of an Airplane

Whenever an airplane changes its flight attitude or position in flight, it rotates about one or more of three axes, which are imaginary lines that pass through the airplane's center of gravity. The axes of an airplane can be considered as imaginary axles around which the airplane turns, much like the axle around which a wheel rotates. At the point where all three axes intersect, each is at a 90° angle to the other two. The axis which extends lengthwise through the fuselage from the nose to the tail is called the longitudinal axis. The axis which extends crosswise, from wingtip to wingtip, is the lateral axis. The axis which passes vertically through the center of gravity, is called the vertical axis (Fig. 17–21).

The airplane's motion about its longitudinal axis resembles the roll of a ship from side to side. In fact, the names used in describing the motion about an airplane's three axes were originally nautical terms. They have been adapted to aeronautical terminology because of the similarity of motion between an airplane and the seagoing ship.

In light of the adoption of nautical terms, the motion about the airplane's longitudinal axis is called "roll"; motion along its lateral

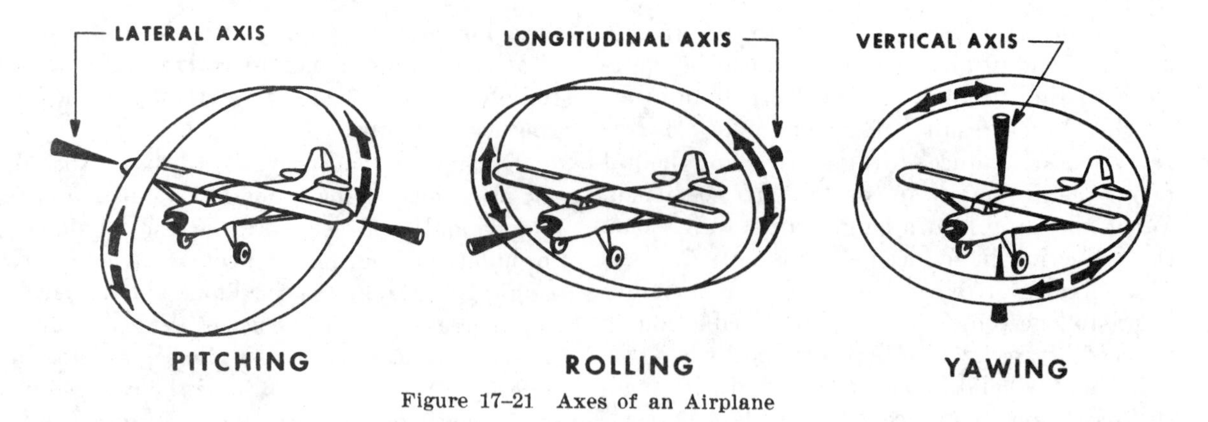

Figure 17–21 Axes of an Airplane

axis is referred to as "pitch." Finally, an airplane moves about its vertical axis in a motion which is termed "yaw"—that is, a horizontal (left and right) movement of the airplane's nose.

The three motions of the airplane—roll, pitch, and yaw—are controlled by three control surfaces. Roll is controlled by the ailerons; pitch is controlled by the elevators; yaw is controlled by the rudder. The use of these controls was explained in the chapter on Use and Effect of Controls.

Moments and Moment Arm

A study of physics shows that a body that is free to rotate will always turn about its center of gravity. In aerodynamic terms, the mathematical measure of an airplane's tendency to rotate about its center of gravity is called a "moment." A moment is said to be equal to the product of the force applied and the distance at which the force is applied. (A moment arm is the distance from a datum [reference point or line] to the applied force.) For airplane weight and balance computations, "moments" are expressed in terms of the distance of the arm times the airplane's weight, or simply, *inch pounds*.

As previously mentioned, airplane designers locate the fore and aft position of the airplane's center of gravity as nearly as possible to the 20 percent point of the mean aerodynamic chord (MAC). If the thrust line is designed to pass horizontally through the center of gravity, it will not cause the airplane to pitch when power is changed, and there will be no difference in moment due to thrust for a power-on or power-off condition of flight. Although designers have some control over the location of the drag forces, they are not always able to make the resultant drag forces pass through the center of gravity of the airplane. However, the one item over which they have the greatest control is the size and location of the tail. The objective is to make the moments (due to thrust, drag, and lift) as small as possible; and, by proper location of the tail, to provide the means of balancing the airplane longitudinally for any condition of flight.

The pilot has no direct control over the location of forces acting on the airplane in flight, except for controlling the center of lift by changing the angle of attack. Such a change however immediately involves changes in other forces. Therefore, the pilot cannot independently change the location of one force without changing the effect of others. For example, a change in airspeed involves a change in lift, as well as a change in drag and a change in the up or down force on the tail. As forces such as turbulence, gusts, etc., act to displace the airplane, the pilot reacts by providing opposing control forces to counteract this displacement.

Some airplanes are subject to changes in the location of the center of gravity with variations of load. Trimming devices are used to counteract the forces set up by fuel burnoff, and loading or off-loading of passengers, cargo, etc. Elevator trim tabs and adjustable horizontal stabilizers comprise the most common devices provided to the pilot for trimming for load variations. Over the wide ranges of balance during flight in large airplanes, the force which the pilot has to exert on the controls would become excessive and fatiguing if means of trimming were not provided.

Design Characteristics

Every pilot who has flown numerous types of airplanes has noted that each airplane handles somewhat differently—that is, each resists or responds to control pressures in its own way. A training type airplane is quick to respond to control applications, while a transport airplane usually feels heavy on the controls and responds to control pressures more slowly. These features can be designed into an airplane to facilitate the particular purpose the airplane is to fulfill by considering certain stability and maneuvering requirements. In the following discussion it is intended to summarize the more important aspects of an airplane's stability; its maneuvering and controllability qualities; how they are analyzed; and their relationship to various flight conditions. In brief, the basic differences between stability, maneuverability, and controllability are as follows:

1. *Stability.* The inherent quality of an airplane to correct for conditions that may disturb its equilibrium, and to return or to continue on the original flightpath. It is primarily an airplane design characteristic.
2. *Maneuverability.* The quality of an airplane that permits it to be maneuvered easily and to withstand the stresses imposed by maneuvers. It is governed by the airplane's weight, inertia, size and location of flight controls, structural strength, and powerplant. It too is an airplane design characteristic.
3. *Controllability.* The capability of an airplane to respond to the pilot's control, especially with regard to flightpath and attitude. It is the quality of the airplane's response to the pilot's control application when maneuvering the airplane, regardless of its stability characteristics.

Basic Concepts of Stability

The flightpaths and attitudes in which an airplane can fly are limited only by the aerodynamic characteristics of the airplane, its propulsive system, and its structural strength. These limitations indicate the maximum performance and maneuverability of the airplane. If the airplane is to provide maximum utility, it must be safely controllable to the full extent of these limits without exceeding the pilot's strength or requiring exceptional flying ability. If an airplane is to fly straight and steady along any arbitrary flightpath, the forces acting on it must be in static equilibrium. The reaction of any body when its equilibrium is disturbed is referred to as stability. There are two types of stability; static and dynamic. We will first consider the static, and in this discussion the following definitions will apply:

1. *Equilibrium.* All opposing forces acting on the airplane are balanced; (i.e., steady unaccelerated flight conditions).
2. *Static Stability.* The initial tendency that the airplane displays after its equilibrium is disturbed.
3. *Positive Static Stability.* The initial tendency of the airplane to return to the original state of equilibrium after being disturbed (Fig. 17–22).
4. *Negative Static Stability.* The initial tendency of the airplane to continue away from the original state of equilibrium after being disturbed (Fig. 17–22).
5. *Neutral Static Stability.* The initial tendency of the airplane to remain in a new condition after its equilibrium has been disturbed (Fig. 17–22).

Static Stability

Stability of an airplane in flight is slightly more complex than just explained, because the airplane is free to move in any direction and must be controllable in pitch, roll, and direction. When designing the airplane, engineers must compromise between stability, maneuverability, and controllability; and the problem is compounded because of the airplane's three-axis freedom.

By comparing this three-axis freedom with the limited flexibility of an automobile, the problem may be better understood. Pitch control of a car is limited to a shock absorbing and leveling problem while roll control is a matter of the lateral spacing of the wheels and the banking of highways around curves. Essentially, then, the car has only one degree of real freedom—directional; in many cases,

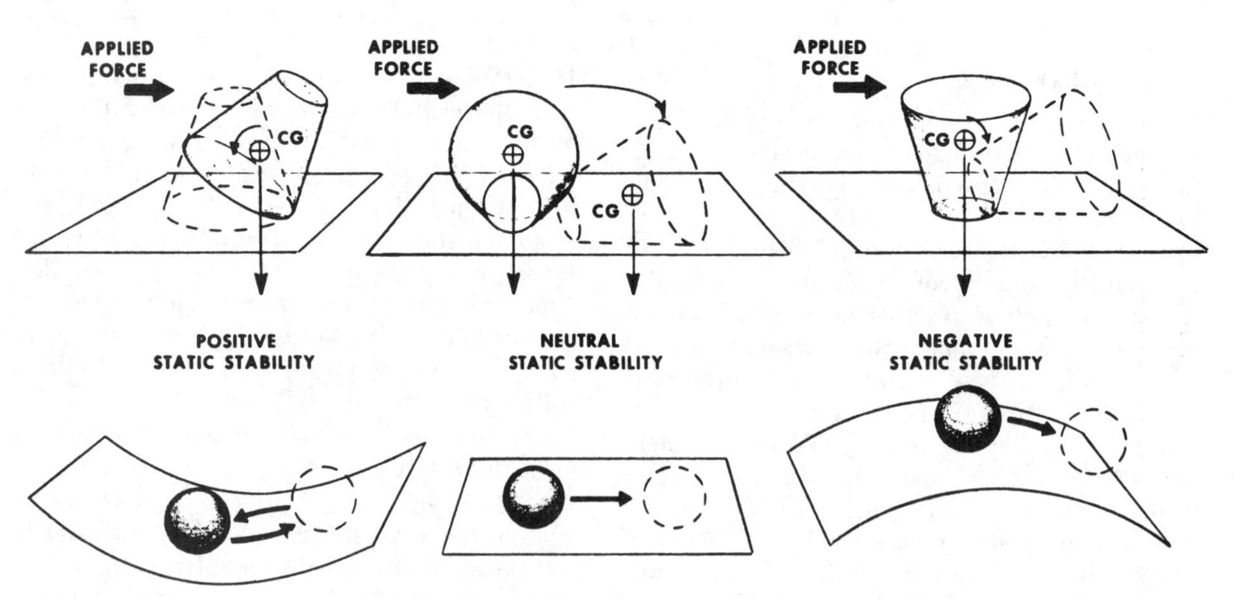

Figure 17–22 Types of Stability

though, the lack of pitch and bank control have caused it to lose directional control. Automobile designers provide "centering" in the steering mechanism or "toe-in" on the wheels to provide directional stability so the car will maintain a straight path when the steering wheel is released. Too much centering or "toe-in" is, of course, objectionable because of the excessive effort required of the driver to turn the car. When power steering was introduced to the driving public, many considered it objectionable because of insufficient centering or "feel." Even now, most automobile models are offered either with or without power steering to satisfy the public.

The conclusions, then, are that too much stability is detrimental to maneuverability, and similarly, not enough stability is detrimental to controllability. In the design of airplanes, *compromise* between the two is the keyword.

Dynamic Stability

We have defined static stability as the *initial* tendency that the airplane displays after being disturbed from its trimmed condition. Occasionally, the initial tendency is different or opposite from the overall tendency, so we must distinguish between the two. Dynamic stability is the *overall* tendency that the airplane displays after its equilibrium is disturbed. The curves of Fig. 17–23 represent the variation of controlled functions versus time. It is seen that the unit of time is very significant. If the time unit for one cycle or oscillation is above 10 seconds' duration, it is called a "long-period" oscillation (phugoid) and is easily controlled. In a longitudinal phugoid oscillation, the angle of attack remains constant when the airspeed increases and decreases. To a certain degree a convergent phugoid is desirable but is not required. The phugoid can be determined only on a statically stable airplane, and this has a great effect on the trimming qualities of the airplane. If the time unit for one cycle or oscillation is less than one or two seconds, it is called a "short-period" oscillation and is normally very difficult, if not impossible, for the pilot to control. This is the type of oscillation that the pilot can easily "get in phase with" and reinforce.

A neutral or divergent, short-period oscillation is dangerous because structural failure usually results if the oscillation is not damped immediately. Short-period oscillations affect airplane and control surfaces alike and reveal themselves as "porpoising" in the airplane, or as in "buzz" or "flutter" in the control surfaces. Basically, the short-period oscillation is a change in angle of attack with no change in airspeed. A short-period oscillation of a control surface is usually of such high frequency that the airplane does not have time to react. Logically, Federal Aviation Regulations require that short-period oscillations be heavily damped (i.e., die out immediately).

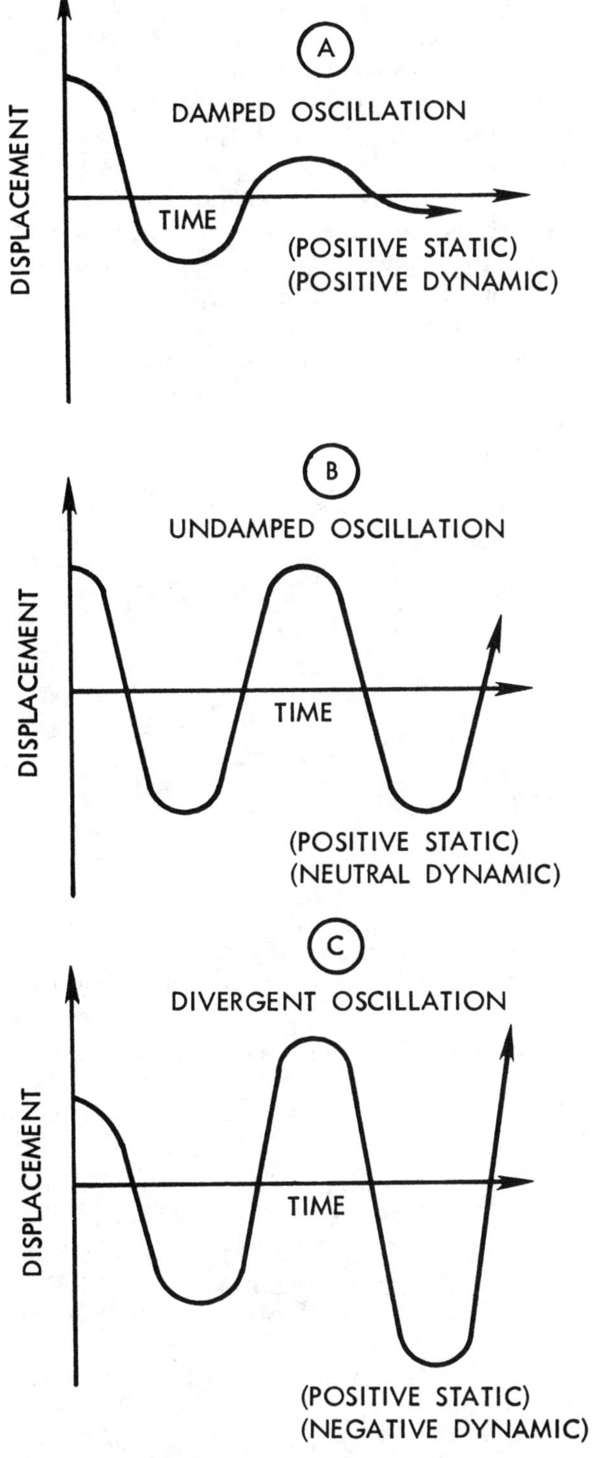

Figure 17–23 Damped Versus Undamped Stability

Flight tests during the airworthiness certification of airplanes are conducted for this condition by inducing the oscillation in the controls for pitch roll, or yaw at the most critical speed (i.e., at V_{ne}, the never-exceed speed). The test pilot strikes the control wheel or rudder pedal a sharp blow and observes the results.

Longitudinal Stability (Pitching)

In designing an airplane a great deal of effort is spent in developing the desired degree of stability around all three axes. But longitudinal stability about the lateral axis is considered to be the most affected by certain variables in various flight conditions.

As we learned earlier, longitudinal stability is the quality which makes an airplane stable about its lateral axis. It involves the pitching motion as the airplane's nose moves up and down in flight. A longitudinally *unstable* airplane has a tendency to dive or climb progressively into a very steep dive or climb, or even a stall. Thus, an airplane with longitudinal instability becomes difficult and sometimes dangerous to fly.

Static longitudinal stability or instability in an airplane, is dependent upon three factors:

1. Location of the wing with respect to the center of gravity;
2. Location of the horizontal tail surfaces with respect to the center of gravity; and
3. The area or size of the tail surfaces.

In analyzing stability it should be recalled that a body that is free to rotate will always turn about its center of gravity.

To obtain static longitudinal stability, the relation of the wing and tail moments must be such that, if the moments are initially balanced and the airplane is suddenly nosed up, the wing moments and tail moments will change so that the sum of their forces will provide an unbalanced but restoring moment which in turn, will bring the nose down again. Similarly, if the airplane is nosed down, the resulting change in moments will bring the nose back up.

We have spoken of the airplane's center of gravity and the airfoil's center of lift in preceding sections. Now let us reexamine the center of lift or as it is sometimes called, the *center of pressure.*

As previously pointed out, the center of pressure in most unsymmetrical airfoils has a tendency to change its fore and aft position with a change in the angle of attack. The center of pressure tends to move *forward* with an increase in angle of attack and to move aft with a decrease in angle of attack. This means that when the angle of attack of an airfoil is increased, the center of pressure (lift) by moving forward, tends to lift the leading edge of the wing still more. This tendency gives the wing an inherent quality of *instability*.

Figure 17–24, shows an airplane in straight-and-level flight. The line CG–CL–T represents the airplane's longitudinal axis from the center of gravity (CG) to a point T on the horizontal stabilizer. The center of lift (or center of pressure) is represented by the point CL.

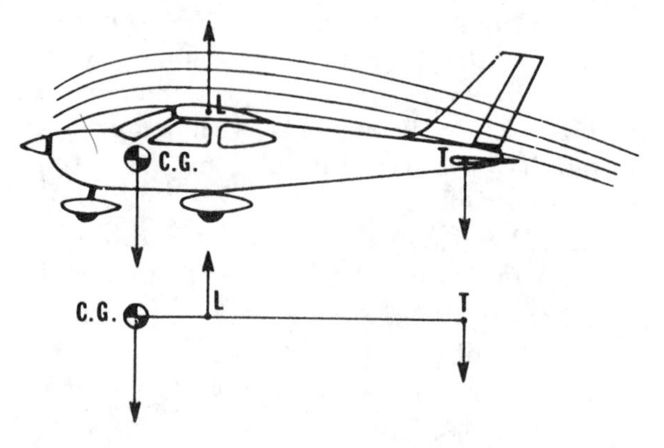

Figure 17–24 Longitudinal Stability

Most airplanes are designed so that the wing's center of lift (CL) is to the rear of the center of gravity. This makes the airplane "nose heavy" and requires that there be a slight downward force on the horizontal stabilizer in order to balance the airplane and keep the nose from continually pitching downward. Compensation for this nose heaviness is provided by setting the horizontal stabilizer at a slight negative angle of attack. The downward force thus produced, holds the tail down, counterbalancing the "heavy" nose. It is as if the line CG–CL–T was a lever with an upward force at CL and two downward forces balancing each other, one a strong force at the CG point and the other, a much lesser force, at point T (downward air pressure on the stabilizer). Applying simple physics principles, it can be seen that if an iron bar were suspended at point CL with a heavy weight hanging on it at the CG, it would take some downward pressure at point T to keep the "lever" in balance.

Even though the horizontal stabilizer may be level when the airplane is in level flight, there is a downwash of air from the wings. This downwash strikes the top of the stabilizer and produces a downward pressure which, at a certain speed, will be just enough to balance the "lever." The faster the airplane is flying, the greater this downwash and the greater the downward force on the horizontal stabilizer (except "T" tails) (Fig. 17–25). In airplanes with fixed position horizontal stabilizers, the airplane manufacturer sets the stabilizer at an angle that will provide the best stability (or balance) during flight at the design cruising speed and power setting (Fig. 17–26).

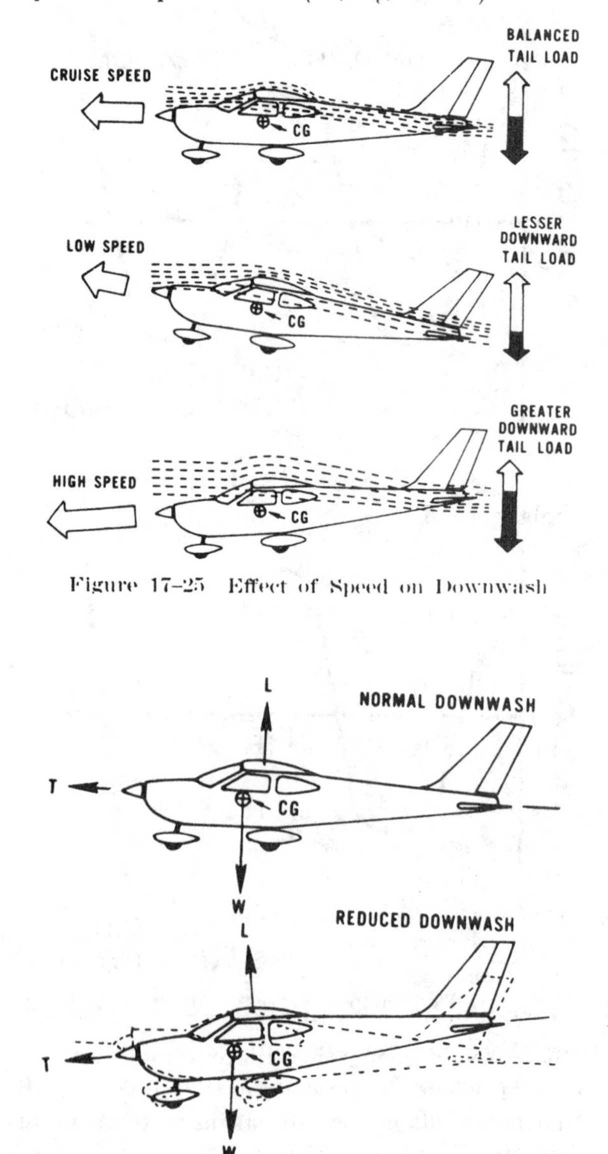

Figure 17–25 Effect of Speed on Downwash

Figure 17–26 Reduced Power Allows Pitch Down

If the airplane's speed decreases, the speed of the airflow over the wing is decreased. As a result of this decreased flow of air over the wing, the downwash is reduced, causing a lesser downward force on the horizontal stabilizer. In turn, the characteristic nose heaviness is accentuated, causing the airplane's nose to pitch down more. This places the airplane in a nose-low attitude, lessening the wing's angle of attack and drag and allowing the airspeed to increase. As the airplane continues in the nose-low attitude and its speed increases, the downward force on the horizontal stabilizer is once again increased. Consequently, the tail is again pushed downward and the nose rises into a climbing attitude.

As this climb continues, the airspeed again decreases, causing the downward force on the tail to decrease until the nose lowers once more. However, because the airplane is dynamically stable, the nose does not lower as far this time as it did before. The airplane will acquire enough speed in this more gradual dive to start it into another climb, but the climb is not so steep as the preceding one.

After several of these diminishing oscillations, in which the nose alternately rises and lowers, the airplane will finally settle down to a speed at which the downward force on the tail exactly counteracts the tendency of the airplane to dive. When this condition is attained the airplane will once again be in balanced flight and will continue in stabilized flight as long as this attitude and airspeed are not changed.

A similar effect will be noted upon closing the throttle. The downwash of the wings is reduced and the force at T in Fig. 17–24 is not enough to hold the horizontal stabilizer down. It is as if the force at T on the lever were allowing the force of gravity to pull the nose down. This, of course, is a desirable characteristic because the airplane is inherently trying to regain airspeed and reestablish the proper balance.

Power or thrust can also have a destabilizing effect in that an increase of power may tend to make the nose rise. The airplane designer can offset this by establishing a "high thrustline" wherein the line of thrust passes above the center of gravity (Figs. 17–27, 17–28). In this case, as power or thrust is increased a moment is produced to counteract the down load on the tail. On the other hand, a very "low thrust line" would tend to add to the nose-up effect of the horizontal tail surface.

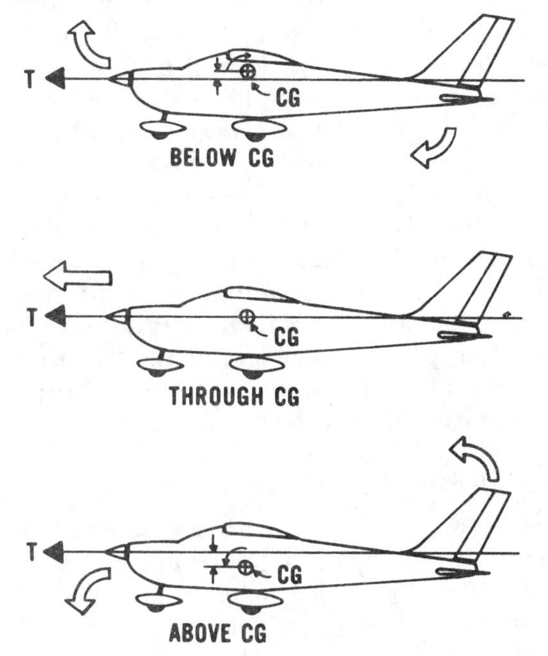

Figure 17–27 Thrust Line Affects Longitudinal Stability

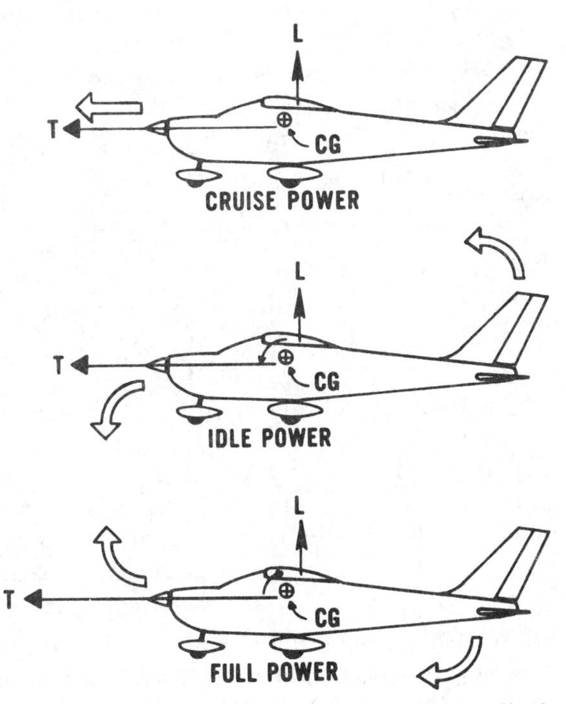

Figure 17–28 Power Changes Affect Longitudinal Stability

It can be concluded, then, that with the center of gravity forward of the center of lift, and with an aerodynamic tail-down force, the result is that the airplane always tries to return to a safe flying attitude.

A simple demonstration of longitudinal stability may be made as follows: Trim the airplane for "hands off" control in level flight. Then momentarily give the controls a slight push to nose the airplane down. If, within a brief period, the nose rises to the original position and then stops, the airplane is statically stable. Ordinarily, the nose will pass the original position (that of level flight) and a series of slow pitching oscillations will follow. If the oscillations gradually cease, the airplane has positive stability; if they continue unevenly the airplane has neutral stability; if they increase the airplane is unstable.

Lateral Stability (Rolling)

Stability about the airplane's longitudinal axis, which extends from nose to tail, is called lateral stability. This helps to stabilize the lateral or rolling effect when one wing gets lower than the wing on the opposite side of the airplane. There are four main design factors which make an airplane stable laterally—dihedral, keel effect, sweepback, and weight distribution. It will be seen in later discussions that these factors also aid in producing yawing or directional stability.

The most common procedure for producing lateral stability is to build the wings with a *dihedral angle* varying from one to three degrees. In other words, the wings on either side of the airplane join the fuselage to form a slight V or angle called "dihedral," and this is measured by the angle made by each wing above a line parallel to the lateral axis.

The basis of rolling stability is, of course, the lateral balance of forces produced by the airplane's wings. Any imbalance in lift results in a tendency for the airplane to roll about its longitudinal axis. Stated another way, dihedral involves a balance of lift created by the wings' angle of attack on each side of the airplane's longitudinal axis.

If a momentary gust of wind forces one wing of the airplane to rise and the other to lower, the airplane will bank. When the airplane is banked without turning, it tends to sideslip or slide downward toward the lowered wing (Fig. 17–29). Since the wings have dihedral, the air strikes the low wing at much greater angle of attack than the high wing. This increases the lift on the low wing and decreases lift on the high wing, and tends to restore the airplane to its original lateral attitude (wings level); that is, the angle of attack and lift on the two wings are again equal.

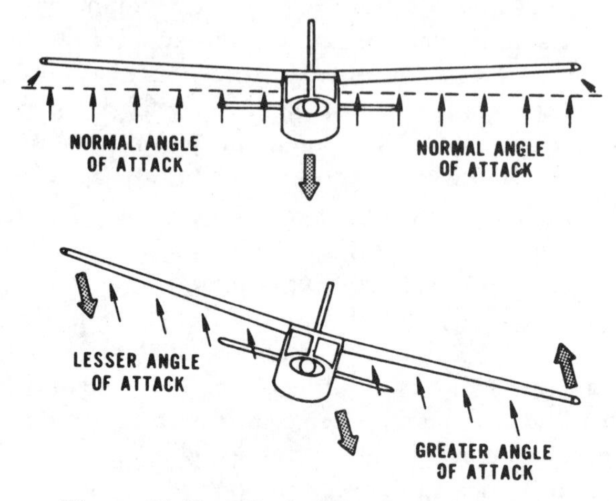

Figure 17–29 Dihedral for Lateral Stability

The effect of dihedral, then, is to produce a rolling moment tending to return the airplane to a laterally balanced flight condition when a sideslip occurs.

The restoring force may move the low wing up too far, so that the opposite wing now goes down. If so, the process will be repeated, decreasing with each lateral oscillation until a balance for wings-level flight is finally reached.

Conversely, excessive dihedral has an adverse effect on lateral maneuvering qualities. The airplane may be so stable laterally that it resists any intentional rolling motion. For this reason, airplanes which require fast roll or banking characteristics usually have less dihedral than those which are designed for less maneuverability.

The contribution of *sweepback* to dihedral effect is important because of the nature of the contribution. In a sideslip the wing into the wind is operating with an effective decrease in sweepback while the wing out of the wind is operating with an effective increase in sweepback. The reader will recall that the swept wing is responsive only to the wind component that is perpendicular to the wing's leading edge. Consequently, if the wing is operating at a positive lift coefficient, the wing into the wind has an increase in lift, and the wing out of the wind has a decrease in lift. In this manner the swept back wing would contribute a positive dihedral effect and the swept forward wing would contribute a negative dihedral effect.

During flight, the side area of the airplane's fuselage and vertical fin react to the airflow in much the same manner as the keel of a ship. That is, it exerts a steadying influence on the airplane laterally about the longitudinal axis.

Such laterally stable airplanes are constructed so that the greater portion of the keel area is above and behind the center of gravity (Fig. 17–30). Thus, when the airplane slips to one side, the combination of the airplane's weight and the pressure of the airflow against the upper portion of the keel area (both acting about the CG) tends to roll the airplane back to wings-level flight.

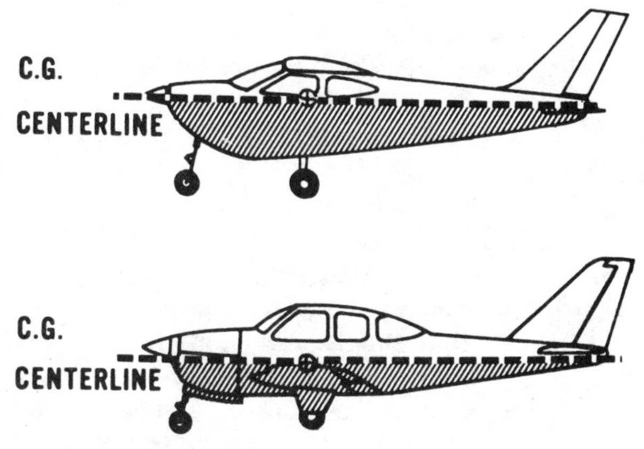

Figure 17–30 Keel Area for Lateral Stability

Vertical Stability (Yawing)

Stability about the airplane's vertical axis (the sideways moment), is called yawing or directional stability.

Yawing or directional stability is the more easily achieved stability in airplane design. The area of the vertical fin and the sides of the fuselage aft of the center of gravity are the prime contributors which make the airplane act like the well known weathervane or arrow, pointing its nose into the relative wind.

In examining a weathervane it can be seen that if exactly the same amount of surface were exposed to the wind in front of the pivot point as behind it, the forces fore and aft would be in balance and little or no directional movement would result. Consequently, it is necessary to have a greater surface aft of the pivot point than forward of it.

Similarly in an airplane, the designer must ensure positive directional stability by making the side surface greater aft than ahead of the center of gravity (Fig. 17–31). To provide more positive stability aside from that provided by the fuselage, a vertical fin is added. The fin acts similar to the feather on an arrow in maintaining straight flight. Like the weathervane and the arrow the farther aft this fin is placed and the larger its size, the greater the airplane's directional stability.

If an airplane is flying in a straight line and a sideward gust of air gives the airplane a slight rotation about its vertical axis (i.e., the right) the motion is retarded and stopped by the fin because while the airplane is rotating to the right, the air is striking the left side of the fin at an angle. This causes pres-

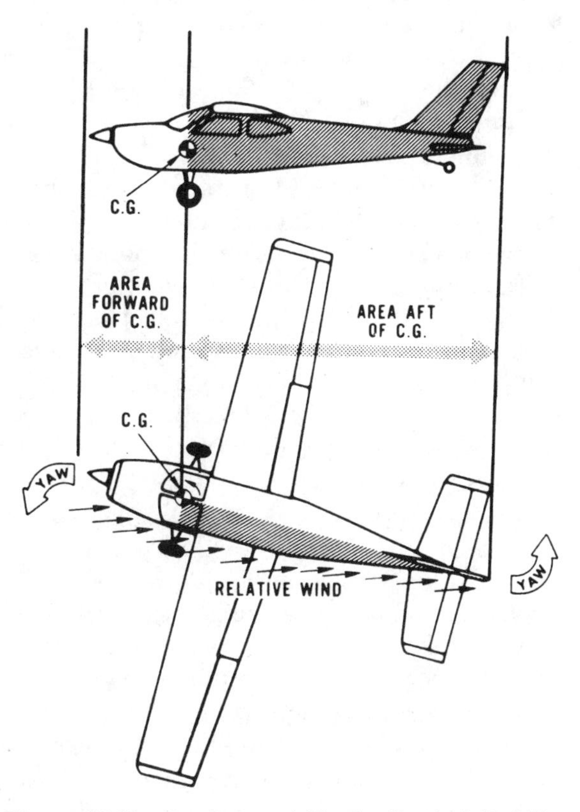

Figure 17–31 Fuselage and Fin for Vertical Stability

sure on the left side of the fin, which resists the turning motion and slows down the airplane's yaw. In doing so it acts somewhat like the weathervane by turning the airplane into the relative wind.

The initial change in direction of the airplane's flightpath is generally slightly behind its change of heading. Therefore, after a slight yawing of the airplane to the right, there is a brief moment when the airplane is still moving along its original path, but its longitudinal axis is pointed slightly to the right.

The airplane is then momentarily skidding sideways, and during that moment (since we assume that although the yawing motion has stopped, the excess pressure on the left side of the fin still persists) there is necessarily a tendency for the airplane to be turned partially back to the left. That is, there is a momentary restoring tendency caused by the fin.

This restoring tendency is relatively slow in developing and ceases when the airplane stops skidding. When it ceases, the airplane will be flying in a direction slightly different from the original direction. In other words, it will not of its own accord return to the original heading; the pilot must reestablish the initial heading.

A minor improvement of directional stability may be obtained through sweepback. Sweepback is incorporated in the design of the wing primarily to delay the onset of compressibility during high speed flight. In lighter and slower airplanes, sweepback aids in locating the center of pressure in the correct relationship with the center of gravity. As we have learned, a longitudinally stable airplane is built with the center of pressure aft of the center of gravity.

Because of structural reasons, airplane designers sometimes cannot attach the wings to the fuselage at the exact desired point. If they had to mount the wings too far forward, and at right angles to the fuselage, the center of pressure would not be far enough to the rear to effect the desired amount of longitudinal stability. By building sweepback into the wings, however, the designers can move the center of pressure toward the rear. The amount of sweepback and the position of the wings then place the center of pressure in the correct location.

The contribution of the *wing* to static directional stability is usually small. The swept wing provides a stable contribution depending on the amount of sweepback but the contribution is relatively small when compared with other components.

Free Directional Oscillations (Dutch Roll)

Dutch roll is a coupled lateral-directional oscillation which is usually dynamically stable but is objectionable in an airplane because of the oscillatory nature. The damping of the oscillatory mode may be weak or strong depending on the properties of the particular airplane.

Unfortunately all air is not smooth. There are bumps and depressions created by gusty updrafts and downdrafts, and by gusts from ahead, behind, or the side of the airplane.

The response of the airplane to a disturbance from equilibrium is a combined rolling-yawing oscillation in which the rolling motion is phased to precede the yawing motion. The yawing motion is not too significant, but the roll is much more noticeable. When the airplane rolls back toward level flight in response to dihedral effect, it rolls back too far and sideslips the other way. Thus, the airplane overshoots each time because of the strong dihedral effect. When the dihedral effect is large in comparison with static directional stability, the Dutch roll motion has weak damping and is objectionable. When the static directional stability is strong in comparison with the dihedral effect, the Dutch Roll motion has such heavy damping that it is not objectionable. However, these qualities tend toward spiral instability.

The choice is then the least of two evils, Dutch Roll is objectionable, and spiral instability is tolerable if the rate of divergence is low. Since the more important handling qualities are a result of high static directional stability and minimum necessary dihedral effect, most airplanes demonstrate a mild spiral tendency. This tendency would be indicated to the pilot by the fact that the airplane cannot be flown "hands off" indefinitely.

In most modern airplanes, excepting high speed swept wing designs, these free directional oscillations usually die out automatically in a very few cycles unless the air continues to be gusty or turbulent. Those airplanes with continuing Dutch Roll tendencies usually are equipped with gyro stabilized yaw dampers. An airplane which has Dutch Roll tendencies is disconcerting, to say the least. Therefore, the manufacturer tries to reach a medium between too much and too little directional stability. Because it is more desirable for the airplane to have "spiral instability" than Dutch Roll tendencies, most airplanes are designed with that characteristic.

Spiral Instability

Spiral instability exists when the static directional stability of the airplane is very strong as compared to the effect of its dihedral in maintaining lateral equilibrium. When the lateral equilibrium of the airplane is disturbed by a gust of air and a sideslip is introduced, the strong directional stability tends to yaw the nose into the resultant relative wind while the comparatively weak dihedral lags in restoring the lateral balance. Due to this yaw, the wing on the outside of the turning moment travels forward faster than the inside wing and as a consequence, its lift becomes greater. This produces an overbanking tendency which, if not corrected by the pilot, will result in the bank angle becoming steeper and steeper. At the same time, the strong directional stability which yaws the airplane into the relative wind is actually forcing the nose to a lower pitch attitude. We then have the start of a slow downward spiral which, if not counteracted by the pilot, will gradually increase into a steep spiral dive. Usually the rate of divergence in the spiral motion is so gradual that the pilot can control the tendency without any difficulty.

All airplanes are affected to some degree by this characteristic although they may be inherently stable in all other normal parameters. This tendency would be indicated to the pilot by the fact that the airplane cannot be flown "hands off" indefinitely.

Much study and effort has gone into development of control devices (wing leveler) to eliminate or at least correct this instability. Advanced stages of this spiral condition demand that the pilot be very careful in application of recovery controls, or excessive loads on the structure may be imposed. Of the inflight structural failures that have occurred in general aviation airplanes, improper recovery from this condition has probably been the underlying cause of more fatalities than any other single factor. The reason is that the airspeed in the spiral condition builds up rapidly, and the application of back elevator force to reduce this speed and to pull the nose up only "tightens the turn," increasing the load factor. The results of the prolonged uncontrolled spiral are always the same; either inflight structural failure, crashing into the ground, or both. The most common causes on record for getting into this situation are: Loss of horizon reference, inability of the pilot to control the airplane by reference to instruments, or a combination of both. Specific instructions on preventing loss of control and the recovery from spirals have been explained in preceding chapters.

Forces in Turns

If an airplane were viewed in straight-and-level flight from the rear (Fig. 17–32), and if the forces acting on the airplane actually could be seen, two forces (lift and weight) would be apparent, and if the airplane were in a bank it would be apparent that lift did not act directly opposite to the weight—it now acts in the direction of the bank. The fact that when the airplane banks, lift acts inward toward the center of the turn, as well as upward, is one of the basic truths to remember in the consideration of turns.

As we learned earlier, an object at rest or moving in a straight line will remain at rest or continue to move in a straight line until acted on by some other force. An airplane, like any moving object, requires a sideward force to make it turn. In a normal turn, this force is supplied by banking the airplane so that lift is exerted inward as well as upward. The force of lift during a turn is separated into two components at right angles to each other. One component which acts vertically and opposite to the weight (gravity) is called the "vertical component of lift." The other which acts horizontally toward the center of

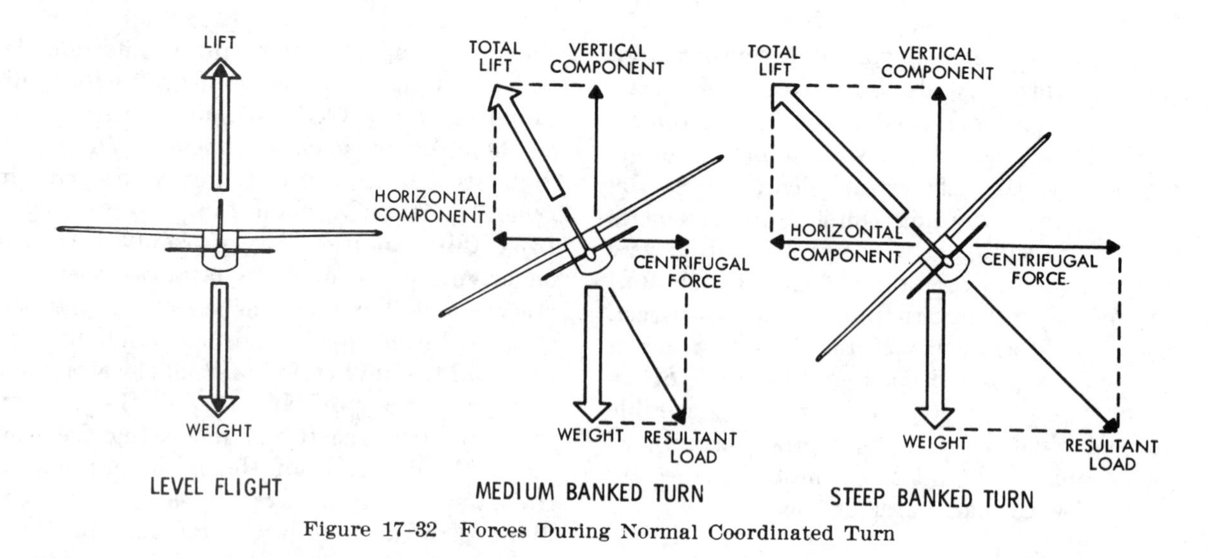

Figure 17–32 Forces During Normal Coordinated Turn

the turn is called the "horizontal component of lift." The horizontal component of lift is the force that pulls the airplane from a straight flightpath to make it turn. Centrifugal force is the "equal and opposite reaction" of the airplane to the change in direction and acts equal and opposite to the horizontal component of lift. This explains why, in a correctly executed turn, the force that turns the airplane is not supplied by the rudder.

An airplane is not steered like a boat or an automobile; in order for it to turn, it must be banked. If the airplane is not banked, there is no force available that will cause it to deviate from a straight flightpath. Conversely, when an airplane is banked, it will turn, provided it is not slipping to the inside of the turn. Good directional control is based on the fact that the airplane will attempt to turn whenever it is banked. This fact should be borne in mind at all times, particularly while attempting to hold the airplane in straight-and-level flight.

Merely banking the airplane into a turn produces no change in the *total* amount of lift developed. However, as was pointed out, the lift during the bank is divided into two components, one vertical and the other horizontal. This division reduces the amount of lift which is opposing gravity and actually supporting the airplane's weight; consequently, the airplane loses altitude unless additional lift is created. This is done by increasing the angle of attack until the vertical component of lift is again equal to the weight. Since the vertical component of lift decreases as the bank angle increases, the angle of attack must be progressively increased to produce sufficient vertical lift to support the airplane's weight. The fact that the vertical component of lift must be equal to the weight to maintain altitude is an important fact to remember when making constant altitude turns.

At a given airspeed, the rate at which an airplane turns depends upon the magnitude of the horizontal component of lift. It will be found that the horizontal component of lift is proportional to the angle of bank; that is, it increases or decreases respectively as the angle of bank increases or decreases. It logically follows then, that as the angle of bank is increased the horizontal component of lift increases, thereby increasing the rate of turn. Consequently, at any given airspeed the rate of turn can be controlled by adjusting the angle of bank.

To provide a vertical component of lift sufficient to hold altitude in a level turn, an increase in the angle of attack is required. Since the drag of the airfoil is directly proportional to its angle of attack, induced drag will increase as the lift is increased. This, in turn, causes a loss of airspeed in proportion to the angle of bank; a small angle of bank results in a small reduction in airspeed and a large angle of bank results in a large reduction in airspeed. Additional thrust (power) must be applied to prevent a reduction in airspeed in level turns; the required amount of additional thrust is proportional to the angle of bank.

To compensate for added lift which would result if the airspeed were increased during a turn, the angle of attack must be decreased, or the angle of bank increased, if a constant altitude were to be maintained. If the angle of bank were held constant and the angle of attack decreased, the rate of turn would decrease. Therefore, in order to maintain a constant rate of turn as the airspeed is increased, the angle of attack must remain constant and the angle of bank increased.

It must be remembered that an increase in airspeed results in an increase of the turn radius and that centrifugal force is directly proportional to the radius of the turn. In a correctly executed turn, the horizontal component of lift must be exactly equal and opposite to the centrifugal force. Therefore, as the airspeed is increased in a constant-rate level turn, the radius of the turn increases. This increase in the radius of turn causes an increase in the centrifugal force, which must be balanced by an increase in the horizontal component of lift. The horizontal component of lift can only be increased by increasing the angle of bank.

In a slipping turn the airplane is not turning at the rate appropriate to the bank being used, since the airplane is yawed toward the outside of the turning flightpath. The airplane is banked too much for the rate of turn, so the horizontal lift component is greater than the centrifugal force (Fig. 17–33). Equilibrium between the horizontal lift component and centrifugal force is reestablished either by decreasing the bank, increasing the rate of turn, or a combination of the two changes.

A skidding turn results from an excess of centrifugal force over the horizontal lift component, pulling the airplane toward the outside of the turn. The rate of turn is too great for the angle of bank. Correction of a skidding turn thus involves a reduction in the rate of turn, an increase in bank, or a combination of the two changes.

To maintain a given rate of turn, the angle of bank must be varied with the airspeed. This becomes particularly important in high-speed airplanes. For instance, at 400 miles per hour, an airplane must be banked approximately 44° to execute a standard rate turn (3° per second). At this angle of bank, only about 79 percent of the lift of the airplane comprises the vertical component of the lift; the result is a loss of altitude unless the angle of attack is increased sufficiently to compensate for the loss of vertical lift.

Forces in Climbs

For all practical purposes, the wing's lift in a steady state normal climb is the same as it is in a steady level flight at the same airspeed. Though the airplane's flightpath has changed when the climb has been established, the angle of attack of the wing with respect to the inclined flightpath reverts to practically the same values, as does the lift. There is an initial momentary change, however, as shown

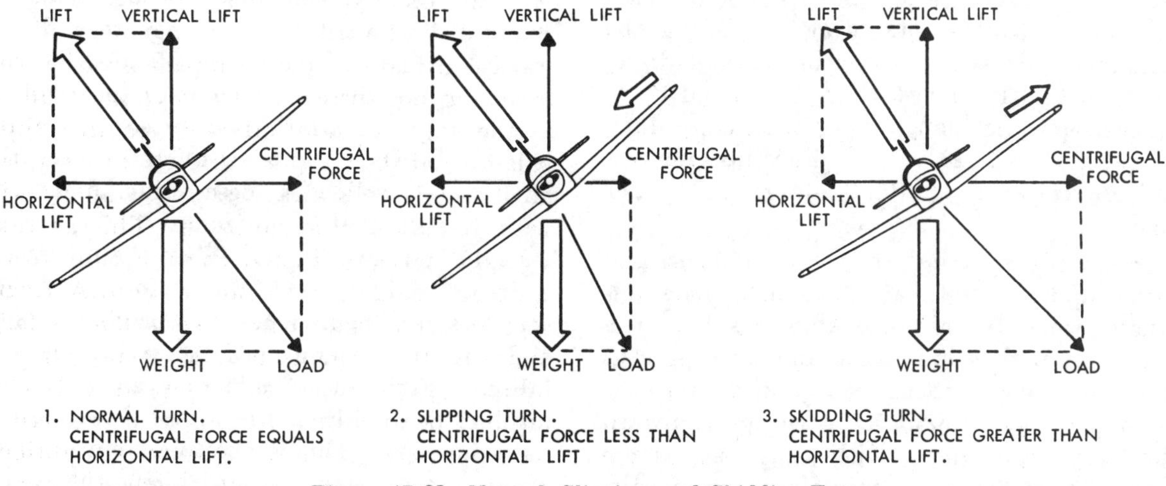

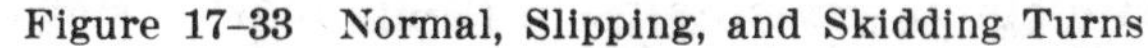
Figure 17–33 Normal, Slipping, and Skidding Turns

in Fig. 17–34. During the transition from straight-and-level flight to a climb, a change in lift occurs when back elevator pressure is first applied. Raising the airplane's nose increases the angle of attack and momentarily increases the lift. Lift at this moment is now greater than weight and starts the airplane climbing. After the flightpath is stabilized on the upward incline, the angle of attack and lift again revert to about the level flight values.

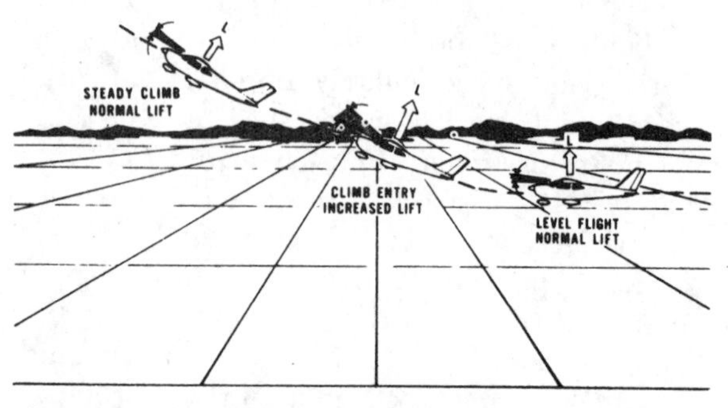

Figure 17–34 Changes in Lift During Climb Entry

If the climb is entered with no change in power setting, the airspeed gradually diminishes because the thrust required to maintain a given airspeed in level flight is insufficient to maintain the same airspeed in a climb. When the flightpath is inclined upward, a component of the airplane's weight acts in the same direction as, and parallel to, the total drag of the airplane, thereby increasing the total *effective* drag. Consequently, the total drag is greater than the power, and the airspeed decreases. The reduction in airspeed gradually results in a corresponding decrease in drag until the total drag (including the component of weight acting in the same direction) equal the thrust (Fig. 17–35). Due to momentum, the change in airspeed is gradual, varying considerably with differences in airplane size, weight, total drag, and other factors.

Generally speaking, the forces of thrust and drag, and lift and weight, again become balanced when the airspeed stabilizes but at a value lower than in straight-and-level flight at the same power setting. Since in a climb the airplane's weight is not only acting downward but rearward along with drag, additional power is required to maintain the same air-

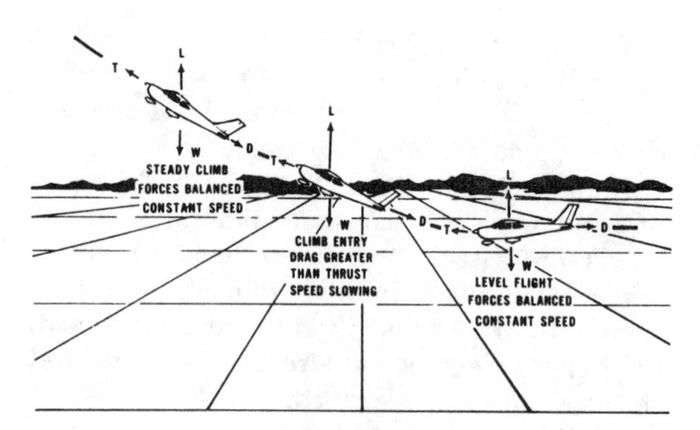

Figure 17–35 Changes in Speed During Climb Entry

speed as in level flight. The amount of power depends on the angle of climb. When the climb is established so steep that there is insufficient power available, a slower speed results. It will be seen then that the amount of reserve power determines the climb performance of the airplane. This is discussed further in the section on Climb Performance.

Forces in Descents

As in climbs, the forces acting on the airplane go through definite changes when a descent is entered from straight-and-level flight. The analysis here is that of descending at the same power as used in straight-and-level flight.

When forward pressure is applied to the elevator control to start descending, or the airplane's nose is allowed to pitch down, the angle of attack is decreased and, as a result, the lift of the airfoil is reduced. This reduction in total lift and angle of attack is momentary and occur during the time the flightpath changes downward. The change to a downward flightpath is due to the lift momentarily becoming less than the weight of the airplane as the angle of attack is reduced. This unbalance between lift and weight causes the airplane to follow a descending flightpath with respect to the horizontal flightpath of straight-and-level flight. When the flightpath is in a steady descent, the airfoil's angle of attack again approaches the original value, and lift and weight will again become stabilized. From the time the descent is started until it is stabilized, the airspeed will gradually increase. This is due to a component of weight now acting forward along the flightpath, similar to the manner it acted rearward in a climb. The overall effect is that of increased power or thrust, which in turn causes the increase in airspeed associated with descending at the same power as used in level flight.

To descend at the same airspeed as used in straight-and-level flight, obviously, the power must be reduced as the descent is entered. The component of weight acting forward along the flightpath will increase as the angle of rate of descent increases and conversely, will decrease as the angle of rate of descent decreases. Therefore, the amount of power reduction required for a descent at the same speed as cruise will be determined by the steepness of the descent.

Stalls

In earlier discussions it was shown that an airplane will fly as long as the wing is creating sufficient lift to counteract the load imposed on it. When the lift is completely lost, the airplane stalls.

Remember, the direct cause of every stall is an excessive angle of attack. There are any number of flight maneuvers which may produce an increase in the angle of attack, but the stall does not occur until the angle of attack becomes excessive.

It must be emphasized that the *stalling speed* of a particular airplane is not a fixed value for all flight situations. However, a given airplane will always stall at the *same angle of attack* regardless of airspeed, weight, load factor, or density altitude. Each airplane has a particular angle of attack where the airflow separates from the upper surface of the wing and the stall occurs. This critical angle of attack varies from 16° to 20° depending on the airplane's design. But each airplane has only *one* specific angle of attack where the stall occurs.

There are three situations in which the critical angle of attack can be exceeded—in low speed flying, in high speed flying, and in turning flight.

The airplane can be stalled in straight-and-level flight by flying too slowly. As the airspeed is being decreased, the angle of attack must be increased to retain the lift required for maintaining altitude. The slower the airspeed becomes the more the angle of attack must be increased. Eventually an angle of attack is reached which will result in the wing not producing enough lift to support the airplane and it will start settling. If the airspeed is reduced further the airplane will stall, since the angle of attack has exceeded the critical angle and the airflow over the wing is disrupted.

It must be reemphasized here that low speed is not necessary to produce a stall. The wing can be brought into an excessive angle of attack at *any* speed. For example, take the case of an airplane which is in a dive with an airspeed of 200 knots when suddenly the pilot pulls back sharply on the elevator control (Fig. 17–36). Because of gravity and centrifugal force, the airplane could not immediately alter its flightpath but would merely change its angle of attack abruptly from quite low to very high. Since the flightpath of the airplane in relation to the oncoming air determines the direction of the relative wind, the angle of attack is suddenly increased, and the airplane would quickly reach the stalling angle at a speed much greater than the normal stall speed.

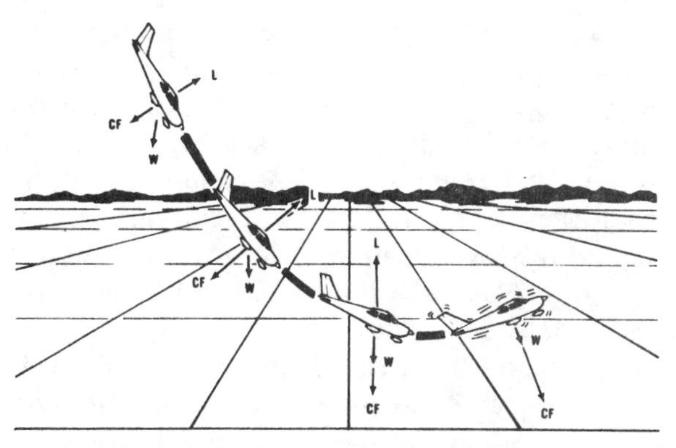

Figure 17–36 Forces Exerted When Pulling Out of a Dive

Similarly, the stalling speed of an airplane is higher in a level turn than in straight-and-level flight (Fig. 17–37). This is because centrifugal force is added to the airplane's weight, and the wing must produce sufficient additional lift to counterbalance the load imposed by the combination of centrifugal force and weight. In a turn, the necessary additional lift is acquired by applying back pressure to the elevator control. This increases the wing's angle of attack, and results in increased lift. As stated earlier, the angle of attack must increase as the bank angle increases to counteract the increasing load caused by centrifugal force. If at any time during a turn the angle of attack becomes excessive, the airplane will stall.

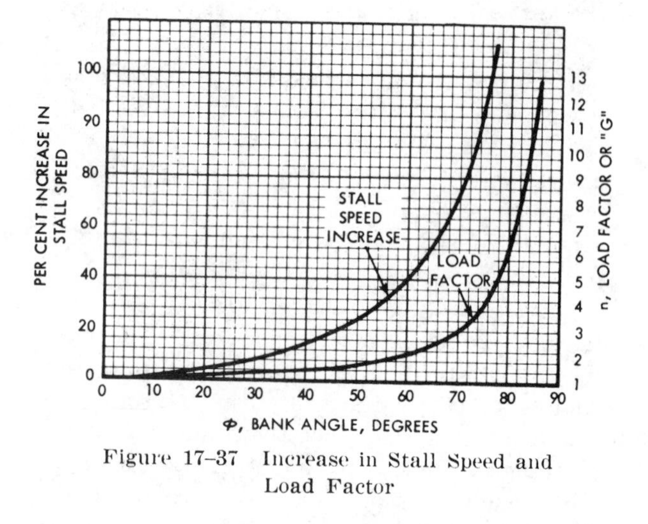

Figure 17–37 Increase in Stall Speed and Load Factor

At this point we should examine the action of the airplane during a stall. In our earlier discussion of pitching (longitudinal) stability, we learned that to balance the airplane aerodynamically, the center of lift is normally located aft of the center of gravity. It was also pointed out that although this made the airplane inherently "nose heavy," downwash on the horizontal stabilizer counteracted this condition. It can be seen then, that at the point of stall when the upward force of the wing's lift and the downward tail force cease, an unbalanced condition exists. This allows the airplane to pitch down abruptly, rotating about its center of gravity. During this nose down attitude the angle of attack decreases and the airspeed again increases; hence, the smooth flow of air over the wing begins again, lift returns, and the airplane is again flying. However, considerable altitude may be lost before this cycle is complete.

Basic Propeller Principles

The airplane propeller consists of two or more blades and a central hub to which the blades are attached. Each blade of an airplane propeller is essentially a rotating wing. As a result of their construction, the propeller blades are like airfoils and produce forces that create the thrust to pull, or push, the airplane through the air.

The power needed to rotate the propeller blades is furnished by the engine. The engine rotates the airfoils of the blades through the air at high speeds, and the propeller transforms the rotary power of the engine into forward thrust.

An airplane moving through the air creates a drag force opposing its forward motion. Consequently, if an airplane is to fly, there must be a force applied to it that is equal to the drag, but acting forward. This force, as we know, is called "thrust."

A cross section of a typical propeller blade is shown in Fig. 17–38. This section or blade element is an airfoil comparable to a cross section of an airplane wing. One surface of the blade is cambered or curved, similar to the upper surface of an airplane wing, while the other surface is flat like the bottom surface of a wing. The chord line is an imaginary line drawn through the blade from its leading edge to its trailing edge. As in a wing, the leading edge is the thick edge of the blade that meets the air as the propeller rotates.

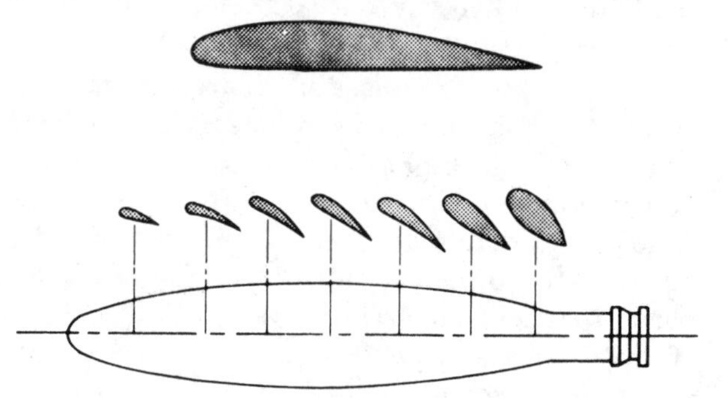

Figure 17–38 Airfoil Sections of Propeller Blade

Blade angle, usually measured in degrees, is the angle between the chord of the blade and the plane of rotation (Fig. 17–39) and is measured at a specific point along the length of the blade. Because most propellers have a flat blade "face," the chord line is often drawn along the face of the propeller blade. Pitch is not the same as blade angle, but because pitch is largely determined by blade angle, the two terms are often used interchangeably. An

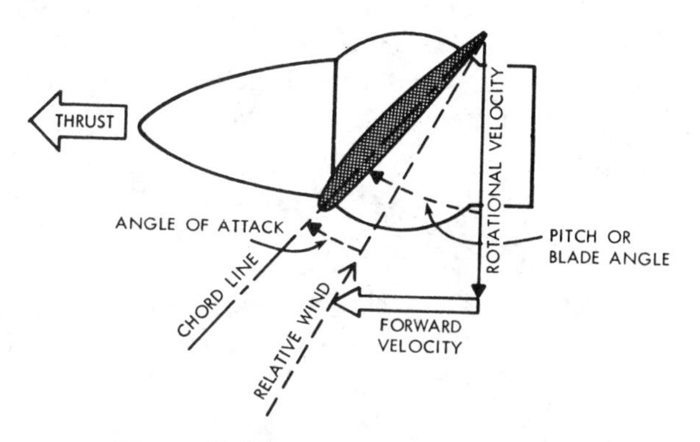

Figure 17–39 Propeller Blade Angle

increase or decrease in one is usually associated with an increase or decrease in the other.

The pitch of a propeller may be designated in inches. A propeller designated as a "74–48" would be 74 inches in length and have an effective pitch of 48 inches. The pitch in inches is the distance which the propeller would screw through the air in one revolution if there were no slippage.

When specifying a fixed-pitch propeller for a new type of airplane, the manufacturer usually selects one with a pitch which will operate efficiently at the expected cruising speed of the airplane. Unfortunately, however, every fixed pitch propeller must be a compromise, because it can be efficient at only a given combination of airspeed and RPM. The pilot does not have it within his power to change this combination in flight.

When the airplane is at rest on the ground with the engine operating, or moving slowly at the beginning of takeoff, the propeller efficiency is very low because the propeller is restrained from advancing with sufficient speed to permit its fixed-pitch blades to reach their full efficiency. In this situation, each propeller blade is turning through the air at an angle of attack which produces relatively little thrust for the amount of power required to turn it.

To understand the action of a propeller, consider first its motion, which is both rotational and forward. Thus, as shown by the vectors of propeller forces in Fig. 17–39, each section of a propeller blade moves downward and forward. The angle at which this air (relative wind) strikes the propeller blade is its angle of attack. The air deflection produced by this angle causes the dynamic pressure at the engine side of the propeller blade to be greater than atmospheric, thus creating thrust.

The shape of the blade also creates thrust, because it is cambered like the airfoil shape of a wing. Consequently, as the air flows past the propeller, the pressure on one side is less than that on the other. As in a wing, this produces a reaction force in the direction of the lesser pressure. In the case of a wing, the air flow over the wing has less pressure, and the force (lift) is upward. In the case of the propeller, which is mounted in a vertical instead of a horizontal plane, the area of decreased pressure is in front of the propeller, and the force (thrust) is in a forward direction. Aerodynamically, then, thrust is the result of the propeller shape *and* the angle of attack of the blade.

Another way to consider thrust is in terms of the mass of air handled by the propeller. In these terms, thrust is equal to the mass of air handled times the slipstream velocity minus the velocity of the airplane. The power expended in producing thrust depends on the rate of air mass movement. On the average, thrust constitutes approximately 80% of the torque (total horsepower absorbed by the propeller). The other 20% is lost in friction and slippage. For any speed of rotation, the horsepower absorbed by the propeller balances the horsepower delivered by the engine. For any single revolution of the propeller, the amount of air handled depends on the blade angle, which determines how big a "bite" of air the propeller takes. Thus, the blade angle is an excellent means of adjusting the load on the propeller to control the engine RPM.

The blade angle is also an excellent method of adjusting the angle of attack of the propeller. On constant-speed propellers, the blade angle must be adjusted to provide the most efficient angle of attack at all engine and airplane speeds. Lift versus drag curves, which are drawn for propellers as well as wings, indicate that the most efficient angle of attack is a small one varying from 2° to 4° positive. The actual blade angle necessary to maintain this small angle of attack varies with the forward speed of the airplane.

Fixed-pitch and ground-adjustable propellers are designed for best efficiency at one rotation and forward speed. They are designed for a given airplane and engine combination. A propeller may be used that provides the maximum propeller efficiency for either takeoff, climb, cruise, or high speed flight. Any change in these conditions results in lowering the efficiency of both the propeller and the engine. Since the efficiency of any machine is the ratio of the useful power output to the actual power input, propeller efficiency is the ratio of thrust horsepower to brake horsepower. Propeller efficiency varies from 50% to 87%, depending on how much the propeller "slips."

Propeller slip is the difference between the geometric pitch of the propeller and its effective pitch (Fig. 17–40). Geometric pitch is the theoretical distance a propeller should advance in one revolution; effective pitch is the distance it actually advances. Thus, geometric or theoretical pitch is based on no slippage, but actual or effective pitch includes propeller slippage in the air.

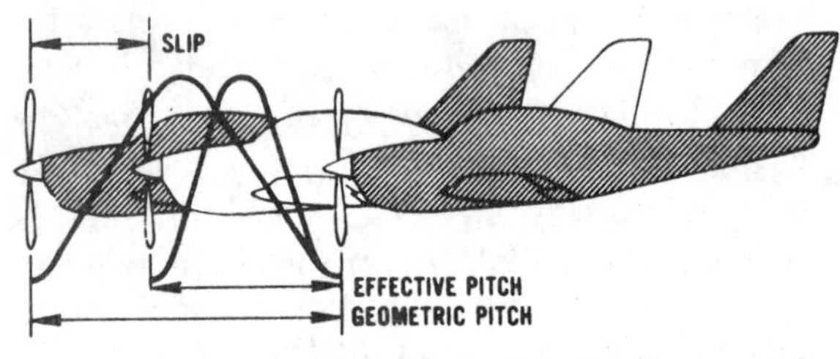

Figure 17–40 Propeller Slippage

If you wonder why a propeller is "twisted," the answer is that the outer parts of the propeller blades, like all things that turn about a central point, travel faster than the portions near the hub (Fig. 17–41). If the blades had the same geometric pitch throughout their lengths, at cruise speed the portions near the hub could have negative angles of attack while the propeller tips would be stalled. "Twisting," or variations in the geometric pitch of the blades, permits the propeller to operate with a relatively constant angle of attack along its length when in cruising flight. To put it another way, propeller blades are twisted to change the blade angle in proportion to the differences in speed of rotation along the length of the propeller and thereby keep thrust more nearly equalized along this length.

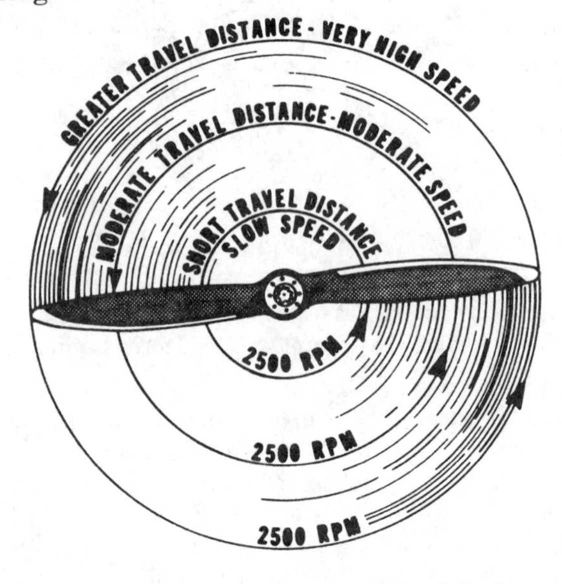

Figure 17–41 Propeller Tips Travel Faster Than Hubs

Usually 1° to 4° provides the most efficient lift/drag ratio, but in flight the propeller angle of attack of a fixed-pitch propeller will vary—normally from 0° to 15°. This variation is caused by changes in the relative airstream which in turn results from changes in aircraft speed. In short, propeller angle of attack is the product of two motions—propeller rotation about its axis and its forward motion.

A constant-speed propeller, however, automatically keeps the blade angle adjusted for maximum efficiency for most conditions encountered in flight. During takeoff, when maximum power and thrust are required, the constant-speed propeller is at a low propeller blade angle or pitch. The low blade angle keeps the angle of attack small and efficient with respect to the relative wind. At the same time, it allows the propeller to handle a smaller mass of air per revolution. This light load allows the engine to turn at high RPM and to convert the maximum amount of fuel into heat energy in a given time. The high RPM also creates maximum thrust; for, although the mass of air handled per revolution is small, the number of revolutions per minute is many, the slipstream velocity is high, and with the low airplane speed, the thrust is maximum.

After liftoff, as the speed of the airplane increases, the constant-speed propeller automatically changes to a higher angle (or pitch). Again, the higher blade angle keeps the angle of attack small and efficient with respect to the relative wind. The higher blade angle increases the mass of air handled per revolution. This decreases the engine RPM, reducing fuel consumption and engine wear, and keeps thrust at a maximum.

After the takeoff climb is established, in an airplane having a controllable pitch propeller, the pilot reduces the power output of the engine to climb power by first decreasing the manifold pressure and then increasing the blade angle to lower the RPM.

At cruising altitude, when the airplane is in level flight and less power is required than is used in takeoff or climb, the pilot again reduces engine power by reducing the manifold pressure and then increasing the blade angle to decrease the RPM. Again, this provides a torque requirement to match the reduced engine power; for, although the mass of air handled per revolution is greater, it is more than offset by a decrease in slipstream velocity and an increase in airspeed. The angle of attack is still small because the blade angle has been increased with an increase in airspeed.

Torque and P Factor

To the pilot, "torque" (the left turning tendency of the airplane) is made up of four elements which cause or produce a twisting or rotating motion around at least one of the airplane's three axes. These four elements are:

1. Torque Reaction from Engine and Propeller.
2. Corkscrewing Effect of the Slipstream.
3. Gyroscopic Action of the Propeller.
4. Asymmetric Loading of the Propeller (P Factor).

Torque Reaction

Torque reaction involves Newton's Third Law of Physics—for every action, there is an equal and opposite reaction. As applied to the airplane, this means that as the internal engine parts and propeller are revolving in one direction, an equal force is trying to rotate the airplane in the opposite direction (Fig. 17–42).

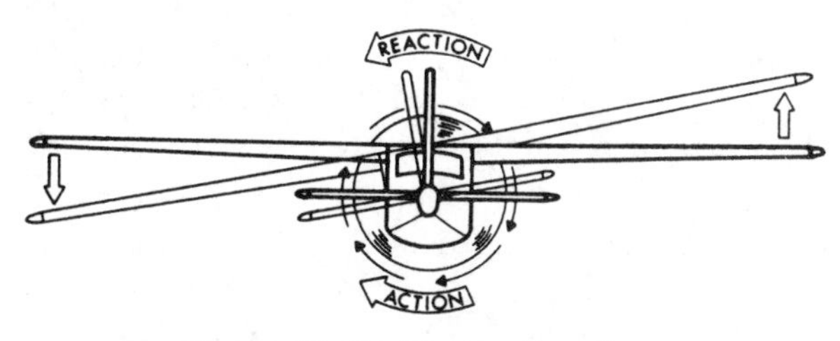

Figure 17–42 Torque Reaction

When the airplane is airborne, this force is acting around the longitudinal axis, tending to make the airplane roll. To compensate for this, some of the older airplanes are rigged in a manner to create more lift on the wing which is being forced downward. The more modern airplanes are designed with the engine offset to counteract this effect of torque.

NOTE.—Most United States built aircraft engines rotate the propeller clockwise, as viewed from the pilot's seat. The discussion here is with reference to those engines.

Generally, the compensating factors are permanently set so that they compensate for this force at cruising speed, since most of the airplane's operating life is at that speed. However, aileron trim tabs permit further adjustment for other speeds.

When the airplane's wheels are on the ground during the takeoff roll, an additional turning moment around the vertical axis is induced by torque reaction. As the left side of the airplane is being forced down by torque reaction, more weight is being placed on the left main landing gear. This results in more ground friction, or drag, on the left tire than on the right, causing a further turning moment to the left. The magnitude of this moment is dependent on many variables. Some of these variables are: (1) size and horsepower of engine, (2) size of propeller and the RPM, (3) size of the airplane, and (4) condition of the ground surface.

This yawing moment on the takeoff roll is corrected by the pilot's proper use of the rudder or rudder trim.

Corkscrew Effect

The high speed rotation of an airplane propeller gives a corkscrew or spiraling rotation to the slipstream. At high propeller speeds and low forward speed (as in the takeoffs. approaches to power-on stalls, etc.), this spiraling rotation is very compact and exerts a strong sideward force on the airplane's vertical tail surface (Fig. 17–43).

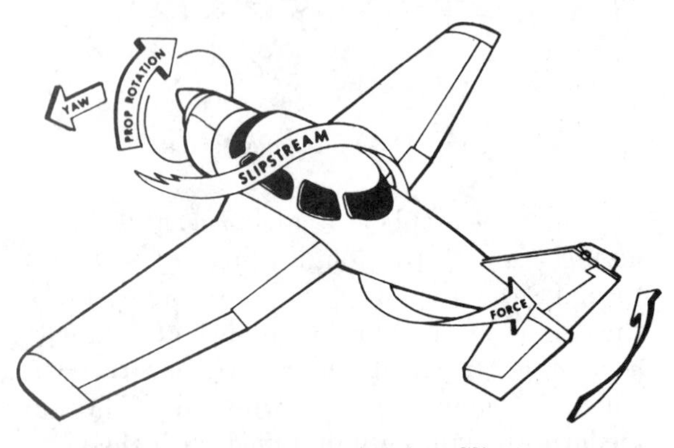

Figure 17–43 Corkscrewing Slipstream

When this spiraling slipstream strikes the vertical fin on the left, it causes a left turning moment about the airplane's vertical axis. The more compact the spiral, the more prominent this force is. As the forward speed increases, however, the spiral elongates and becomes less effective.

The corkscrew flow of the slipstream also causes a rolling moment around the longitudinal axis.

Note that this rolling moment caused by the corkscrew flow of the slipstream is to the right, while the rolling moment caused by torque reaction is to the left—in effect one may be counteracting the other. However, these forces vary greatly and it is up to the pilot to apply proper correction action by use of the flight controls at all times. These forces must be counteracted regardless of which is the most prominent at the time.

Gyroscopic Action

Before the gyroscopic effects of the propeller can be understood, it is necessary to understand the basic principle of a gyroscope.

All practical applications of the gyroscope are based upon two fundamental properties of gyroscopic action—rigidity in space, and precession. The one in which we are interested for this discussion is precession.

Precession is the resultant action, or deflection, of a spinning rotor when a deflecting force is applied to its rim. As can be seen in Fig. 17–44, when a force is applied, the resulting force takes effect 90° ahead of and in the direction of rotation.

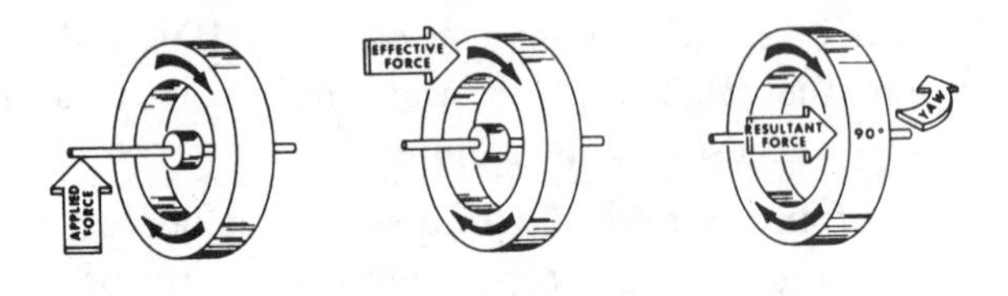

Figure 17–44 Gyroscopic Precession

The rotating propeller of an airplane makes a very good gyroscope and thus has similar properties. Any time a force is applied to deflect the propeller out of its plane of rotation, the resulting force is 90° ahead of and in the direction of rotation and in the direction of application, causing a pitching moment, a yawing moment, or a combination of the two depending upon the point at which the force was applied.

This element of torque effect has always been associated with and considered more prominent in tailwheel-type airplanes, and most often occurs when the tail is being raised during the takeoff roll (Fig. 17–45). This change in pitch attitude has the same effect as applying a force to the top of the propeller's plane of rotation. The resultant force acting 90° ahead causes a yawing moment to the left around the vertical axis. The magnitude of this moment depends on several variables, one of which is the abruptness with which the tail is raised (amount of force applied). However, precession, or gyroscopic action, occurs when a force is applied to any point on the rim of the propeller's plane of rotation; the resultant force will still be 90° from the point of application in the direction of rotation. Depending on where the force is applied, the airplane is caused to yaw left or right, to pitch up or down, or a combination of pitching and yawing.

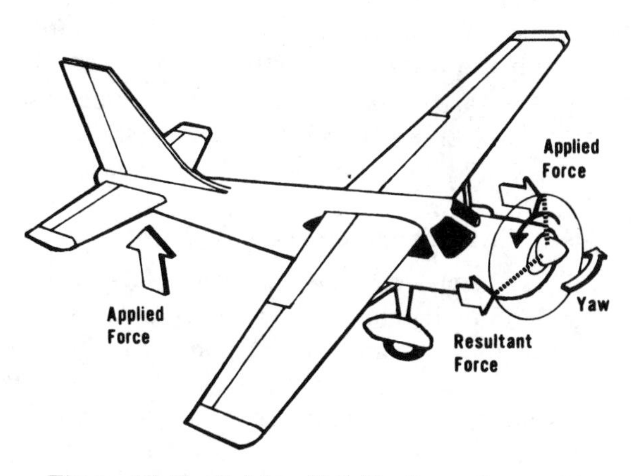

Figure 17–45 Raising Tail Produces Gyroscopic Precession

It can be said that as a result of gyroscopic action—any yawing around the vertical axis results in a pitching moment, and any pitching around the lateral axis results in a yawing moment.

To correct for the effect of gyroscopic action, it is necessary for the pilot to properly use elevator and rudder to prevent undesired pitching and yawing.

Asymmetric Loading (P Factor)

As in the past, it has been explained that when an airplane is flying with a high angle of attack, the "bite" of the downward moving blade is greater than the "bite" of the upward moving blade; thus moving the center of thrust to the right of the prop disc area—causing a yawing moment toward the left around the vertical axis. That explanation is correct; however, to prove this phenomenon, it would be necessary to work wind vector problems on each blade, which gets quite involved when considering both the angle of attack of the airplane and the angle of attack of each blade.

This asymmetric loading is caused by the resultant velocity which is generated by the combination of the velocity of the propeller blade in its plane of rotation and the velocity of the air passing horizontally through the propeller "disc." With the airplane being flown at positive angles of attack, the right (viewed from the rear) or downswinging blade, is passing through an area of resultant velocity which is greater than that affecting the left or upswinging blade. Since the propeller blade is an airfoil, increased velocity means increased lift. Therefore, the downswinging blade having more "lift" tends to pull (yaw) the airplane's nose to the left.

Simply stated, when the airplane is flying at a high angle of attack, the downward moving blade has a higher resultant velocity; therefore creating more lift than the upward moving blade (Fig. 17–46). This might be easier to visualize if we were to mount the propeller shaft perpendicular to the ground (like a helicopter). If there was no air movement at all, except that generated by the propeller itself, identical sections of each blade would have the same airspeed. But now, let's start air moving horizontally across this vertically mounted propeller. Now the blade proceeding forward into the flow of air will have a higher airspeed than the blade retreating with the airflow. Thus the blade proceeding into the horizontal airflow is creating more lift, or thrust, moving the center of thrust toward that blade. Now, let's visualize ROTATING the vertically mounted propeller shaft to shallower angles relative to the moving air (as on an airplane). This unbalanced thrust then will become proportionately smaller and continues getting smaller until it reaches the value of zero when the propeller shaft is exactly horizontal in relation to the moving air.

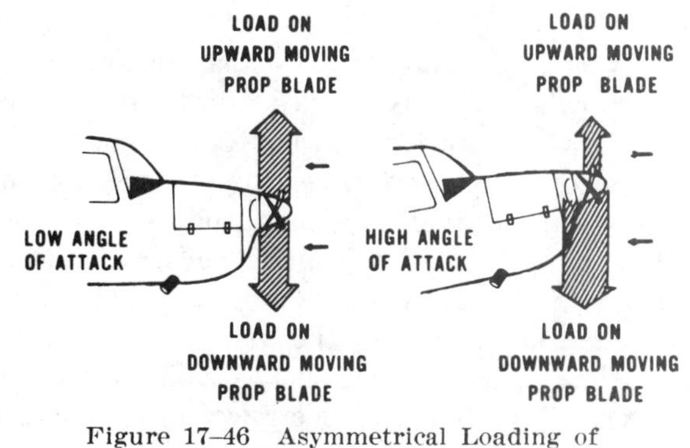

Figure 17–46 Asymmetrical Loading of Propeller (P-Factor)

Each of these four elements of torque effects vary in values with changes in flight situations. In one phase of flight, one of these elements may be more prominent than another; whereas, in another phase of flight, another element may be more prominent. The relationship of these values to each other will vary with different airplanes—depending on the AIRFRAME, ENGINE, AND PROPELLER combinations as well as other design features.

To maintain positive control of the airplane in all flight conditions, the pilot must apply the flight controls as necessary to compensate for these varying values.

Load Factors

The preceding sections only briefly considered some of the practical points of the principles of flight. As we know to become a pilot, a detailed technical course in the science of aerodynamics is not necessary. However, with responsibilities for the safety of passengers, the *competent* pilot must have a well-founded concept of the forces which act on the airplane, and the advantageous use of these forces, as well as the operating limitations of the particular airplane. Any force applied to an airplane to deflect its flight from a straight line produces a stress on its structure; the amount of this force is termed "load factor."

A load factor is the ratio of the total airload acting on the airplane to the gross weight of the airplane. For example, a load factor of 3 means that the total load on an airplane's structure is three times its gross weight. Load factors are usually expressed in terms of "G"; that is, a load factor of 3 may be spoken of as 3 G's, a load factor of 4 as 4 G's, etc.

It is interesting to note that in subjecting an airplane to 3 G's in a pullup from a dive, one will be pressed down into the seat with a force equal to three times the person's weight. Thus, an idea of the magnitude of the load factor obtained in any maneuver can be determined by considering the degree to which one is pressed down into the seat. Since the operating speed of modern airplanes has increased significantly, this effect has become so pronounced that it is a primary consideration in the design of the structure for all airplanes.

With the structural design of airplanes planned to withstand only a certain amount of overload, a knowledge of load factors has become essential for all pilots. Load factors are important to the pilot for two distinct reasons:

1. Because of the obviously dangerous overload that is possible for a pilot to impose on the aircraft structures; and
2. Because an increased load factor increases the stalling speed and makes stalls possible at seemingly safe flight speeds.

Load Factors in Airplane Design

The answer to the question "how strong should an airplane be" is determined largely by the use to which the airplane will be subjected. This is a difficult problem, because the maximum possible loads are much too high for use in efficient design. It is true that any pilot can make a very hard landing or an extremely sharp pullup from a dive which would result in abnormal loads. However, such extremely *abnormal* loads must be dismissed somewhat if we are to build airplanes that will take off quickly, land slowly, and carry a worthwhile payload.

The problem of load factors in airplane design then reduces to that of determining the highest load factors which can be expected in normal operation under various operational situations. These load factors are called "limit load factors." For reasons of safety, it is required that the airplane be designed to withstand these load factors without any structural damage. Although Federal Aviation Regulations require that the airplane structure be capable of supporting one and one-half times these limit load factors without failure, it is accepted that parts of the airplane may bend or twist under these loads and that some structural damage may occur.

This 1.5 value is called the "factor of safety" and provides, to some extent, for loads higher than those expected under normal and reasonable operation. However, this strength reserve is not something which pilots should willfully abuse; rather it is there for their protection when they encounter unexpected conditions.

The above considerations apply to all loading conditions, whether they be due to gusts, maneuvers, or landings. The gust load factor requirements now in effect are substantially the same as those which have been in existence for years. Hundreds of thousands of operational hours have proven them adequate for safety. Since the pilot has little control over gust load factors (except to reduce the airplane's speed when rough air is encountered), the gust loading requirements are substantially the same for most general aviation type airplanes regardless of their operational use. Generally speaking, the gust load factors control the design of airplanes which are intended for strictly nonacrobatic usage.

An entirely different situation exists in airplane design with maneuvering load factors. It is necessary to discuss this matter separately with respect to: (1) Airplanes which are designed in accordance with the Category System (i.e., Normal, Utility, Acrobatic); and (2) Airplanes of older design which were built to requirements which did not provide for operational categories.

Airplanes designed under the Category System are readily identified by a placard in the cockpit which states the operational category (or categories) in which the airplane is certificated. The maximum safe load factors (limit load factors) specified for airplanes in the various categories are as follows:

Category	*Limit Load*	
Normal[1]	3.8	−1.52
Utility (mild acrobatics, including spins)	4.4	−1.76
Acrobatic	6.0	−3.0

To the limit loads given above a safety factor of 50 percent is added.

[1] For airplanes with gross weight of more than 4,000 pounds, the limit load factor is reduced.

There is an upward graduation in load factor with the increasing severity of maneuvers. The Category System provides for obtaining the maximum utility of an airplane. If normal operation alone is intended, the required load factor (and consequently the weight of the airplane) is less than if the airplane is to be employed in training or acrobatic maneuvers as they result in higher maneuvering loads.

Airplanes which do not have the category placard are designs which were constructed under earlier engineering requirements in which no operational restrictions were specifically given to the pilots. For airplanes of this type (up to weights of about 4,000 pounds) the required strength is comparable to present-day utility category airplanes, and the same types of operation are permissible. For airplanes of this type over 4,000 pounds, the load factors decrease with weight so that these airplanes should be regarded as being comparable to the normal category airplanes designed under the Category System, and they should be operated accordingly.

Load Factors in Steep Turns

In a constant altitude, coordinated turn in any airplane, the load factor is the result of two forces—centrifugal force and gravity (Fig. 17–47). For any given bank angle, the rate of turn varies with the airspeed; the higher the speed, the slower the rate of turn. This compensates for added centrifugal force, allowing the load factor to remain the same.

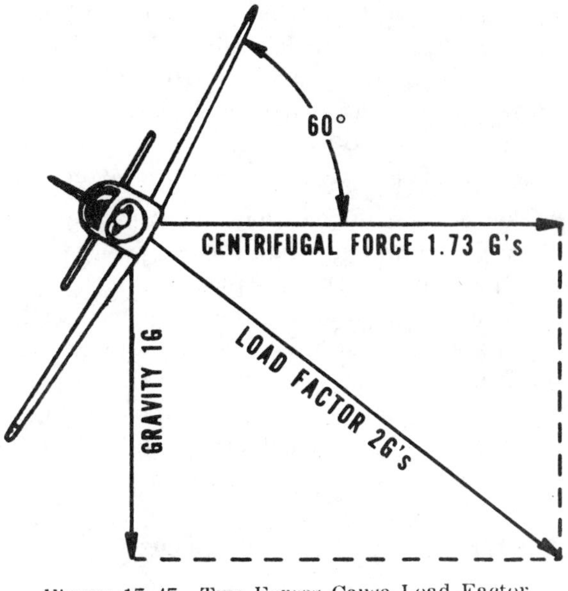

Figure 17–47 Two Forces Cause Load Factor During Turns

Fig. 17–48 reveals an important fact about turns—that the load factor increases at a terrific rate after a bank has reached 45° or 50°. The load factor for any airplane in a 60° bank is 2 G's. The load factor in an 80° bank is 5.76 G's. The wing must produce lift equal to these load factors if altitude is to be maintained.

It should be noted how rapidly the line denoting load factor rises as it approaches the 90° bank line, which it reaches only at infinity. The 90° banked, constant altitude turn mathe-

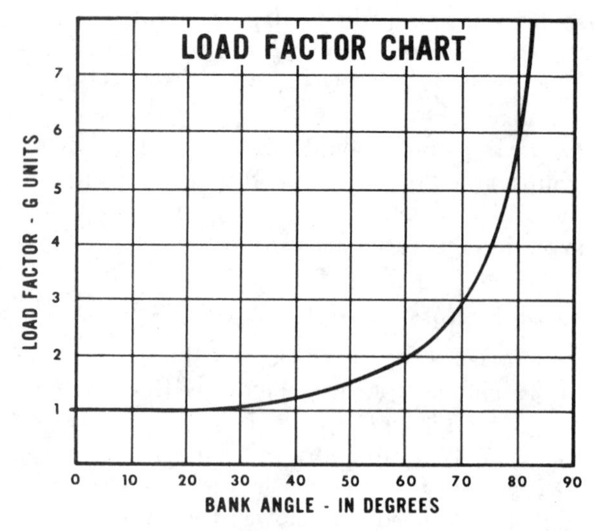

Figure 17–48 Angle of Bank Changes Load Factor

matically is not possible. True, an airplane may be banked to 90° but not in a coordinated turn; an airplane which can be held in a 90° banked slipping turn is capable of straight knife-edged flight. At slightly more than 80° the load factor exceeds the limit of 6 G's, the limit load factor of an *acrobatic airplane.*

For a coordinated, constant altitude turn, the approximate maximum bank for the average general aviation airplane is 60°. This bank and its resultant necessary power setting reach the limit of this type of airplane. An additional 10° bank will increase the load factor by approximately 1G (Fig. 17–48), bringing it close to the yield point established for these airplanes.

Load Factors and Stalling Speeds

Any airplane, within the limits of its structure, may be stalled at any airspeed. When a sufficiently high angle of attack is imposed, the smooth flow of air over an airfoil breaks up and separates, producing an abrupt change of flight characteristics and a sudden loss of lift which results in a stall.

A study of this effect has revealed that the airplane's stalling speed increases in proportion to the square root of the load factor. This means that an airplane with a normal unaccelerated stalling speed of 50 knots can be stalled at 100 knots by inducing a load factor of 4 G's. If it were possible for this airplane to withstand a load factor of 9, it could be stalled at a speed of 150 knots. Therefore, a competent pilot should be aware of the following:

1. The danger of inadvertently stalling the airplane by increasing the load factor, as in a steep turn or spiral; and
2. That in intentionally stalling an airplane above its design maneuvering speed, a tremendous load factor is imposed.

Reference to the charts in Figs. 17–48 and 49 will show that by banking the airplane to just beyond 72° in a steep turn produces a load factor of 3, and the stalling speed is increased significantly. If this turn is made in an airplane with a normal unaccelerated stalling speed of 45 knots, the airspeed must be kept above 90 knots to prevent inducing a stall. A similar effect is experienced in a quick pullup, or any maneuver producing load factors above 1 G. This has been the cause of accidents resulting from a sudden, unexpected loss of control, particularly in a steep turn or abrupt application of the back elevator control near the ground.

Since the load factor squares as the stalling speed doubles, it may be realized that tremendous loads may be imposed on structures by stalling an airplane at relatively high airspeeds.

The maximum speed at which an airplane may be stalled safely is now determined for all new designs. This speed is called the "design maneuvering speed," (V_A) and is required to be entered in the FAA approved flight manual of all recently designed airplanes. For older general aviation airplanes this speed will be approximately 1.7 times the normal stalling speed. Thus, an older airplane which normally stalls at 60 knots must never be stalled at above 102 knots. 60 kts. × 1.7=102 kts.). An airplane with a normal stalling speed of 60 knots will undergo, when stalled at 102 knots, a load factor equal to the square of the increase in speed or 2.89 G's (1.7×1.7=2.89 G's). (The above figures are an approximation to be considered as a guide and are not the exact answers to any set of problems. The design maneuvering speed should be determined from the particular airplane's operating limitations when provided by the manufacturer.)

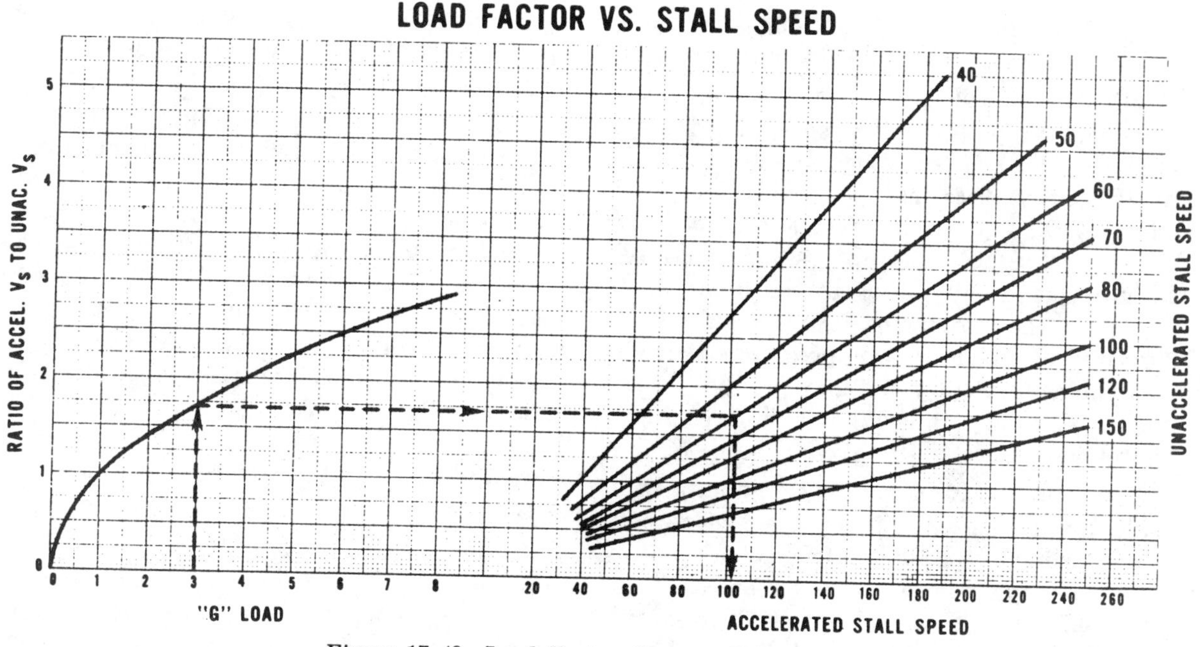

Figure 17–49 Load Factor Changes Stall Speed

Since the leverage in the control system varies with different airplanes and some types employ "balanced" control surfaces while others do not, the pressure exerted by the pilot on the controls cannot be accepted as an index of the load factors produced in different airplanes. In most cases, load factors can be judged by the experienced pilot from the feel of seat pressure. They can also be measured by an instrument called an "accelerometer," but since this instrument is not common in general aviation training airplanes, the development of the ability to judge load factors from the feel of their effect on the body is important. A knowledge of the principles outlined above is essential to the development of this ability to estimate load factors.

A thorough knowledge of load factors induced by varying degrees of bank, and the significance of design maneuvering speed (V_A) will aid in the prevention of two of the most serious types of accidents:

1. Stalls from steep turns or excessive maneuvering near the ground; and
2. Structural failures during acrobatics or other violent maneuvers resulting from loss of control.

Load Factors and Flight Maneuvers

Critical load factors apply to all flight maneuvers except unaccelerated straight flight where a load factor of 1 G is always present. Certain maneuvers considered in this section are known to involve relatively high load factors.

Turns. Increased load factors are a characteristic of all banked turns. As noted in the section Load Factors in Steep Turns and particularly Figs. 17–48 and 17–49, load factors become significant both to flight performance and to the load on wing structure as the bank increases beyond approximately 45°.

The yield factor of the average light plane is reached at a bank of approximately 70°–75°, and the stalling speed is increased by approximately one-half at a bank of approximately 63°.

Stalls. The normal stall entered from straight level flight, or an unaccelerated straight climb, will not produce added load factors beyond the 1 G of straight-and-level flight. As the stall occurs, however, this load factor may be reduced toward zero, the factor at which nothing seems to have weight; and the pilot has the feeling of "floating free in space." In the event recovery is effected by snapping the elevator control forward, negative load factors, those which impose a down load on the wings and raise the pilot from the seat, may be produced.

During the pullup following stall recovery, significant load factors sometimes are induced. Inadvertently these may be further increased during excessive diving (and consequently high airspeed) and abrupt pullups to level flight. One usually leads to the other, thus increasing the load factor. Abrupt pullups at high diving speeds may impose critical loads on airplane structures and may produce recurrent or secondary stalls by increasing the angle of attack to that of stalling.

As a generalization, a recovery from a stall made by diving only to cruising or design maneuvering airspeed, with a gradual pullup as soon as the airspeed is safely above stalling, can be effected with a load factor not to exceed 2 or 2.5 G's. A higher load factor should never be necessary unless recovery has been effected with the airplane's nose near or beyond the vertical attitude, or at extremely low altitudes to avoid diving into the ground.

Spins. Since a stabilized spin is not essentially different from a stall in any element other than rotation, the same load factor considerations apply as those which apply to stall recovery. Since spin recoveries usually are effected with the nose much lower than is common in stall recoveries, higher airspeeds and consequently higher load factors are to be expected. The load factor in a proper spin recovery will usually be found to be about 2.5 G's.

The load factor during a spin will vary with the spin characteristics of each airplane but is usually found to be slightly above the 1 G of level flight. There are two reasons this is true:

1. The airspeed in a spin is very low; usually within 2 knots of the unaccelerated stalling speeds; and
2. The airplane pivots, rather than turns, while it is in a spin.

High-Speed Stalls. The average light plane is not built to withstand the repeated application of load factors common to high-speed stalls. The load factor necessary for these maneuvers produces a stress on the wings and tail structure, which does not leave a reasonable margin of safety in most light airplanes.

The only way this stall can be induced at an airspeed above normal stalling involves the imposition of an added load factor, which may be accomplished by a severe pull on the elevator control. A speed of 1.7 times stalling speed (about 102 knots in a light airplane with a stalling speed of 60 knots) will produce a load factor of 3 G's. Further, only a very narrow margin for error can be allowed for acrobatics in light airplanes. To illustrate how rapidly the load factor increases with airspeed, a high-speed stall at 112 knots in the same airplane would produce a load factor of 4 G's.

Chandelles and Lazy Eights. It would be difficult to make a definite statement concerning load factors in these maneuvers as both involve smooth, shallow dives and pullups. The load factors incurred depend directly on the speed of the dives and the abruptness of the pullups.

Generally, the better the maneuver is performed, the less extreme will be the load factor induced. A chandelle or lazy eight, in which the pullup produces a load factor greater than 2 G's will not result in as great a gain in altitude, and in low-powered airplanes it may result in a net loss of altitude.

The smoothest pullup possible, with a moderate load factor, will deliver the greatest gain in altitude in a chandelle and will result in a better overall performance in both chandelles and lazy eights. Further, it will be noted that recommended entry speed for these maneuvers is generally near the manufacturer's design maneuvering speed, thereby allowing maximum development of load factors without exceeding the load limits.

Rough Air. All certificated airplanes are designed to withstand loads imposed by gusts of considerable intensity. Gust load factors increase with increasing airspeed and the strength used for design purposes usually corresponds to the highest level flight speed. In extremely rough air, as in thunderstorms or frontal conditions, it is wise to reduce the speed to the design maneuvering speed. Regardless of the speed held, there may be gusts that can produce loads which exceed the load limits.

Most airplane flight manuals now include turbulent air penetration information. Operators of modern airplanes, capable of a wide range of speeds and altitudes, are benefited by this added feature both in comfort and safety. In this connection it is to be noted that the maximum "never exceed" placard dive speeds are determined for smooth air only. High-speed dives or acrobatics involving speed above the known maneuvering speed should never be practiced in rough or turbulent air.

In summary, it must be remembered that load factors induced by intentional acrobatics, abrupt pullups from dives, high-speed stalls, and gusts at high airspeeds all place added stress on the entire structure of an airplane.

Stress on the structure involves forces on any part of the airplane. There is a tendency for the uninformed to think of load factors only in terms of their effect on spars and struts. Most structural failures due to excess load factors involve rib structure within the leading and trailing edges of wings and tail group. The critical area of fabric-covered airplanes is the covering about one-third of the chord aft on the top surface of the wing.

The cumulative effect of such loads over a long period of time may tend to loosen and weaken vital parts so that actual failure may occur later when the airplane is being operated in a normal manner.

Vg Diagram

The flight operating strength of an airplane is presented on a graph whose horizontal scale is based on load factor (Fig. 17–19). The diagram is called a V-g diagram—velocity versus "g" loads or load factor. Each airplane has its own V-g diagram which is valid at a certain weight and altitude.

The lines of maximum lift capability (curved lines) are the first items of importance on the V-g diagram. The subject airplane in the illustration is capable of

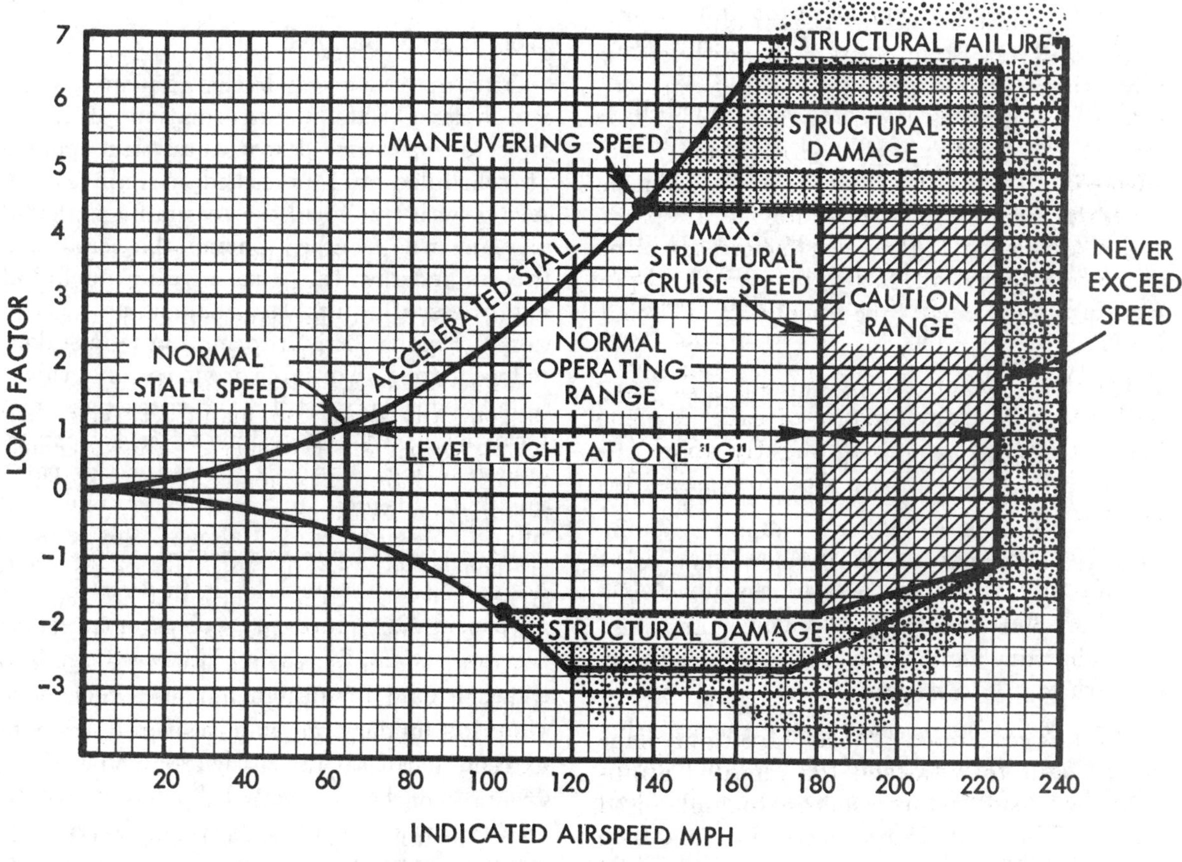

Figure 17–50 Typical Vg Diagram

developing no more than one positive "g" at 62 m.p.h., the wing level stall speed of the airplane. Since the maximum load factor varies with the square of the airspeed, the maximum positive lift capability of this airplane is 2 "g" at 92 m.p.h., 3 "g" at 112 m.p.h., 4.4 "g" at 137 m.p.h., etc. Any load factor above this line is unavailable aerodynamically; i.e., the subject airplane cannot fly above the line of maximum lift capability (it will stall). Essentially the same situation exists for negative lift flight with the exception that the speed necessary to produce a given negative load factor is higher than that to produce the same positive load factor.

If the subject airplane is flown at a positive load factor greater than the positive limit load factor of 4.4, structural damage will be possible. When the airplane is operated in this region, objectionable permanent deformation of the primary structure may take place and a high rate of fatigue damage is incurred. Operation above the limit load factor must be avoided in normal operation.

There are two other points of importance on the V-g diagram. Point A is the intersection of the positive limit load factor and the line of maximum positive lift capability. The airspeed at this point is the minimum airspeed at which the limit load can be developed aerodynamically. Any airspeed greater than point A provides a positive lift capability sufficient to damage the airplane; any airspeed less than point A does NOT provide positive lift capability sufficient to cause damage from excessive flight loads. The usual term given to the speed at point A is the "maneuvering speed," since consideration of subsonic aerodynamics would predict minimum usable turn radius to occur at this condition. The maneuver speed is a valuable reference point since an airplane operating below this point cannot produce a damaging positive flight load. Any combination of maneuver and gust cannot create damage due to excess airload when the airplane is below the maneuver speed.

Point B is the intersection of the negative limit load factor and line of maximum negative lift capability. Any airspeed greater than point B provides a negative lift capability sufficient to damage the airplane; any airspeed less than point B does not provide negative lift capability sufficient to damage the airplane from excessive flight loads.

The limit airspeed (or redline speed) is a design reference point for the airplane—the subject airplane is limited to 225 m.p.h. If flight is attempted beyond the limit airspeed structural damage or structural failure may result from a variety of phenomena.

Thus, the airplane in flight is limited to a regime of airspeeds and g's which do not exceed the limit (or redline) speed, do not exceed the limit load factor, and cannot exceed the maximum lift capability. The airplane must be operated within this "envelope" to prevent structural damage and ensure that the anticipated service life of the airplane is obtained. The pilot must appreciate the V-g diagram as describing the allowable combination of airspeeds and load factors for safe operation. Any maneuver, gust, or gust plus maneuver outside the structural envelope can cause structural damage and effectively shorten the service life of the airplane.

Weight and Balance

Often a pilot regards the airplane's weight and balance data as information of interest only to engineers, dispatchers, and operators of scheduled and nonscheduled air carriers. Along with this idea, the reasoning is that the airplane was weighed during the certification process and that this data is valid indefinitely, regardless of equipment changes or modifications. Further, this information is mistakenly reduced to a workable routine or "rule of thumb" such as: "If I have three passengers, I can load only 100 gallons of fuel; four passengers—70 gallons." Admittedly, this rule of thumb is adequate in many cases, but as the subject "Weight and Balance" suggests, we are concerned not only with the weight of the airplane but also the location of its center of gravity. The importance of the C.G. should have become apparent in the discussions of stability, controllability, and performance. If all pilots understood and respected the effect of C.G. on an airplane, then one type of accident would be eliminated from our records: "PRIMARY CAUSE OF ACCIDENT—AIRPLANE CENTER OF GRAVITY OUT

OF REARWARD LIMITS AND UNEQUAL LOAD DISTRIBUTION RESULTING IN AN UNSTABLE AIRPLANE. PILOT LOST CONTROL OF AIRPLANE ON TAKEOFF AND CRASHED."

The reasons airplanes are so certificated are obvious when one gives it a little thought. For instance, it is of added value to the pilot to be able to carry extra fuel for extended flights when the full complement of passengers is not to be carried. Further, it is unreasonable to forbid the carriage of baggage when it is only during spins that its weight will adversely affect the airplane's flight characteristics. Weight and balance limits are placed on airplanes for two principal reasons:

1. Because of the effect of the weight on the airplane's primary structure and its performance characteristics; and
2. Because of the effect the location of this weight has on flight characteristics, particularly in stall and spin recovery and stability.

Effects of Weight on Flight Performance

The takeoff/climb and landing performance of an airplane are determined on the basis of its maximum allowable takeoff and landing weights. A heavier gross weight will result in a longer takeoff run and shallower climb, and a faster touchdown speed and longer landing roll. Even a minor overload may make it impossible for the airplane to clear an obstacle which normally would not have been seriously considered during takeoffs under more favorable conditions.

The detrimental effects of overloading on performance are not limited to the immediate hazards involving takeoffs and landings. Overloading has an adverse effect on all climb and cruise performance which leads to overheating during climbs, added wear on engine parts, increased fuel consumption, slower cruising speeds, and reduced range.

The manufacturers of modern airplanes furnish weight and balance data with each airplane produced. Generally, this information may be found in the FAA approved Airplane Flight Manual or pilot's operating handbook. With the advancements in airplane design and construction in recent years has come the development of "easy to read charts" for determining weight and balance data. Increased performance and load carrying capability of these airplanes require strict adherence to the operating limitations prescribed by the manufacturer. Deviations from the recommendations can result in structural damage or even complete failure of the airplane's structure. Even if an airplane is loaded well within the maximum weight limitations, it is imperative that weight distribution be within the limits of center of gravity location. The preceding brief study of aerodynamics and load factors points out the reasons for this precaution. The following discussion is background information into some of the reasons why weight and balance conditions are important to the safe flight of an airplane.

The pilot is often completely unaware of the weight and balance limitations of the airplane being flown and of the reasons for these limitations. In some airplanes it is not possible to fill all seats, baggage compartments, and fuel tanks, and still remain within approved weight or balance limits. As an example, in several popular four-place airplanes the fuel tanks may not be filled to capacity when four occupants and their baggage are carried. In a certain two-place airplane, no baggage may be carried in the compartment aft of the seats when spins are to be practiced.

Effects of Weight on Airplane Structure

The effect of additional weight on the wing structure of an airplane is not readily apparent. Airworthiness requirements prescribe that the structure of an airplane certificated in the normal category (in which acrobatics are prohibited) must be strong enough to withstand a load factor of 3.8 to take care of dynamic loads caused by maneuvering and gusts. This means that the primary structure of the airplane can withstand a load of 3.8 times the approved gross weight of the airplane without structural failure occurring. If this is accepted as indicative of the load factors which may be imposed during operations for which the airplane is intended, a 100-pound overload imposes a potential structural overload of 380 pounds. The same consideration is even more impressive in the case of utility and acrobatic category airplanes, which have load factor requirements of 4.4 and 6.0 respectively.

Structural failures which result from overloading may be dramatic and catastrophic, but more often they affect structural components progressively in a manner which is difficult to detect and expensive to repair. One of the most serious results of habitual overloading is that its results tend to be cumulative, and may result in structural failure later during completely normal operations. The additional stress placed on structural parts by overloading is believed to accelerate the occurrence of metallic fatigue failures.

A knowledge of load factors imposed by flight maneuvers and gusts will emphasize the consequences of an increase in the gross weight of an airplane. The structure of an airplane about to undergo a load factor of 3 G's, as in the recovery from a steep dive, must be prepared to withstand an added load of 300 pounds for each 100-pound increase in weight. It should be noted that this would be imposed by the addition of about 16 gallons of unneeded fuel in a particular airplane. The FAA certificated civil airplane has been analyzed structurally, and tested for flight at the maximum gross weight authorized and within the speeds posted for the type of flights to be performed. Flights at weights in excess of this amount are quite possible and often are well within the performance capabilities of an airplane. Nonetheless, this fact should not be allowed to mislead the pilot, as he may not realize that loads for which the airplane was not designed are being imposed on all or some part of the structure.

In loading an airplane with either passengers or cargo, the structure must be considered. Seats, baggage compartments, and cabin floors are designed for a certain load or concentration of load and no more. As an example, a light-plane baggage compartment may be placarded for 20 pounds because of the limited strength of its supporting structure even though the airplane may not be overloaded or out of center of gravity limits with more weight at that location.

Effects of Weight on Stability and Controllability

The effects that overloading has on stability also are not generally recognized. An airplane which is observed to be quite stable and controllable when loaded normally, may be discovered to have very different flight characteristics when it is overloaded. Although the distribution of weight has the most direct effect on this, an increase in the airplane's gross weight may be expected to have an adverse effect on stability, regardless of location of the center of gravity.

The stability of many certificated airplanes is completely unsatisfactory if the gross weight is exceeded.

Effect of Load Distribution

The effect of the position of the center of gravity on the load imposed on an airplane's wing in flight is not generally realized, although it may be very significant to climb and cruising performance. Contrary to the beliefs of some pilots, an airplane with forward loading is "heavier" and consequently, slower than the same airplane with the center of gravity further aft.

Fig. 17–51 illustrates the reason for this. With forward loading, "nose-up" trim is required in most airplanes to maintain level cruising flight. Nose-up trim involves setting the tail surfaces to produce a greater down load on the aft portion of the fuselage, which adds to the wing loading and the total lift required from the wing if altitude is to be

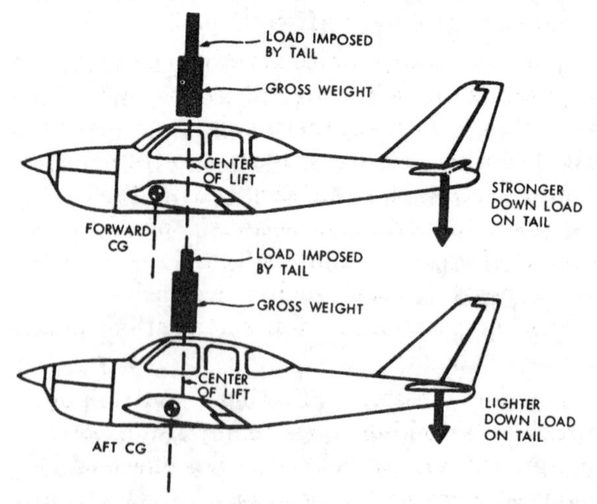

Figure 17–51 Load Distribution Affects Balance

maintained. This requires a higher angle of attack of the wing, which results in more drag and, in turn, produces a higher stalling speed.

With aft loading and "nose-down" trim, the tail surfaces will exert less down load, relieving the wing of that much wing loading and lift required to maintain altitude. The required angle of attack of the wing is less, so the drag is less, allowing for a faster cruise speed. Theoretically, a neutral load on the tail surfaces in cruising flight would produce the most efficient overall performance and fastest cruising speed, but would also result in instability. Consequently, modern airplanes are designed to require a down load on the tail for stability and controllability.

Remember that a zero indication on the trim tab control is not necessarily the same as "neutral trim" because of the force exerted by downwash from the wings and the fuselage on the tail surfaces.

The effects of the distribution of the airplane's useful load have a significant influence on its flight characteristics, even when the load is within the center of gravity limits and the maximum permissible gross weight. Important among these effects are changes in controllability, stability, and the actual load imposed on the wing.

Generally speaking, an airplane becomes less controllable, especially at slow flight speeds, as the center of gravity is moved further aft. An airplane which cleanly recovers from a prolonged spin with the center of gravity at one position may fail completely to respond to normal recovery attempts when the center of gravity is moved aft by 1 or 2 inches.

It is common practice for airplane designers to establish an aft center of gravity limit that is within 1 inch of the maximum which will allow normal recovery from a one-turn spin. When certificating an airplane in the utility category to permit intentional spins, the aft center of gravity limit is usually established at a point several inches forward of that which is permissible for certification in the normal category.

Another factor affecting controllability which is becoming more important in current designs of large airplanes is the effect of long moment arms to the positions of heavy equipment and cargo. The same airplane may be loaded to maximum gross weight within its center of gravity limits by concentrating fuel, passengers, and cargo near the design center of gravity; or by dispersing fuel and cargo loads in wingtip tanks and cargo bins forward and aft of the cabin.

With the same total weight and center of gravity, maneuvering the airplane or maintaining level flight in turbulent air will require the application of greater control forces when the load is dispersed. This is true because of the longer moment arms to the positions of the heavy fuel and cargo loads which must be overcome by the action of the control surfaces. An airplane with full outboard wing tanks or tip tanks tends to be sluggish in roll when control situations are marginal, while one with full nose and aft cargo bins tends to be less responsive to the elevator controls.

The rearward center of gravity limit of an airplane is determined largely by considerations of stability. The original airworthiness requirements for a type certificate specify that an airplane in flight at a certain speed will dampen out vertical displacement of the nose within a certain number of oscillations. An airplane loaded too far rearward may not do this; instead when the nose is momentarily pulled up, it may alternately climb and dive becoming steeper with each oscillation. This instability is not only uncomfortable to occupants but it could even become dangerous by making the airplane unmanageable under certain conditions.

The recovery from a stall in any airplane becomes progressively more difficult as its center of gravity moves aft. This is particularly important in spin recovery, as there is a point in rearward loading of any airplane at which a "flat" spin will develop. A flat spin is one in which centrifugal force, acting through a center of gravity located well to the rear, will pull the tail of the airplane out away from the axis of the spin, making it impossible to get the nose down and recover.

An airplane loaded to the rear limit of its permissible center of gravity range will handle differently in turns and stall maneuvers and have different landing characteristics than when it is loaded near the forward limit.

The forward center of gravity limit is determined by a number of considerations. As a safety measure, it is required that the trimming device, whether tab or adjustable stabilizer, be capable of holding the airplane in a normal glide with the power off. A conventional airplane must be capable of a full stall, power-off landing in order to ensure minimum landing speed in emergencies. A tailwheel type airplane loaded excessively nose heavy will be difficult to taxi, particularly in high winds. It can be nosed over easily by use of the brakes, and it will be difficult to land without bouncing since it tends to pitch down on the wheels as it is slowed down and flared for landing. Steering difficulties on the ground may occur in nosewheel-type airplanes, particularly during the landing roll and takeoff.

1. The CG position influences the lift and angle of attack of the wing, the amount and direction of force on the tail, and the degree of deflection of the stabilizer needed to supply the proper tail force for equilibrium. The latter is very important because of its relationship to elevator control force.
2. The airplane will stall at a higher speed with a forward CG location. This is because the stalling angle of attack is reached at a higher speed due to increased wing loading.
3. Higher elevator control forces normally exist with a forward CG location due to the increased stabilizer deflection required to balance the airplane.
4. The airplane will cruise faster with an aft CG location because of reduced drag. The drag is reduced because a smaller angle of attack and less downward deflection of the stabilizer are required to support the airplane and overcome the nose-down pitching tendency.
5. The airplane becomes less and less stable as the CG is moved rearward. This is because when the angle of attack is increased it tends to result in additional increased angle of attack. Therefore, the wing contribution to the airplane's stability is now decreased, while the tail contribution is still stabilizing. When the point is reached that the wing and tail contributions balance, then neutral stability exists. Any CG movement further aft will result in an unstable airplane.
6. A forward CG location increases the need for greater back elevator pressure. The elevator may no longer be able to oppose any increase in nose-down pitching. Adequate elevator control is needed to control the airplane throughout the airspeed range down to the stall.

Weight and Balance Control

The aircraft manufacturer and the Federal Aviation Administration have major roles in designing and certificating the aircraft with a safe and workable means of controlling weight and balance. If the prototype aircraft has weight and balance control problems which are potentially dangerous or complicated, design changes must be made to ensure the airworthiness of the aircraft.

When an airplane is placarded requiring that it must be flown solo from a specified seat, that a given fuel tank is to be emptied first, or that a compartment or seat is to be left empty under certain conditions, the pilot may rest assured that the placard is necessary for some well-founded reason. Such placards must be maintained in the airplane and observed.

Weight and balance control is a matter of serious concern to all pilots as well as many other people who are involved in the flight. The pilot has to personally assume the responsibility because he has control over both the loading and the fuel management, the two variable factors which can change both total weight and CG location. Weight and balance information is available to the pilot in the form of aircraft records, operating handbooks, and placards in baggage compartments and on fuel tank caps. It is the aircraft owner or operator's responsibility to make certain that up-to-date information is available in the aircraft for the pilot's use.

It must be stressed that the empty weight and moment given in most manufacturers' handbooks are for the basic airplane prior to the installation of additional optional equipment. When the owner later adds such items as radio navigation equipment, auto pilot, de-icers, etc., the empty weight and the moment are changed. These changes must be recorded in the airplane's weight and balance data and used in all computations. In addition, the actual weight of occupants, baggage, fuel, and other useful load should be used rather than the sample weights given in the manufacturers' handbooks.

The owner or operator of the aircraft should ensure that maintenance personnel make appropriate entries in the aircraft maintenance records when repairs or modifications have been accomplished, and when optional equipment (radios, etc.) have been installed, or removed. Weight changes must be accounted for and proper notations made in the weight and balance records. Without such notations, the pilot has no foundation upon which to base his calculations and decisions.

The airplane's latest Weight and Balance Loading Form and/or its maintenance record will list the empty weight, the useful load, and the empty center of gravity location. If it is possible to load the airplane out of center of gravity limits, it will also include a specific listing of the most forward and the most rearward allowable limits. This information should be consulted when a pilot proposes to load and fly an airplane with which he or she is not thoroughly familiar.

Although seemingly complex at times, all weight and balance problems are based on the following moment equation:

$$\text{Moment} = \text{Weight} \times \text{Arm}$$

This equation is the basic equation used to find the center of gravity location of an airplane and/or its components. By rearrangement of this equation to the forms, $\text{Weight} = \text{Moment} \div \text{Arm}$, and $\text{Arm} = \text{Moment} \div \text{Weight}$, with any two known values, the third value can be found.

Weight and balance computations are sometimes simplified by two graphic aids—the loading graph and the center of gravity moment envelope. The loading graph (Fig. 17–52) is typical of those found in many general aviation airplane owner's manuals. This graph, in effect, multiplies weight by arm giving moment, then divides the moment by a reduction factor, giving an index number. Weight values appear along the left side of the graph. The moment/1,000 or index numbers are along the bottom. In this chart, each line representing a useful load item is labeled. To determine the moment of any load item, find the weight along the left margin, then project a line directly to a point of intersection with the appropriate load item line. The CG moment envelope (Fig. 17–53) allows the pilot to bypass the computation of a CG number. It gives an acceptable range of index numbers for any airplane weight from minimum to maximum. If the lines from total weight and total moment intersect within the envelop, the airplane is within weight and balance limits.

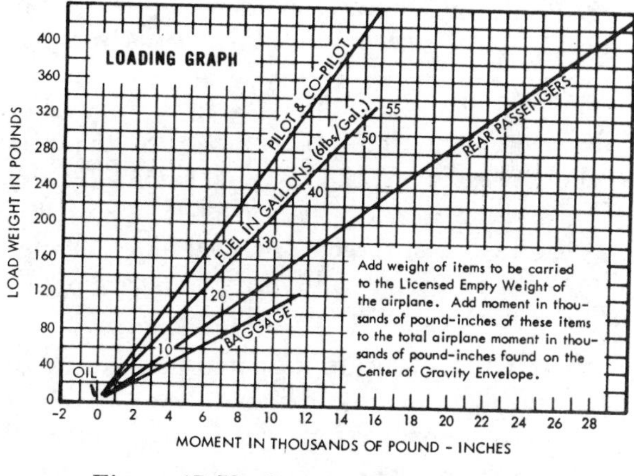

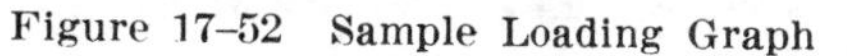
Figure 17–52 Sample Loading Graph

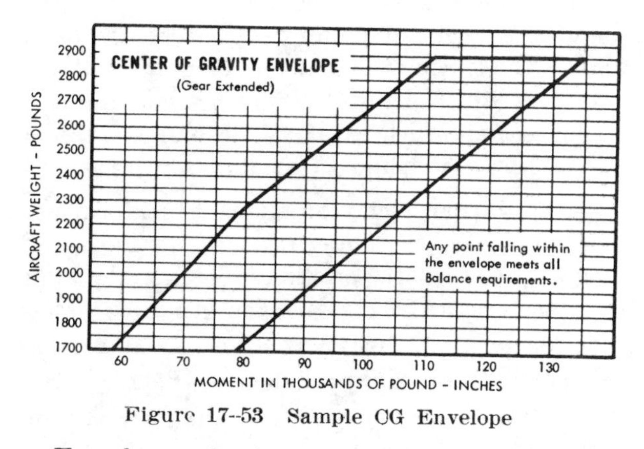

Figure 17–53 Sample CG Envelope

For those who have a need for greater appreciation of weight and balance control as it relates to safety of flight, a more comprehensive text on the subject is contained in the FAA Advisory Circular 91–23A, *Pilot's Weight and Balance Handbook.*

Airplane Performance

The clean, sleek appearance and the splendid performance characteristics of modern airplanes, even the most inexpensive training airplanes, reflect the demands of modern travel, business, and industry. The present-day airplane is clean and sleek because aerodynamic cleanness of line, and efficiency of design, result in greater range, speed, and payload at the least operating cost. Although this is especially important in the functions of large jet transports, it is also a principal factor in the operation of executive and personal type airplanes.

"Performance" is a term used to describe the ability of an airplane to accomplish certain things which make it useful for certain purposes. For example, the ability of the airplane to land and take off in a very short distance is an important factor to the pilot who operates in and out of confined fields. The ability to carry heavy loads, fly at high altitudes at fast speeds, or travel long distances is essential performance for operators of airline and executive type airplanes.

The chief elements of performance are the takeoff and landing distance, rate of climb, ceiling, payload, range, speed, maneuverability, stability, and fuel economy. Some of these factors are often directly opposed: for example, high speed versus shortness of landing distance; long range versus great payload; and high rate of climb versus fuel economy. It is the preeminence of one or more of these factors which dictates differences between airplanes and which explains the high degree of specialization found in modern airplanes.

The various items of airplane performance result from the combination of airplane and powerplant characteristics. The aerodynamic characteristics of the airplane generally define the power and thrust *requirements* at various conditions of flight while powerplant characteristics generally define the power and thrust *available* at various conditions of flight. The matching of the aerodynamic configuration with the powerplant is accomplished by the manufacturer to provide maximum performance at the specific design condition, e.g., range, endurance, climb, etc.

Straight-and-Level Flight

All of the principal items of flight performance involve steady-state flight conditions and equilibrium of the airplane. For the airplane to remain in steady level flight, equilibrium must be obtained by a lift equal to the airplane weight and a powerplant thrust equal to the airplane drag. Thus, the airplane drag defines the *thrust required* to maintain steady level flight.

All parts of the airplane that are exposed to the air contribute to the drag, though only the wings provide lift of any significance. For this reason, and certain others related to it, the total drag may be divided into two parts, the wing drag (induced) and the drag of everything but the wings (parasite).

The total power required for flight then can be considered as the sum of induced and parasite effects; that is, the total drag of the airplane. Parasite drag is the sum of pressure and friction drag which is due to the airplane's basic configuration and, as defined, is independent of lift. Induced drag is the undesirable but unavoidable consequence of the development of lift.

While the parasite drag predominates at high speed, induced drag predominates at low speed (Fig. 17–18). For example, if an airplane in a steady flight condition at 100 knots is then accelerated to 200 knots, the *parasite* drag becomes four times as great but the power required to overcome that drag is eight times the original value. Conversely, when the airplane is operated in steady level flight at twice as great a speed, the *induced* drag is one-fourth the original value and the power required to overcome that drag is only one-half the original value.

The wing or induced drag changes with speed in a very different way, because of the changes in the angle of attack. Near the stalling speed the wing is inclined to the relative wind at nearly the stalling angle, and its drag is very strong. But at ordinary flying speeds, with the angle of attack nearly zero, the wing cuts through the air almost like a knife, and the drag is minimal. After attaining a certain high speed, the angle of attack changes very little with any further increase in speed and the drag of the wing increases in direct proportion to any further increase in speed. This does not consider the factor of compressibility drag which is involved at speeds beyond the top speed of most general aviation airplanes.

To sum up these changes, for a typical, moderately powered airplane: As the speed increases from stalling speed to top speed, the induced drag decreases and parasite drag increases. As a result *the total drag decreases for the first part of the range and then increases again.*

When the airplane is in steady, level flight, the condition of equilibrium must prevail. The unaccelerated condition of flight is achieved with the airplane trimmed for lift equal to weight and the powerplant set for a thrust to equal the airplane drag.

The maximum level flight speed for the airplane will be obtained when the power or thrust required equals the maximum power or thrust available from the powerplant (Fig. 17–54). The minimum level flight airspeed is not usually defined by thrust or power requirement since conditions of stall or stability and control problems generally predominate.

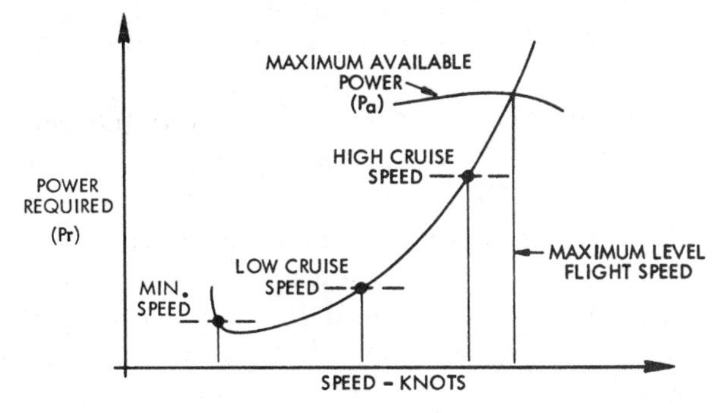

Figure 17–54 Power Versus Speed

Climb Performance

Increasing the power by advancing the throttle produces a marked difference in the rate of climb. Climb depends upon the *reserve* power or thrust. Reserve power is the available power over and above that required to maintain horizontal flight at a given speed. Thus, if an airplane is equipped with an engine which produces 200 total available horsepower and the airplane requires only 130 horsepower at a certain level flight speed, the power available for climb is 70 horsepower.

Although we sometimes use the terms "power" and "thrust" interchangeably, erroneously implying that they are synonymous, it is well to distinguish between the two when discussing climb performance. Work is the product of a force moving through a distance and is usually independent of time. Work is measured by several standards, the most common unit is called a "foot-pound." If a 1-pound mass is raised 1 foot, a work unit of 1 foot-pound has been performed. The common unit of mechanical power is horsepower; one horsepower is work equivalent to lifting 33,000 pounds a vertical distance of 1 foot in 1 minute. The term, "power," implies work *rate* or units of work per unit of time, and as such is a function of the speed at which the force is developed. "Thrust," also a function of work, means the force which imparts a change in the velocity of a mass. This force is measured in pounds but has no element of time or rate. It can be said then, that during a steady climb, the *rate* of climb is a function of excess thrust.

When the airplane is in steady level flight or with a slight angle of climb, the vertical component of lift is very nearly the same as the actual total lift. Such climbing flight would exist with the lift very nearly equal to the weight. The net thrust of the powerplant may be inclined relative to the flightpath but this effect will be neglected here for the sake of simplicity. Although the weight of the airplane acts vertically, a component of weight will act rearward along the flightpath (Fig. 17–55).

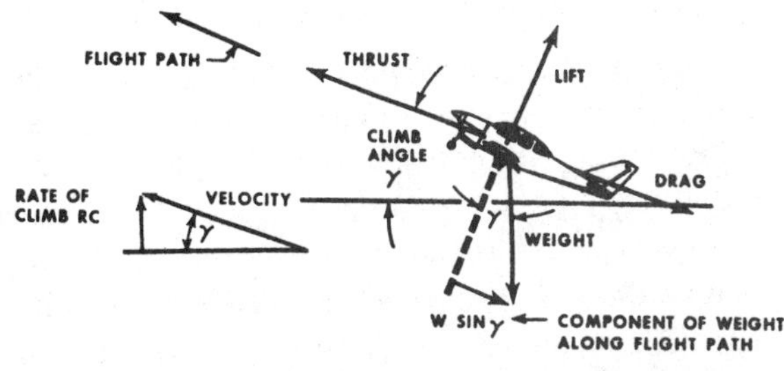

Figure 17–55 Weight Has Rearward Component

If it is assumed that the airplane is in a steady climb with essentially a small inclination of the flightpath, the summation of forces along the flightpath resolves to the following:

Forces forward = Forces aft

The *basic* relationship neglects some of the factors which may be of importance for airplanes of very high climb performance. (For example, a more detailed consideration would account for the inclination of thrust from the flightpath, lift not being equal to weight, a subsequent change of induced drag, etc.) However, this *basic* relationship will define the principal factors affecting climb performance.

This relationship means that, for a given weight of the airplane, *the angle of climb* depends on the difference between thrust and drag, or the excess thrust (Fig. 17–56). Of course, when the excess thrust is zero, the inclination of the flightpath is zero and the airplane will be in steady, level flight. When the thrust is greater than the drag, the excess thrust will allow a climb angle depending on the value of excess thrust. On the other hand, when the thrust is less than the drag, the deficiency of thrust will allow an angle of descent.

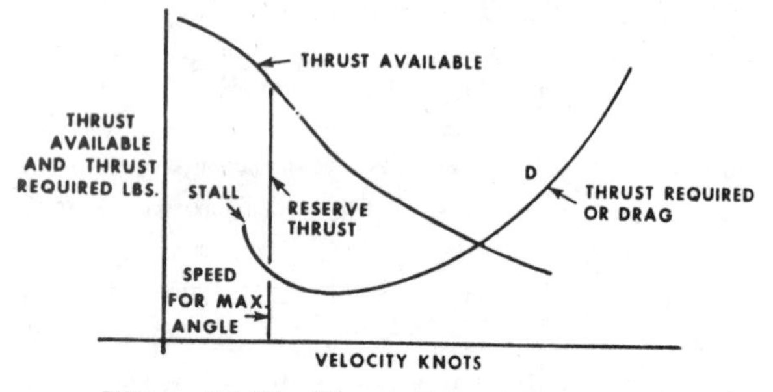

Figure 17–56 Thrust Versus Climb Angle

The most immediate interest in the climb angle performance involves obstacle clearance. The most obvious purpose for which it might be used is to clear obstacles when climbing out of short or confined airports.

The maximum *angle of climb* would occur where there exists the greatest difference between thrust available and thrust required; i.e., for the propeller powered airplane, the maximum excess thrust and angle of climb will occur at some speed just above the stall speed. Thus, if it is necessary to clear an obstacle after takeoff, the propeller powered airplane will attain maximum angle of climb at an airspeed close to—if not at—the takeoff speed.

Of greater general interest in climb performance are the factors which affect the *rate of climb*. The vertical velocity of an airplane depends on the flight speed and the inclination of the flightpath. In fact, the rate of climb is the vertical component of the flightpath velocity.

For rate of climb, the maximum rate would occur where there exists the greatest difference between power available and power required (Fig. 17–57). The above relationship means that, for a given weight of the airplane, the *rate of climb* depends on the difference between the power available and the power required, or the excess power. Of course, when the excess power is zero, the rate of climb is zero and the airplane is in steady level flight. When power available is greater than the power required, the excess power will allow a rate of climb specific to the magnitude of excess power.

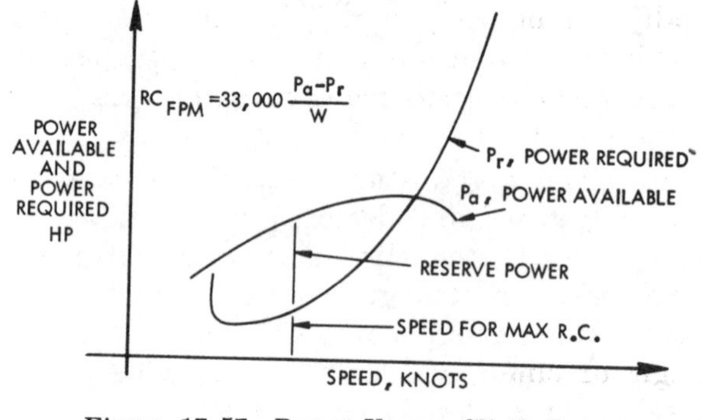

Figure 17–57 Power Versus Climb Rate

It can be said, then, that during a steady climb, the *rate* of climb will depend on *excess power* while the *angle* of climb is a function of *excess thrust*.

The climb performance of an airplane is affected by certain variables. The conditions of the airplane's maximum climb angle or maximum climb rate occur at specific speeds, and variations in speed will produce variations in climb performance. Generally, there is sufficient latitude in most general aviation airplanes that small variations in speed from the optimum do not produce large changes in climb performance, and certain operational considerations may require speeds slightly different from the optimum. Of course, climb performance would be most critical with high gross weight, at high altitude, in obstructed takeoff areas, or during malfunction of a powerpiant. Then, optimum climb speeds are *necessary*.

Weight has a very pronounced effect on airplane performance. If weight is added to the airplane, it must fly at a higher angle of attack to maintain a given altitude and speed. This increases the induced drag of the wings, as well as the parasite drag of the airplane. Increased drag means that additional power is needed to overcome it, which in turn means that less reserve power is available for climbing. Airplane designers go to great effort to minimize the weight since it has such a marked effect on the factors pertaining to performance.

A change in the airplane's weight produces a twofold effect on climb performance. First, the weight affects both the climb angle and the climb rate. In addition, a change in weight will change the drag and the power required. This alters the reserve power available. Generally, an increase in weight will reduce the maximum rate of climb but the airplane must be operated at some increase of climb speed to achieve the smaller peak climb rate.

An increase in altitude also will increase the power required and decrease the power available. Hence, the climb performance of an airplane is affected greatly by altitude. The speeds for maximum rate of climb, maximum angle of climb, and maximum and minimum level flight airspeeds vary with altitude. As altitude is increased, these various speeds finally converge at the *absolute* ceiling of the airplane. At the absolute ceiling, there is no excess of power and only one speed will allow steady level flight. Consequently, the absolute ceiling of the airplane produces zero rate of climb. The *service ceiling* is the altitude at which the airplane is unable to climb at a rate greater than 100 feet per minute. Usually, these specific performance reference points are provided for the airplane at a specific design configuration.

In discussing performance, it frequently is convenient to use the terms "power loading" and "wing loading." Power loading is expressed in pounds per horsepower and is obtained by dividing the total weight of the airplane by the rated horsepower of the engine. It is a significant factor in the airplane's takeoff and climb capabilities. Wing loading is expressed in pounds per square foot and is obtained by dividing the total weight of the airplane in pounds by the wing area (including ailerons) in square feet. It is the airplane's wing loading that determines the landing speed. These factors are discussed in subsequent sections of this chapter.

Range Performance

The ability of an airplane to convert fuel energy into flying distance is one of the most important items of airplane performance. In flying operations, the problem of efficient range operation of an airplane appears in two general forms: (1) to extract the maximum flying distance from a given fuel load or (2) to fly a specified distance with a minimum expenditure of fuel. A common denominator for each of these operating problems is the "specific range"; that is, nautical miles of flying distance per pound of fuel. Cruise flight operations for maximum range should be conducted so that the airplane obtains maximum specific range throughout the flight.

The specific range can be defined by the following relationship:

$$\text{specific range} = \frac{\text{nautical miles}}{\text{lbs. of fuel}}$$

or

$$\text{specific range} = \frac{\text{nautical miles/hr.}}{\text{lbs. of fuel/hr.}}$$

or

$$\frac{\text{knots}}{\text{fuel flow}}$$

If maximum specific range is desired, the flight condition must provide a maximum of speed versus fuel flow.

The general item of *range* must be clearly distinguished from the item of *endurance* (Fig. 17–58). The item of range involves consideration of flying *distance*, while endurance involves consideration of flying *time*. Thus, it is appropriate to define a separate term, "specific endurance."

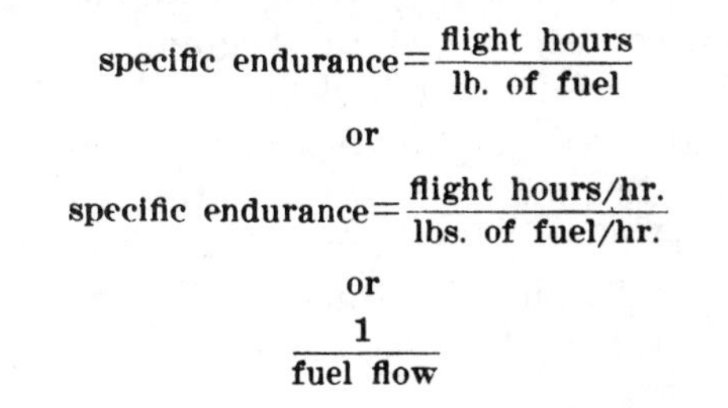

$$\text{specific endurance} = \frac{\text{flight hours}}{\text{lb. of fuel}}$$

or

$$\text{specific endurance} = \frac{\text{flight hours/hr.}}{\text{lbs. of fuel/hr.}}$$

or

$$\frac{1}{\text{fuel flow}}$$

If maximum endurance is desired, the flight condition must provide a minimum of fuel flow.

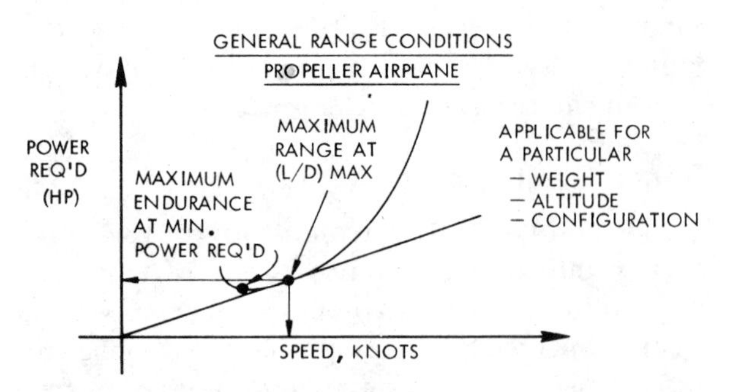

Figure 17–58 Power Versus Range

While the peak value of specific range would provide maximum range operation, long range cruise operation is generally recommended at some slightly higher airspeed. Most long range cruise operations are conducted at the flight condition which provides 99 percent of the absolute maximum specific range. The advantage of such operation is that 1 percent of range is traded for 3 to 5 percent higher cruise speed. Since the higher cruise speed has a great number of advantages, the small sacrifice of range is a fair bargain. The values of specific range versus speed are affected by three principal variables: (1) airplane gross weight, (2) altitude, and (3) the external aerodynamic configuration of the airplane. These are the source of range and endurance operating data included in the performance section of the airplane's flight handbook.

"Cruise control" of an airplane implies that the airplane is operated to maintain the recommended long—range cruise condition throughout the flight. Since fuel is consumed during cruise, the gross weight of the airplane will vary and optimum airspeed, altitude, and power setting can also vary. Generally, "cruise control" means the control of the *optimum* airspeed, altitude, and power setting to maintain the 99 percent maximum specific range condition. At the beginning of cruise flight, the relatively high initial weight of the airplane will require specific values of airspeed, altitude, and power setting to produce the recommended cruise condition (Fig. 17–59). As fuel is consumed and the airplane's gross weight decreases, the optimum airspeed and power setting may decrease, or, the optimum altitude may increase. In addition, the optimum specific range will increase. Therefore, the pilot must provide the proper cruise control technique to ensure that optimum conditions are maintained.

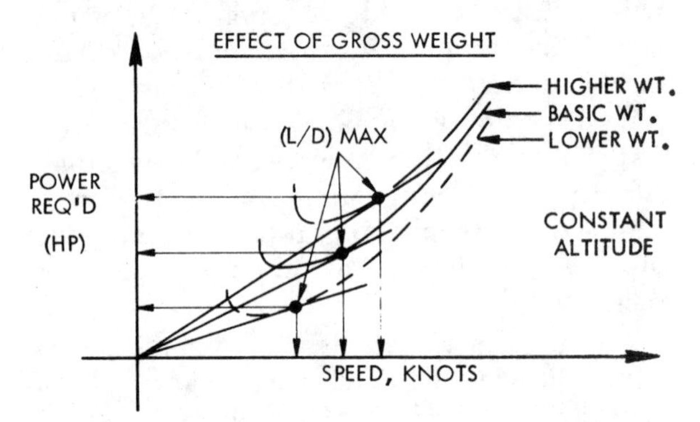

Figure 17–59 Weight Versus Cruise Speed

Total range is dependent on both fuel available and specific range. When range and economy of operation are the principal goals, the pilot must ensure that the airplane will be operated at the recommended long range cruise condition. By this procedure, the airplane will be capable of its maximum design operating radius, or can achieve flight distances less than the maximum with a maximum of fuel reserve at the destination.

The propeller driven airplane combines the propeller with the reciprocating engine for propulsive power. In the case of the reciprocating engine, fuel flow is determined mainly by the shaft *power* put into the propeller rather than *thrust*. Thus, the fuel flow can be related directly to the power required to maintain the airplane in steady, level flight. This fact allows for the determination of range through analysis of power required versus speed—variation of fuel flow versus speed.

The maximum *endurance* condition would be obtained at the point of minimum power required since this would require the lowest fuel flow to keep the airplane in steady, level flight. Maximum *range* condition would occur where the proportion between speed and power required is greatest (Fig. 17–58). The maximum *range* condition is obtained at maximum lift-drag ratio (L/D max) and it is important to note that for a given airplane configuration, the maximum lift-drag ratio occurs at a particular angle of attack and lift coefficient, and is unaffected by weight or altitude.

The flight condition of maximum lift-drag ratio is achieved at one particular value of lift coefficient for a given airplane configuration. Hence, a variation of gross weight will alter the values of airspeed, power required, and specific range obtained at the maximum lift-drag ratio.

The variations of speed and power required must be monitored by the pilot as part of the cruise control procedure to maintain the maximum lift-drag ratio. When the airplane's fuel weight is a small part of the gross weight and the airplane's range is small, the cruise control procedure can be simplified to essentially maintaining a constant speed and power setting throughout the time of cruise flight. On the other hand, the long range airplane has a fuel weight which is a considerable part of the gross weight, and cruise control procedures must employ scheduled airspeed and power changes to maintain optimum range conditions.

The effect of altitude on the range of the propeller driven airplane may be understood by inspection of Fig. 17–60. A flight conducted at high altitude will have a greater true airspeed and the power required will be proportionately greater than when conducted at sea level. The drag of the airplane at altitude is the same as the drag at sea level but the higher true airspeed causes a proportionately greater power required. Note that the straight line that is tangent to the sea level power curve is also tangent to the altitude power curve.

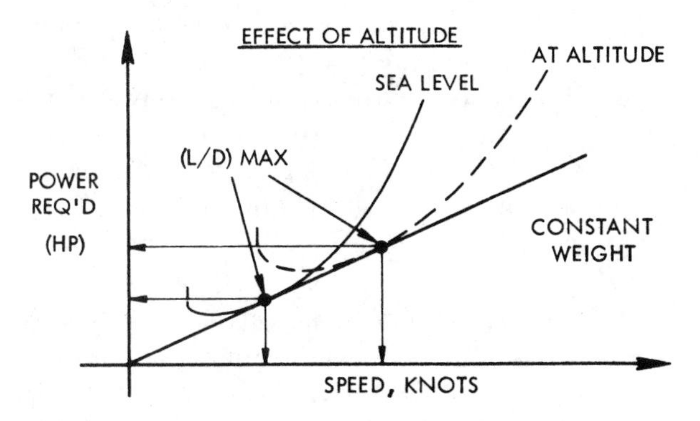

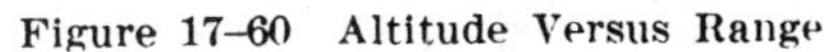
Figure 17–60 Altitude Versus Range

The effect of altitude on specific range also can be appreciated from the previous relationships. If a change in altitude causes identical changes in speed and power required, the proportion of speed to power required would be unchanged. The fact implies that the specific range of the propeller driven airplane would be unaffected by altitude. Actually, this is true to the extent that specific fuel consumption and propeller efficiency are the principal factors which could cause a variation of specific range with altitude. If compressibility effects are negligible, *any variation of specific range with altitude is strictly a function of engine-propeller performance.*

The airplane equipped with the reciprocating engine will experience very little, if any, variation of specific range with altitude at low altitudes. There is negligible variation of brake specific fuel consumption for values of brake horsepower below the maximum cruise power rating of the engine which is the lean range of engine operation. Thus, an increase in altitude will produce a decrease in specific range only when the increased power requirement exceeds the maximum cruise power rating of the engine. One advantage of supercharging is that the cruise power may be maintained at high altitude and the airplane may achieve the range at high altitude with the corresponding increase in true airspeed. The principal differences in the high altitude cruise and low altitude cruise are the true airspeeds and climb fuel requirements.

Takeoff and Landing Performance

The majority of pilot-caused airplane accidents occur during the takeoff and landing phase of flight. Because of this fact, the pilot must be familiar with all the variables which influence the takeoff and landing performance of an airplane and must strive for exacting, professional techniques of operation during these phases of flight.

Takeoff and landing performance is a condition of accelerated and decelerated motion. For instance, during takeoff the airplane starts at zero speed and accelerates to the takeoff speed to become airborne. During landing, the airplane touches down at the landing speed and decelerates to zero speed.

The important factors of takeoff or landing performance are:

(1) The takeoff or landing *speed* which will generally be a function of the stall speed or minimum flying speed.

(2) The rate of *acceleration* and *deceleration* during the takeoff or landing roll. The acceleration and deceleration experienced by any object vary directly with the unbalance of force and inversely as the mass of the object.

(3) The takeoff or landing roll *distance* is a function of both acceleration/deceleration and speed.

Takeoff Performance

The minimum takeoff distance is of primary interest in the operation of any airplane because it defines the runway requirements. The minimum takeoff distance is obtained by taking off at some minimum safe speed which allows sufficient margin above stall and provides satisfactory control and initial rate of climb. Generally, the liftoff speed is some fixed percentage of the stall speed or minimum control speed for the airplane in the takeoff configuration. As such, the lift-off will be accomplished at some particular value of lift coefficient and angle of attack. Depending on the airplane characteristics, the lift-off speed will be anywhere from 1.05 to 1.25 times the stall speed or minimum control speed.

To obtain minimum takeoff distance at the specific lift-off speed, the forces which act on the airplane must provide the maximum acceleration during the takeoff roll. The various forces acting on the airplane may or may not be under the control of the pilot, and various techniques may be necessary in certain airplanes to maintain takeoff acceleration at the highest value.

The powerplant thrust is the principal force to provide the acceleration and, for minimum takeoff distance, the output thrust should be at a maximum. Lift and drag are produced as soon as the airplane has speed, and the values of lift and drag depend on the angle of attack and dynamic pressure.

In addition to the important factors of proper technique, many other variables affect the takeoff performance of an airplane. Any item which alters the takeoff speed or acceleration rate during the takeoff roll will affect the takeoff distance.

For example, the effect of *gross weight* on takeoff distance is significant and proper consideration of this item must be made in predicting the airplane's takeoff distance. Increased gross weight can be considered to produce a threefold effect on takeoff performance: (1) higher lift-off speed, (2) greater mass to accelerate, and (3) increased retarding force (drag and ground friction). If the gross weight increases, a greater speed is necessary to produce the greater lift necessary to get the airplane airborne at the takeoff lift coefficient. As an example of the effect of a change in gross weight, a 21 percent increase in takeoff weight will require a 10 percent increase in lift-off speed to support the greater weight.

A change in gross weight will change the net accelerating force, and change the mass which is being accelerated. If the airplane has a relatively high thrust-to-weight ratio, the change in the net accelerating force is slight and the principal effect on acceleration is due to the change in mass.

The takeoff distance will vary at least as the square of the gross weight. For example, a 10 percent increase in takeoff gross weight would cause:

(1) a 5 percent increase in takeoff velocity.

(2) at least a 9 percent decrease in rate of acceleration.

(3) at least a 21 percent increase in takeoff distance.

For the airplane with a high thrust-to-weight ratio, the increase in takeoff distance might be approximately 21 to 22 percent, but for the airplane with a relatively low thrust-to-weight ratio, the increase in takeoff distance would be approximately 25 to 30 percent. Such a powerful effect requires proper consideration of gross weight in predicting takeoff distance.

The effect of *wind* on takeoff distance is large, and proper consideration also must be provided when predicting takeoff distance. The effect of a headwind is to allow the airplane to reach the lift-off speed at a lower groundspeed while the effect of a tailwind is to require the airplane to achieve a greater groundspeed to attain the lift-off speed.

A headwind which is 10 percent of the takeoff airspeed will reduce the takeoff distance approximately 19 percent (Fig. 17–63). However, a tailwind which is 10 percent of the takeoff airspeed will increase the takeoff distance approximately 21 percent. In the case where the headwind speed is 50 percent of the takeoff speed, the takeoff distance would be approximately 25 percent of the zero wind takeoff distance (75 percent reduction).

The effect of wind on landing distance is identical to the effect on takeoff distance. Figure 17–61 illustrates the general effect of wind by the percent change in takeoff or landing distance as a function of the ratio of wind velocity to takeoff or landing speed.

The effect of *proper takeoff speed* is especially important when runway lengths and takeoff distances are critical. The takeoff speeds specified in the airplane's flight handbook are generally the minimum safe speeds at which the airplane can become airborne. Any attempt to take off below the recommended speed could mean that the airplane may stall, be difficult to control, or have a very

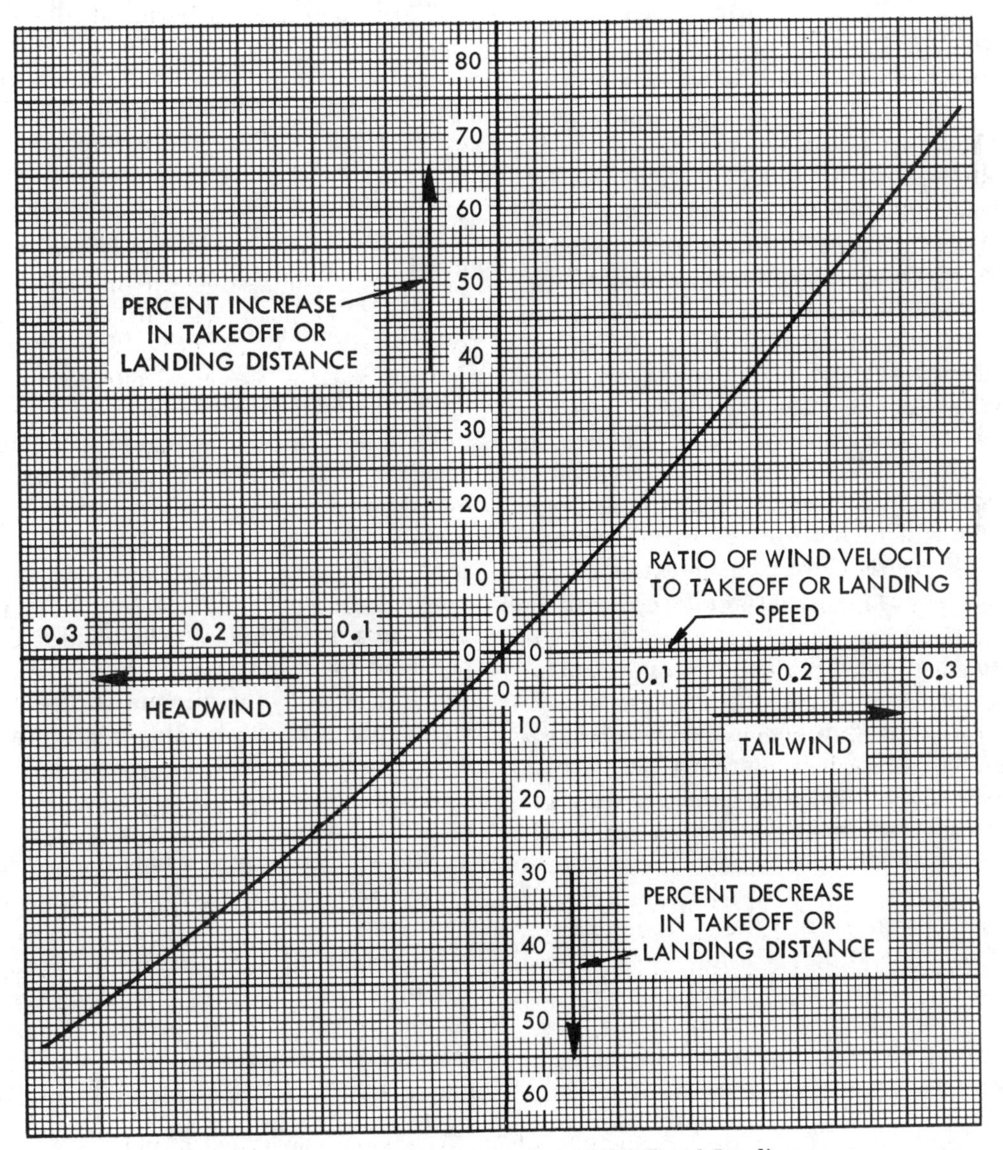

Figure 17–61 Effect of Wind on Takeoff and Landing

low initial rate of climb. In some cases, an excessive angle of attack may not allow the airplane to climb out of ground effect. On the other hand, an excessive airspeed at takeoff may improve the initial rate of climb and "feel" of the airplane, but will produce an undesirable increase in takeoff distance. Assuming that the acceleration is essentially unaffected, the takeoff distance varies as the square of the takeoff velocity.

Thus, 10 percent excess airspeed would increase the takeoff distance 21 percent. In most critical takeoff conditions, such an increase in takeoff distance would be prohibitive and the pilot must adhere to the recommended takeoff speeds.

The effect of *pressure altitude* and *ambient temperature* is to define primarily the *density altitude* and its effect on takeoff performance. While subsequent corrections are appropriate for the effect of temperature on certain items of powerplant performance, *density altitude* defines specific effects on takeoff performance. An increase in density altitude can produce a two-fold effect on takeoff performance: (1) greater takeoff speed and (2) decreased thrust and reduced net accelerating force. If an airplane of given weight and configuration is operated at greater heights above standard sea level, the airplane will still require the same dynamic pressure to become airborne at the takeoff lift coefficient. Thus, the airplane at altitude will take off at the same *indicated* airspeed as at sea level, but because of the reduced air density, the true airspeed will be greater.

The effect of density altitude on powerplant thrust depends much on the type of powerplant. An increase in altitude above *standard sea level* will bring an immediate decrease in power output for the unsupercharged reciprocating engine. However, an increase in altitude above standard sea level will not cause a decrease in power output for the supercharged reciprocating engine until the altitude exceeds the critical operating altitude. For those powerplants which experience a decay in thrust with an increase in altitude, the effect on the net accelerating force and acceleration rate can be approximated by assuming a direct variation with density. Actually, this assumed variation would closely approximate the effect on airplanes with high thrust-to-weight ratios.

Proper accounting of pressure altitude (field elevation is a poor substitute) and temperature, is mandatory for accurate prediction of takeoff roll distance.

The most critical conditions of takeoff performance are the result of some combination of high gross weight, altitude, temperature, and unfavorable wind. In all cases, it behooves the pilot to make an accurate prediction of takeoff distance from the performance data of the Airplane's Flight Handbook, regardless of the runway available, and to strive for a polished, professional takeoff technique.

In the prediction of takeoff distance from the handbook data, the following primary considerations must be given:

(1) Pressure altitude and temperature—to define the effect of density altitude on distance.

(2) Gross weight—a large effect on distance.

(3) Runway slope and condition—the effect of an incline and retarding effect of snow, ice, etc.

(4) Wind—a large effect due to the wind or wind component along the runway.

Landing Performance

In many cases, the landing distance of an airplane will define the runway requirements for flying operations. The minimum landing distance is obtained by landing at some minimum safe speed which allows sufficient margin above stall and provides satisfactory control and capability for a go-around. Generally, the landing speed is some fixed percentage of the stall speed or minimum control speed for the airplane in the landing configuration. As such, the landing will be accomplished at some particular value of lift coefficient and angle of attack. The exact values will depend on the airplane characteristics but, once defined, the values are independent of weight, altitude, wind, etc.

To obtain minimum landing distance at the specified landing speed, the forces which act on the airplane must provide maximum deceleration during the landing roll. The various forces acting on the airplane during the landing roll may require various techniques to maintain landing deceleration at the peak value.

A distinction should be made between the techniques for minimum landing distance and an ordinary landing roll with considerable excess runway available. Minimum landing distance will be obtained with the actual landing speed by creating a continuous peak deceleration of the airplane; that is, extensive use of the brakes for maximum deceleration. On the other hand, an ordinary landing roll with considerable excess runway may allow extensive use of aerodynamic drag to minimize wear and tear on the tires and brakes. If aerodynamic drag is sufficient to cause deceleration of the airplane, it can be used in deference to the brakes in the early stages of the landing roll; i.e., brakes and tires suffer from continuous hard use but airplane aerodynamic drag is free and does not wear out with use. The use of aerodynamic drag is applicable only for deceleration to 60 or 70 percent of the touchdown speed. At speeds less than 60 to 70 percent of the touchdown speed, aerodynamic drag is so slight as to be of little use, and braking must be utilized to produce continued deceleration of the airplane. Since the objective during the landing roll is to decelerate, the powerplant thrust should be the smallest possible positive value (or largest possible negative value in the case of thrust reversers).

In addition to the important factors of proper technique, many other variables affect the landing performance of an airplane. Any item which alters the landing speed or deceleration rate during the landing roll will affect the landing distance.

The effect of *gross weight* on landing distance is one of the principal items determining the landing distance of an airplane. One effect of an increased gross weight is that the airplane will require a greater speed to support the airplane at the landing angle of attack and lift coefficient.

As an example of the effect of a change in gross weight, a 21 percent increase in landing weight will require a 10 percent increase in landing speed to support the greater weight.

When minimum landing distances are considered, braking friction forces predominate during the landing roll and, for the majority of airplane configurations, braking friction is the main source of deceleration.

The minimum landing distance will vary in direct proportion to the gross weight. For example, a 10 percent increase in gross weight at landing would cause:

(1) a 5 percent increase in landing velocity,

(2) a 10 percent increase in landing distance.

A contingency of this is the relationship between weight and braking friction force.

The effect of *wind* on landing distance is large and deserves proper consideration when predicting landing distance. Since the airplane will land at a particular airspeed independent of the wind, the principal effect of wind on landing distance is due to the change in the groundspeed at which the airplane touches down. The effect of wind on deceleration during the landing is identical to the effect on acceleration during the takeoff.

A headwind which is 10 percent of the landing airspeed will reduce the landing distance approximately 19 percent but a tailwind which is 10 percent of the landing speed will increase the landing distance approximately 21 percent. Figure 17–61 illustrates this general effect.

The effect of *pressure altitude* and *ambient temperature* is to define *density altitude* and its effect on landing performance. An increase in density altitude will increase the landing speed but will not alter the net retarding force. Thus, the airplane at altitude will land at the same indicated airspeed—as at sea level but, because of the reduced density, the true airspeed (TAS) will be greater. Since the airplane lands at altitude with the same weight and dynamic pressure, the drag and braking friction throughout the landing roll have the same values as at sea level. As long as the condition is within the capability of the brakes, the net retarding force is unchanged and the deceleration is the same as with the landing at sea level. Since an increase in altitude does not alter deceleration, the effect of density altitude on landing distance would actually be due to the greater TAS (true airspeed).

The minimum landing distance at 5,000 feet would be 16 percent greater than the minimum landing distance at sea level. The approximate increase in landing distance with altitude is approximately 3½ percent for each 1,000 feet of altitude. Proper accounting of density altitude is necessary to accurately predict landing distance.

The effect of proper *landing speed* is important when runway lengths and landing distances are critical. The landing speeds specified in the airplane's flight handbook are generally the minimum safe speeds at which the airplane can be landed. Any attempt to land at below the specified speed may mean that the airplane may stall, be difficult to control, or develop high rates of descent. On the other hand, an excessive speed at landing may improve the controllability slightly (especially in crosswinds), but will cause an undesirable increase in landing distance.

Thus, a 10 percent excess landing speed would cause a 21 percent increase in landing distance. The excess speed places a greater working load on the brakes because of the additional kinetic energy to be dissipated. Also, the additional speed causes increased drag and lift in the normal ground attitude and the increased lift will reduce the normal force on the braking surfaces. The deceleration during this range of speed immediately after touchdown may suffer and it will be more likely that a tire can be blown out from braking at this point.

The most critical conditions of landing performance are the result of some combination of high gross weight, high density altitude, and unfavorable wind. These conditions produce the greatest landing distance and provide critical levels of energy dissipation required of the brakes. In all cases, it is necessary to make an accurate prediction of minimum landing distance to compare with the available runway. A polished, professional landing technique is necessary because the landing phase of flight accounts for more pilot-caused airplane accidents than any other single phase of flight.

In the prediction of minimum landing distance from the handbook data, the following considerations must be given:

(1) Pressure altitude and temperature—to define the effect of density altitude.

(2) Gross weight—which defines the CAS or EAS for landing.

(3) Wind—a large effect due to wind or wind component along the runway.

(4) Runway slope and condition—relatively small correction for ordinary values of runway slope, but a significant effect of snow, ice, or soft ground.

Importance of Performance Data

The performance or operational information section of the airplane's flight handbook contains the operating data for the airplane; that is, the data pertaining to takeoff, climb, range, endurance, descent, and landing (Figs. 17–62 through 17–68). The ordinary use of these data in flying operations is mandatory for safe and efficient operation. Considerable knowledge and familiarity of the airplane can be gained through study of this material. Complete familiarity of the airplane's performance characteristics can be obtained only through extensive analysis and study of the complete handbook.

It must be emphasized that the information and the data furnished in the airplane flight handbooks by the various manufacturers has not always been standardized. Some provide the data in tabular form, while others use graphs. In addition, the performance data may be presented on the basis of standard atmospheric conditions, pressure altitude, or density altitude. Thus, the performance information in airplane operating handbooks has little or no value unless the user recognizes those variations and makes the necessary adjustments.

To be able to make practical use of the airplane's capabilities and limitations, it is essential to understand the significance of the operational data. The pilot must be cognizant of the basis for the performance data, as well as the meanings of the various terms used in expressing performance capabilities and limitations.

Since the characteristics of the atmosphere in which the airplane operates has a predominant effect on performance, we should first review some of the dominating factors—pressure and temperature.

TAKE-OFF DATA

TAKE-OFF DISTANCE FROM HARD SURFACE RUNWAY WITH FLAPS UP

GROSS WEIGHT POUNDS	IAS AT 50' MPH	HEAD WIND KNOTS	AT SEA LEVEL & 59° GROUND RUN	AT SEA LEVEL & 59° TOTAL TO CLEAR 50 FT OBS	AT 2500 FT. & 50°F GROUND RUN	AT 2500 FT. & 50°F TOTAL TO CLEAR 50 FT OBS	AT 5000 FT. & 41°F GROUND RUN	AT 5000 FT. & 41°F TOTAL TO CLEAR 50 FT OBS	AT 7500 FT. & 32°F GROUND RUN	AT 7500 FT. & 32°F TOTAL TO CLEAR 50 FT OBS
2300	68	0	865	1525	1040	1910	1255	2480	1565	3855
		10	615	1170	750	1485	920	1955	1160	3110
		20	405	850	505	1100	630	1480	810	2425
2000	63	0	630	1095	755	1325	905	1625	1120	2155
		10	435	820	530	1005	645	1250	810	1685
		20	275	580	340	720	425	910	595	1255
1700	58	0	435	780	520	920	625	1095	765	1370
		10	290	570	355	680	430	820	535	1040
		20	175	385	215	470	270	575	345	745

NOTES: 1. Increase distance 10% for each 25°F above standard temperature for particular altitude.
2. For operation on a dry, grass runway, increase distances (both "ground run" and "total to clear 50 ft. obstacle") by 7% of the "total to clear 50 ft. obstacle" figure.

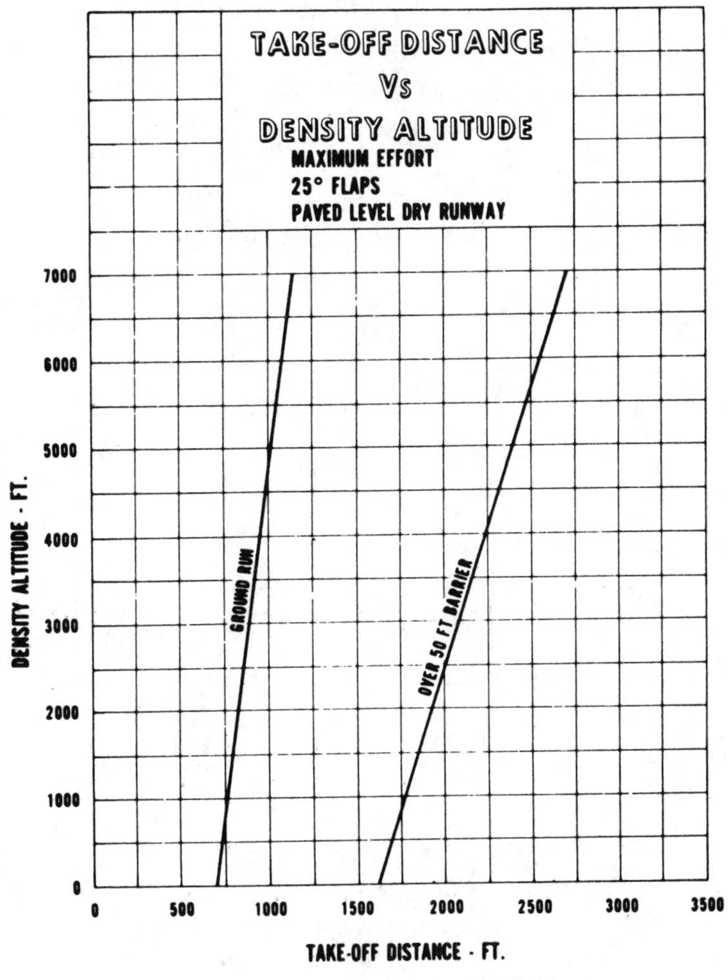

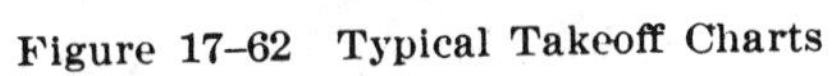
Figure 17–62 Typical Takeoff Charts

MAXIMUM RATE-OF-CLIMB DATA

GROSS WEIGHT POUNDS	AT SEA LEVEL & 59°F			AT 5000 FT. & 41°F			AT 10,000 FT. & 23°F			AT 15,000 FT. & 5°F		
	IAS MPH	RATE OF CLIMB FT/MIN	GAL. OF FUEL USED	IAS MPH	RATE OF CLIMB FT/MIN	FROM S.L. FUEL USED	IAS MPH	RATE OF CLIMB FT/MIN	FROM S.L. FUEL USED	IAS MPH	RATE OF CLIMB FT/MIN	FROM S.L. FUEL USED
2300	82	645	1.0	81	435	2.6	79	230	4.8	78	22	11.5
2000	79	840	1.0	79	610	2.2	76	380	3.6	75	155	6.3
1700	77	1085	1.0	76	825	1.9	73	570	2.9	72	315	4.4

NOTES: 1. Flaps up, full throttle, mixture leaned for smooth operation above 3000 ft.
2. Fuel used includes warm up and take-off allowance.
3. For hot weather, decrease rate of climb 20 ft./min. for each 10°F above standard day temperature for particular altitude.

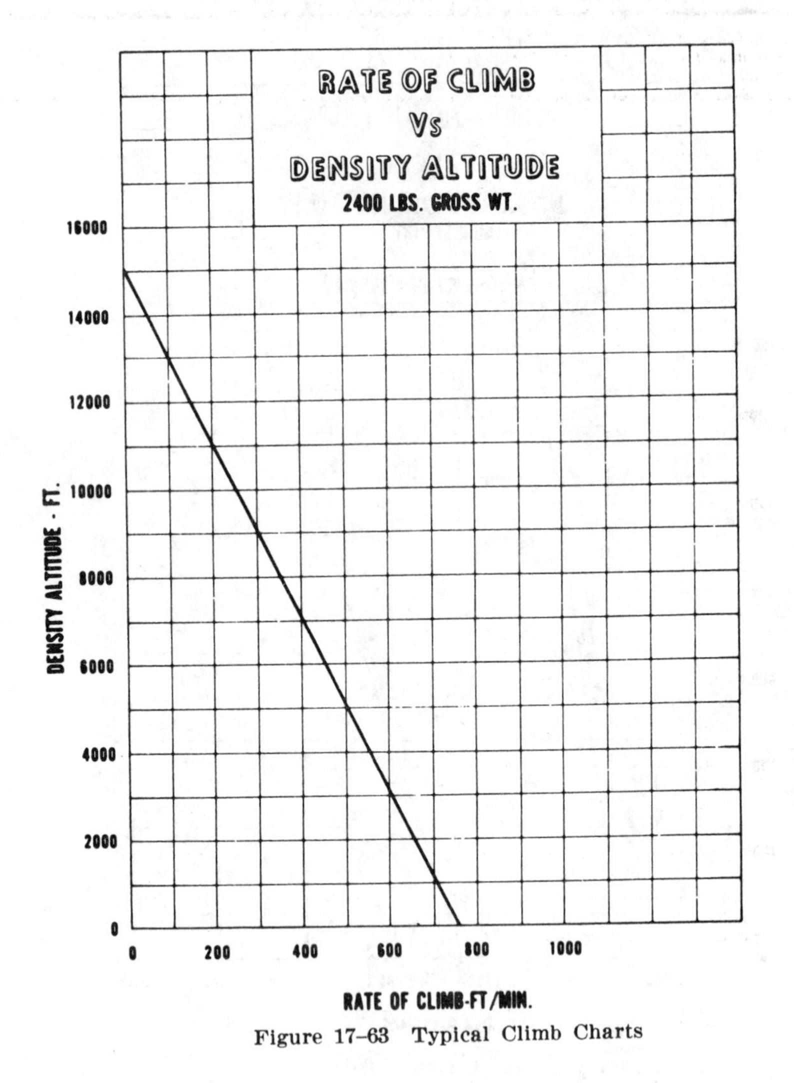

Figure 17–63 Typical Climb Charts

LANDING DATA

LANDING DISTANCE ON HARD SURFACE RUNWAY
NO WIND – 40° FLAPS – POWER OFF

GROSS WEIGHT LBS.	APPROACH IAS MPH	@ S.L. & 59° F		@ 2500 ft. & 50° F		@ 5000 ft. & 41° F		@ 7500 ft. & 32° F	
		GROUND ROLL	TOTAL TO CLEAR 50' OBS.	GROUND ROLL	TOTAL TO CLEAR 50' OBS.	GROUND ROLL	TOTAL TO CLEAR 50' OBS.	GROUND ROLL	TOTAL TO CLEAR 50' OBS.
2300	69	520	1250	560	1310	605	1385	650	1455

NOTES: 1. Reduce landing distance 10% for each 5 knot headwind.
2. For operation on a dry, grass runway, increase distances (both "ground roll" and "total to clear 50 ft. obstacle") by 20% of the "total to clear 50 ft. obstacle" figure.

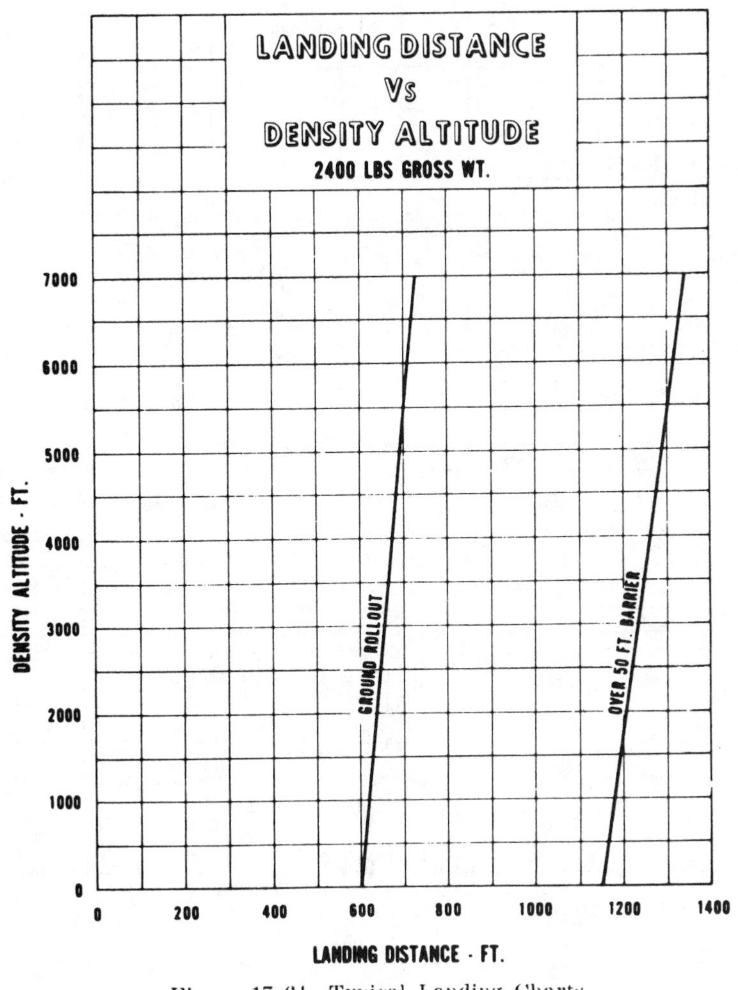

Figure 17–64 Typical Landing Charts

CRUISE & RANGE PERFORMANCE

GROSS WEIGHT-2200 LBS
STANDARD CONDITIONS
ZERO WIND
LEAN MIXTURE

ALTITUDE	RPM	PERCENT POWER	TRUE AIR SPEED—MPH	GALLONS/ HOUR	ENDURANCE HOURS	RANGE MILES
2500	2600	81	136	9.3	3.9	524
	2500	73	129	8.3	4.3	555
	2400	65	122	7.5	4.8	586
	2300	58	115	6.6	5.4	617
	2200	52	108	6.0	6.0	645
4500	2600	77	135	8.8	4.0	539
	2500	69	129	7.9	4.5	572
	2400	62	121	7.1	5.0	601
	2300	56	113	6.4	5.5	628
	2200	51	106	5.7	6.1	646
6500	2700	81	140	9.3	3.8	530
	2600	73	134	8.3	4.2	559
	2500	66	126	7.5	4.7	587
	2400	60	119	6.8	5.2	611
	2300	54	112	6.1	5.7	632
8500	2700	77	139	8.8	4.0	547
	2600	70	132	7.9	4.4	575
	2500	63	125	7.2	4.9	599
	2400	57	118	6.5	5.3	620
	2300	52	109	5.9	5.8	635
10500	2700	73	138	8.3	4.2	569
	2600	66	130	7.6	4.6	590
	2500	60	122	6.9	5.0	610
	2400	55	115	6.3	5.4	625
	2300	50	106	5.7	5.9	631

NOTES:

1. Range and endurance data include allowance for take-off and climb.
2. Fuel consumption is for level flight with mixture leaned. See Section III for proper leaning technique. Continuous operations at powers above 75% should be with full rich mixture.
3. Speed performance is without wheel fairings. Add 2 MPH for wheel fairings.
4. For temperatures other than standard, add or subtract 1%power for each 10° F. below or above standard temperature respectively.

Power Setting Table- Model [illegible] Series, 180 HP Engine

Press. Alt	Std Alt Temp °F	108 HP 60% Power RPM	117 HP 65% Power RPM	126 HP 70% Power RPM	135 HP 75% Power RPM	Press. Alt
SL	59	2290	2370	2440	2500	SL
1,000	55	2310	2390	2460	2520	1,000
2,000	52	2330	2410	2480	2540	2,000
3,000	48	2350	2430	2500	2560	3,000
4,000	45	2370	2450	2520	2580	4,000
5,000	41	2390	2470	2540	2600	5,000
6,000	38	2410	2490	2560	2620	6,000
7,000	34	2430	2510	2580	2640	7,000
8,000	31	2450	2530	2600	–	8,000
9,000	27	2470	2550	2620	–	9,000
10,000	23	2490	2570	–	–	10,000
11,000	19	2510	2590	–	–	11,000
12,000	16	2530	–	–	–	12,000

Figure 17–65 Typical Cruise Charts

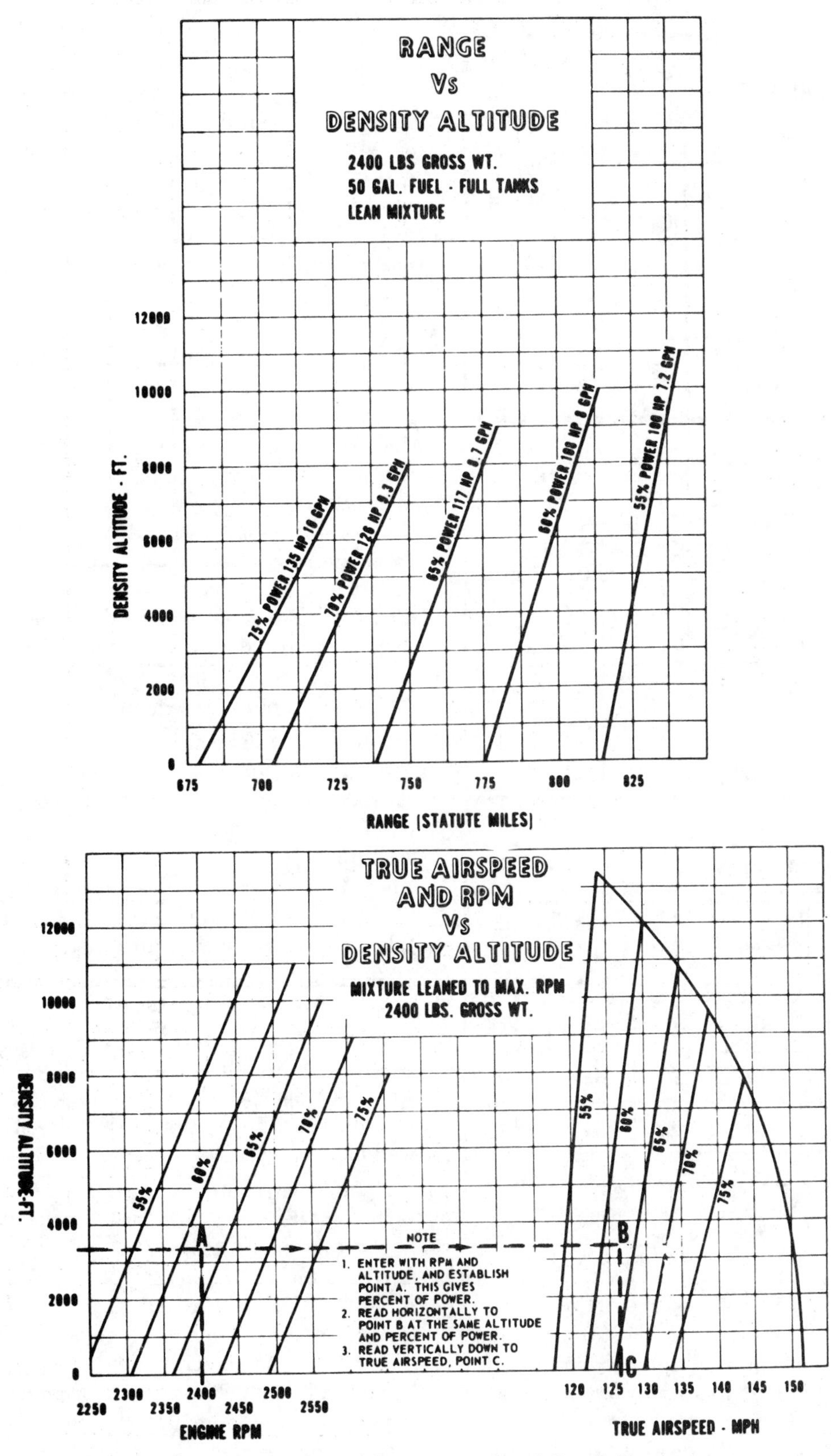

Figure 17–66 Typical Range and Speed Charts

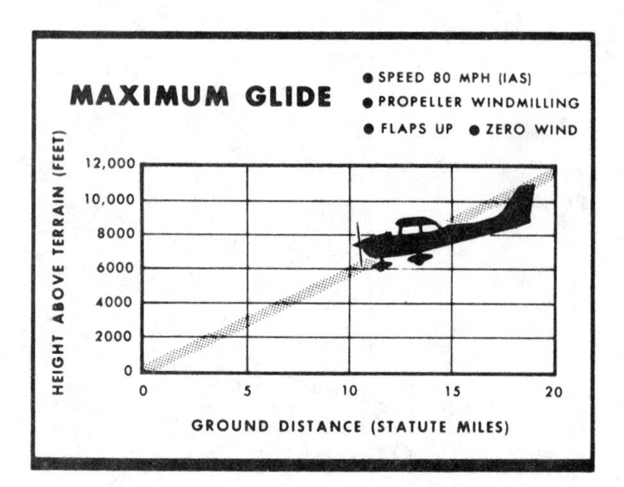

Figure 17–67 Typical Glide Chart

AIRSPEED CORRECTION TABLE										
FLAPS UP										
IAS-MPH	50	60	70	80	90	100	110	120	130	140
CAS-MPH	53	60	69	78	87	97	107	117	128	138
FLAPS DOWN										
IAS-MPH	40	50	60	70	80	90	100			
CAS-MPH	40	50	61	72	83	94	105			

STALL SPEEDS – MPH CAS				
Gross Weight 1600 lbs.	ANGLE OF BANK			
CONDITION	0°	20°	40°	60°
Flaps UP	55	57	63	78
Flaps 20°	49	51	56	70
Flaps 40°	48	49	54	67
POWER OFF — AFT CG				

Figure 17–68 Typical Airspeed Charts

Atmospheric Pressure

Atmospheric pressure is the force exerted by the weight of the atmosphere above a unit area. The weight of the air exerts a force on the earth or on any object placed in the atmosphere. *Pressure* is a measure of this *force per unit area:* i.e., grams-per-square centimeter, pounds-per-square foot, etc. Pressure at any point on the earth's surface is the weight per unit area of the column of air above that point. In other words, if the pressure is 14 pounds-per-square inch, a column of air having one square inch cross section extending to the top of the atmosphere weighs 14 pounds. (The average pressure at the surface of the earth is approximately 14.7 pounds per square inch.)

We have mentioned that at any point in the atmosphere, pressure is the weight of the air *above* that point. Since pressure is the weight of the air above, and since less and less air lies above a point as it moves upward through the atmosphere, pressure must decrease with increasing altitude. The greater pressure at low altitude compresses the air more than does the lesser pressure at higher altitude. Therefore, the rate of decrease (lapse rate) in pressure with height becomes less with increasing altitude. For example, from sea level to 1,000 feet, pressure drops about one inch of mercury; but from 19,000 to 20,000 feet, pressure drops only about six-tenths of an inch.

The rate of decrease of pressure with height, however, is not always constant. Like most substances, air contracts as it cools and expands as it becomes warmer. Therefore, when a sample of air cools, it occupies less space; when heated, it occupies more. As a result, the rate of pressure decrease with height in cold air is greater than in warm air.

Since air is a gas, it may be compressed or permitted to expand. When air is compressed, a given volume contains more air, hence its density, or weight, is increased. Conversely, when air is permitted to expand, a given volume contains less air, thus its density, or weight, is decreased.

Heat is a property of all matter. From early studies of science, we learned that heat is the motion of molecules. *Heat* is then defined as *the total energy of motion of molecules.* We also learned that dense air has more molecules than less dense air. The two might have the same average motion, and thus have the same temperature, but the total energy, and consequently the degree of heat is greater in the dense air with more molecules. We cannot measure heat directly, but we can measure temperature with the thermometer.

A general gas law defines the relationship of pressure, temperature, and density when there is no change of state or heat transfer. Simply stated, this would be "density varies directly with pressure, inversely with temperature." Consequently, the higher we fly the less dense the air becomes due to less pressure and temperature.

Atmospheric pressure is continually changing. It varies with both time and location. These pressure changes are caused primarily by changes in the air density (weight of air per unit volume) produced by variations in the distribution of temperature.

Standard Atmosphere

In order to provide a common denominator for comparison of the performance of various aircraft, a *standard* atmosphere has been adopted. The set of standard conditions presently used in the U.S. is known as the *International Standard Atmosphere* (ISA) and has been adopted by most of the nations and airlines of the world—the International Civil Aviation Organization.

The standard atmosphere actually represents the mean or average properties of the atmosphere; that is, it represents the year-round average of the pressure-height temperature soundings observed over a period of years.

In the standard atmosphere, sea level pressure is 29.92″ Hg and 15° C. (59° F.); the standard lapse rates (decrease) for pressure are approximately 1″ Hg per 1,000 feet increase in altitude and 2° C. (3.5° F.) per 1,000 feet increase (up to the tropopause). Figure 17–62 illustrates the variation of the most important properties of the air throughout the standard atmosphere.

Since all airplane performance is compared and evaluated in the environment of the standard atmosphere, all of the airplane performance instrumentation is calibrated for the standard atmosphere. Thus, certain corrections must apply to the instrumentation, as well as the airplane performance, if the actual operating conditions do not fit the standard atmosphere. In order to account properly for the nonstandard atmosphere certain related terms must be defined.

STANDARD ATMOSPHERE

Altitude (ft)	Pressure (in. Hg)	Temp. (°C.)	Temp. (°F.)	Density-slugs per cubic foot
0	29.92	15.0	59.0	.002378
1,000	28.86	13.0	55.4	.002309
2,000	27.82	11.0	51.9	.002242
3,000	26.82	9.1	48.3	.002176
4,000	25.84	7.1	44.7	.002112
5,000	24.89	5.1	41.2	.002049
6,000	23.98	3.1	37.6	.001988
7,000	23.09	1.1	34.0	.001928
8,000	22.22	-0.9	30.5	.001869
9,000	21.38	-2.8	26.9	.001812
10,000	20.57	-4.8	23.3	.001756
11,000	19.79	-6.8	19.8	.001701
12,000	19.02	-8.8	16.2	.001648
13,000	18.29	-10.8	12.6	.001596
14,000	17.57	-12.7	9.1	.001545
15,000	16.88	-14.7	5.5	.001496
16,000	16.21	-16.7	1.9	.001448
17,000	15.56	-18.7	-1.6	.001401
18,000	14.94	-20.7	-5.2	.001355
19,000	14.33	-22.6	-8.8	.001310
20,000	13.74	-24.6	-12.3	.001267

Figure 17–69 Properties of Standard Atmosphere

Pressure Altitude

Pressure altitude is the altitude in the standard atmosphere corresponding to a particular pressure level. The airplane altimeter is essentially a sensitive barometer calibrated to indicate altitude in the standard atmosphere. If the altimeter is set for 29.92″ Hg (Standard Datum Plane) the altitude indicated is the pressure altitude—the altitude in the standard atmosphere corresponding to the sensed pressure.

The Standard Datum Plane is a theoretical level where the weight of the atmosphere is 29.92″ of mercury as measured by a barometer. As atmospheric pressure changes, the Standard Datum Plane may be below, at, or above sea level. Pressure altitude is important as a basis for determining aircraft performance as well as for assigning flight levels to aircraft operating at high altitude (above 18,000 feet).

The pressure altitude can be determined by either of two methods: (1) by setting the barometric scale of the altimeter to 29.92 and reading the indicated altitude, or (2) by applying a correction factor to the elevation according to the reported "altimeter setting" as shown in Figure 17–63.

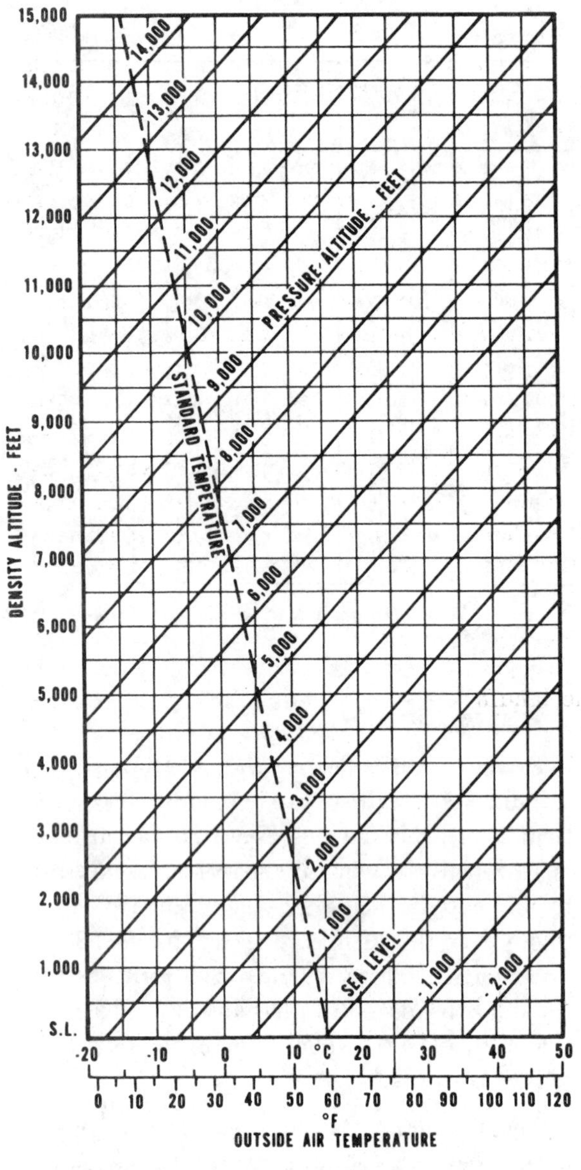

Figure 17–70 Density Altitude Chart

Density Altitude

The more appropriate term for correlating aerodynamic performance in the nonstandard atmosphere is density altitude—the altitude in the standard atmosphere corresponding to a particular value of air density.

Density altitude is pressure altitude corrected for *nonstandard* temperature. Under standard atmospheric conditions, air at each level in the atmosphere has a specific density, and under *standard* conditions, pressure altitude and density altitude identify the same level. Density altitude, then, is the vertical distance above sea level in the standard atmosphere at which a given density is to be found.

Since density varies directly with pressure, and inversely with temperature, a given pressure altitude may exist for a wide range of temperature by allowing the density to vary. However, a known density occurs for any one temperature and pressure altitude. The density of the air, of course, has a pronounced effect on airplane and engine performance. Regardless of the actual altitude at which the airplane is operating, its performance will be as though it were operating at an altitude equal to the existing density altitude.

For example, when set at 29.92″ the altimeter may indicate a pressure altitude of 5,000 feet. According to the airplane flight handbook, the ground run on takeoff may require a distance of 790 feet under standard temperature conditions. However, if the temperature is 20° C. above standard, the expansion of air raises the density level. Using temperature correction data from tables or graphs, or by deriving the density altitude with a computer, it may be found that the density level is above 7,000 feet, and the ground run may be closer to 1,000 feet.

The computation of density altitude must involve consideration of pressure (pressure altitude) and temperature. Since airplane performance data at any level is based upon air density under standard day conditions, such performance data apply to air density levels that may not be identical with altimeter indications. Under conditions higher or lower than standard, these levels cannot be determined directly from the altimeter.

Density altitude can be computed by applying the pressure altitude and outside air temperature at flight level to a navigation computer. Density altitude can also be determined by referring to the table and chart as shown in Fig. 17–71 and Fig. 17–70.

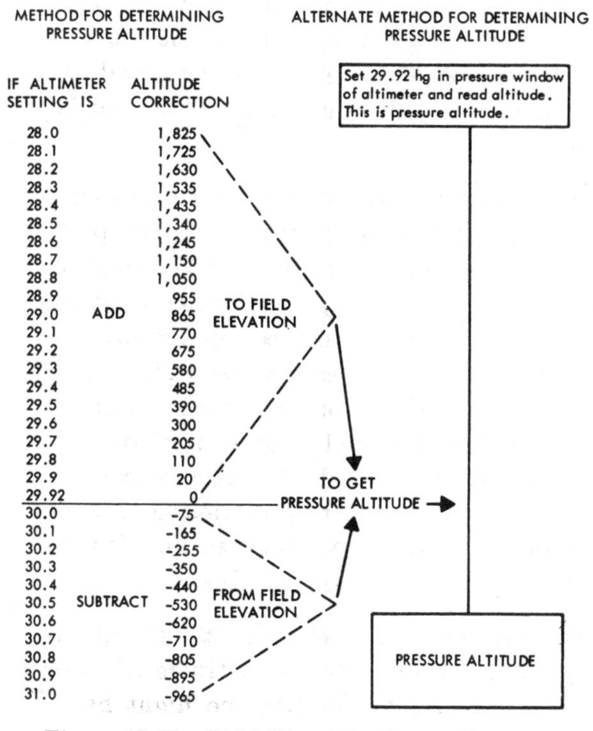

Figure 17–71 Field Elevation Versus Pressure Altitude

Airplane Performance Speeds

True Airspeed (TAS)—the speed of the airplane in relation to the air mass in which it is flying.

Indicated Airspeed (IAS)—the speed of the airplane as observed on the airspeed indicator. It is the airspeed without correction for indicator, position (or installation), or compressibility errors.

Calibrated Airspeed (CAS)—the airspeed indicator reading corrected for position (or installation), and instrument errors. (CAS is equal to TAS at sea level in standard atmosphere.) The color coding for various design speeds marked on airspeed indicators may be IAS or CAS.

Equivalent Airspeed (EAS)—the airspeed indicator reading corrected for position (or installation), or instrument error, and for adiabatic compressible flow for the particular altitude. (EAS is equal to CAS at sea level in standard atmosphere.)

V_{SO}—the calibrated power-off stalling speed or the minimum steady flight speed at which the airplane is controllable in the landing configuration.

V_{S1}—The calibrated power-off stalling speed or the minimum steady flight speed at which the airplane is controllable in a specified configuration.

V_Y—the calibrated airspeed at which the airplane will obtain the maximum increase in altitude per unit of time. This best rate-of-climb speed normally decreases slightly with altitude.

V_X—the calibrated airspeed at which the airplane will obtain the highest altitude in a given horizontal distance. This best angle-of-climb speed normally increases slightly with altitude.

V_{LE}—the maximum calibrated airspeed at which the airplane can be safely flown with the landing gear extended. This is a problem involving stability and controllability.

V_{LO}—the maximum calibrated airspeed at which the landing gear can be safely extended or retracted. This is a problem involving the air loads imposed on the operating mechanism during extension or retraction of the gear.

V_{FE}—the highest calibrated airspeed permissible with the wing flaps in a prescribed extended position. This is a problem involving the air loads imposed on the structure of the flaps.

V_A—the calibrated design maneuvering airspeed. This is the maximum speed at which the limit load can be imposed (either by gusts or full deflection of the control surfaces) without causing structural damage.

V_{NO}—the maximum calibrated airspeed for normal operation or the maximum structural cruising speed. This is the speed at which exceeding the limit load factor may cause permanent deformation of the airplane structure.

V_{NE}—the calibrated airspeed which should NEVER be exceeded. If flight is attempted above this speed, structural damage or structural *failure* may result.

☆U.S. GOVERNMENT PRINTING OFFICE 1984 421-018/4782